Mechanisms of Taste Transduction

Mechanisms of Taste Transduction

Sidney A. Simon
Professor
Department of Neurobiology
Duke University Medical Center
Durham, North Carolina

Stephen D. Roper
Professor
The Rocky Mountain Taste and Smell Center
University of Colorado Health Sciences Center
and
Colorado State University
Fort Collins, Colorado

CRC Press is an imprint of the
Taylor & Francis Group, an **informa** business

CRC Press
Taylor & Francis Group
6000 Broken Sound Parkway NW, Suite 300
Boca Raton, FL 33487-2742

Visit the Taylor & Francis Web site at
http://www.taylorandfrancis.com

and the CRC Press Web site at
http://www.crcpress.com

ISBN 13: 9780849353413 (hbk)
ISBN 13: 9780367449827 (pbk)

DEDICATION

To those who taught us.

"We have drunk from wells we did not dig and have been warmed by fires we did not build."

— Anonymous

PREFACE

Information about a field of science grows towards an asymptote, sometimes smoothly, sometimes broken by new paradigms and surges of information. Ideally, a textbook should be published when the field nears its asymptote. However, during the actual development of a field of inquiry, one never knows where this asymptote lies and, consequently, when it is appropriate to stop and summarize a body of knowledge. We decided to edit a book devoted entirely to taste now because we believed it would be useful for researchers to have a single reference book encompassing many aspects of taste while this is still possible. Thus, we have included topics ranging from the molecular biology of taste, to taste psychophysics, to the design of new foods.

The field of taste, as with most branches of biological science, is strongly interdisciplinary. Consequently, readers may encounter descriptions of several fields in a single chapter. We have attempted, however, to organize the chapters in a manner that describes information transfer from the peripheral sensory organs to the higher processing centers in the brain. After finishing this book the reader will be left with the impression that although great strides have been made, the field of taste is still in its embryonic phase.

As with any major undertaking, editing this book had its frustrations and rewards. By far our biggest reward was the pleasure of working with and learning from the many contributors. In addition to each contributor, we would especially like to acknowledge Ms. Lisa Bethage, biomedical illustrator, for her expert assistance and contribution of the cover drawing.

Sid Simon
Durham, North Carolina

Steve Roper
Ft. Collins, Colorado

THE EDITORS

Sidney A. Simon, Ph.D., is Professor of Neurobiology and Biochemical Engineering and Associate Professor of Anesthesiology at Duke University in Durham, North Carolina.

Dr. Simon received his B.S. (Physics) at I.I.T. in 1965, M.S. (Physics) Arizona State University in 1968, and Ph.D. (Materials Science) from Northwestern University in 1973. He then served as a postdoctoral fellow at Duke University in the Department of Physiology and Pharmacology until 1975 whereupon he was given a faculty appointment as Assistant Professor. In 1980 he was promoted to Associate Professor and in 1988 to Full Professor.

Dr. Simon is a member of the Biophysical Society, the American Society for Chemoreception Sciences and the Society for Neuroscience. He has received grants from the National Institutes of Health and the Smokeless Tobacco Research Council. Dr. Simon also consults for industry in regard to taste and drug delivery.

Dr. Simon has published over 100 research papers. His present interests include mechanisms of chemical activation of trigeminal nerves, the basis of cell adhesion, and the authorship of the plays and sonnets attributed to William Shakespeare.

Stephen Roper, Ph.D., is Professor of Anatomy and Neurobiology at Colorado State University, Adjunct Professor of Physiology at the University of Colorado School of Medicine, and is one of the founding members of the Rocky Mountain Taste and Smell Center, an NIH-funded research center at the University of Colorado Health Sciences Center.

Dr. Roper obtained his B.A. at Harvard College in 1967 and his doctoral degree at University College, University of London, England in 1970. After completing his doctorate, Dr. Roper spent 2 years with Professor Stephen Kuffler at Harvard Medical School in Boston as a postdoctoral trainee. From Boston, Dr. Roper took a position as Assistant Professor of Anatomy at the University of Colorado Medical School in Denver where he rose to the rank of Professor in two departments, Anatomy and Physiology. In 1985, Dr. Roper assumed the position of chairman of the Department of Anatomy and Neurobiology at Colorado State University and Director of the CSU Electron Microscopy Center and served in that capacity from 1985 to 1990.

Dr. Roper is a member of the Association for Chemoreception Sciences. He served as a councillor and as a member of the awards committee in that organization. Dr. Roper is also a member of the Society for Neuroscience, having served on the Chapters committee and presently on the Public Information Committee in that society. In 1992, Roper was awarded the Frito-Lay Award by the Association for Chemoreception Sciences in honor of distinguished research in taste. He has been the recipient of numerous research grants from the National Institutes of Health and has published several research reports, reviews, and book chapters in the field of taste. Dr. Roper also distinguished himself as teacher by receiving a Kaiser Permanente Teaching Award from the University of Colorado Medical School.

CONTRIBUTORS

Myles H. Akabas
Departments of Physiology, Cellular Biophysics, and Medicine
Columbia University
New York, New York

P. Avenet
Department of Biology
Synthelabo Recherche
Bagneux, France

John A. DeSimone
Department of Physiology
Virginia Commonwealth University
Richmond, Virginia

V. G. Dethier
Department of Biology
University of Massachusetts
Amherst, Massachusetts

Grant E. DuBois
The CocaCola Company
Atlanta, Georgia

Robert P. Erickson
Departments of Psychology and Neurobiology
Duke University
Durham, North Carolina

Marion E. Frank
Department of Biostructure and Function
University of Connecticut Health Center
Farmington, Connecticut

Hank I. Frier
Central Research Division
Pfizer, Inc.
Groton, Connecticut

Masaya Funakoshi
Office of the President
Asahi University
Gifu, Japan

Gerard L. Heck
Department of Physiology
Virginia Commonwealth University
Richmond, Virginia

David L. Hill
Department of Psychology
University of Virginia
Charlottesville, Virginia

Michael S. Kellogg
The Nutra Sweet Company
Deerfield, Illinois

Sue C. Kinnamon
The Rocky Mountain Taste and Smell Center
University of Colorado Health Sciences Center
and
Department of Anatomy and Neurobiology
Colorado State University
Fort Collins, Colorado

Bryce L. Munger
Department of Anatomy
University of Tasmania
Hobart, Tasmania
Australia

Yuzo Ninomiya
Department of Oral Physiology
Asahi University School of Dentistry
Gifu, Japan

Bruce Oakley
Department of Biology
University of Michigan
Ann Arbor, Michigan

Klaus Reutter
Institute of Anatomy
University of Tubingen
Tübingen, Germany

Stephen D. Roper
The Rocky Mountain Taste and Smell Center
University of Colorado Health Sciences Center
and
Department of Anatomy and Neurobiology
Colorado State University
Fort Collins, Colorado

Susan S. Schiffman
Departments of Psychiatry and Psychology
Duke University
Durham, North Carolina

Sidney A. Simon
Department of Neurobiology
Duke University Medical Center
Durham, North Carolina

David V. Smith
Department of Otolaryngology
Head and Neck Surgery
University of Cincinnati
College of Medicine
Cincinnati, Ohio

Robert E. Stewart
Program in Neuroscience
University of Virginia
Charlottesville, Virginia

Susan P. Travers
Departments of Oral Biology and Psychology
The Ohio State University
Columbus, Ohio

Carol A. Venanzi
Department of Chemical Engineering, Chemistry, and Environmental Science
New Jersey Institute of Technology
Newark, New Jersey

Thomas J. Venanzi
Department of Chemistry
College of New Rochelle
New Rochelle, New York

D. Eric Walters
Department of Biological Chemistry
The Chicago Medical School
North Chicago, Illinois

Y. Wang
Department of Neurobiology
Duke University
Durham, North Carolina

Martin Witt
Institute of Anatomy
University of Tubingen
Tübingen, Germany

TABLE OF CONTENTS

FOOD INTAKE

MORPHOLOGY OF TASTE RECEPTORS

DEVELOPMENT OF PERIPHERAL TASTE SYSTEM

TRANSDUCTION MECHANISMS

Food Intake

Chapter 1

THE ROLE OF TASTE IN FOOD INTAKE: A COMPARATIVE VIEW

V. G. Dethier

TABLE OF CONTENTS

0-8493-5341-6/93/$0.00 + $.50

I. INTRODUCTION

Even though the relation of taste to food intake is manifest to all who have observed animals eating and have reflected on their own eating habits, many aspects of the complex integration of chemosensory input with antecedent or concomitant physiological activities remain unclear. In human beings the situation is complicated further by the psychological, ethnic, and cultural refinements that have evolved. These complexities add to the difficulty of analyzing fundamental biological mechanisms.

Studies of evolutionary origins together with comparisons of ancient lineages and aquatic forerunners of mammals provide opportunities for clarification. This chapter examines these origins and traces developments from prokaryotes to invertebrates and fishes. It addresses in annelids, mollusks, and insects some of the behavioral and physiological phenomena studied in mammals. Information relating to mammals and other vertebrates has been compiled in volumes on nutrition and taste edited by Kare and Maller[124,125] and Kare et al.[123]

II. EVOLUTION AND DISTINCTIVENESS OF GUSTATION

The eminency of the chemical senses in the economy of animals is indicated by the occurrence of chemical sensitivity in one form or another in all animals. The capacity of a cell to react with molecules in its ambient environment without necessarily transporting them into itself for metabolic purposes is an essential characteristic of these receptors. Gustation and olfaction evolved from this 3 billion-year-old potential (cf. References 64 and 66).

The conclusions of Adler,[4-6] Armitage and Socklett,[9] and Vyas et al.[212] regarding chemoreceptive mechanisms in *Escherichia coli* probably represent fairly accurately the situation as it existed in primitive prokaryotes. In metazoans the development of a diversification of neurons, segregation in which like units were gathered together, and compartmentalization of functional assemblies within ganglia provided clear anatomical and functional division of chemoreceptive systems into two major and distinct categories, gustatory and olfactory.

While the identification of the categories is reasonably clear in aerial vertebrates[170] and invertebrates,[72] it becomes imprecise in aquatic animals. Laverack[134] suggested that "receptor organs responding to a high concentration of substances are often associated with tactile sensitivity and needing contact with material corresponding to taste, while receptors needing few molecules for stimulation and that may receive them in solution were effectively olfactory receptors." For fish and aquatic crustacea, Atema[10] proposed that taste was the sense concerned with food and required local reflex activity and that other chemoreception (olfaction) involved more complex behavior and was concerned with mate and prey detection. Altner and Prillinger[7] wrote that "a clear distinction between olfaction . . . and taste . . . is not convincing." Ache[3] proposed a separation based on the degree of complexity of chemoreceptive projections in the central nervous system. Seelinger,[195] writing about aerial crustacea, noted that taste is usually restricted to oral appendages and olfaction to others. Laverack[135] later proposed as definitions the following: taste, a combined sense with both chemical and mechanical sensors active in the same end organ, and olfaction, a sense in which chemical signals alone are received by a population of receptors without simultaneous contact being required.

Another distinction is made on the basis of receptor cell morphology. While this criterion may be distinctive when applied to vertebrates, it fails in the case of invertebrates. In crustacea (e.g., Florida spiny lobster, American lobster, European crayfish) basically identical chemoreceptive cells occur on antennules, legs, and mouthparts. All are modified ciliary cells, bipolar, with the soma at the periphery sending axons, without synapsing, directly into the central nervous system.[54] A similar situation occurs in insects where chemoreceptive cells are located on the antennae, mouthparts, legs, and ovipositors. Those on the antennae are designated as olfactory, those on the mouthparts as gustatory. The end organs (sensilla) are hair-like cuticular structures consisting of a variable number of chemosensory and accessory cells. The antennal sensilla are multiporous, each pore connected to a filament of a branched dendrite. The sensilla on the mouthparts are uniporous and the dendrite is unbranched.

It is probable, however, that the most significant distinction between the two senses lies, as Ache[3] proposed, in the organization of the central projections. This is the one criterion that holds across phyla.[66] It is the one on which the present discussion is based.

Two patterns of central organization are apparent in every animal in which olfaction and gustation are highly developed. In the first instance receptors are densely distributed in large localized sheets of epithelium and extend their axons to discrete, concentrated synaptic areas, glomerular neuropil. In the second instance, the axons project into organized centers that are less compact and nonglomerular. This delineation of distinctive central organization constitutes the most rigorous criterion for defining the separateness of the two senses. None of the other criteria that are diagnostic in some animals are contradicted.

III. RELATION TO FEEDING

In vertebrates the presence of "taste" buds in the oral cavity, and especially on the tongue, was always taken as presumptive evidence of an ability to taste. Corroborating evidence for this conclusion came readily from casual observations of eating behavior. It was supported by observations of loss of taste following pathological degeneration of taste buds. The presence of taste buds came to be associated with eating, food selection, palatability, and pleasure.

This narrow view lost some of its focus, however, when it was discovered that taste buds in different papillae and in different parts of the oropharyngeal cavity have different functions. Those in the fungiform papillae may serve ends different from those in the circumvallate papillae.[39] Additionally, those on the epiglottis are not normally exposed to gustatory stimuli. They are bathed by tracheal and laryngeal glandular secretions rather than by chemical stimuli originating in food. Among the functions with which some of these end organs are concerned are the following visceral activities: salivation, chewing, sucking, swallowing, preabsorptive insulin release, and such protective reflexes as gagging, coughing, spitting, regurgitating, and reflex apnea.[104,127a,205a,207a,209a] Other aversive responses (in the rat, for example) include oral gapes and somatic motor sequences (e.g., head shaking, chin rubbing, fore limb flailing, paw pushing, face washing, fluid rejection, and locomotion).[104,193] Clearly, taste buds are not restricted to the tasks of assessing food quality, initiating ingestion, or stimulating rejection. They trigger, in addition, responses preparatory to digestion and actions providing protection.

A still broader view of taste bud function became necessary when some fishes were found to possess taste buds not only within the oral cavity and pharynx but in the general body surface as well.[40,109] Subsequent studies suggested that these extraoral taste buds were more probably concerned with orientation to and localization of sapid substances (as is sometimes true of olfaction) than with ingestion.[12,129]

In vertebrates generally the nerves that subserve what we recognize as taste are the facial (VII), glossopharyngeal (IX), and vagus (X).[105,106,155] These three constitute the neural Ceberus that guards the entrance to the lower regions of the digestive tract. The anatomical substrate for the integration of the sensory information lies within the medulla, the specific relative anatomy of which has been described by Finger and Morita,[90] Morita and Finger,[163] and Finger.[89]

Once these relationships are understood it is clear that taste is spatially associated with organs of ingestion in the majority of animals. Filter feeders, e.g., oysters and Daphnia, are exceptions; they sweep in suspended particulate food more or less continuously and nonselectively. Rejection of unsuitable material occurs after ingestion.

To understand the degree to which taste is necessary and sufficient for feeding, it is helpful to supplement knowledge of the characteristics of gustatory receptors with behavioral information about the sequential steps in feeding. These may be divided into two phases, appetitive and consummatory. Appetitive behavior involves arousal, foraging, and recognition. These activities may be mediated by vision, thermoreception (in leeches, blood-sucking insects, and snails), mechanical vibration (water ripples detected by leeches), olfaction, and contact chemoreceptions (taste buds on the general body surface and barbels in some fish). Taste plays little or no direct role in these activities. Assessment of quality and concentration of prospective food is accomplished primarily by taste (but detection of texture and shape may also be important).

Consummatory activity begins with those patterns (biting, sucking, rasping, handling, etc.) that place food into the oral cavity, is followed by passing food into the buccal cavity or pharynx, and ceases with satiation. The sense of taste is actively concerned with initiating and maintaining the first three of these processes. Termination occurs as a result of postingestive events in various regions of the digestive tract, homeostatic conditions, and the release of neural transmitters. One or more of these consummatory activities has been investigated in a variety of animals, especially the following: the medicinal leech *(Hirudo medicinalis)*, gastropod mollusks *(Aplysia californica, A. punctata, A. juliana, A. vicarria, A. dactylomela, A. diplans, A. fasciata, Ariolimax californica, A. reticularis, Limax maximus, Helisoma trivolvis, Pleurobranchaea californica, Navanax inermis)*, insects (the blowfly *Phormia regina*, the locusts *Locusta migratoria, Schistocerca gregaria*, various caterpillars including *Manduca sexta, Diacrisia virginica, Spodoptera* spp., and others), hummingbirds (c.f. Reference 77), rats, rabbits, hamsters, and human beings. Not all phases of feeding have been investigated in each species, but taken together the comparative information reveals underlying themes and illuminates the role of taste in this enterprise.

IV. ANNELIDS

Attention to the feeding behavior of leeches has focused principally on the central control of ingestion, swallowing, and termination of feeding. Details of behavior differ among different species and especially between sanguivorous and predaceous species. A general picture emerges, however, from studies of the medicinal leech. This species feeds principally on mammals but will also attack birds, reptiles, fish, and amphibia.

Feeding has been described by Mann,[148] Sawyer,[186] and in a series of papers by Lent and his associates (Dickinson and Lent,[78] Lent and Dickinson,[138,139] Marshall and Lent,[149] Lent, Dickinson, and Marshall,[140] Lent et al.[142]). "Hungry" leeches rest at the surface of water where photic and mechanical stimuli, specifically moving shadows and local ripples, initiate appetitive behavior. This takes the form of orientational swimming. When a warm area or a host-specific chemical is encountered, consummatory behavior begins.

All the events of ingestion are centered in the anterior sucker. This organ, which delimits the oral cavity, is bounded anteriorly by the lip of the prostomium. Labial chemoreceptors are situated on the lip. The oral cavity is separated from the buccal cavity (in which the jaws are housed) by a tissue fold, the velum. The buccal cavity leads to the pharynx.

Heat alone or chemical stimulation of receptors on the prostomium evokes biting. Host specificity is mediated by labial chemoreceptors on the dorsal lip. As salivation occurs, and blood flows into the buccal cavity, the mixed fluid is pumped by rhythmic movements of the pharynx into the crop (Lent et al.[141]; Marshall and Lent,[149]). This process of ingestion is initiated by whole blood. A mixture of NaCl and arginine also constitutes an adequate stimulus.[80] Continuous stimulation is required to maintain ingestion and is so dominant that many kinds of noxious stimuli are ineffective as inhibitors.[140] Some compounds, however, are able to terminate earlier phases of feeding. Among these are carboxylic acids, alkaloids, ammonia, indole, and skatol.[120]

The relevant chemoreceptive organs resemble in structure the taste buds of vertebrates. They are aggregates of ciliated bipolar sensory cells[8] with basal processes connecting with the first two cephalic nerves.[143,214] Their role in feeding resembles that of vertebrate taste buds, but detailed information concerning their response characteristics is lacking.

Stimulation of the sensory receptors, thermal and chemical alike, activates serotonin effectors in the first suboesophageal ganglion and Retzius cells in the segmental ganglia.[138,140] The feeding pattern is orchestrated by this release of serotonin. As feeding progresses, distension of the body wall stimulates mechanoreceptors which in turn hyperpolarize the serotonergic neurons. Inhibitory information is integrated with incoming excitation from lip chemoreceptors.[138,139] Generation of the pattern then ceases and feeding stops.[142]

V. MOLLUSKS

The patterns of feeding in gastropod mollusks resemble in many respects those characteristic of leeches.[132] Herbivorous sea hares of the genus *Aplysia* exemplify the sequence of events (for a comprehensive discussion consult Kandel[121]). Although all species studied feed on seaweeds, there are specific feeding preferences. Some *Aplysia* prefer red algae, others, green algae.[41-43] In all species the appetitive phase of feeding, consisting of arousal, searching, and orientation, is a variable response leading to a more stereotyped consummatory response.[131,133]

Orientation and head waving are elicited by chemical stimulation of the anterior tentacles and rhinophores (posterior tentacles).[11,118,174] The organs are functionally analogous to olfactory organs. Receptors (gustatory) on the mouth and lips respond to higher concentrations of compounds and their stimulation initiates biting.[113,114,207] Swallowing is activated by the presence of food on the inner surfaces of the mouth or buccal cavity. Distasteful material entering the buccal cavity can cause reverse peristalsis resulting in rejection.[131] The presence of bulk in the digestive system leads to satiation.[206] Denervation of the gut (in the sea slug *Pleurobranchaea*) causes hyperphagia.[48]

Mechanostimulation of the mouth also can provide information sufficient for exciting feeding behavior. The information passes to a neuron (C2) that generates diverse outputs. This histaminergic afferent contributes to the flexibility and adaptability of feeding.[45]

The feeding behavior of the carnivorous snail *Navanax inermis* also begins with locomotion that is unoriented until the mucous trail of prey is encountered. Specific chemical recognition is required for orientation to begin.[169] The receptors involved in tracking are restricted to the anterolateral folds of the head shield.[166] As soon as the head comes into contact with the prey, specific chemicals from that source stimulate receptors on the lips and pharynx. The pharynx is then protracted.

Pharyngeal expansion following protraction is initiated by sensory feedback from the antecedent response. Mechanoreceptors in the lip and anterior wall of the pharynx provide this information.[203,204] The remaining sequence of actions leading to passage of food into the midgut is mediated by additional mechanical events and differential inhibition of motor neurons.

Feeding behavior in another gastropod, the carnivorous opisthobranch *Pleurobranchaea californica*, resembles that of *Aplysia*. It too depends on chemoreceptors in the anterior tentacles and rhinophores to effect successful orientation to prey.[137,174] Details of the consummatory behavior have been reported by Davis et al.,[49] Gillette and Davis,[101] and Kovac and Davis.[130] As in leeches and *Aplysia*, distension of the gut causes feeding to cease.

Studies of the feeding behavior of the pond snail *Helisoma trivolvis*[122,126,127,154] and of the land slug *Limax maximus*[98-100,175,177] have been concerned primarily with neural circuitry. Feeding sequences are essentially the same in all the gastropods studied, especially *Aplysia*,[128,179] *Pleurobranchaea*,[102] *Lymnaea*,[150-152] and *Helisoma*.[103]

The role of taste has been investigated less intensively. Motor neuron activity patterns are triggered by chemical and tactile receptors in the lips. The afferent neurons connect with the cerebral ganglion which in turn is connected to the buccal ganglion by the buccal nerves. In *Limax*, fructose and food-plant extracts (which initiate feeding) cause buccal nerves to become active; aversive stimuli, such as quinidine sulfate and citric acid, inhibit these neurons.[51-53] The higher the concentration of stimulating chemicals the larger the meal. Food-aversion learning occurs in this species but appears to be mediated by olfaction rather than taste.[99]

VI. INSECTS

Whereas studies of feeding behavior of annelids and mollusks have focused on the participation of central neuronal pattern generators, motor neural activity, and feedback systems, investigations of insects have been concerned primarily with the identity and characteristics of chemoreceptors and how they effect food choice. Thus, although there have been studies of satiation, emphasis has been placed on the initiation of feeding, selection of food, regulation of meal size, nutrition, homeostasis, and the role of learning. Even with this selective interest an enormous volume of literature has accumulated. The subject has been reviewed most extensively by Barton Brown,[13] Bernays and Simpson,[27] and Bernays.[22] Other recent reviews include those of Miller and Miller[156] and the symposium *Insect-Plants '89* edited by Szentesi and

Jermy.[209] The number of experiments is great but comprehensive studies have been restricted to a few species, notably blowflies, caterpillars, and locusts.[27,60] Similarly, the microstructure of feeding, that is, meal size, frequency, etc., has been investigated in detail only in locusts,[196] caterpillars,[37,38,178] and aphids.[153]

Appetitive behavior of these species is directed by a variety of nongustatory stimuli, among them visual and olfactory. Recognition and qualitative assessment of food is accomplished by mechanical and olfactory stimulation and, in the case of blowflies and locusts, by contact chemoreceptors. In the blowfly these are situated on the tarsi; in locusts, on tarsi and maxillary and labial palpi.

The sequence of motor events involved in feeding follows the same general pattern as that executed in annelids and mollusks. The first bites by phytophagous species (e.g., locusts and caterpillars) release leaf sap which passes into the preoral cavity (cf. Snodgrass[202]). There it stimulates gustatory sensilla on the inside surface of the clypeolabrum. In sucking species, as exemplified by the blowfly, stimulation of tarsal contact chemoreceptors orients the walking insect to food, arrests locomotion, and triggers extension of the proboscis (labium and palpi). Extension brings taste receptors on the aboral and oral surfaces into contact with the food. If the food is appropriate, sucking ensues. Fluid passes into the functional "mouth cavity" (cibarium) where cibarial chemoreceptors serve as the final sensory checkpoint before swallowing occurs.

The activities of these various populations of gustatory receptors provide information concerning the nature (quality) of the stimulus and its intensity (concentration). In this respect they are essential for identifying the kind of food, monitoring the breadth of diet, initiating acceptance or rejection, regulating the size of meals, providing a means for intensity discrimination (Weber fraction), adjusting intake to need, and allowing for nutritional compensation.

VII. DIET SELECTION

Diet selection has been studied most intensively in phytophagous species. Breadth of diet among the various species ranges from strict monophagy to polyphagy. The involvement of taste in initiating feeding and exercising choice is most dramatically demonstrated in monophagous and oligophagous species by ablation experiments.[35,192,213] When all gustatory receptors of these species are removed, caterpillars, which heretofore starved rather than accept a nonhost plant, now accept a wide variety of leaves. The experiment also reveals that although taste normally stimulates ingestion by nonstarved insects it is not necessary.

Food-plants present the chemical senses with extremely complex arrays of compounds. The chemical profile of each species of plant is unique. It must be assessed by the chemoreceptors in order that a code indicating acceptability might be sent to the central nervous system. There is some evidence that key token stimuli (e.g., glucosinolates in cruciferous plants) alone are able to stimulate biting and ingestion by cruciferae-feeders.[81,145,167,168,187,211] Despite an intensive search few other token stimuli have been discovered.

Generally speaking food stimuli are not directly related to nutritional value, hence the appellation "token" (cf. Reference 108). As Rozin and Vollmecke[185] have pointed out in another context, people eat foods not nutrients. In an evolutionary sense, however, token stimuli in plants are associated with the nutritional and ecological value of the plant to the responding insect. That is to say, each insect has adapted to those token stimuli which occur in plants that meet its specific needs.

In addition to compounds that stimulate feeding there are deterrent compounds that constrain the spectrum of potentially available food. The importance of deterrents in this respect has been pointed out by many investigators.[14,24,25,81,91,111,189] It is clear that a balance of positive and negative factors ultimately determines the choice of food.[15,25,32,33,145,168,190]

The nature of the receptors that effect feeding behavior reveals the complexity of gustatory involvement. Electrophysiological studies have not on the whole identified receptors that are narrowly responsive to token stimuli. Consequently, attention is now focused on the probability of ensemble coding involving receptors that are rather broadly tuned. As Pfaffmann[171-173] and Erickson[83-85] suggested for vertebrates, a set of broadly but distinctively tuned receptors together can generate an across-fiber code that reflects the chemical profile of a food (see Chapters 13 and 14).

The view that insect chemoreceptors were either specialists or generalistists was superceded by the discovery that there is in fact a continuum of breadth of response spectra. The constituent sensitivities may vary from species to species. There are also differences within any given complement of receptors. In the tobacco hornworm, for example, the following receptors have been identified: glucose, inositol, salt, and deterrent.[188] Other receptor types that have been reported in various other caterpillars are sorbitol, alkaloid, glycoside, phlorizin, adenosine, populin, and salicin.[191] The designations refer to the compound or class of compounds to which each receptor is *most* sensitive.

The capability of these receptors acting together to encode quality has been investigated in flies and caterpillars.[57,59,61,73,188,191] Across-fiber patterning has also been invoked to explain intensity discrimination.[147]

VIII. INTENSITY DISCRIMINATION

Gustatory receptors provide information not only about the identity and quality of food and whether or not it should be ingested and in what quantity, but also about the concentration of the stimulating compounds. Input/output relationships in some phytophagous insects have been studied by Blom.[32,33] Information transmission by the tarsal "sugar" receptors of the blowfly exemplifies another approach to the relation between concentration and volume intake.[200] The volume of a single drink is directly proportional to the number of impulses, prior to adaptation, of the receptors on the tarsi and labellum.

Comparable analyses suggest that the coding mechanism underlying difference thresholds (Weber fraction) derives from the events of transduction at the receptor membranes.[67,68] The similarity of the curves describing the relationship $\Delta I/I$ for *E. coli*, flies, rats, and human beings implies that the ability to discriminate depends ultimately on fundamental relations between the stimulus and the receptor, relations that are based on the Law of Mass Action. Central neural information from other sources within the organism sharpen the discrimination.

IX. DEPRIVATION

Gustatory information is also involved in changes in eating related to deprivation. As deprivation increases in the blowfly the concentration of sugar required to initiate ingestion decreases. No convincing accompanying decrease in receptor sensitivity (in the blowfly) has been demonstrated. On the other hand, changes in receptor activity related to variation in internal nutritional states have been found in locusts.[23,24] In the rat, peripheral changes in salt sensitivity associated with salt deprivation have been found by Contreras.[46] In nonspecific deprivation in the fly, however, immediate changes in behavioral thresholds can be attributed to an integration of gustatory sensory input and a decreasing inhibition from mechanoreceptors associated with the gut.[38,71,74,87,96,97] The importance of inhibitory mechanoreception generated by stretch of gut or body wall appears to be a general mechanism for terminating feeding in annelids, mollusks, and insects (see also Maddrell[146]).

The same kind of balance between gustatory excitation and mechanosensory inhibition is responsible for decreasing finickiness as deprivation increases; that is, a fly will accept higher concentrations of aversive solutions as it becomes "hungrier".[164] For example, the acceptance of a mixture of quinine HCl and sugar depends upon the stimulating effectiveness of the sugar in relation to the sum of stretch-generated inhibition and the negative sensory effect of quinine. Thus, the behavioral response to quinine (and other unacceptable compounds) in a mixture is determined by a change in the central nervous system's set point for responding to sensory excitation by sugar. These findings raise the question of how unacceptable compounds bring about behavioral rejection.

X. DETERRENCY

Compounds that deter or reduce feeding may act on gustatory receptors in any of the following three ways: (1) by stimulating "deterrent" receptors which specifically initiate motor patterns inhibiting ingestion; (2) by inhibiting the activity of receptors (e.g., "sugar" receptors) that normally initiate motor patterns leading to ingestion and swallowing; and (3) causing receptors that are normally responsive to food to fire erratically or in bursts. All three modes of action have been demonstrated by combined electrophysiological and behavioral studies.

Deterrent receptors, analogous to the bitter receptors of vertebrates, have been reported in a number of leaf-feeding insects.[107,112,190a] A deterrent cell in the maxillary palpi that is involved in the rejection of unacceptable plants such as canna by the tobacco hornworm has been studied in detail. This cell responds vigorously to extract of canna. It also responds to triterpene limonin, a compound that is a feeding deterrent for many insects. On the other hand, it does not respond to a number of alkaloids (caffeine, salicin, nicotine) that reportedly deter feeding by the hornworm. Other cells do respond. Receptors in the preoral cavity also contribute to rejection.[34] In contrast to glucose and inositol cells, which have short latencies and rises in spike frequency, the deterrent cell has a five- to tenfold longer latency and rise.

Many compounds that deter feeding do so by shutting down receptors that normally signal acceptance. Reversible inhibition of sugar, salt, and water receptors has been demonstrated in flies, beetles, and caterpillars.[30,63,65,157-161] Alkaloids not only have an immediate inhibitory effect, they have a perseverating inhibition that may last as long as one minute.[69,70]

A wide variety of other compounds also inhibit receptors. Monovalent salts not only stimulate salt cells, they inhibit sugar receptors.[110,160,162] Divalent salts, sometimes in millimolar concentrations, inhibit sugar and sometimes water and salt receptors.[88,159,162] Heavy metal ions cause rejection and depress neural responses to gustatory stimuli in insects and vertebrates alike. Aliphatic organic compounds, of which more than 200 were tested on the blowfly, cause rejection by suppressing the activity of water, sugar, and salt receptors (for a review see Dethier[65]).

The first two mechanisms of deterrency achieve the same behavioral result by decreasing the *effective* intensity of the stimulating components of a mixture. In the first case the central nervous system integrates inhibitory and excitatory sensory signals. In the second case there is a direct inhibition of the excitatory receptor.

The third mode of action by deterrent compounds, abnormal firing of gustatory receptors, usually initiates actions that are aversive in nature, as, for example, excessive regurgitation, vigorous wiping of mouthparts, and evasive locomotion — actions analogous to the protective responses of rats and hamsters when presented with nociceptive stimuli.

XI. COMPENSATORY FEEDING AND SPECIFIC HUNGERS

There is such a diversity of potential food in the world that most animals can choose from Lucullan banquets. This latitude is available especially for omnivores and phytophagous species. On the other hand, not all foods are equally nutritious, and, as has been pointed out, taste is not an infallible guide to nutrient value. Obtaining a balanced diet from day to day requires more information than taste alone can provide. Furthermore, there are critical periods in life cycles when compensation must be made for changing metabolic needs. The ability to meet these challenges undoubtedly appeared early in evolutionary time. It has been observed in contemporary mollusks[44] and is common among modern insects.

Although there are satisfactory models for explaining the interaction of gustatory stimulation and postingestive feedback during states of satiation and "hunger", knowledge regarding mechanisms effecting compensation for nutritional imbalance is scanty. As Simpson and Simpson[199] have pointed out in their detailed review of the entomological literature, compensation can be achieved a number of ways, specifically by altering dietary selection, consumption, or utilization efficiency. Taste can be involved in the first two of these.

Many investigations demonstrating that insects do make adjustments in response to deficiencies in their feeding regimens have been undertaken with locusts, caterpillars, and aphids. Clearest insight to the physiological mechanisms involved have been provided by studies of locusts.[1,2,198] Fifth instar nymphs of *Locusta migratoria*, for example, respond to dilution of the protein component of their diet by altering feeding behavior. Specific nutrient feedback from the blood influences gustatory sensitivity to key amino acids. Changes in receptor sensitivity as a means of compensating for deficiencies in diet have also been recorded in the caterpillar *Spodoptera*.[31] It is also possible that there may be a learning component to the response in the form of altered preferences (e.g., induction or aversion learning).

Insofar as specific hungers are concerned, only one case has been studied in detail. This is the change of food preference by the female blowfly from carbohydrate to protein during vitellogenesis.[56] The basis of this switch has been investigated by Belzer[16-19] and by Rachman.[176] The nature of the gustatory receptor contribution to this adjustment is not clear.

XII. EXPERIENCE

All the mechanisms and behavioral outcomes discussed thus far are based on the simplifying assumption that the relation between taste and feeding derives from a rigid input/output system modulated only by changing conditions of the internal milieu. Until relatively recently the influence of experience was ignored or minimized. During the past two decades, however, it has been demonstrated in invertebrates, principally in phytophagous insects, that experience plays a significant role in modifying qualitative details of feeding (for comprehensive reviews see Szentesi and Jermy[208]).

The changes that experience can impose on feeding behavior have been characterized as habituation, sensitization and pseudoconditioning, and associative learning. Habituation can occur either to deterrents or to compounds (e.g., sucrose) that stimulate feeding. In the first instance unacceptable food is eaten more readily. In the second instance food that is highly stimulating becomes less acceptable as eating progresses. In all cases it is difficult to demonstrate that the decrement reflects central rather than peripheral neural change. The most convincing evidence has been obtained with flies[55] and with locusts.[21]

Changes in which there has been an increment in feeding may involve either associative or nonassociative learning. One class of nonassociative learning is the phenomenon of central excitatory state (CES). In the blowfly a perseverating increase in response to sucrose following repeated stimulation has been shown to be of central origin.[75] This is probably a widespread phenomenon and one that augments the stimulating effect of taste in driving feeding after initial ingestion.

Feeding on a particular food can also induce a predilection for that food. The phenomenon was originally termed "induction" because its nature in terms of the various categories of learning was not known.[117] Some later authors have referred to it as imprinting. Whatever its nature the involvement of taste has not been elucidated.

The antithesis of induction is food-aversion learning, a well-known phenomenon in vertebrates.[20,28,29,86,92,94,95,144,181,184,215] It occurs in mollusks, particularly *Limax*,[50,98] and *Pleurobranchaea*,[165] and in insects.[26,62,116,136]

XIII. EPILOGUE

Although the evolutionary development of the chemosensory systems that became associated with feeding followed two different courses, one in invertebrates and one in vertebrates, the purposes they serve and the general principles on which they operate are essentially the same. The goals that had to be achieved were the following: detection and discrimination among foods, selection of nutritious diets, initiating and sustaining ingestion, adjusting to fluctuating metabolic needs, and terminating ingestion.

In all organisms, taste exercises tight control over detection and discrimination. It is sufficient but not necessary for initiating ingestion. It may control swallowing. It regulates the quantity of food eaten. In matters of homeostatic control, nutrient balance, size of meals, and termination of feeding, taste does not act alone. This is clear from studies of invertebrates, especially insects. It is more dramatically illustrated by experiments with mammals.

Taste obviously drives ingestion by mammals;[173,216] however, the force of control may be out of proportion to what is required by homeostasis.[205] Rats will drink their body weight of a glucose-saccharine mixture overnight with few caloric or osmotic consequences.[201,210] Rats also overeat on "supermarket" diets[194] and on highly palatable fat diets.[47] Human beings behave similarly. In a situation where two pumps simultaneously deliver equal amounts into mouth and stomach, subjects eat in accordance with oral stimulation.[119]

Clearly the homeostatic aspect of eating does not depend exclusively on gustatory stimulation. Epstein[82] demonstrated that rats with intragastric tubes would feed themselves by manipulating a lever and maintain constancy of intake, respond to deprivation, and maintain a normal dirunal cycle of feeding. The control is effected by a complexity of internal factors.

The complexity of feeding behavior in mammals is enhanced by highly evolved neural and neurohumoral refinements and by a greatly increased capacity for learning. In human beings, hedonistic, social, and cultural influences have greatly modified feeding behavior.[182,183,185] What began evolutionarily as a rather limited chemosensitivity associated with identifying and initiating feeding became not only a highly refined physiological capability integrated with homeostatic feedback, but eventually an instrument of individualistic and social activity enriched by hedonic reward and tradition.

REFERENCES

1. **Abisgold, J. D. and Simpson, S. J.,** The physiology of compensation by locusts for changes in dietary protein, *J. Exp. Biol.*, 129, 329, 1987.
2. **Abisgold, J. D. and Simpson, S. J.,** The effect of dietary protein levels and haemolymph composition on the sensitivity of the maxillary palp chemoreceptors of locusts, *J. Exp. Biol.*, 135, 215, 1988.
3. **Ache, B. W.,** Chemoreception and thermoreception, in *Biology of Crustacea*, Vol. 3, Bliss, D. H., Atwood, H. L., and Sandman., D. C., Eds., Academic Press, New York, 1982, 369.
4. **Adler, J.,** The sensing of chemicals by bacteria, *Sci. Am.*, 234, 40, 1976.
5. **Adler, J.,** How motile bacteria sense and respond to chemicals, in *Olfaction and Taste IX*, Roper S. D. and Atema, J., Ann. N.Y. Acad. Sci., 510, 95–97, 1987.
6. **Adler, J.,** The sense of smell in bacteria, in *R. H. Wright Lectures on Olfaction*, Colbow, K., Ed., Simon Fraser University, Burnaby, Canada, 1990.
7. **Altner, H. and Prillinger, L.,** Ultrastructure of invertebrate chemo- thermo- and hygroreceptors and its functional significance, *Int. Rev. Cytol.*, 67, 69, 1980.
8. **Apathy, St. V.,** Analyse der äusserm Körperform der Hirudineen, *Mittheil. Zool. Sta. Neapel.*, 8, 153, 1888.
9. **Armitage, J. P. and Socklett, R. E.,** Sensory transduction in flagellate bacteria, in *Olfaction and Taste IX*, Roper, S. D. and Atema, J., Eds., Ann. N.Y. Acad. Sci., 510, 9, 1987.
10. **Atema, J.,** Functional separation of smell and taste in fish and Crustacea, in *Olfaction and Taste VI*, LeMagnen, J. and MacLeod, P., Eds., Information Retrieval, London, 1977, 165.

11. **Audesirk, T. E.,** Chemoreception in *Aplysia californica*. I. Behavioral localization of distance chemoreceptors used in food-finding, *Behav. Biol.*, 15, 45, 1975.
12. **Bardach, J. E., Todd, J. H., and Crickmer, R.,** Orientation by taste in fish of the genus *Ictalurus*, *Science*, 155, 1276, 1967.
13. **Barton Brown, L.,** Regulatory mechanisms in insect feeding, *Adv. Insect Physiol.*, 11, 1, 1975.
14. **Beck, S. D.,** Resistance of plants to insects, *Annu. Rev. Entomol.*, 10, 207, 1965.
15. **Beck, S. D. and Hanec, W.,** Effect of amino acids on feeding behavior of the European corn borer, *Pyrausta nubilalis* (Hubn.), *J. Insect Physiol.*, 2, 85, 1958.
16. **Belzer, W. R.,** Patterns of selective protein ingestion by the blowfly *Phormia regina*, *Physiol. Ent.*, 3, 169, 1978.
17. **Belzer, W. R.,** Factors conducive to increased protein feeding by the blowfly *Phormia regina*, *Physiol. Ent.*, 3, 251, 1978.
18. **Belzer, W. R.,** Recurrent nerve inhibition of protein feeding in the blowfly *Phormia regina*, *Physiol. Ent.*, 3, 259, 1978.
19. **Belzer, W. R.,** Abdominal stretch in the regulation of protein ingestion by the black blowfly, *Phormia regina*, *Physiol. Ent.*, 4, 7, 1979.
20. **Bermudez-Rattoni, F., Frothman, D. L., Sanchez, M. A., Perez, J. I., and Garcia, J.,** Odor and taste aversions conditioned in anesthetized rats, *Behav. Neurosci.*, 102, 726, 1988.
21. **Bernays, E. A.,** Antifeedants in crop pest management, in *Natural Products for Innovative Pest Management*, Whitehead, D. L. and Bowers, W. S., Eds., Pergamon Press, Oxford, 1983, 586.
22. **Bernays, E. A.,** Regulation of feeding behavior, in *Comprehensive Insect Physiology, Biochemistry and Pharmacology*, Vol. 4, Kerkut, G. A. and Gilbert, L. I., Pergamon Press, Oxford, 1985, 1.
23. **Bernays, E. A., Blaney, W. M., and Chapman, R. F.,** Changes in chemoreceptive sensilla on the maxillary palps of *Locusta migratoria* in relation to feeding, *J. Exp. Biol.*, 57, 745, 1972.
24. **Bernays, E. A. and Chapman, R. F.,** The control of changes in peripheral sensilla associated with feeding in *Locusta migratoria*, *J. Exp. Biol.*, 57, 755, 1972.
25. **Bernays, E. A. and Chapman, R. F.,** Deterrent chemicals as a basis of oligophagy in *Locusta migratoria* (L.), *Ecol. Entomol.*, 2, 1, 1977.
26. **Bernays, E. A. and Lee, J. E.,** Food aversion learning in the polyphagous grasshopper *Schistocerca americana*, *Physiol. Entomol.*, 13, 131, 1988.
27. **Bernays, E. A. and Simpson, S. J.,** Control of food intake, *Adv. Insect Physiol.*, 16, 59, 1982.
28. **Bernstein, I. L.,** Learned taste aversions in children receiving chemotherapy, *Science*, 200, 1302, 1978.
29. **Bernstein, I. L. and Webster, M. M.,** Learned taste aversions in humans, *Physiol. Behav.*, 25, 363, 1980.
30. **Blades, D. and Mitchell, B. K.,** Effect of alkaloids on feeding by *Phormia regina*, *Entomol. Exp. Appl.*, 41, 299, 1986.
31. **Blaney, W. M., Simmonds, M. S. J., and Simpson, S. J.,** Dietary selection behavior: comparisons between locusts and caterpillars, in *Insect-Plants '89*, Szentesi, A. and Jermy, T., Eds., Akadémiai Kiadó, Budapest, 1991, 47.
32. **Blom, F.,** Sensory activity and food intake: a study of input-output relationships in two phytophagous insects, *Neth. J. Zool.*, 28, 277, 1978.
33. **Blom, F.,** Sensory input behavioral output relationships in the feeding activity of some lepidopterous larvae, *Entomol. Exp. Appl.*, 24, 258, 1978.
34. **deBoer, G., Dethier, V. G., and Schoonhoven, L. M.,** Chemoreceptors in the preoral cavity of the tobacco hornworm, *Manduca sexta*, and their possible function in feeding behavior, *Entomol. Exp. Appl.*, 21, 287, 1977.

35. **deBoer, G. and Hanson, F. E.,** Differentiation of roles of chemosensory organs in food discrimination among host and non-host plants in larvae of the tobacco hornworm, *Manduca sexta, Physiol. Entomol.*, 12, 387, 1987.
36. **Bowdan, E.,** Microstructure of feeding by tobacco hornworm caterpillars, *Manduca sexta, Entomol. Exp. Appl.*, 47, 127, 1988.
37. **Bowdan, E.,** The effect of deprivation on the microstructure of feeding by the tobacco hornworm caterpillar, *J. Insect Behav.*, 1, 31, 1988.
38. **Bowdan, E. and Dethier, V. G.,** Coordination of a dual inhibitory system regulating feeding behavior in the blowfly, *J. Comp. Physiol.*, A, 158, 713, 1986.
39. **Bradley, R. M.,** Diversity of taste bud function, in *The Beidler Symposium on Taste and Smell*, Miller, I. J., Ed., Book Service Assoc., Winston-Salem, NC, 1988, 33.
40. **Caprio, J.,** Peripheral fibers and chemoreceptor cells in fishes, in *Sensory Biology of Aquatic Animals*, Atema, J., Fay, R. R., Popper, A. N., and Tavolga, W. N., Eds., Springer-Verlag, New York, 1988, 313.
41. **Carefoot, T. H.,** Growth and nutrition of three species of opisthobranch mollusks, *Comp. Biochem. Physiol.*, 2, 627, 1967.
42. **Carefoot, T. H.,** Growth and nutrition of *Aplysia punctata* feeding on a variety of marine algae, *J. Marine Biol. Assoc. U.K.*, 47, 565, 1967.
43. **Carefoot, T. H.,** A comparison of absorption and utilization of food energy in two species of tropical *Aplysia, J. Exp. Marine Biol. Ecol.*, 5, 47, 1970.
44. **Carefoot, T. H. and Switzer-Dunlop, M.,** Effect of amino acid imbalance in artificial diets on food choice and feeding rates in two species of terrestrial snails, *Copaea nemoralis* and *Achatina fulica, J. Molluscan Stud.*, 55, 323, 1989.
45. **Chiel, H. J., Weiss, K. R., and Kupfermann, I.,** Multiple roles of a histaminergic afferent neuron in the feedling behavior of *Aplysia, TINS*, 13, 223, 1990.
46. **Contreras, R. J.,** Peripheral neural changes associated with sodium deprivation, in *Biological and Behavioral Aspects of Salt Intake*, Kare, M. R., Fregly, M. J., and Bernard, R. A., Eds., Academic Press, New York, 1980, 300.
47. **Corbit, J. D. and Stellar, E.,** Palatability, food intake, and obesity in normal and hyperphagic rats, *J. Comp. Physiol. Psychol.*, 58, 63, 1964.
48. **Croll, R. P., Albuquerque, T., and Fitzpatrick, L.,** Hyperphagia resulting from gut denervation in the sea slug, *Pleurobranchaea, Behav. Neural Biol.*, 47, 212, 1987.
49. **Davis, W. J., Siegler, M. V. S., and Mpitsos, G. J.,** Distributed neuronal oscillators and efference copy in the feeding system of *Pleurobranchaea, J. Neurophysiol.*, 36, 258, 1973.
50. **Delaney, K. and Gelperin, A.,** Post-ingestive food aversion learning to amino acid deficient diets by the terrestrial slug *Limax maximus, J. Comp. Physiol.*, A, 159, 281, 1986.
51. **Delaney, K. and Gelperin, A.,** Cerebral interneurons controlling fictive feeding in *Limax maximus*. I. Anatomy and criteria for reidentification, *J. Comp. Physiol.*, A, 166, 297, 1990.
52. **Delaney, K. and Gelperin, A.,** Cerebral interneurons controlling fictive feeding in *Limax maximus*. II. Intiation and modulation of fictive feeding, *J. Comp. Physiol.*, A, 166, 311, 1990.
53. **Delaney, K. and Gelperin, A.,** Cerebral interneurons controlling fictive feeding in *Limax maximus*. III. Integration of sensory inputs, *J. Comp. Physiol.*, A, 166, 327, 1990.
54. **Derby, C. D. and Atema, J.,** Chemoreceptor cells in aquatic vertebrates: peripheral mechanisms of chemical signal processing in Decapod Crustaceans, in *Sensory Biology of Aquatic Organisms*, Atema, J., Fay, R. R., Popper, A. N., and Tavolga, W. N., Springer-Verlag, New York, 1988, 365.
55. **Dethier, V. G.,** Adaptation to chemical stimulation of the tarsal receptors of the blowfly, *Biol. Bull.*, 103, 179, 1952.
56. **Dethier, V. G.,** Behavioral aspects of protein ingestion by the blowfly *Phormia regina* Meigen, *Biol. Bull.*, 121, 456, 1961.

57. **Dethier, V. G.,** Electrophysiological studies of gustation in lepidopterous larvae. II. Taste spectra in relation to food-plant discrimination, *J. Comp. Physiol.*, 82, 103, 1973.
58. **Dethier, V. G.,** The specificity of the labellar chemoreceptors of the blowfly and the response to natural foods, *J. Insect Physiol.*, 20, 1859, 1974.
59. **Dethier, V. G.,** The importance of stimulus patterns for host-plant recognition and acceptance, *Symp. Biol. Hung.*, 16, 67, 1976.
60. **Dethier, V. G.,** *The Hungry Fly*, Harvard University Press, Cambridge, MA, 1976.
61. **Dethier, V. G.,** Gustatory sensing of complex mixed stimuli by insects, in *Olfaction and Taste VI*, LeMagnen, J. and Macleod, P., Information Retrieval, London, 1977, 323.
62. **Dethier, V. G.,** Food-aversion learning in two polyphagous caterpillars, *Diacrisia virginica* and *Estigmene congrua*, *Physiol. Entomol.*, 5, 321, 1980.
63. **Dethier, V. G.,** Mechanism of host-plant recognition, *Entomol. Exp. Appl.*, 31, 49, 1982.
64. **Dethier, V. G.,** Chemoreception and behavior from an evolutionary and comparative perspective, in *Mechanisms in Insect Olfaction*, Payne, T. L., Birch, M. C., and Kennedy, C. E. J., Eds., Oxford University Press, Oxford, 1986, 1.
65. **Dethier, V. G.,** Discriminative taste inhibitors affecting insects, *Chem. Senses*, 12, 251, 1987.
66. **Dethier, V. G.,** Five hundred million years of olfaction, in *R. H. Wright Lectures on Olfaction*, Colbow, K., Ed., Simon Fraser University, Burnaby, Canada, 1990, 1.
67. **Dethier, V. G. and Bowdan, E.,** Relations between differential threshold and sugar receptor mechanisms in the blowfly, *Behav. Neurosci.*, 98, 791, 1984.
68. **Dethier, V. G. and Bowdan, E.,** An interspecific comparison of difference thresholds for sucrose, in *The Beidler Symposium in Taste and Smell*, Miller, I., Ed., Book Services Assoc., Winston-Salem, NC, 1988, 193.
69. **Dethier, V. G. and Bowdan, E.,** The effect of alkaloids on sugar receptors and feeding behavior of the blowfly, *Physiol. Entomol.*, 14, 127, 1989.
70. **Dethier, V. G. and Bowdan, E.,** Inhibition by alkaloids of proboscis extension of the blowfly *Phormia regina*, *Physiol. Entomol.*, in press.
71. **Dethier, V. G. and Bodenstein, D.,** Hunger in the blowfly, *Z. Tierpsychol.*, 15, 129, 1958.
72. **Dethier, V. G. and Chadwick, L. E.,** Chemoreception in insects, *Physiol. Rev.*, 28, 220, 1948.
73. **Dethier, V. G. and Crnjar, R. M.,** Candidate codes in the gustatory system of caterpillars, *J. Gen. Physiol.*, 79, 549, 1982.
74. **Dethier, V. G. and Gelperin, A.,** Hyperphagia in the blowfly, *J. Exp. Biol.*, 47, 191, 1967.
75. **Dethier, V. G., Solomon, R. L., and Turner, L. H.,** Sensory input and central excitation and inhibition in the blowfly, *J. Comp. Physiol. Psychol.*, 60, 303, 1965.
76. **Deutsch, J. S., Gonzalez, M. F., and Young, M. G.,** Two factors control meal size, *Brain Res. Bull.*, 2 (Suppl. 4), 55, 1980.
77. **Diamond, J. M., Karasov, W. H., Phan, D., and Carpenter, F. L.,** Digestive physiology is a determinant of foraging bout frequencies in hummingbirds, *Nature*, 320, 62, 1986.
78. **Dickinson, M. H. and Lent, C. M.,** Feeding behavior of the medicinal leech, *Hirudo medicinalis*, *J. Comp. Physiol.*, A, 154, 449, 1984.
79. **Dorfman, K. A.,** Feeding Behavior in *Manduca sexta:* The Roles of Nutrition and Experience, Ph. D. dissertation, University of Massachusetts, Amherst, 1982.
80. **Eliot, E. J.,** Chemosensory stimuli in feeding behavior of the leech *Hirudo medicinalis*, *J. Comp. Physiol.*, A, 159, 391, 1986.
81. **Emden, H. F. van,** Aphids as phytochemists, in *Phytochemical Ecology*, Harborne, J. B., Ed., Academic Press, London, 1972, 25.

82. **Epstein, A. N.,** Feeding without oropharyngeal sensations, in *The Chemical Senses and Nutrition*, Kare, M. R. and Maller, O., Eds., The Johns Hopkins University Press, Baltimore, 1967, 263.
83. **Erickson, R.,** Sensory neural patterns and gustation, in *Olfaction and Taste I*, Zotterman, Y., Ed., Pergamon Press, New York, 1963, 205.
84. **Erickson, R.,** Neural coding of taste quality, in *The Chemical Senses and Nutrition*, Kare, M. R. and Maller, O., Eds., The Johns Hopkins University Press, Baltimore, 1967, 313.
85. **Erickson, R.,** Stimulus coding in topographic and nontopographic afferent modalities: on the significance of the activity of individual sensory neurons, *Psychol. Rev.*, 75, 447, 1968.
86. **Etscorn, F.,** Effects of a preferred vs a non-preferred CS in the establishment of a taste aversion, *Physiol. Psychol.*, 1, 5, 1973.
87. **Evans, D. R. and Barton Browne, L.,** Physiology of hunger in the blowfly, *Am. Midland Nat.*, 64, 282, 1960.
88. **Evans, D. R. and Mellon, DeF.,** Electrophysiological studies of a water receptor associated with the taste sensilla of the blowfly, *J. Gen. Physiol.*, 45, 487, 1962.
89. **Finger, T. E.,** Organization of chemosensory systems within the brains of bony fishes, in *Sensory Biology of Aquatic Animals*, Atema, J., Fay, R. R., Popper, A. N., and Tavolga, W. N., Eds., Springer-Verlag, New York, 1988, 339.
90. **Finger, T. E. and Morita, Y.,** Two gustatory systems: facial and vagal gustatory nuclei have different brainstem connections, *Science*, 227, 776, 1985.
91. **Fraenkel, G.,** Evaluation of our thoughts on secondary plant substances, *Entomol. Exp. Appl.*, 12, 473, 1969.
92. **Garcia, J. and Brett, L. P.,** Conditioned responses to food odor and taste in rats and wild predators, in *Chemical Senses and Nutrition*, Kare, M. R. and Maller, O., Eds., Academic Press, New York, 1977, 277.
93. **Garcia, J., Hankins, W. G., and Rusiniak, K. W.,** Behavioral regulation of the milieu interne in man and rat, *Science*, 185, 824, 1974.
94. **Garcia, J., Kimmeldorf, D. J., and Koelling, R. A.,** Conditioned aversion to saccharine resulting from exposure to gamma radiation, *Science*, 122, 157, 1955.
95. **Garcia, J. and Koelling, R. A.,** A comparison of aversions induced by X-rays, toxin, and drugs in the rat, *Radiation Res.*, 7, 439, 1967.
96. **Gelperin, A.,** Stretch receptors in the foregut of the blowfly, *Science*, 157, 208, 1967.
97. **Gelperin, A.,** Abdominal sensory neurons providing negative feedback to the feeding behavior of the blowfly, *Z. Vergl. Physiol.*, 72, 17, 1971.
98. **Gelperin, A.,** Rapid food-aversion learning in a terrestrial mollusk, *Science*, 189, 567, 1975.
99. **Gelperin, A.,** A taste for learning, *Am. Zool.*, 30, 549, 1990.
100. **Gelperin, A., Chang, J. J., and Reingold, S. C.,** Feeding motor program in *Limax*: neuromuscular correlates and control by sensory input, *J. Neurobiol.*, 9, 285, 1978.
101. **Gillette, R. and Davis, W. J.,** The role of the metacerebral giant neuron in the feeding behavior of *Pleurobranchaea*, *J. Comp. Physiol.*, A, 116, 129, 1977.
102. **Gillette, R., Kovac, M. P., and Davis, W. J.,** Control of feeding motor output by paracerebral neurons in the brain of *Pleurobranchaea californica*, *J. Neurophysiol.*, 47, 885, 1982.
103. **Granzow, B. and Kater, S. B.,** Identified higher-order neurons, controlling the feeding motor program of *Helisoma*, *Neuroscience*, 2, 1049, 1977.
104. **Grill, H. J. and Norgren, R.,** The taste reactivity test. I. Mimetic responses to gustatory stimuli in neurologically normal rats, *Brain Res.*, 143, 263, 1978.
105. **Hanamori, T. and Smith, D. V.,** Gustatory innervation in the rabbit: central distribution of sensory and motor components of the chorda tympani, glossopharyngeal, and superior laryngeal nerves, *J. Comp. Neurol.*, 282, 1, 1989.

106. **Hanamori, T., Miller, I. J., and Smith, D. V.,** Gustatory responsiveness of fibers in the hamster glossopharyngeal nerve, *J. Neurophysiol.*, 60, 478, 1988.
107. **Hanson, F. E. and Peterson, S. C.,** Sensory coding in *Manduca sexta* for deterrence by a non-host plant, *Canna generalis, Symp. Biol. Hung.*, 30, 29, 1990.
108. **Hassett, C. C., Dethier, V. G., and Gans, J.,** A comparison of nutritive values and taste thresholds of carbohydrates for the blowfly, *Biol. Bull.*, 99, 446, 1950.
109. **Herrick, C. J.,** The cranial nerves and cutaneous sense organs of North American siluroid fishes, *J. Comp. Neurol.*, 11, 177, 1901.
110. **Hodgson, E. S.,** Electrophysiological studies of arthropod chemoreception. II. Responses of labellar chemoreceptors of the blowfly to stimulation by carbohydrates, *J. Insect Physiol.*, 1, 240, 1957.
111. **Hsaio, T. H.,** Chemical influence on feeding behavior of *Leptinotarsa* beetles, in *Experimental Analysis of Insect Behavior,* Barton Browne, L., Eds., Springer-Verlag, Berlin, 1974, 237.
112. **Ishikawa, S.,** Electrical response and function of a bitter receptor associated with the maxillary sensilla of the silkworm, *Bombyx mori.* L., *J. Cell. Physiol.*, 67, 1, 1966.
113. **Jahan-Parwar, B.,** Behavioral and electrophysiological studies on chemoreception in *Aplysia, Am. Zool.*, 12, 525, 1972.
114. **Jahan-Parwar, B.,** Central projection of chemosensory pathways of *Aplysia, Physiologist,* 15, 180, 1972.
115. **Jermy, T.,** Feeding inhibitors and food preferences in chewing phytophagous insects, *Entomol. Exp. Appl.*, 9, 1, 1966.
116. **Jermy, T.,** The role of experience in the host selection of phytophagous insects, in *Perspectives in Chemoreception and Behavior,* Chapman, R. F., Bernays, E. A., and Stoffolano, J. G., Eds., Springer-Verlag, New York, 1987, 143.
117. **Jermy, T., Hanson, F. E., and Dethier, V. G.,** Induction of specific food preference in lepidopterous larvae, *Entomol. Exp. Appl.*, 11, 211, 1968.
118. **Jordan, H.,** Das Wahrnehmen der Nahrung bei *Aplysia limacina* und *Aplysia depilans, Biol. Zentrabbl.*, 37, 2, 1917.
119. **Jordan, H. A.,** Voluntary intragastric feeding: oral and gastric contributions to food intake and hunger in man, *J. Comp. Physiol. Psychol.*, 68, 498, 1969.
120. **Kaiser, F.,** Beiträge zur Bewegungsphysiologie der Hirudineen, *Zool. Jahrh. Abt. Allg. Zool. Physiol.*, 65, 59, 1954.
121. **Kandel, E. R.,** *Behavioral Biology of Aplysia,* W. H. Freeman, San Francisco, 1979, 313.
122. **Kaneko, C. R. S., Kater, S. B., and Merikel, I.,** Centrally programmed feeding in *Helisoma* controlled by electrically coupled network. I. Premotor neuron identification and characteristics, *Brain Res.*, 146, 1, 1978.
123. **Kare, M. R., Fregly, M. J., and Bernard, R. A., Eds.,** *Biological and Behavioral Aspects of Salt Intake,* Academic Press, New York, 1980.
124. **Kare, M. R. and Maller, O., Eds.,** *The Chemical Senses and Nutrition,* The Johns Hopkins University Press, Baltimore, 1967.
125. **Kare, M. R. and Maller, O., Eds.,** *The Chemical Senses and Nutrition,* Academic Press, New York, 1977.
126. **Kater, S. B.,** Feeding in *Helisoma trivolvis:* the morphological and physiological bases of a fixed action pattern, *Am. Zool.*, 14, 1017, 1974.
127. **Kater, S. B. and Fraser-Rowell, C. H.,** Integration of sensory and centrally programmed components in generation of cyclical feeding activity of *Helisoma trivolvis, J. Neurophysiol.*, 36, 142, 1973.
127a. **Kawamura, Y. and Yamamoto, T.,** Studies on neural mechanisms of the gustatory-salivary reflex in rabbits, *J. Physiol. (London)*, 285, 35, 1978.
128. **Kirk, M. D.,** Premotor neurons in the feeding system of *Aplysia californica, J. Neurobiol.*, 20, 497, 1989.

129. **Kiyohara, S.**, Anatomical studies of the facial taste system in teleost fish, in *The Beidler Symposium on Taste and Smell*, Miller, J., Ed., Book Services Assoc., Winston-Salem, NC, 1988.
130. **Kovac, M. P. and Davis, W. J.**, Behavioral choice: neural mechanisms in *Pleurobranchaea*, *Science*, 198, 632, 1977.
131. **Kupfermann, I.**, Feeding behavior in *Aplysia*. A simple system for the study of motivation, *Behav. Biol.*, 10, 1, 1974.
132. **Kupfermann, I.**, Dissociation of the appetitive and consummatory phases of feeding behavior in *Aplysia*. A lesion study, *Behav. Biol.*, 10, 89, 1974.
133. **Kupfermann, I. and Weiss, K. R.**, Activity of an identified serotonergic neuron in free moving Aplysia correlates with behavioral arousal, *J. Brain Res.*, 241, 334, 1982.
134. **Laverack, M. S.**, On the receptors of marine invertebrates, *Oceanogr. Marine Biol. Annu. Rev.*, 6, 249, 1968.
135. **Laverack, M. S.**, The diversity of chemoreceptors, in *Sensory Biology of Aquatic Animals*, Atema, J., Fay, R. R., Popper, A. N., and Tavolga, W. N., Eds., Springer-Verlag, New York, 287.
136. **Lee, J. C. and Bernays, E. A.**, Declining acceptability of a food plant for the polyphagous grasshopper *Schistocerca americana:* the role of food aversion learning, *Physiol. Entomol.*, 13, 291, 1988.
137. **Lee, R. M., Robbins, M. R., and Palovcik, R.**, *Pleurobranchaea* behavior: food finding and other aspects of feeding, *Behav. Biol.*, 12, 297, 1974.
138. **Lent, C. M. and Dickinson, M. H.**, Serotonin integrates the feeding behavior of the medicinal leech, *J. Comp. Physiol.*, A, 154, 457, 1984.
139. **Lent, C. M. and Dickinson, M. H.**, On the termination of ingestive behavior by the medicinal leech, *J. Exp. Biol.*, 131, 1, 1987.
140. **Lent, C. M., Dickinson, M. H., and Marshall, C. G.**, Serotonin and leech feeding behavior. Obligatory neuromodulation, *Am. Zool.*, 24, 1241, 1989.
141. **Lent, C. M., Fliegner, K. H., Freedman, E., and Dickinson, M. H.**, Ingestive behavior and physiology of the medicinal leech, *J. Exp. Biol.*, 137, 513, 1988.
142. **Lent, C. M., Zundel, D., Freedman, E., and Groome, J. R.**, Serotonin in the leech central nervous system: anatomical correlates and behavioral effects, *J. Comp. Physiol.*, A, 168, 191, 1991.
143. **Livanow, N. A.**, Klass Piyavot (Hirudinae), *Rukovodstvo Po Zool.*, 2, 203, 1940.
144. **Logue, A. W.**, Conditioned food aversion learning in humans, *Ann. N.Y. Acad. Sci.*, 443, 316, 1985.
145. **Ma, W. C.**, Dynamics of feeding responses in *Pieris brassicae* Linn. as a function of chemosensory input: a behavioral, ultrastructural, and electrophysiological study, *Meded. Landbouwhogesch. Wageningen*, 72, 1972.
146. **Maddrell, S. H. P.**, Control of ingestion in *Rhodnius prolixus*, *Nature*, 198, 210, 1963.
147. **Maes, F. W.**, A neural coding model for sensory intensity discrimination, to be applied to gustation, *J. Comp. Physiol.*, A, 155, 263, 1984.
148. **Mann, K. H.**, *Leeches (Hirudinae). Their Structure, Physiology, Ecology, and Embryology*, Pergamon Press, London, 1962.
149. **Marshall, C. G. and Lent, C. M.**, Excitability and secretory activity in salivary glands of jawed leeches (Hirudinea: Gnalhobdellida), *J. Exp. Biol.*, 137, 89, 1988.
150. **McCrohan, C. R.**, Initiation of feeding motor output by an identified interneurone in the snail, *Lymnaea stagnalis*, *J. Exp. Biol.*, 113, 351, 1984.
151. **McCrohan, C. R.**, Properties of ventral cerebral neurones involved in the feeding system of the snail *Lymnaea stagnalis*, *J. Exp. Biol.*, 108, 257, 1984.
152. **McCrohan, C. R. and Audesirk, T. E.**, Initiation, maintenance and modification of patterned buccal motor output by the cerebral giant cells of *Lymnaea stagnalis*, *Comp. Biochem. Physiol.*, 87A, 969, 1987.

153. **McLean, D. L. and Kinsey, M. G.,** Probing behavior of the pea aphid *Acrythosiphon pisum*. II. Comparisons of salivation and ingestion in host and non-host plant leaves, *Ann. Entomol. Soc. Am.*, 61, 730, 1968.
154. **Merickel, M. B., Eyman, E. D., and Kater, S. B.,** Analysis of a network of electrically coupled neurons producing rhythmic activity in the snail *Helisoma trivolvis*, *I.E.E.E. Trans. Biomed. Engr.*, 24, 277, 1977.
155. **Miller, I. J. and Smith, D. V.,** Quantitative taste bud distribution in the hamster, *Physiol. Behav.*, 32, 275, 1984.
156. **Miller, J. R. and Miller, T. A.,** *Insect-Plant Interactions*, Springer-Verlag, New York, 1986.
157. **Mitchell, B. K. and Harrison, G. D.,** Effects of *Solanum* glycoalkaloids on chemosensilla in the Colorado potato beetle. A mechanism of feeding deterrence?, *J. Chem. Ecol.*, 11, 73, 1985.
158. **Mitchell, B. K. and Sutcliffe, J. F.,** Sensory inhibition as a mechanism of feeding deterrence: effects of three alkaloids on leaf beetle feeding, *Physiol. Entomol.*, 9, 57, 1984.
159. **Morita, H.,** Initiation of spike potentials in contact chemosensory hairs of insects. III. D.C. stimulation and generator potentials of labellar chemoreceptor of *Calliphora*, *J. Cell. Comp. Physiol.*, 54, 189, 204.
160. **Morita, H., Hidaka, T., and Shiraishi, A.,** Excitatory and inhibitory effects of salts on the sugar receptor of the fleshfly, *Mem. Fac. Sci. Kyushu Univ.* E4, 123, 1966.
161. **Morita, H., Enomoto, K.-I., Nakashima, M., Shimada, I., and Kijima, H.,** The receptor sites for sugars in chemoreception of the fleshfly and blowfly, in *Olfaction and Taste VI*, LeMagnen, J. and MacLeod, P., Eds., Information Retrieval, London, 1977, 39.
162. **Morita, H. and Yamashita, S.,** Further studies on the receptor potential of the blowfly, *Mem. Fac. Sci. Kyushu Univ.*, E4, 83, 1966.
163. **Morita, Y. and Finger, T. E.,** Reflex connections of the facial and gustatory systems in the brainstem of the bullhead catfish *Ictalurus nebulosus*, *J. Comp. Neurol.*, 231, 547, 1985.
164. **Moss, C. F. and Dethier, V. G.,** Central nervous system regulation of finicky feeding by the blowfly, *Behav. Neurosci.*, 97, 541, 1983.
165. **Mpitsos, G. J. and Collins, S. D.,** Learning: rapid aversive conditioning in the gastropod mollusk *Pleurobranchaea*, *Science*, 180, 317, 1975.
166. **Murray, M. J.,** The Biology of a Carnivorous Mollusk: Anatomical, Behavioral, Electrophysiological Observations on Navanax inermis, Ph.D. dissertation, University of California, Berkeley, 1971.
167. **Nayer, J. K. and Thorsteinson, A. J.,** Further investigations into the chemical basis of insect-host relationships in an oligophagous insect, *Plutella maculipennis* (Curtis), *Can. J. Zool.*, 41, 923, 1963.
168. **Nielsen, J. K.,** Host plant selection of monophagous and oligophagous flea beetles feeding on crucifers, *Entomol. Exp. Appl.*, 24, 562, 1978.
169. **Paine, R. T.,** Food recognition and predation on opisthobranchs by *Navanax inermis* (Gastropoda: Opisthobranchia), *Veliger*, 6, 1, 1963.
170. **Parker, G. H.,** *Smell, Taste and Allied Senses in Vertebrates*, J. B. Lippincott, Philadelphia, 1922.
171. **Pfaffmann, C.,** Gustatory afferent impulses, *J. Cell. Comp. Physiol.*, 17, 243, 1941.
172. **Pfaffmann, C.,** Physiological and behavioral processes of the sense of taste, in *Taste and Smell in Vertebrates*, Wolstenholme, G. E. W. and Knight, J., Eds., J. & A. Churchill, London, 1955, 31.
173. **Pfaffmann, C.,** The sensory and motivating properties of the sense of taste, in *Nebraska Symposium on Motivation*, Jones, M. R., Ed., Nebraska Press, Lincoln, 1961.

174. **Preston, R. J. and Lee, R. M.,** Feeding behavior in *Aplysia californica*. Role of chemical and tactile stimuli, *J. Comp. Physiol. Psychol.*, A, 82, 368, 1973.

175. **Prior, D. J. and Gelperin, A.,** Autoactive molluscan neuron: reflex function and synaptic modulation during feeding in the terrestrial slug, *Limax maximus*, *J. Comp. Physiol.*, A, 114, 217, 1977.

176. **Rachman, N. J.,** Physiology of feeding preference patterns of female black blowflies (*Phormia regina* Meigen). I. The role of carbohydrate reserves, *J. Comp. Physiol.*, 139, 59, 1980.

177. **Reingold, S. C. and Gelperin, A.,** Feeding motor program in *Limax*. III. Modulation by sensory inputs in intact animals and isolated nervous systems, *J. Exp. Biol.*, 85, 1, 1980.

178. **Reynolds, S. E., Yoemans, M. R., and Timmins, W. A.,** The feeding behavior of caterpillars (*Manduca sexta*) on tobacco and on artificial diet, *Physiol. Entomol.*, 11, 39, 1986.

179. **Rosen, S. C., Miller, M. W., Weiss, K. R., and Kupfermann, I.,** Control of buccal motor programs in *Aplysia* by identified neurons in the cerebral ganglion, *Soc. Neurosci. Abstr.*, 13, 1061, 1987.

180. **Rozin, P.,** Specific hunger for thiamin: Recovery from deficiency and thiamin preference, *J. Comp. Physiol. Psychol.*, 59, 98, 1965.

181. **Rozin, P.,** Specific aversion as a component of specific hungers, *J. Comp. Physiol. Psychol.*, 64, 237, 1967.

182. **Rozin, P.,** The selection of foods by rats, humans, and other animals, *Adv. Study Behav.*, 4, 21, 1976.

183. **Rozin, P.,** Human food selection: the interaction of biology, culture and individual preference, in *The Psychobiology of Human Food Selection*, Barker, L. M., Ed., AVI, Bridgeport, CN, 1982, 225.

184. **Rozin, P. and Kalat, J. W.,** Specific hungers and poison avoidance as adaptive specializations of learning, *Psychol. Rev.*, 78, 459, 1971.

185. **Rozin, P. and Vollmecke, T. A.,** Food likes and dislikes, *Annu. Rev. Nutr.*, 6, 433, 1986.

186. **Sawyer, R. T.,** *Leech Biology and Behavior*, Vols. 1–3, Clarendon Press, Oxford, 1986.

187. **Schoonhoven, L. M.,** Chemosensory bases of host plant selection, *Annu. Rev. Entomol.*, 13, 115, 1968.

188. **Schoonhoven, L. M.,** Gustation and food plant selection in some lepidopterous larvae, *Entomol. Exp. Appl.*, 12, 535, 1969.

189. **Schoonhoven, L. M.,** Secondary plant substances and insects, *Recent Adv. Phytochem.*, 4, 197, 1972.

190. **Schoonhoven, L. M.,** Plant recognition by lepidopterous larvae, in *Insect Plant Relationships*, van Emden, H. F., Ed., Blackwell Scientific, Oxford, 1973.

190a. **Schoonhoven, L. M.,** Chemical mediators between plants and phytophagous insects, in *Semiochemicals: Their Role in Pest Control*, Nordlund, D. A., Ed., John Wiley & Sons, New York 1981, 31.

191. **Schoonhoven, L. M.,** What makes a caterpillar eat?, The sensory code underlying feeding behavior, in *Perspectives in Chemoreception and Behavior*, Chapman, R. F., Bernays, E. A., and Stoffolano, J. G., Springer-Verlag, New York, 1987, 67.

192. **Schoonhoven, L. M. and Dethier, V. G.,** Sensory aspects of host-plant discrimination by lepidopterous larvae, *Arch. Neerl. Zool.*, 16, 497, 1966.

193. **Schwartz, G. J. and Grill, H. J.,** Relationships between taste reactivity and intake in the neurologically intact rat, *Chem. Senses*, 9, 249, 1984.

194. **Sclafani, A. and Springer, D.,** Dietary obesity in adult rats. Similarities to hypothalamic and human obesity syndromes, *Physiol. Behav.*, 17, 461, 1976.

195. **Seelinger, G.,** Response characteristics and specificity of chemoreceptors in *Hemilepistus reaumuri* (Crustacea: Isopoda), *J. Comp. Physiol.*, 152, 219, 1983.

196. **Simpson, S. J.,** Patterns in feeding: a behavioral analysis using *Locusta migratoria* nymphs, *Physiol. Entomol.*, 7, 325, 1982.
197. **Simpson, S. J. and Abisgold, J. D.,** Compensation by locusts for changes in dietary nutrients: behavioral mechanisms, *Physiol. Entomol.*, 10, 443, 1985.
198. **Simpson, C. L., Simpson, S. J., and Abisgold, J. D.,** The role of various amino acids in the protein compensatory response of *Locusts migratoria, in Insect-Plant 89*, Szentesi, A. and Jermy, T., Eds., Akadémiai Kiadó, Budapest, 1991, 39.
199. **Simpson, S. J. and Simpson, C. L.,** The mechanisms of nutritional compensation by phytophagous insects, in *Insect-Plant Interactions II*, Bernays, E. A., Ed., CRC Press, Boca Raton, FL, 1990, 111.
200. **Smith, D. V., Bowdan, E., and Dethier, V. G.,** Information transmission in tarsal sugar receptors of the blowfly, *Chem. Senses*, 1, 81, 1983.
201. **Smith, J. C., Williams, D. P., and Jue, S. S.,** Rapid oral mixing of glucose and saccharine by rats, *Science*, 191, 304, 1976.
202. **Snodgrass, R. E.,** *Principles of Insect Morphology*, McGraw-Hill, New York, 1935.
203. **Spray, D. C. and Bennett, M. V. L.,** Pharyngeal sensory neurons and feeding behavior in the opisthobranch mollusk *Navanax, Neurosci. Abstr.*, 1, 570, 1975.
204. **Spray, D. C. and Bennett, M. V. L.,** Proprioceptive inputs to large buccal motor neurons controlling pharyngeal expansion in *Navanax, Fed. Proc.*, 34, 418, 1975.
205. **Stellar, J. R. and Stellar, E.,** *The Neurobiology of Motivation and Reward*, Springer-Verlag, New York, 1985.
205a. **Storey, A. T. and Johnson, P.,** Laryngeal water receptors initiating apnea in the lamb, *Exp. Neurol.*, 47, 42, 1975.
206. **Susswein, A. J. and Kupfermann, I.,** Localization of bulk stimuli underlying satiation in *Aplysia, J. Comp. Physiol.*, 101, 309, 1975.
207. **Susswein, A. J. and Kupfermann, I.,** The stimulus control of biting in *Aplysia, J. Comp. Physiol.*, 108, 75, 1976.
207a. **Sweazey, R. D. and Bradley,** Central connections of the lingual-tonsillar branch of the glassopharyngeal nerve and the superior laryngeal nerve in lamb, *J. Comp. Neurol.*, 245, 471, 1986.
208. **Szentesi, A. and Jermy, T.,** The role of experience in host plant choice by phytophagous insects, in *Insect-Plant Interactions III*, Bernays, E. A., Ed., CRC Press, FL, Boca Raton, 1990, 39.
209. **Szentesi, A. and Jermy, T., Ed.,** *Insect-Plants '89. Symp. Biol. Hung.*, Vol. 39, Akadémiai Kiadó, Budapest, 1991.
209a. **Travers, J. B. and Norgren, R.,** Electromyographic analysis of the ingestion and rejection of rapid stimuli in the rat, *Behav. Neurosci.*, 100, 544, 1986.
210. **Valenstein, E. S., Cox, V. C., and Kakolewski, J. W.,** Polydipsia elicited by the synergistic action of a saccharin and glucose solution, *Science*, 157, 552, 1967.
211. **Verschaffelt, E.,** The cause determining the selection of food in some herbivorous insects, *Proc. Roy. Acad. Amsterdam*, 13, 536, 1910.
212. **Vyas, N. K., Vyas, M. N., and Quiocho, F. A.,** Sugar and signal-transducer binding sites of the *Escherichia coli* galactose chemoreceptor protein, *Science*, 242, 1290, 1988.
213. **Waldbauer, G. P. and Fraenkel, G.,** Feeding on normally rejected plants by maxillectomized larvae of the tobacco hornworm, *Protoparce sexta* (Lepidoptera, Sphingidae), *Ann. Entomol. Soc. Am.*, 54, 477, 1961.
214. **Whitman, C. O.,** The metamerism of *Clepsine, Festschr. Z.* 70 *Geburtstage R. Keuckarts.*, 385, 1892.
215. **Wiggins, L. L., Frank, R. A., and Smith, D. V.,** Generalization of learned taste aversions in rabbits: similarities among gustatory stimuli, *Chem. Senses*, 14, 103, 1989.
216. **Young, P. T.,** The role of hedonic processes in motivation, in *Nebraska Symposium on Motivation*, Jones, M. R., Ed., University of Nebraska Press, Lincoln, 1955, 193.

Morphology of Taste Receptors

Chapter 2

MORPHOLOGY OF VERTEBRATE TASTE ORGANS AND THEIR NERVE SUPPLY

Klaus Reutter and Martin Witt

TABLE OF CONTENTS

0-8493-5341-6/93/$0.00 + $.50

I. INTRODUCTION

Taste buds (TBs) are the peripheral gustatory organs belonging to the gustatory (chemosensory) system which occurs in all groups of vertebrates. Due to the different habitat of different vertebrates, these organs are situated at different positions of the animal body. In fish, TBs occur not only within the oral cavity, but may also occur in the outer skin. In amphibians, in some developmental stages TBs are found within the oral cavity and at the head's skin, also. In air-living animals such as reptiles, birds, and mammals, including man, TBs are restricted to the moist environment of the oral cavity. In each case, TBs are intraepithelially positioned, and their cellular components derive from epithelial cells or, at least, from the ectoderm or the neural crest, respectively.

At first glance, TBs seem to be organized similarly throughout the classes of vertebrates. Normally, they are of ovoid shape and consist of slender cells which run parallel to the longitudinal axis of the bud. But, on the other hand, in frogs the taste organs are flat and disk-like and therefore called taste disks (TDs). The apices of TBs also differ: In fish and frogs, the microvillar end structures of TB cells form together the so-called receptor area*, whereas in birds, reptiles, and especially in mammals (including man) the microvilli project into a relatively deep taste pit. In addition, there are different specialized structures such as gland cells, in and around the TBs, which might fulfill auxiliary functions during the chemosensory events.

Taste organs were first described in the middle of the last century. Waller (1849)[1] investigated the "taste disk" of the frog and Leydig (1851)[2] found the "Geschmacksbecher" in fish. In mammals, the TBs were first and independently described by Lovén (1867 preliminary report in Swedish, 1868)[3] and Schwalbe (1867, 1868)[4,5] and simultaneously named "Schmeckbecher", "Geschmackszwiebel", and "Geschmacksknospe". Since these early times numerous papers concerned with TBs from representatives of different vertebrate groups have appeared. The older work is extensely reviewed by Oppel,[6] von Ebner,[7] Herrick,[8] Kolmer,[9] and Boeke.[10] Newer and therefore more electron microscopic data are compiled and compared by Graziadei,[11] Jeppsson,[12] Murray,[13-15] Murray and Murray,[16,17] Reutter,[18] and Roper.[19] This paper will concentrate on the more recent findings in TB morphology.

* The often used term *receptor field* should be replaced by *receptor area* because receptor field is easily mixed up with *receptive field* which means a cluster of several neighboring TB-bearing papillae (2-4) which is innervated by the branches of the same nerve fiber (Oakley, B., Receptive fields of cat taste fibers, *Chem. Senses Flavour*, 1, 431, 1975).

II. TASTE ORGANS IN DIFFERENT VERTEBRATE GROUPS

A. TASTE BUDS IN FISH

Most of the work done in the last two decades in fish TB is concerned with the one of numerous teleost species, and the respective literature has been reviewed several times.[18-28]

Beside all the differences between teleostean species there seems to exist one basic type of TBs in bony fishes. Such TBs are typically ovoid or pear-shaped; their longitudinal axis is oriented vertically to the surface of the epidermis of the outer skin or the epithelium of the oral cavity, the pharynx, esophagus, or gill rakers. Normally TB-bearing epithelia are stratified and squamous and not keratinized on the apical surface.

In the scanning electron microscope (SEM), a TB is visualized by its receptor area. This consists of the apical endings of TB cells containing large and small receptor villi (Figure 1).[11,18,29-39] As a rule, these villi become visible only after removal of the mucus lying on the TB surface. At very exposed positions the TBs are situated in epidermal hillocks and therefore elevated over the average level of the epidermis, whereas in other cases they are sunken. Fish TBs may be classified as type I (elevated), type II (slightly elevated), or type III (not elevated or sunken).[29,40] Possibly the elevated TBs may be deflected by food or other particles and thus may also serve as mechanoreceptors.[18,40,41]

As seen in longitudinal sections (light microscopy, LM and transmission electron microscopy, TEM), TBs rest on papillae of the corium (dermis) of different heights which project into the thinner epithelium. As intraepithelial formations, the greater part of a TB consists of its sensory epithelium respective to the TB cells. At the borderline between the TB and the normal epidermis, marginal cells are located. If the TB belongs to the elevated type, the marginal cells, to some extent, form the TB's hillock. The basal pole of a TB is formed by up to five basal cells. These are situated directly on the basal membrane which demarcates the TB from the connective tissue of the corium papilla. Basal cells are oriented transversely to the longitudinal axis of the TB. The TB's nerve fiber plexus is located between the basal cells and the basal parts of the sensory epithelium's cells. The nerve fibers of the plexus derive from the TB's nerve. It is located in the corium papilla and consists of several poorly myelinated nerve fibers. The fibers penetrate the basal membrane at one or two distinct holes where they lose their myelin sheaths. Another essential structure of the corium papilla is a capillary blood vessel which reaches in a loop-like manner the base of the TB. They are seen in corrosion casts of resin (mercox)-injected papillary blood vessels.[41] The gap between the blood vessel and the basal lamina normally is filled by transversely oriented fibrocytes. Fibrocytes, reticular and collagenous connective fibers, and, in some cases, melanocytes are common structures of the corium papilla.

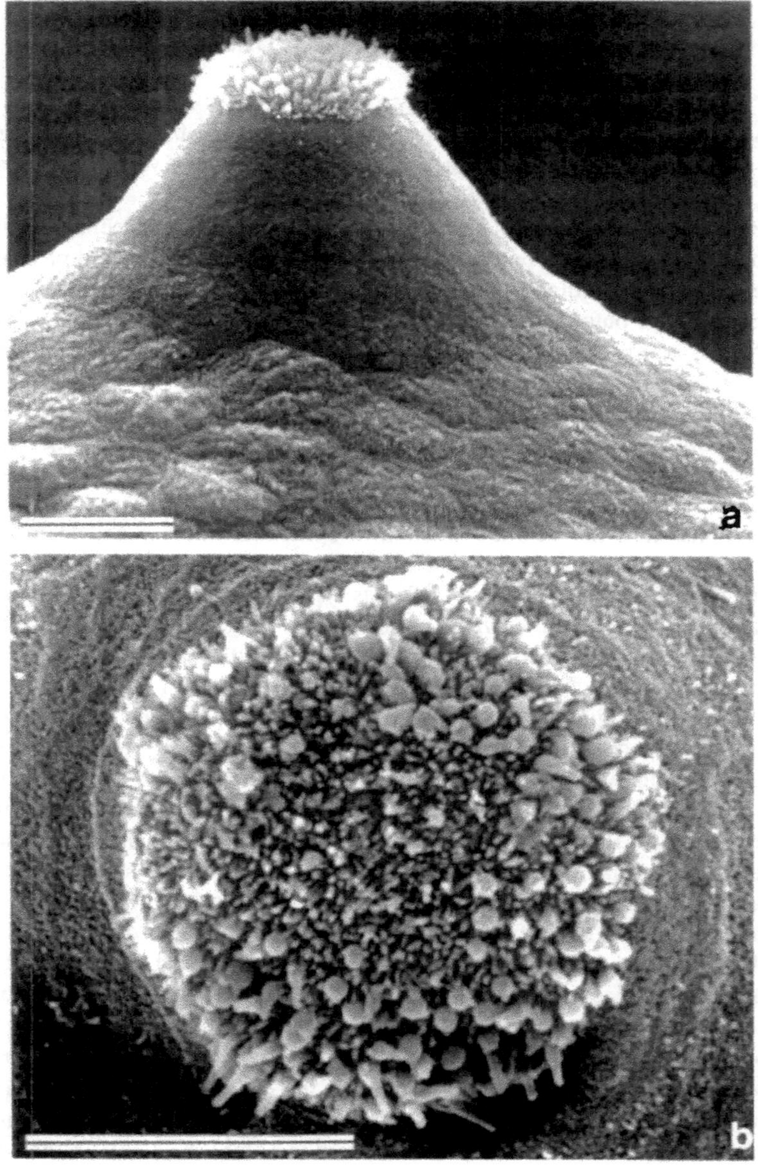

FIGURE 1. Scanning electron micrographs of taste buds from the teleost fish, *Ictalurus* = *Ameiurus nebulosus*. (a) Lateral view of a barbel piece with a TB situated in an epidermal hillock. On top of the hillock is the TBs receptor field. (b) Vertical view to the receptor field consisting of small and large receptor villi. (a) ×1800; bar: 10 μm. (b) ×4350; bar: 10 μm.

B. TASTE ORGANS IN AMPHIBIANS

In contrast to the more or less uniformly organized TBs of fish the taste organs of amphibians present themselves in two distinctly different forms. In the Caudata (Urodela) the taste organs are typically bud-like, more or less roundish (or ovoid) in shape, and, to some extent, similar to the TBs in fish. In contrast, the taste organ of the Salientia (Anura) and especially of frogs (*Rana* sp.) is a flat and disk-like sensory epithelium situated at the distal end of a relatively large fungiform papilla.

1. Taste Buds in Caudata

Most of the electron microscopic work done in Caudata was performed using the mudpuppy, *Necturus*,[42-45] and the axolotl, *Ambystoma*.[46-48] In these animals TBs are situated for the most part in the dorsal surface epithelium of the tongue, but some are also seen in the palatal mucosa and, in the axolotl, at the gill arches. In the SEM, the places where TBs are identified appear as narrow mounds or hillocks which, at their apices, bear the TB receptor area.[43,47,48] The receptor area consists of different microvillar structures, i.e., the apical endings of the sensory epithelium. In a longitudinal section, the sensory epithelium consists of two types of elongated cells and more roundish and basally situated cells called basal cells. The sensory epithelium is up to 200 μm in width and up to 120 μm in length (*Necturus;* References 42,43). The underlying connective tissue may form a small corium papilla on which the TB rests. Nerve fibers ascend to the TB and penetrate the basal lamina which demarcates the epithelial from the underlying connective tissue. It also contains blood capillaries. Within the TB, nerve fibers are situated more basally and do not form an extensive nerve fiber plexus.[42,43,46] The border between the sensory epithelium of the bud and the nonspecialized squamous and stratified epithelium is formed by marginal cells (Randzellen; Reference 46).

2. Taste disks in Salientia

TDs in Salientia have diameters of up to 200 μm and therefore are easily identified. They are located on top of fungiform papillae which occur on the dorsal side of the tongue. In other regions of the oral cavity, as in its floor and palatal epithelia, flat sensory epithelia are found as nonpapillary taste organs.[49,50] TDs were subject to numerous physiological and morphological investigations. SEM and TEM studies were most often concerned with the TDs of real frogs (Ranidae).[50-58,179,180] However, some papers also report about the smaller TDs of toads (Bufonidae, Discoglossidae, Aglossidae).[59-63,180]

In longitudinal sections (LM, TEM) the sensory epithelium of a TD shows three layers of cells: superficial stratum, intermediate stratum, and deep or basal stratum. These strata especially contain the nuclei of cells which belong to different types of cells with different functions, including the sensory cells which with their apical processes reach the TD surface, the receptor area (see below). As clearly seen in the SEM, the sensory epithelium of a frog's TD is lined by a ring of ciliated cells. The border of the TD and also of the sides of the papilla are formed by marginal cells which are nonsensory epithelial cells. The sensory epithelium rests on connective tissue, the dermal core of the fungiform papilla. The basal lamina between the sensory epithelium and the connective tissue is penetrated several times by unmyelinated nerve fibers. Then these fibers lie in the basal stratum and form synaptic contacts with (especially) the sensory cells (see below). Blood capillaries are common in the dermal core which also contains muscle fibers arranged obliquely or longitudinally to its axis.

On the whole, the smaller TDs of toads are similar to frog TDs. However, it should be noted that the outer ring of ciliated cells is lacking and that there are some differences with regard to cellular specializations (see below).

C. TASTE BUDS IN REPTILES

There exist only a few papers dealing with the ultrastructure of reptile TBs.[64-68] One reason for this might be that reptiles are not very popular laboratory animals.

In the red-eared turtle, TBs lie on the rostral part of its nonprotusible tongue.[64] In tortoises and lizards, TBs occur on tongues and oral epithelia of the maxillae and mandibles. Snakes have TBs exclusively in the stratified squamous epithelium along the dental arch.[66]

TBs of the turtle are recognized by their roundish taste pores in the center of which the apical processes of the TB cells are situated: normal and, less common, branched or brush-like microvilli. The receptor areas are not elevated and therefore TB-bearing epidermal mounds do not occur (SEM).[64] Longitudinally sectioned TBs (TEM) of turtles[64] and tortoises as well as snakes and lizards[65] are barrel-like or ovoid in shape and differ in height and width from 35 μm × 30 μm (turtle, lizard, snake) to 120 μm × 65 μm (tortoise). These TB cells are more or less parallelly oriented; they form a somewhat depressed receptor area, but not a pronounced taste pit. Nerve fiber profiles are regularly found in the basal part of the bud, but a nerve fiber plexus is lacking. The TBs are positioned in the middle of the stratified squamous epithelium, and therefore "ascending" and TB-bearing dermal papillae are not present. In shape, reptilian TBs resemble those of mammals.

D. TASTE BUDS IN BIRDS

Only a few SEM[69] and TEM[68-73] studies on bird TBs have been performed. As birds and reptiles belong to the same superclass (subphylum), the sauropsida, TBs of both classes should be expected to have a similar organization. However, the few examples of bird TBs showed a great variability between different species. This may arise because birds in different habitats often eat different foods and have different mechanisms of food intake. Unfortunately, the few studies on bird TBs were performed mostly with grain-eating birds (e.g., pigeon [*Columba*],[69] canary, parrot [*Serinus* and *Melopsittacus*],[70] chicken, [*Gallus*],[71,72] and quail [*Coturnix*]).[73,74] Birds' TBs appear at the basal posterior part of the tongue, around and in the proximal part of the pharynx, and in the distal palatal mucosa. The middle and distal parts of the avian tongue are TB-free.[71-73]

Within the pore of canary TBs, lots of short microvilli are present, whereas parrot TBs contain some longer microvilli.[70] In these cases the TB sensory epithelium is positioned directly under the epidermal surface, as to be seen in the LM and the TEM on longitudinally cut TBs. This is in contrast to the TBs of the chicken[71,72] and the quail.[73,74] In these species, the TBs are deeply positioned within the extremely thick (up to 250 μm) stratified epithelium so that the apical microvilli of the TB cells do not reach the epidermal surface, but project into a tube-like, curved, and narrow TB-pit (Fig. 11 d). As TBs are regularly associated with serous glands, the TBs "channel" opens also into the excretory duct of the glands.[71-74] The TB itself is about 70 μm high and 35 μm wide and of elongate-ovoid shape. The TB epithelium is located on top of an often high, narrow cup-like papilla of the dermis which projects into the base of the epithelium. Nerve fibers may form a plexus-like situation at the distal end of the papilla and penetrate the basal lamina. The latter always demarcates the dermal papilla. Within the TB a distinct nerve fiber plexus is lacking. Blood capillaries regularly underlie the TB.[73]

E. TASTE BUDS IN MAMMALS

In the last three decades numerous SEM and TEM TB investigations were done on various species of mammals. Most of the work was concentrated on popular laboratory animals such as rodents (mouse,[75-81] rat,[82-85] hamster,[86-88] guinea pig[89]) and rabbits (reviews in References 12–17 and 90–99). There also exist some studies of TBs on monkey,[100-104,184] human,[105,106,184-186] dog,[107] and camel.[108]

In mammals, TBs occur throughout the oral cavity, in the middle and the lower parts of the pharynx, in the epiglottis as part of the larynx, and at the entrance of the esophagus. TBs occur in greater numbers on the dorsal surface of the tongue, where they are located in three types of gustatory papillae. The numerous (200 and more) fungiform papillae are located predominantly at the tip and anterior two thirds of the tongue and bear 1 to 20 TBs in their apical epithelium. There are always only two foliate papillae on the posterior lateral sides of the tongue. These papillae consist of several foliae which contain TBs in great numbers in the epithelia of their lateral sides. The (circum)vallate papillae are situated at the posterior dorsal border between the oral and pharyngeal parts of mammals tongue. These papillae are the largest and contain numerous TBs at the lateral sides. They also vary in number: 1 in the mouse and rat, 8 to 12 in humans, and large numbers in cow and sheep. In this regard, neither cows nor sheep have foliate papillae. In all species, the numerous filiform and conical papillae which are present throughout the surface of the tongue do not bear TBs (data from Mistretta).[109]

Mammalian TBs possess taste pores. At the dorsal surfaces of fungiform papillae, the taste pores become easily visible, whereas the TB regions of vallate and foliate papillae must be cut free as they are within the side walls of the deep trenches between the vallate papilla and the neighboring epithelium or between the foliae of a foliate papilla (SEM).[11,95,107,108,110,111] The mammalian TB pores are encircled by microridges bearing desquamating epithelial cells and are about 4 μm[95] to about 10 μm[108] in diameter. Within the TB pores the "taste hairs" — the microvillar apical endings of the sensory epithelium of different shape — are present. To some extent, the villi may be club-shaped and protrude into the pore.

Longitudinally sectioned TBs are typically ovoid, although they are sometimes of more roundish shape. They might be located on top of calyx-like papillae of the dermis and then project into a relatively thick epithelium (as it is especially the case in fungiform papillae). On the other hand, they might be as high as the TB-bearing epithelium itself and then placed flat on the basal lamina between the dermis and the TB respective to the epithelium, as in vallate and foliate papillae. At the TB's bottom there is a hole in the basal lamina through which the nerve fibers enter the bud; this is named "basal pore" of the TB.[79] From the basal pore the TB nerve fibers derive, as especially seen in vallate papillae. The connective tissue underlying the TBs also contains capillary blood vessels as especially seen in SEMs of resin microcorrosion casts done in the tongue of the kitten,[111] dog,[112] and rat.[113]

The TB itself or its sensory epithelium is of variable size. In rabbits, for example, it varies between 50 to 60 μm (height) and 30 to 70 μm (width).[99] The elongated slender TB cells extend from the TB base up to the taste pit which contains the microvillar "taste hairs". The taste hairs project into an electron-dense mucus of the taste pit and may reach the epithelial surface. The apical margin of the pit reaches the epithelial surface and forms the TB pore.

The excretory ducts of the von Ebner's lingual glands open at the bases of the furrows, crypts, or trenches of the vallate and foliate papillae. The seromucous secretions of the glands are thought to have three functions: (1) removal of detritus and general cleaning of the crypt of the papilla, (2) assistance in the process of taste by circulation of tastants, and (3) circulation of the von Ebner glands' secretions and other oral fluids which bathe the pores of the TBs.[115] This is likely, as kinocilia-bearing cells occur in vallate papillae between the opening of the glands' duct and the TB near the bottom of the trench. These cells likely serve to propel secretions from the von Ebner's glands (man[115]; rat[116]).

III. TASTE ORGAN CELLS

A. TASTE BUD CELLS AND THEIR SYNAPSES IN FISH

The sensory epithelium of the (teleost) fish TB contains two main types of elongated cells running parallel to the longitudinal axis of the bud, from its base to the apically situated receptor area: light cells and dark cells. (They also contain intermediate and degenerative cells.) TB cells terminate with large and small receptor villi which form together the receptor area. The nerve fiber plexus is interposed between the basal processes of the TB cells and basal cells which form the base of the TB. The interfacial zone between the TB and the normal or unspecialized epithelium cells is marked by the marginal cells.

This is the teleost-TB-cell nomenclature which now is preferred.[25] In previous work we used "light sensory cells" and "dark sensory cells" (e.g., References 18, 23, 24, 117). These terms contradict the "sensory cell", "gustatory cell" or "receptor cell", and the "sustentacular cell" or "supporting cell" nomenclature as favored and discussed extensely by Whitear,[118] Jakubowski,[41] and Jakubowski and Whitear.[28,119] Unfortunately, this nomenclature suggests that synaptic contacts only occur on proven gustatory cells, i.e. those which normally appear lighter than the other cells. In some cases, synapses are found in dark cells, too, and therefore these cells must also be sensory. This does not exclude that dark cells possibly have additional functions, such as supportive or nutritive ones. In morphology, where we often are unsure about what a darker cell really represents, it seems to be more convenient to simply use the term "dark cell", keeping in mind that this cell might be sensory, as well. On the other hand, the light cells are not problematic: they are unanimously considered to be the main sensory cells. The discussed controversy resembles, to some extent, the one in mammalian TB nomenclature.

The following description of fish TB cell types is in general agreement with morphological data from other laboratories,[34,36,38,39,41,117-136] although we are aware that not everyone will agree with these interpretations.

Light cells — (Figures 2 to 4) These cells appear clear and electron-lucent when fixed with glutaraldehyde and osmium tetroxide and stained with lead citrate and uranyl acetate.[25] These cells are elongated, slender, and wider only in the nuclear region. Apically, light cells terminate into one club-shaped or conoid microvillus, named large receptor villus (Figure 2). The bases of light cells are divided into several processes. Cut transversely, these cells are roundish. The nucleus of a light cell is somewhat elongated, often lobed, and relatively light. Its nucleoplasm shows compartments of hetero- and euchromatin. As seen in Figure 4a, the nucleus may contain a paracrystalline inclusion. The nuclear envelope is rich in nuclear pores and has a well-developed perinuclear cisterna (Figure 4a). The cytoplasm is rich in organelles. The supranuclear region is rich in Golgi systems, whereas the apical third of the cell is Golgi-free. Vesicular structures of different sizes occur around the Golgi systems. Rough endoplasmic reticulum is rare, but smooth endoplasmic reticulum is common. Profiles of the smooth endoplasmic reticulum occur throughout the cell body, including the apical receptor villi (Figure 2). Free ribosomes and polysomes are rich in the nuclear region. Mitochondria are

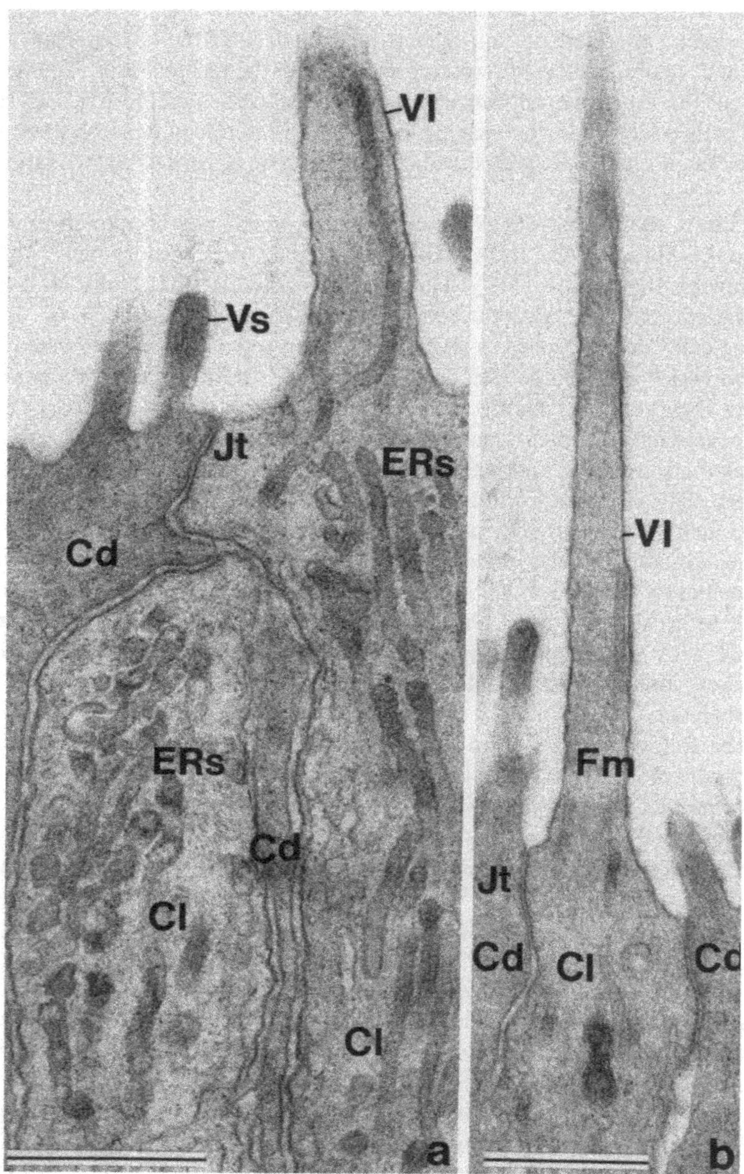

FIGURE 2. Longitudinal sections through the apical parts of light and dark TB cells from the teleost *Silurus glanis*. The large receptor villi belong to the light cells and vary in shape and length. (a) A small lobar process of a dark cell lies between two light cells. (b) Note tight junctions between light and dark cells. (a) ×26250; bar: 0.5 μm; (b) ×21750; bar: 0.5 μm.

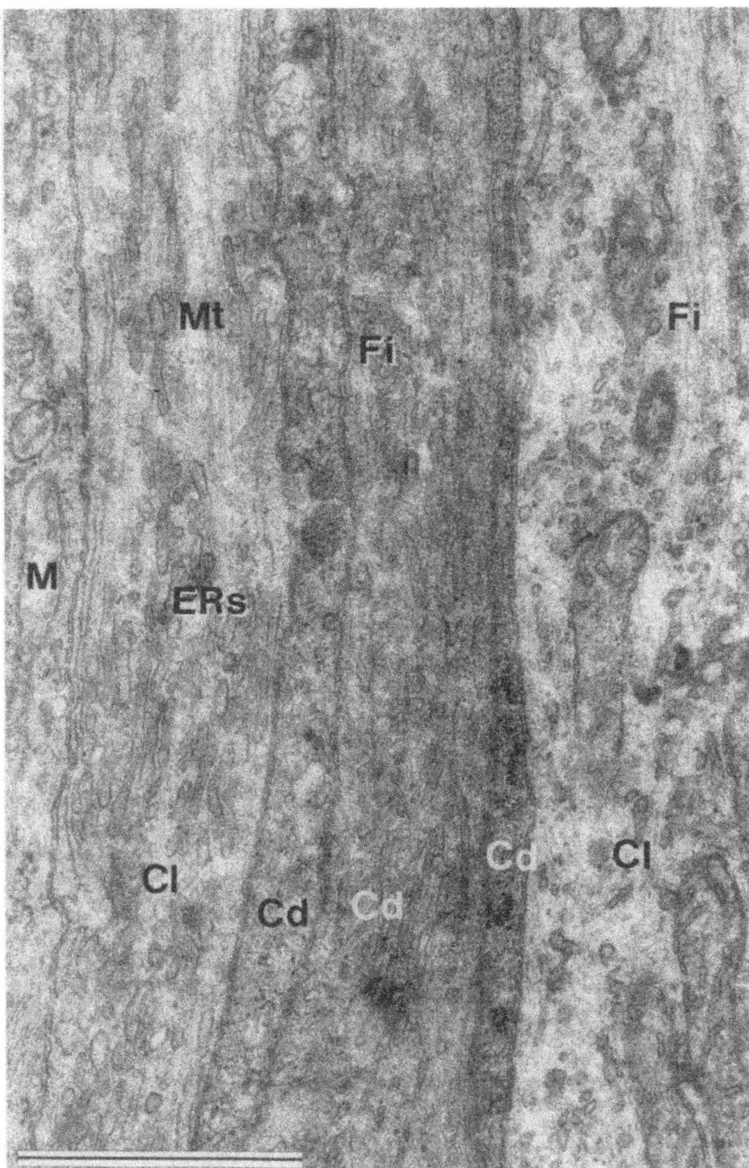

FIGURE 3. Longitudinal section through the supranuclear part of a barbel TB from *Plotosus lineatus* (teleost). Parallel running light and dark cells are close together. One dark cell is cut as a lamella-like structure. ×37500; bar: 1 μm.

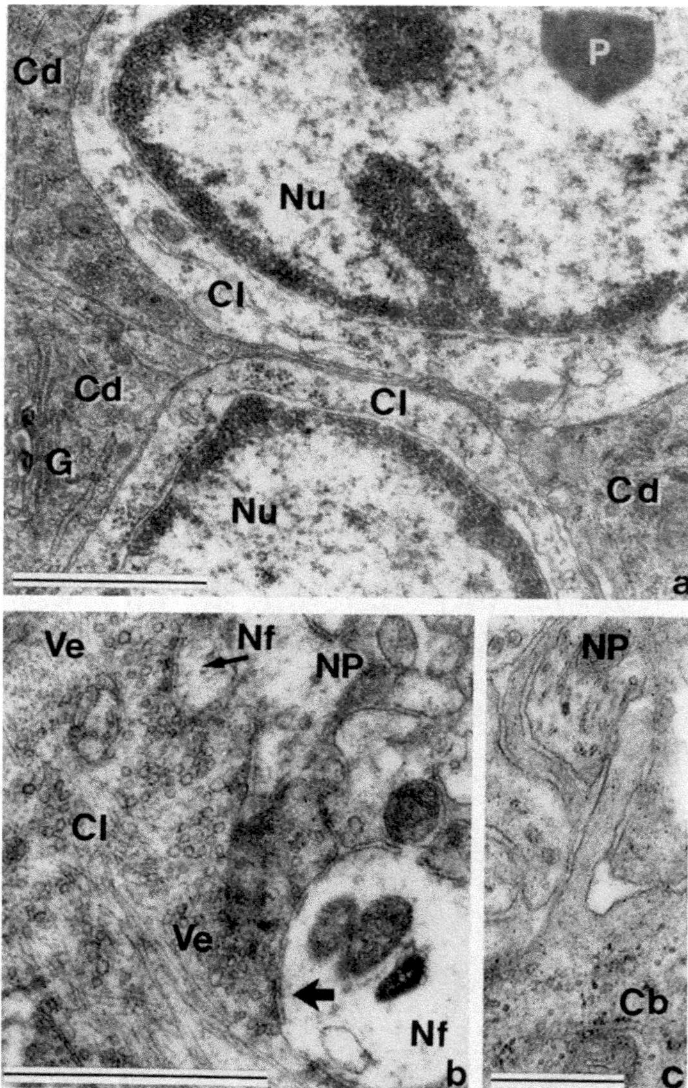

FIGURE 4. Nuclear region and basal part of teleostean TBs. (a) Cross section of a *Plotosus* TB. Light sensory cells are roundish in shape, whereas the dark cells possess thin and sheet-like processes. Note the paracrystalline inclusion in the nucleus of the upper light cell. (b) Region of the nerve fiber plexus of a *Plotosus* TB in longitudinal section. A basal process of a light cell is in synaptic contacts (arrows) to nerve fibers. (c) Part of a *Silurus*-basal cell with a spine-like process projecting into the nerve fiber plexus. (a) ×25900; bar: 1 μm; (b) ×35000; bar: 1 μm; (c) ×36400; bar: 0.5 μm.

abundant and occur in all the parts of the cell. In the supranuclear and apical region they run parallel to the longitudinal axis of the cell. Intermediate filaments (tonofilaments) and microtubules are found throughout the cell. Lysosomes, multivesicular bodies, centrioles, and glycogen granules are rare. The basal processes of light cells contain clear vesicles (30 to 60 nm in diameter).

Dark cells (Figures 2 to 4) — These cells are somewhat more electron-dense and rich in organelles. They are slender and reach from the TB's base up to the receptor area where they end in several undivided small receptor villi. In cross section, dark cells are often star-shaped as they possess lobed processes which ensheathe the neighboring light cells (Figure 4). In longitudinal sections, dark cells appear as sheet- or lamella-like structures, mostly interposed with the light cells, especially at the apical part of the bud (Figure 3). The bases of dark cells are divided into processes. These are intermingled with those of the light cells and the fibers of the nerve fiber plexus. The processes pass down to the basal cells and to the basal lamina. The nuclei of dark cells are darker than the nuclei of light cells and ovoid in shape. Hetero- and euchromatin areas are distinct; paracrystalline inclusions may occur. Nuclear pores and perinuclear cisternae are typical signs of the nuclear envelope. The cytoplasm contains rough endoplasmic reticulum and also elongated and tubular smooth endoplasmic reticulum. Golgi systems lie perinuclearly, and vesicles of different sizes and mitochondria are numerous. Intermediate filaments (tonofilaments) are also numerous and normally organized into bundles which run longitudinally through the cell. Microtubules and centrioles are seldom found, in contrast to lysosomes, polyribosomes, and glycogen granula. The basal processes are rich in clear vesicles (30 to 60 nm in diameter).

Other cell types — Within a TB regularly occur cells of indistinct type which might be differentiating into light or dark cells. These intermediate cells contain organelles typical of both light and dark cells. Normally these cells are located more in the basal and nuclear region of the TB. Degenerative cells are cells with pycnotic nuclei, relatively light cytoplasm, swollen mitochondria, and lobed vesicles of irregular sizes. These cells are frequently observed.

Receptor area (Figures 1,2) — The receptor area is built up exclusively by the apical endings of the TB cells. Large receptor villi are conical in shape and measure up to 2.5 μm in length and about 0.3 μm in width. They belong individually to the light cells. The large villi contain parallel running microfilaments which form a relatively dense core. For a short distance, the core passes down into its light cell. The small receptor villi occur in greater number than the large ones. Up to ten small villi belong to one dark cell. A small villus is cylindrical, about 0.5 μm long, and about 0.1 μm wide. They also contain filaments. Both large and small receptor villi are covered by a fuzzy coat (the glycocalyx). The receptor area is covered by mucous substances which are identified by frozen preparations in the SEM[137,138] or in TEM-histochemical tests, such as lectin-binding studies.[130,139,140] Besides obvious protective functions this mucous surface coat possibly fulfils special functions (e.g., ion reservoir or modulator of tastants) comparable to those in the olfactory organ, named "perireceptor events".[141,194]

Basal cells (Figure 4) — TBs contain up to five basal cells that lie directly on the basal lamina surrounding the corium (dermal) papilla. They lie, transversely oriented to the longitudinal axis of the TB, in depressions of the basal lamina. Basal cells are disk-like and do not form hemidesmosomes with the basal lamina. The apical plasmalemma is irregularly impressed by nerve fibers and by the basal processes of light and dark cells. The nuclei of basal cells are elliptic or lobed and are also oriented transversely; the nucleoplasm shows distinct areas of heterochromatin and euchromatin. The nuclear envelope contains some nuclear pores and well-developed perinuclear cisternae. In the relatively light cytoplasm of basal cells, small, roundish, sometimes elongated mitochondria are present, as are both rough endoplasmic reticulum and smooth endoplasmic reticulum. Golgi systems, free ribosomes, glycogen granules, intermediate filaments, microtubules, and a centriole are also regularly found. Finally, basal cells contain multivesicular bodies, small vesicles (30 to 60 nm in diameter), and large, sometimes dense-cored vesicles (60 to 80 nm). In some species of fish, basal cells have spine-like processes (about 0.5 μm long) which extend into the nerve fiber plexus and interdigitate with its structures.[24,28,41,123,138,142] The basal cells are rich in serotonin[117,143,144] and neuron-specific enolase[142] (NSE) and consequently are possibly related to Merkel cells, which belong to the class of paraneurons.[145] Basal cells have not been described in all species of fish so far (cf. References 28, 126, 128, 130). Finally, it should be noted that this type of basal cell has nothing to do with basally situated stem or regenerative cells, as they occur in the mammalian TB.

Nerve fiber plexus — This structure consists of unmyelinated nerve fibers intermingled with themselves and the basal processes of light and dark cells. Therefore, it is difficult to decide how the individual cell processes are arranged exactly and what belongs to a single nerve fiber. The nerve fiber plexus is the place of synaptic contacts within the TB. In fish TBs, nerve fibers normally are restricted to the plexus region and do not occur within the sensory epithelium of the bud.

TBs of rockling (*Ciliata*),[28] trout (*Salmo*),[36] killifish (*Fundulus*),[39] or bass (*Dicentrarchus*)[136] contain only some unmyelinated nerve fibers in the respective TB region. Therefore a typical nerve fiber plexus seems not to be obligatory in all fish species.

Marginal cells — These cells are not true TB cells. They form the border or margin between a TB and the neighboring stratified squamous epithelium. As elongated and flattened ones the marginal cells follow the shape of the TB. In the upper region of the epithelium there lie more marginal cells, and around TBs situated in an epithelial hillock they contribute to form this elevation. The superficial marginal cells are rich in Golgi systems and secretory vesicles of different sizes. These cells contribute to the surface mucus of the epithelium and possibly to that of the receptor field, too (TEM-histochemical findings).[139,140] Thus, marginal cells may be also important in view of "perireceptor events".[141] On the other hand, marginal cells situated next to the TBs base are thought to be the regenerative or stem cells of the TB. This is the place where mitoses occur, where ^{3}H-thymidine is preferentially incorporated, and where the renewal of the TB sensory epithelium takes place (cell renewal is temperature-dependent and takes 10 to 12 days at 30°C[146]).

Synaptic connections (Figure 4) — In fish TBs, the morphology of synapses is not always clear and well understood because the pre- and postsynaptic membrane specializations often are poorly defined. Small clear synaptic vesicles (30 to 60 nm in diameter) and some dense-cored vesicles (60 to 100 nm) sporadically lie on the presynaptic side. The presynaptic web is often unclear; the synaptic cleft, however, is distinctly organized; the postsynaptic (subsynaptic) membrane often lacks postsynaptic densities. On the whole, these types of synapses may be characterized as afferent.

Synapses occur at the subnuclear part and at the basal processes of light and dark cells, and also at basal cells. Light cells (presynaptic side) are synaptically connected to fibers of the nerve fiber plexus (Figure 4b). These synapses are relatively rich in (presynaptic) small vesicles. Dark cells respective to their processes (presynaptic side) are seldom interconnected with nerve fiber profiles and contain numerous small and a few large dense-cored vesicles. Both types of TB cells may be synaptically connected by their processes with the basal cells.[24,28,36,117,123] Basal cells might function like interneurons (see Chapter 11).[18,117]

Efferent synapses are identified by subsurface cisternae opposite a nerve fiber profile. They typically occur in the sensory cells of the acoustico-lateralis system (e.g., References 147, 148). In fish TBs, such subsurface cisternae as parts of the smooth endoplasmic reticulum, situated close to the plasmalemma of an innervated cell directly opposite a nerve fiber (presynaptic side, with clear vesicles), have been described several times.[28,122,123] Obviously, these synapses are lacking in most species or even in individuals of one species.[28] All in all, the general occurrence and the significance of efferent synapses in fish TBs is uncertain and not well understood.

Besides the synapses within a TB, there exist numerous "normal" intercellular contacts. The plasmalemmata of the TB cells are interdigitated, especially in the supra- and perinuclear region of the cells. Desmosomes occur between neighboring dark cells, between dark and light cells, between the processes of dark and light cells and the basal cells, and between light and dark cells and marginal cells. Hemidesmosomes lie between the bases of elongated processes of dark cells and the basal lamina. (Processes of light cells do not have hemidesmosomes.) Tight junctions occur just below the receptor area where the dark cells closely ensheathe the light cells. Tight junctions are clearly visible in freeze-fracture preparations.[149]

B. CELLS OF THE TASTE ORGAN AND THEIR SYNAPSES IN AMPHIBIANS

1. Caudata

Part of information given in the preceding section on fish TB cells is, to some extent, of general character and therefore also valid for this and the other sections. This is the case especially in view of the electron density of cell types. Provided that fixation and contrast of the tissues are the same, the light and dark characters of cells in different vertebrate groups are the consequences of their more or less identical organelle content. Therefore, cells rich in filaments appear darker than those rich in smooth endoplasmic reticulum.

The sensory epithelium of Urodelan TBs consists of light cells, dark cells, and basal cells.[42-44,47] The apical endings of light and dark cells project into the taste pore[43] and form there the receptor area. Marginal cells are described under the term Randzellen.[46] Nerve fibers occur at the basal part of the TB, but they are not densely arranged in a plexus-like manner. (Figure 11).

Most of Urodelan work has been done on *Necturus*. These animals have unusually large cells with big nuclei, and TBs.[42] This property is useful in electrophysiological studies, especially in intracellular recordings. Single cell studies, i.e., by use of the patch-clamp technique, have also been done on isolated *Necturus* TB cells.[150] From this prospective, there exists a special experimental interest in the cellular organization of *Necturus* TB cells.

Light cells — Light cells occur especially in the center of the TB and comprise about 20 to 30% of the 50 to 100 TB cells.[42,44] Their electron-lucent cytoplasm contains large amounts of smooth endoplasmic reticulum and numerous mitochondria. In the apical cell region, the elongated mitochondria are aligned in the vertical axis of the cell. Light cells end apically with short and stubby microvilli, and their nuclei are larger and more regular than the nuclei in dark cells.

Dark cells — Dark cells occupy the more outer region of the TB and comprise 50 to 60% of its cells. The most striking feature is the presence of secretory vesicles in the apical third of the cell. These vesicles or granules are arranged in clusters which are separated by bundles of delicate filaments.[42] Rough endoplasmic reticulum are abundant, as are elongated mitochondria. Dark cells terminate apically with long and branched microvilli.

Receptor area/taste pore — In the *Necturus* TB three different types of microvillar structures occur.[43] At the apices of light cells there are the aforementioned short and unbranched microvilli and, as endings of a special light cell, stereocilia (long and slender nonmotile microvilli) which are more than 3 μm long. Dark cells terminate in long and branched microvilli. As the stereocilia-like processes project above the mucus-covered receptor area, they are highly exposed to taste stimuli and possibly a favored chemoreceptive environment.

In *Ambystoma*[47,151] the receptor area consists of small and large microvilli: A dark cell terminates in several small microvilli, and a light cell (= gustatory cell[151]) ends in one large microvillus. Large villi contain a core of parallel-oriented (micro)filaments. Thus, these relations resemble the ones in fish.[151] A peculiarity of dark cells respective to their microvilli is the apical secretion of bilaminar and leaf-shaped membrane fragments which are about 25 to 100 nm long and, similar to the unit membrane itself, about 9 nm thick.[47,151] These membrane fragments are incorporated into the surface mucus of the receptor field and possibly consist of lipoproteins which aid to wash the surface of the receptor area clear (Reference 151; cf. References 160, 209).

Basal cells — About 10% of TB cells are basal cells. They are located at the base of the TB, sitting on the basal lamina. At the periphery of the TB they radiate outward and therefore are somewhat elongated. Centrally located basal cells appear roundish. Basal cells have large nuclei which occupy most of the cell. The cytoplasm contains small elongated mitochondria, bundles of tonofilaments, polyribosomes, clear vesicles (90 to 150 nm in diameter), and large numbers of dense-cored vesicles (90 to 180 nm). Basal cells have spiny processes (0.25 μm wide and 2 to 3 μm long) which project into the adjacent light and dark cells. They contain microfilaments, possibly actin filaments. By use of immunohistochemical techniques it was demonstrated that some *Necturus* basal cells contain serotonin.[152] On the other hand, in *Triturus* basal cells the dense-cored vesicles are rich in serotonin (fluorescence microscopy)[153]. Thus, this monoamine is a putative neurotransmitter in the TB. Interestingly, also the basal cells of *Triturus* show spine- (finger)like processes and are related to Merkel cells.[153]

Synaptic connections — As mentioned above, in the Urodelan TB a distinct nerve fiber plexus does not occur. In *Necturus*[44] there are several unmyelinated nerve fibers in the basal region of the TB which interwind with basal cells and with the basal processes of dark and light cells and form synaptic contacts there. Most of the synapses (65%) lie between basal cells and nerve fibers. Synapses are not found between nerve fibers and the spines of basal cells. Pre- and postsynaptic densities are often symmetrical and the synapses are rather small (0.5 μm in diameter). If the basal cell is the presynaptic side, it contains both large dense-cored vesicles and small clear vesicles next to the active side of the synapses. There also occur putative efferent synapses between nerve fibers and basal cells with clear vesicles only on the nerve fiber's side (presynaptic). There also exist synapses containing vesicles on both sides of their active zone: clear ones on the side of the nerve fiber and large clear and dense-cored ones on the basal cells' side. Possibly these synapses are bidirectional. Synapses between basal cells and light cells and between basal cells and dark cells are also present. By virtue of the vesicular arrangement, these synapses would also appear to be bidirectional.

Synapses between light cells and nerve fibers, and dark cells and nerve fibers, are found in the lower third of the TB and represent about 18% of all TB synapses. As the basal processes of both types of cells interdigitate intensely with themselves and with the nerve fibers, it is rarely possible to identify the cell from which a synaptic process originates. Synapses are found on light and dark cell processes and therefore both forms of cells are considered to be taste receptor cells.[44] The basal cells, "interposed between afferent fibers, may serve as interneurons which mediate some form of peripheral integration in chemosensory signal processing" (Reference 44; compare References 24, 117, 138).

2. Salientia

In the final, adult state of their development, taste organs (taste discs, TDs) of frogs and their relatives consist of up to eight different types of cells (Figure 11). No other group of vertebrates possesses such a diversity of different cells within their taste organs. But, interestingly, these large and complex TDs of frogs are developed just before, during, and after metamorphosis of the tadpoles. The TDs are the successor-organs of smaller and more conventional taste organs. Tadpoles possess TBs which resemble in other vertebrates. In the SEM, the receptor areas of tadpole TBs are found on top of conic and slender papillae within the oral cavity, named *oral premetamorphic papillae*.[154,155] During formation of the tongue and TD-bearing fungiform papillae, these papillae and their TBs are reduced.

The TBs of the tadpole consist of two main types of cells: receptor cells and supporting cells.[155] Basal cells (Merkel-like basal cells) are lacking. With regard to their electron density and organelle content, the receptor cells correspond to light cells, and the supporting cells to dark TB cells. In the TD of the adult frog, the aforementioned eight different types of cells lie within the three different layers of the sensory epithelium. In the *basal stratum*, there occur two types of basal cells, basal stem cells (or regenerative cells) and basal cells which resemble Merkel cells (Merkel cell-like basal cells), and the basal processes of cells of the middle layer. The nuclei and the perikarya of two morphologically slightly different sensory cells and of the wing cells lie in the *intermediate stratum*. All these cells reach with their mostly slender apical processes the TD's receptor area. The *superficial stratum* is mainly occupied by the mucus cells, and between them lie the processes of cells situated in the intermediate layer. The lateral lining of the TD is formed by epithelial cells named marginal cells. In contrast to the situation in toads, the TD of Ranidae is encircled by a ring of ciliated cells. The nerve fibers are situated more basally and form there a small nerve fiber plexus.

To some extent this nomenclature for Salientian TD cells attempts to standardize the diverse nomenclatures or names which have been used over the years in TD histology and electron microscopy.

Sensory cells[50,53,55,63] — This type of cell is also described under several names: taste cell,[51,58,60,62,159,160] receptor cell,[59,156] taste receptor cell,[56,57] gustatory cell,[151,155] rod cell,[59,157] and dark cell (type I).[158]

As sensory cells contain numerous organelles, they are indeed somewhat darker than the surrounding cells (Figure 5). The main portion of the cell which includes the nucleus is located in the middle layer of the sensory epithelium. The long dendritic process (or processes[53]) of the cell reaches the receptor area and ends there with one rod-like and apically flat microvillus-like process. The process is about 0.5 μm wide and up to 0.8 μm high. As revealed by freeze-fracture techniques, its plasmalemma contains numbers of relatively large intramembranous particles, possibly a sign of chemoreceptive function.[56] Inside the process occur 500 to 1500 parallel-oriented microfilaments (actin filaments;[57,59,80] cf. Figure 5) which "stick" in the apical cytoplasm of its respective dendritic process. Such a process contains elongated mitochondria, intermediate filaments, tubular profiles of the smooth endoplasmic reticulum, some granulated vesicles, ribosomes, and Golgi systems. Its perinuclear cytoplasm is rich in mitochondria, rough endoplasmic reticulum, and free ribosomes. Clear and dense-cored vesicles occur within the basal processes of a sensory cell.[50,53,59,62]

In addition to this form of sensory cell, a second type is present. In contrast to the apical termination of the first type of cell, this sensory cell ends in a brush of slender and relatively long microvilli which normally protrude above the level of the rod-like processes.[55,63] These cells are associated with nerve fiber profiles which reach just below the receptor field. Synaptic connections between these structures are not seen.

Wing cells[59,63,159] — These cells were also named satellite supporting cells (in contrast to glandular supporting cells)[55] and sustentacular, flat cells.[59] The greater part of the cell body, including the nucleus, lies in the *intermediate stratum* of the sensory epithelium. The cell possesses very thin, sheet-like processes which extend to the receptor field and which envelop or ensheathe the thin dendrite-like processes of the sensory cells and the mucus cells. The apical processes of wing cells are rich in mitochondria; they are intensely interdigitated with the neighboring mucus cells (see below), but not with the processes of the sensory cells. Desmosomes, and, apically, tight junctions are present between all these neighboring structures. The basal process (or processes) of a wing cell contains bundles of intermediate filaments. Wing cells may form hemidesmosomes with the basal lamina which underlies the sensory epithelium.[55] Some of the basal processes may contain dense-cored vesicles in the vicinity of nerve fibers, but they form no synaptic contacts there.[63] The cytoplasm of wing cells is inconstant but relatively rich in organelles and therefore may be stained either lightly or darkly.

Mucus cells[57,59,63,157,159] — This type of cell also was described under many names: glandular supporting cell,[55] glandular cell,[156] supporting cell,[51,58,60,151] sustentacular cell,[62] associate cell,[50,53,60] and light cell (type

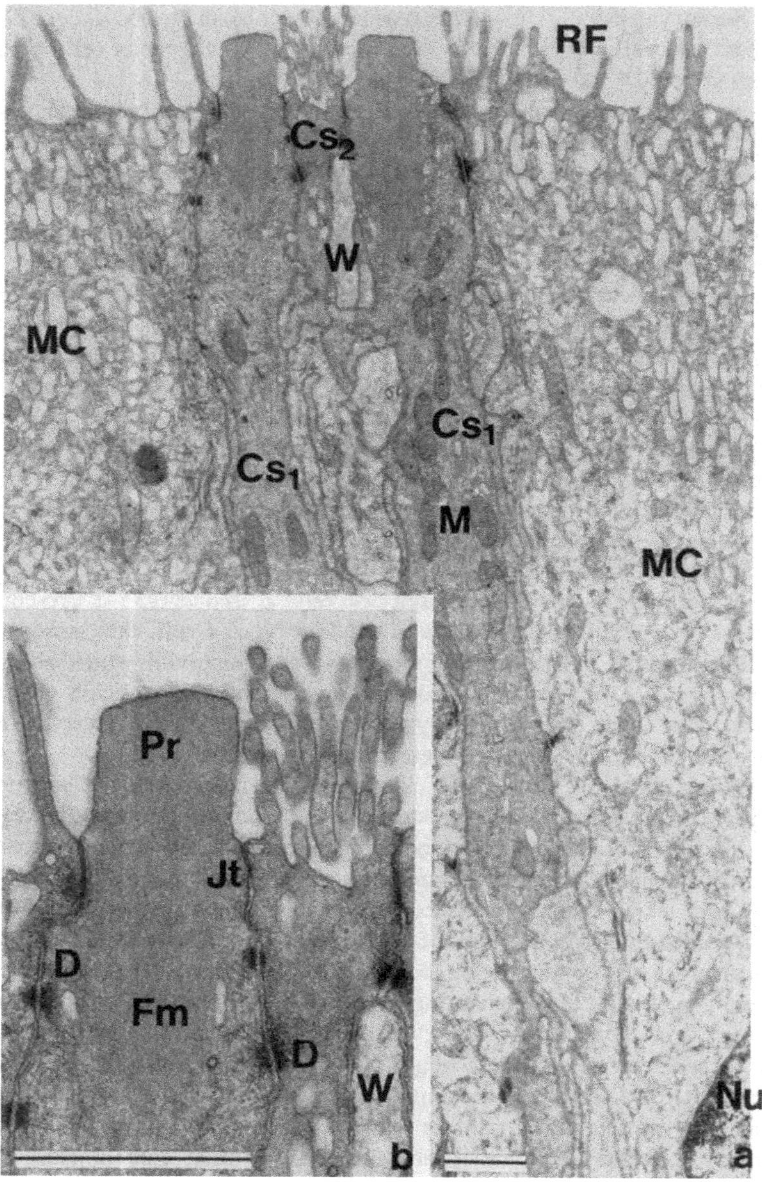

FIGURE 5. Apical part of an amphibian taste disc (*Bombina orientalis*, Salientia) in longitudinal section. (a) The slender processes of sensory cells are close together between surrounding mucus cells. (b) Detail from (a) showing a sensory cell with a rod-like process and a sensory cell with a bunch of microvilli. The villus on the left belongs to a mucus cell. (a) ×30750; bar: 1 μm; (b) ×41000; bar: 1 μm.

II).[158] Within the sensory epithelium, mucus cells occupy the greater part of the superficial stratum (Figure 5) and of the receptor area. These large and more or less cuboid cells have roundish or somewhat lobed nuclei which altogether appear as an apical row of nuclei within the epithelium. Mucus cells have short basal processes. In cross-section, these cells appear pentagonal or hexagonal and are interdigitated with the surrounding thin processes of the wing cells (References 57, 169; see above). The mucus cells ensheathe the processes of sensory cells together with the wing cells. The apical plasmalemma of a mucus cell is rich in microridges from which microvilli arise and thus contribute to the receptor area. Mucous cells possess organelles which are typically found in secretory glandular cells. The supranuclear area of the cell is extensely filled with electron-lucent vesicles of about 100 to 250 nm in diameter. Golgi systems are numerous and some mitochondria as well as bundles of intermediate filaments are regularly found. The subnuclear region is free of vesicles, whereas free ribosomes, intermediate filament bundles, and mitochondria are present. The base and basal processes of mucous cells lack synaptic vesicles. The function of mucus cells seems to be the production of a serous secretion which serves to maintain a rapid flow of fluids on top of the receptor area. These fluids might be actively moved by the cilia — the ciliated cells situated on the TD's border. Mucus cells are important for the events which happen at the receptor field (see below).

Ciliated cells — There are, closely together, one to four rows of ciliated cells around the TD (cf. Receptor area, below). These cells are quite normal kinocilia-bearing epithelial cells. They apparently function to transport the surface mucus on top of the receptor field and to remove taste stimuli.

Marginal cells — These cells are nongustatory epithelial cells which form the margin of the TD and partially cover the lateral sides of the fungiform papilla.

Receptor area — As mentioned above, the surface of the TD consists of the apical endings of the sensory cells, the wing, and the mucus cells. In frogs, it is laterally limited by a ring of ciliated cells, whereas in toads, cilia-bearing cells are often situated in the deep furrow which demarcates the fungiform papilla (SEM).[55-57,63,159,172,179-181] After removal of the surface mucus the cellular arrangement becomes visible. Pentagonal or hexagonal fields correspond to mucus cells with microridges; these terminate in microvillar processes. Between the mucus cells lie slim and longish areas which form together a polygonal network; in the meshes of the net lie the mucus cells. The net is formed by the apical processes of the wing cells, especially their microvilli. The sensory cells are enclosed by the wing cells and to be recognized by either brush-like microvilli or by one large rod-like microvillus. The surface mucus of the receptor field (TEM)[151] has been studied intensely by use of light and electron microscopical histochemistry. It contains acid and PAS-positive mucopolysaccharides (now glycoproteins).[156,173,174] As lectin histochemical studies show, the mucus (and TD) cells contains α-D-galactosamine, β-D-glucosamine, and galactose.[175] The surface mucus is not homogeneously arranged and composed. There are three "microenvironments" as distinctly different mucus compartments: one on top of the mucus cells, one on top of the wing cells and one around the apices of sensory cells. As the latter is rich in calcium, lipids,[159] in a surfactant-like material (Reference 160; cf. References 151,209), and as it shows a special affinity to lanthanum,[172] this microenvironment may play a key role in the perireceptor events.[141,159,172]

Basal cells — In Salientian TDs two forms of basal cells occur, both situated in the *basal stratum* of the sensory epithelium: basal stem — or regenerative — cells in the more central part of the basal layer and highly specialized basal cells which resemble Merkel cells. These cells are named basal cells, too,[50,58] or Merkel-like basal cells[63] or, in short, Merkel cells.[55,60,62] Merkel-like basal cells occupy the more lateral parts of basal cell layer.

Basal stem cells — These cells are of flat or roundish shape and are directly located at the basal lamina. They have no long processes. Their cytoplasm corresponds to the one of normal basally situated epithelial cells. It seems likely that these basal cells are mitotically active and might be the source of the renewal of the cells of the sensory epithelium.[55]

Merkel-like basal cells — There are about 10 to 20 cells located in a ring-like manner at the base of the sensory epithelium. These cells have one long, finger-like and undivided process which runs radially to the center of the TD where it ends in a foot-like plate.[55] They are oriented in a wheel- or rosette-like manner, as seen in cross-sections through a fungiform papilla (fluorescence microscopy after induced amine fluorescence, see below). In an ideal, longitudinally-cut section of the TD, the Merkel-like basal cell is club shaped. Its transversely oriented nucleus lies in the lateral and larger portion of the cytoplasm. It contains abundant perinuclear Golgi systems, lysosomes and multivesicular bodies, roundish mitochondria, free ribosomes, and intermediate filaments. The cytoplasm of the cell process is rich in microtubules and intermediate filaments which run parallel within the process. Near the nerve fibers, dense-cored vesicles (60 to 100 nm in diameter) are accumulated.[50,55,63,161] The most striking morphological criterion for identifying the Merkel-like basal cells are the short spine- or finger-like processes of about 1.5 μm in length. They project between the neighboring structures, such as nerve fibers, cell processes, and marginal cells.[55,63,161,162] These spines resemble the ones regularly and typically found in epidermal Merkel cells.[163]

Merkel-like basal cells are rich in the biogenic monoamine, serotonin (5-hydroxytryptamine, 5-HT), and are capable of taking up the respective "false" neurotransmitters such as 5,6-dihydroxytryptamine (5,6-DHT) and 5,7-dihydroxytryptamine (5,7,-DHT).[54,165-170] These amines show formaldehyde-induced fluorescence[54,165-170] and are stored within the dense-cored vesicles of the cell.[167-169] As demonstrated immunohistochemically, the basal Merkel cell is the only TD cell which contains NSE. The enzyme is one of the marker substances of Merkel cells and other paraneurons.[58,145,171,177]

Paraneurons as derivatives of the neural crest, do not contain keratin filaments. However, the basal Merkel cells react with antibodies against keratin which usually occurs in epithelial (ectodermal) cells.[58] Therefore these basal cells seem to be of epidermal (ectodermal) origin. But on the whole, they are similar to cutaneous Merkel cells, morphologically, and, to some extent, histochemically as well. They may serve as mechanoreceptors[162,170] or as paracrine[161,162] or neuroendocrine[169] cells (see Chapter 11).

Synaptic connections — In the connective tissue stalk of the fungiform papilla there occur two types of nerve fibers: thicker ones about 4 to 5 μm in diameter and a myelin sheath consisting of about 40 lamellae and thinner fibers. The thinner nerve fibers measure about 3 to 4 μm and have about 20 myelin lamellae. Both types of fibers lose their myelin sheaths just below the basal lamina which underlies the sensory epithelium.[55] There the nerve fibers are arranged in form of an inconstant and small nerve fiber plexus.[161] From here the nerve fibers penetrate the basal lamina. The thicker fibers exclusively contact the Merkel-like basal cells, whereas the thinner ones wind themselves especially through the basal stratum and make synaptic contacts there.[55]

Synapses are located between:

1. The somata of sensory cells and, to a greater extent, their basal processes (presynaptic sides) and nerve fibers. These synapses are clearly organized and show numerous dense-cored vesicles and abundantly some clear vesicles, too, on the sensory cell's side. These synapses are afferent.[50,54,58,59,62,155,157,158,162,164,166,168]
2. The basal Merkel cells (presynaptic side) and nerve fibers. The synaptic membrane specializations are developed and the basal Merkel cells cytoplasm is rich in dense-cored vesicles. This synapse is afferent.[63,161,169] (There exist narrow contacts between the basal processes of sensory cells, of wing cells and of mucus cells with the basal Merkel cells, but no real synapses.[161])
3. The nerve fibers (presynaptic side) and the sensory cells. There occur numerous clear vesicles on the nerve fibers' side. These synapses are regarded to be afferent.[50,59,157] Efferent synapses which possess a subsurface (subsynaptic) cisterna exist, too[162,176] (see above).
4. The sensory cells and nerve fibers with dense-cored vesicles on the sensory cell's side and clear vesicles on the nerve fiber's side. Membrane specializations are scarce. Such synapses are reported to be reciprocal (bidirectional) synapses.[155]
5. Closely related nerve fibers. Synaptic clear vesicles occur on both sides. These synapses seem to be reciprocal, too.[50,157]

On the whole, most of these synapses are characterized by their contents of dense-cored vesicles, and this is in agreement with the above-mentioned fluorescence histochemical data which show serotonin to be the main transmitter substance in the Salientian TD. Nevertheless, there exist also some clear vesicles, especially within the nerve fibers. These vesicles may be cholinergic as the respective nerve fibers show positive cholinesterase activities in the light and electron microscope.[164]

Serotonin and acetylcholine are not the only putative neurotransmitters within the Salientian taste organ. As immunohistochemically found, frog TDs contain nerve fibers rich in substance P (SP).[178] In the bullfrog, the TD is supplied by different nerve fibers in which the following bioactive peptides were proved: SP together with calcitonin gene-related peptide (CGRP); tyrosine hydroxylase (TH); neurofilament protein (NFP); vasoactive intestinal polypeptide (VIP) together with peptide HI (PHI); and gastrin-releasing peptide (GRP). The functional significance of these diverse peptide-containing nerve fibers is not clear yet. It is suggested that they "may be involved in the maintenance of the morphology and in the regulation of the function of cellular elements in the taste organ".[170]

C. TASTE BUD CELLS AND THEIR SYNAPSES IN REPTILES

In turtles,[64] tortoises, lizards, and snakes[65] there occur up to five different cell types which have been classified as types 1, 2, 3, A, B,[64] or as types I, II, III, and basal cells.[65] It is difficult to compare these types and to unify the nomenclature. There are also dark cells, light cells, intermediate and degenerate cells in reptiles. Cells belonging to these three types reach from the base of the TB up to the taste pit (pore). Basally, there may occur two different types of basal cells. Marginal cells are not especially mentioned and a nerve fiber plexus does not exist.

Type 1 resp. I cells — Cells of this type are dark cells. They are long and slender and terminate apically with some microvilli which are about 1 μm long and 0.2 μm wide. The cytoplasm is rich in organelles and contains numerous electron-dense granules (100 to 300 nm in diameter), profiles of rough endoplasmic reticulum, ribosomes, Golgi systems in the perinuclear region, filamentous structures, and glycogen. These cells have no synapses.

Type 2 resp. II cells — Cells of this type are light cells. They are also long and slender and terminate apically with (or without) several long and slender microvilli that resemble stereocilia. The smooth endoplasmic reticulum is well developed and shows parallel-oriented profiles or lamellae, especially in the supranuclear cell region. Perinuclear Golgi systems, rough endoplasmic reticulum, ribosomes, and microtubules are common. The basal processes contain synaptic vesicles and contact nerve fibers. Following the type 1 nomenclature,[64] there exist synapses at the basal processes and the cells are considered to be sensory cells. Following type II nomenclature,[65] there exist contacts to nerve fibers but no well-developed synapses. The type I cells seem to be intermediate cells which differentiate into type III cells.

Type 3 cells[64] — These cells appear to be dark cells. They end apically in some shorter microvilli, contain swollen mitochondria and irregular vacuoles belonging to the smooth endoplasmic reticulum. These cells contain a bundle of (intermediate) filaments which is located centrally and runs parallel to the longitudinal axis of the cell. These cells appear to be degenerating cells.

Type III cells[65] — These cells stain slightly darker than type II cells. They reach the receptor area with numerous short microvilli. The cytoplasm is rich in organelles and contains well-developed Golgi systems, especially in the supranuclear region, and numerous dense-cored vesicles (80 to 150 nm in diameter). These vesicles probably contain monoamines (fluorescence microscopy tests). Typical synapses are located at the basal processes. Type III cells are considered to be sensory cells.

Taste pore/receptor area — The branched brush-like microvilli, as seen in the SEM, belong to type 2 cells.[64] All the other and more numerous microvilli are shorter and more irregularly arranged; they belong to cell types 1 and 3 (I to III).[65] There is no information about the mucus surface cover on the receptor area. Interestingly, the receptor area of the turtle (*Testudo hermanni*) TB contains single large microvillar processes, 1.5 μm long and 0.5 μm wide, which are part of the so-called receptor cells. The surrounding supporting cells bear several small microvilli (Figure 3 in Reference 68).

Basal cells — Type A[64] and basal cells[65] are identical. These cells resemble the basal cells in lower vertebrates (Merkel-like basal cells, see above). They are rich in dense-cored vesicles (100 nm in diameter)[64] and seem to contain monoamines.[65] The basal cells are in contact with nerve fibers, but typical synapses are lacking.

Basal cells — Type B[64] are small roundish cells. They are fixed to the basal lamina on the TB's bottom by hemidesmosomes. Their cytoplasm stains relatively dark, is rich in ribosomes, rough endoplasmic reticulum, and in intermediate filaments. The cells engulf the nerve fibers, but there are no synaptic contacts. These basal cells seem to be stem cells.

Synaptic connections — Real synapses occur at the basal processes of type 2 (light) and type III (dark) cells. Type 2 cells processes (presynaptic side) contain numerous clear vesicles (60 to 70 nm in diameter). The pre- and postsynaptic densities are well developed and the synaptic cleft is distinct. The postsynaptic nerve fiber is afferent. Type III cell processes are rich in dense-cored vesicles (80 to 150 nm in diameter) and also contain clear vesicles. At synaptic sites, vesicles are accumulated on the cell's (presynaptic) side; membrane specializations are developed. These synapses are afferent.

In the tortoise[65] there also occur axo-axonal synapses within the TB's base, some of which show clear vesicles on both nerve fiber sides. These synapses seem to be bidirectional.

D. TASTE BUD CELLS AND THEIR SYNAPSES IN BIRDS

Apart from the great variability of avian TBs with regard to their size and especially to their position within the epithelium (including the problem of how the TBs elongated pit or channel is connected with the oral cavity and how taste stimuli reach the TB cells), there seems to exist some uniformity in the TB cells in different birds. There are light and dark cells, intermediate cells, and basal cells (Figure 11).[71-74] A second nomenclature uses the terms type I, type II, and basal cells.[70] Fortunately, the cellular characteristics of both classifications can be related to each other. The taste pore may lead into the taste pit or into an elongated channel-like structure. Marginal cells (perigemmal cells)[72] and a small nerve fiber plexus (located on top of the dermal papilla and therefore outside the actual base of the TB) may be described.

Light cells (type II) **(Figure 6)** — These cells are long and slender and reach the TB pore (or TB channel) with several microvilli of about 1.0 to 1.5 μm in length and about 0.1 μm in width. Some of them may be divided. The cytoplasm is rich in tubular profiles of smooth endoplasmic reticulum. Golgi systems lie in the perinuclear zone, round mitochondria are common, and microtubules and intermediate filaments are regularly found. Clear vesicles (80 nm in diameter) and some dense-cored vesicles (200 nm) loosely aggregate adjacent to nerve fibers.[72] In cross-section, light cells are round in shape.

Dark cells (type I) (Figure 6) — The supranuclear part of dark cells forms a thin process which reaches the TB pore with several shorter, curved, and sometimes divided microvilli. The cytoplasm is rich in free ribosomes (polysomes) and contains some rough endoplasmic reticulum. In the perinuclear cytoplasm, Golgi systems occur, and intermediate filaments may form a dense network around the nucleus.[72] The nucleus is irregularly lobed. Mitochondria are relatively small and roundish. Clear vesicles are occasionally found. In cross-section, the cells are star-like in shape. They possess thin elongated and thin processes which ensheathe the neighboring (light) cells. There exist numerous interdigitations and desmosomes between dark and light cells and to the dark cells' processes and the ensheathed cells, respectively. Tight junctions are located just below the taste pore, as seen in the longitudinally cut TB.

Intermediate cells — These cells occur in abundance. They stain intermediate compared to light and dark cells and, to some extent, posses characteristics of light and dark cells. Some of these cells may be in development; others may be degenerating cells.

Basal cells — These cells lie in the basal zone of the TB. They are of more or less round shape and relatively large. Basal cells stain dark, in fact, often darker than the dark TB cells. They are rich in organelles, especially in rough endoplasmic reticulum and free ribosomes. These cells seem to be stem cells as in the basal cell region; occasional mitoses were found.[72,73] In any case, basal cells have nothing to do with Merkel-like basal cells (see above).

Synaptic connections — As mentioned above, some nerve fibers form a small and plexus-like structure situated at the apical part of the dermal papilla, just below the TB's base. Unmyelinated fibers penetrate the basal

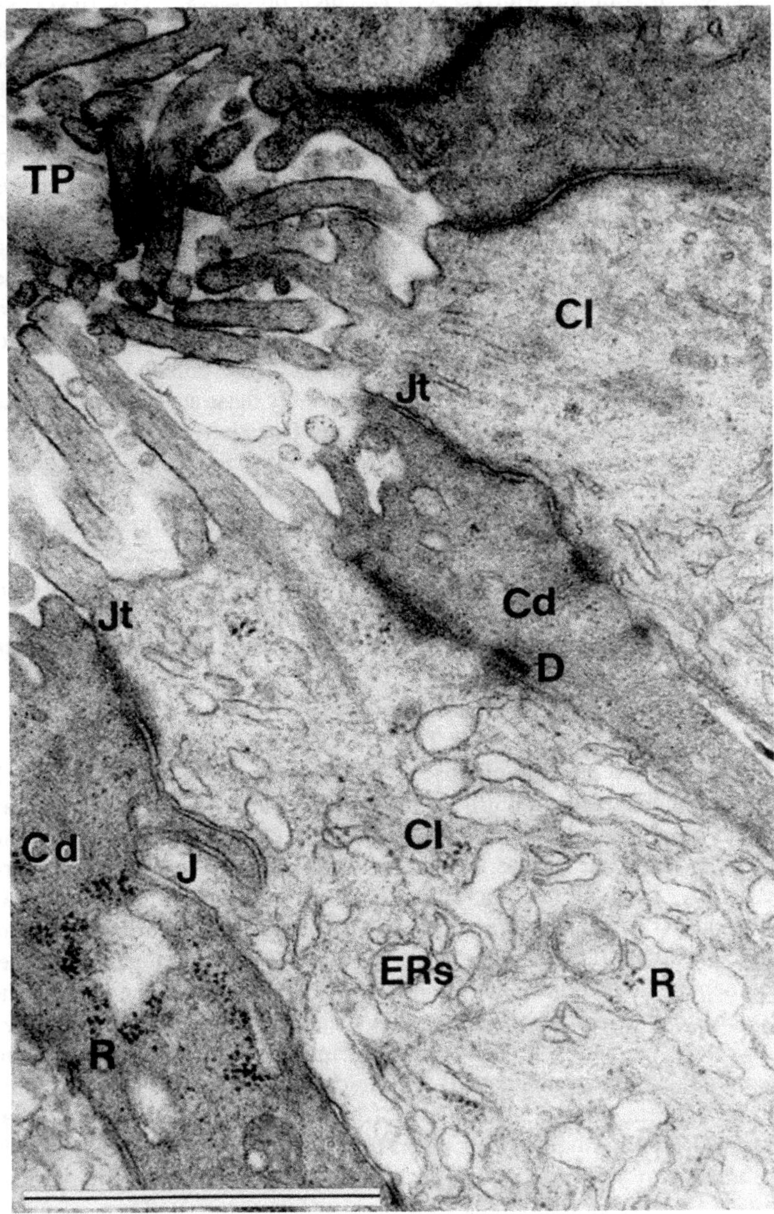

FIGURE 6. Cross section through the middle part of a quail *(Coturnix coturnix japonica)* TB. Light and dark cells converge to the narrow TB pore (channel). Some of the receptor villi are divided. ×48000; bar: 1 μm. (Courtesy of Christoph Sprißler, Tübingen.)

lamina and intermingle with TB cells and their processes. Nerve fibers occur especially in the basal part of the TB, but are also found in all the other parts of the organ. Some nerve fibers reach to the apical end of TB cells and therefore are in distinct relationship with the TB pore. Often only a thin lamella of a TB cell is located between the nerve fiber itself and the TB pore.[73] Synaptic contacts seem to occur throughout the TB, but well-developed synapses are rare. Putative afferent synapses are infrequently seen between light cells (presynaptic side) and nerve fibers. There are only a few clear vesicles (80 nm in diameter) and dense-cored vesicles (100 nm) in the light cell; membrane specializations are scarce. In some cases there are also clear and dense-cored vesicles in the nerve fiber.[73] Afferent synapses between dark cells and nerve fibers may occur, too. Some nerve fibers which are in close contact with light or dark cells contain clear and dense-cored vesicles in close relationship to membrane specializations. Possibly these are efferent synapses, but submembranous cisternae do not exist.[73] Especially in the basal part of a TB there also exist axo-axonal contacts with clear and dense-cored vesicles on one or both sides. Membrane specializations are lacking.[72,73]

The data given above represent observations from the perihatching chick[72] and from newly hatched quails.[73,74] Therefore it is expected that synaptic connections and their specializations may not be fully developed and might change on their way to the adult state. It seems likely that the putative efferent nerve fibers might have inductive or trophic functions in TB cell development.

E. TASTE BUD CELLS AND THEIR SYNAPSES IN MAMMALS

Since the first electron microscopical description of a mammalian TB by Engström and Rytzner (rabbit)[90] a variety of cell names have been used and attempts have been made to classify the TB cells exactly — the same situation as in other vertebrate groups. Due to the technical standards of early electron microscopy, the first TB descriptions were closely related to the terms and classifications developed earlier by light microscopists. Engström and Rytzner described all the TB cells as "taste cells",[90] Trujillo-Cenóz named them "neuroepithelial cells",[91] and Nemetschek-Ganssler and Ferner distinguished between "dichte Zellen" (dense cells), "vakuolisierte Zellen" (vacuolized cells), and "perigemmale Zellen" (perigemmal cells).[92] As a consequence of continuous improvement of the technical equipment in TEM (including fixation and embedding), structural differences between the TB cells have become clearer and have led to several classifications of the observed cells. The attempt to subject TB cells to one nomenclature is justified. But the question is, whether mammalian TBs are organized in a uniform manner. If not, different TB cell classifications may be the result of species differences.

In the past, the following classifications have been applied to TB cells of the following species:

Cells	Species	Ref.
Sustentacular, chemosensory, and basal cells	Monkey	101
Dark, intermediate, and light cells	Rabbit	12
	Mouse	78,79,81
	Rat	84
Dark, light, and peripheral (perigemmal) cells	Hamster	88
	Rat	186
Dark, light, and peripheral cells	Hamster	86,87
Pregustatory, gustatory type Ia,Ib,Ic cells	Rabbit	93
Type I and type II cells	Rabbit	94
Type I, type II, and basal cells	Rat	82
Type I, type II, type III, type IV, and basal cells	Rabbit	13–17, 96–99, 110, 187, 239
	Rat	83
	Guinea pig	89
	Monkey	103–104
	Man	105
One type of gustatory cell: light/dark cells are affected by fixation	Monkey	100,102,182
	Man	106

It should be noted that peripheral, perigemmal, or marginal cells are regarded as most as normal epithelial cells and therefore not named TB cells.

Murray's nomenclature,[14,16,96,187] as developed over the years especially for the TB of the rabbit, possesses the most advantages. Cell types in this classification scheme are well defined and thus the criteria can be used to describe the cells of other mammalian species, too. The application of this nomenclature does not exclude the description of cell variants or of further cell types; it also allows for missing a special TB cell type in a specific animal.

Besides the cell types I-IV, the marginal cells, the TB pore and pit, and the nerve fibers within the bud are of great interest (Figure 11).

Type I cells (Figures 7 to 9) — These cells stain relatively electron-dense and therefore are dark cells. They are very slender and occur especially at the margin of the TB and line the taste pore, but they are also located in its center. Type I cells extend from the base of the TB to the taste pit. They are the most predominant cells in the bud and comprise 55 to 75% of its cells. The most significant organelles of type I cells are dark granules; they allow one to distinguish these cells from others. The granules occur especially in the apical region and, to some extent, in the perinuclear region. The membrane-bound granules measure 100 to 400 nm in diameter. In vallate and foliate TBs, their content corresponds in density and homogeneity to the quality of the dense substrate in the taste pit.[14] However, this is not the case in fungiform TBs.[82] Each cell terminates apically in one brush of microvilli (Figures 7,8) which projects into the taste pit. A brush consists of about 30 to 40 microvilli, each 0.1 to 0.2 μm wide and up to 2 μm long. They contain filamentous cores which might also be microtubules. Type I cells have irregular cytoplasmic processes which, in thin lamellae, surround type II and type III cells. The cytoplasm also contains strands of intermediate filaments and microtubules. In the supranuclear zone, Golgi systems as well as cisternae

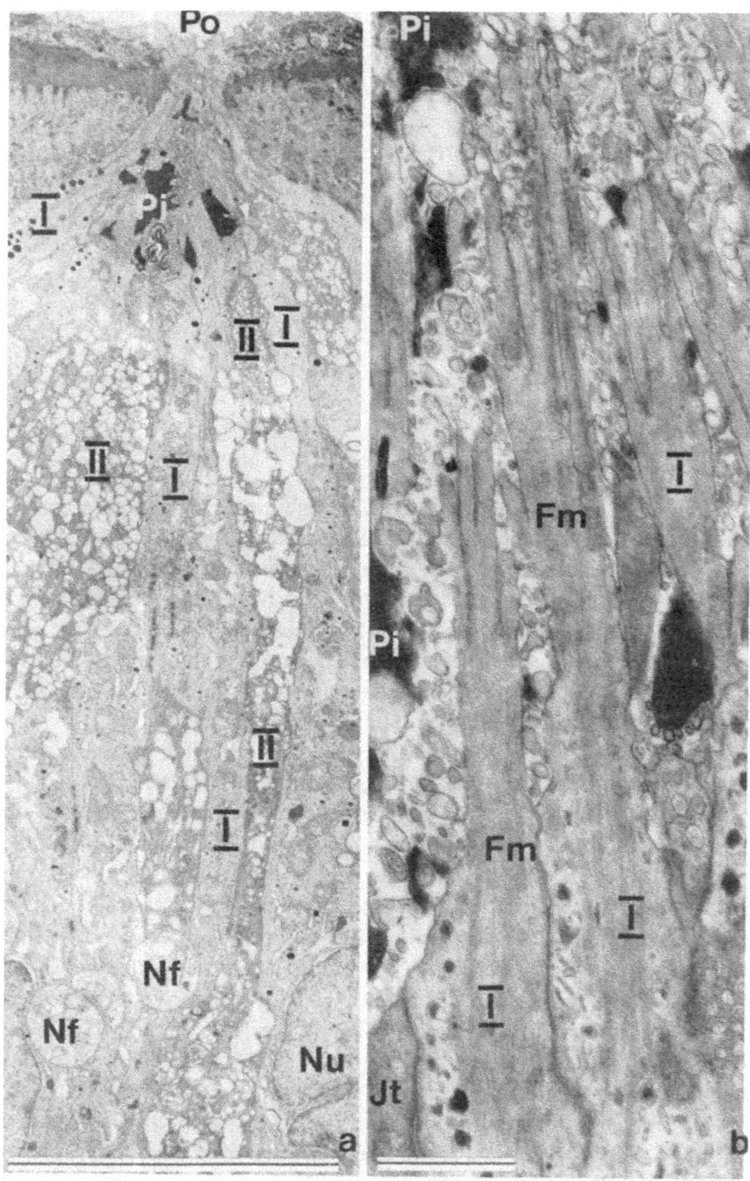

FIGURE 7. Longitudinal sections through TBs from the circumvallate papilla of the swine *(Sus scrofa domestica).* (a) Overview: represented is the supranuclear region up to the (not exactly medially cut) TB pore. (b) Dark granules containing type I cells (dark cells) end in slender necklike processes with a terminal brush of microvilli. The microvilli project into the more basal part of the taste pit which is filled with intensely contrasted mucous substance and abundant cross and obliquely cut microvilli and vesicles. (a) ×4440; bar: 10 μm; (b) ×18600; bar: 1 μm.

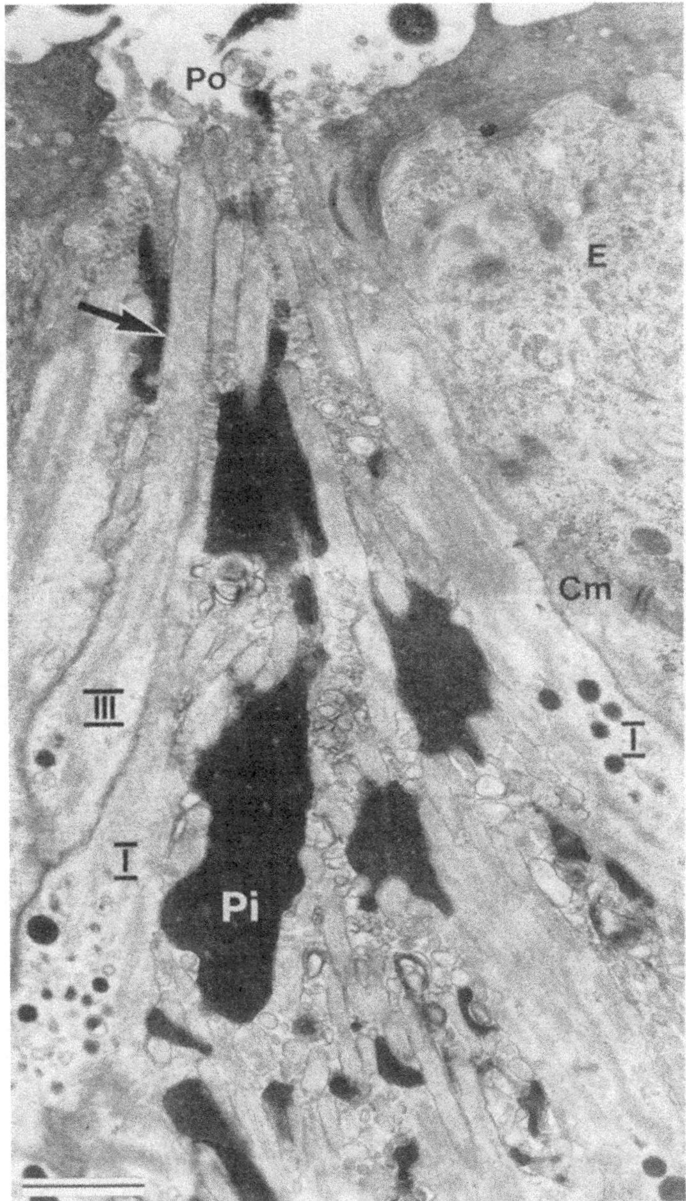

FIGURE 8. Longitudinal section through a swine TB pit in which, near the microvilli of other TB cells, the apical undivided process of a type III cell (arrow) is located. The process reaches up to the TB pore. × 16400; bar: 1 μm.

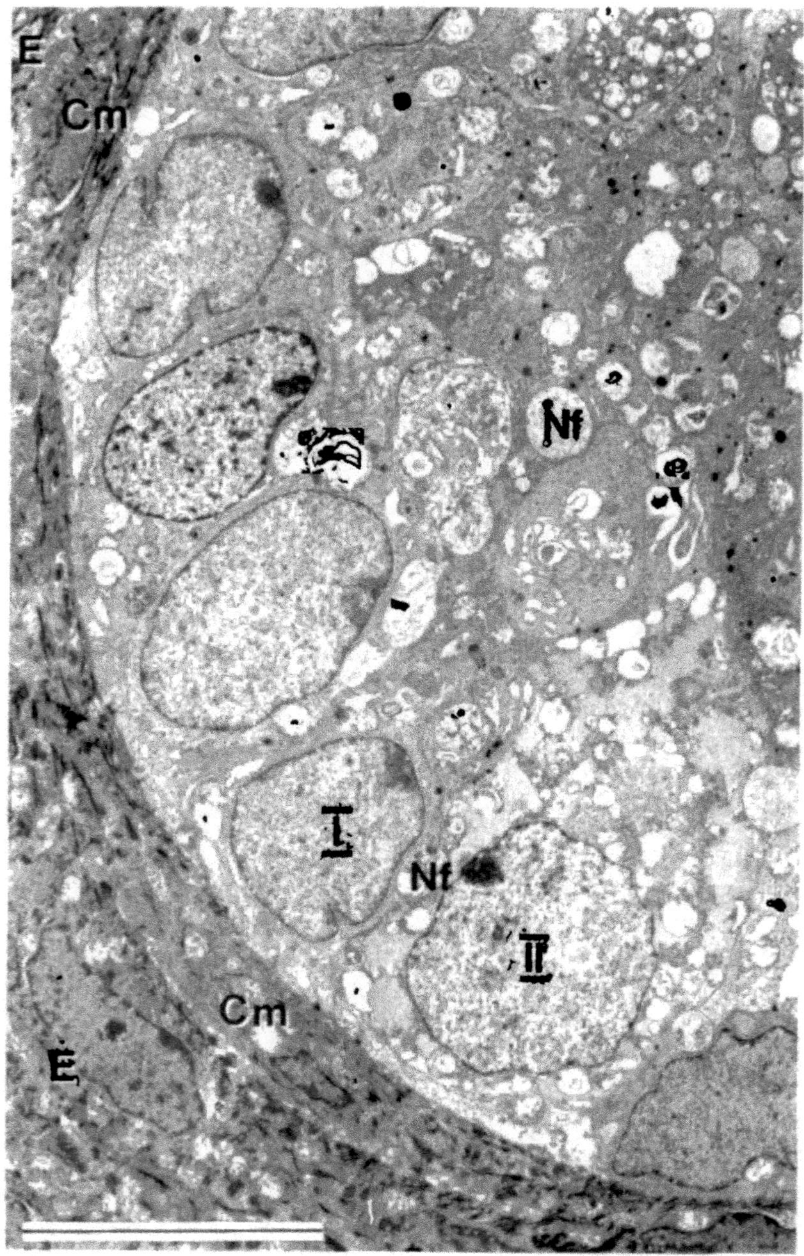

FIGURE 9. Detail from a cross section through the middle part of a swine TB. Cells of different types, including degenerative ones, as well as nerve fibers are close together. The border to unspecialized epithelial cells is formed by a small zone of marginal cells. ×4000; bar: 10 μm.

of smooth endoplasmic reticulum and lysosomes occur. Compartments of the rough endoplasmic reticulum and free ribosomes, mostly as polysomes, are abundant. The elongated and relatively small nucleus of the cell is irregularly shaped and often deeply indented.

Type I cells are considered to have a secretory function as they are believed to produce the dark substance of the taste pit. As they ensheathe the neighboring type II and III cells, type I cells possibly function like Schwann cells in terms of isolation and nutrition. They may also act as phagocytes. Since type I cells are in distinct relation to nerve fibers (but do not form synapses), an active gustatory role cannot be ruled out (Reference 96; see below). On the whole, type I cells presumably fulfil supportive functions.

Type II cells (Figures 7, 9, 10) — These cells stain less electron-dense and therefore are light cells. Type II cells are slender, elongated cells and reach from the TB base to the taste pit. Type II cells are less numerous (15 to 30%)[14] and occur in both the lateral and the more central parts of the TB. Apically, type II cells terminate either with or without well-developed, but shorter and wider microvilli than type I cells. Their cytoplasm is rich in vesicular structures of different size and profiles than smooth endoplasmic reticulum. The latter are often swollen and highly vacuolated. The rough endoplasmic reticulum is sparsely developed; free ribosomes and tonofilaments are scarce, too. Mitochondria are numerous and relatively large. In the perinuclear region, Golgi systems and vesicles of different sizes, including lysosomes, are numerous. Multivesicular bodies occur. The nuclei of type II cells are the largest ones in the TB, more round and less electron-dense than in type I cells. Type II cells are wider than type I cells and, when seen in cross-section, round in shape and more or less isolated from each other by the thin processes of type I cells.

Especially at their basal parts, type II cells show synaptic contacts with nerve fibers, including unique dense-cored vesicles and accumulations of clear vesicles, membrane specializations, and, in some cases, subsynaptic cisternae.

In view of their synapse-like contacts to nerve fibers, type II cells are considered to be chemosensory cells which are innervated efferently, too (see below).

Type III cells (Figure 8) — With regard to their staining, these cells are similar to light cells and therefore type II cells, too. But, in contrast to type II cells, a key feature of type III cells is that they contain numerous dense-cored vesicles (80 to 140 nm in diameter). These have nothing to do with the dark secretory granules of type I cells. Type III cells comprise the smallest fraction of TB cells (5 to 15%).[187] A type III cell is slender and reaches from near the basal lamina to the TB pore. It extends through the dense substance of the taste pit up to the pore and ends as a blunt microvillus-like process, without smaller microvilli. This process therefore shows smooth outlines; it contains filamentous materials. The apical part of the cell's cytoplasm, up to the base of the process, contains dense-cored vesicles (Figure 8). The cell is rich in smooth and rough endoplasmic reticulum and, in view of the other

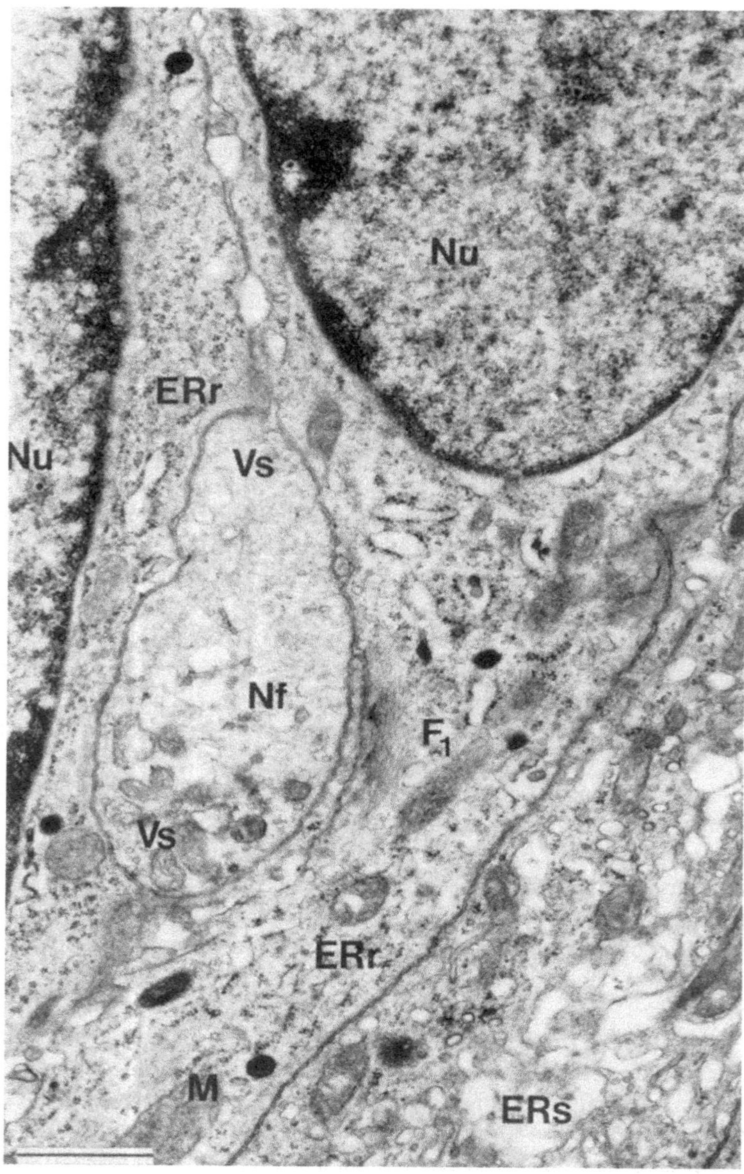

FIGURE 10. Longitudinal section through the nuclear region of a swine TB. A nerve fiber is located between two TB cells and deeply impressed into these cells. The cells contain only a few rare synaptic vesicles; more vesicles lie in the nerve fiber. Synaptic specializations are not seen. ×18000; bar: 1 μm.

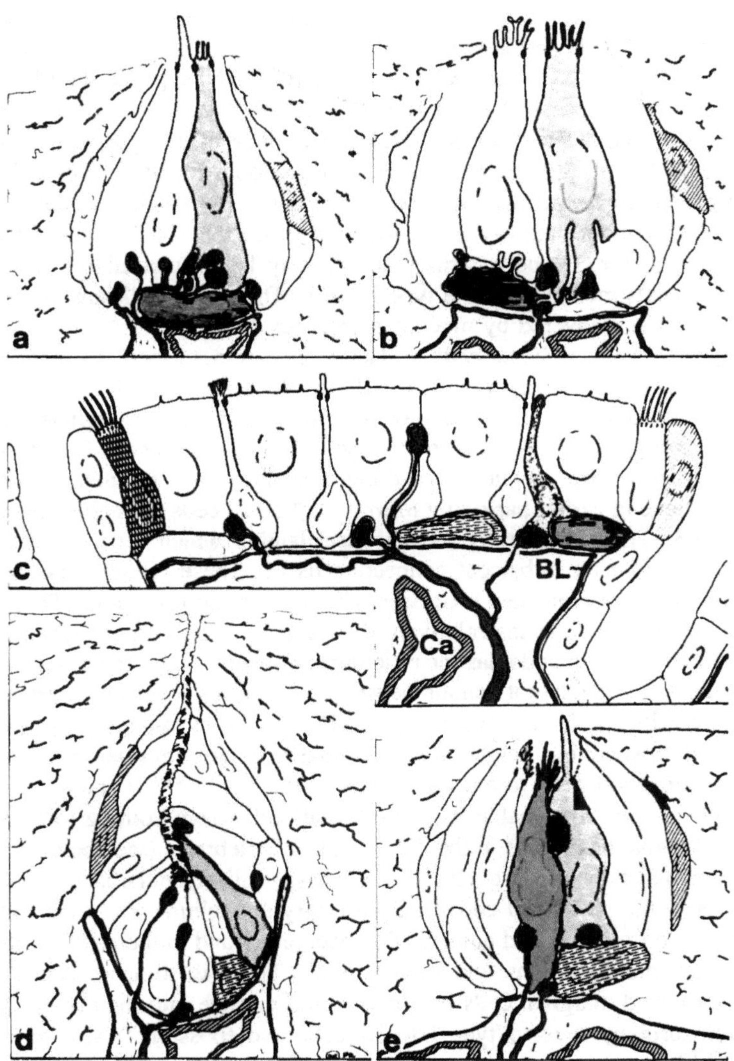

FIGURE 11. Schematic drawings of longitudinally cut taste organs which belong to representatives of different vertebrate groups. In each organ a distinct cell type is represented by only one cell. For this drawing the respective findings of several authors were used. (a) TB of the bullhead, a teleost fish, with light, dark, basal, and marginal cells and a nerve fiber plexus. (b) TB of the newt, an urodelan amphibian, with light, dark, basal, and marginal cells. (c) TD of the frog, a salientian amphibian, with two kinds of sensory cells, basal (Merkel), basal (stem), wing, ciliated, mucous, and marginal cells. (d) TB of the quail, a gallinaceous bird, with light, dark, basal, and marginal cells. (e) TB of a mammal, with type I, II, III, basal (stem), and marginal cells. A reptile TB is not depicted, but it is similar to a mammalian TB. Nerve fibers in black.

organelles, resembles type II cells. The nucleus might be somewhat darker than in type II cells and is larger than in other TB cells. In cross-section, a type III cell is roundish and enveloped by the lamellar processes of type I cells. The basal process of a type III cell is rich in dense-cored vesicles. They accumulate together with clear vesicles at regions of synapse-like contacts with nerve fibers. Subsynaptic cisternae are seldom found.[99] Type III cells are regarded to be primary gustatory receptor cells which may provide a specific and direct pathway in gustatory transduction. Their apical process protrude into the TB pore. This pathway may be influenced by the more diffuse signals mediated by the type II cells. Their microvilli are located in the taste pit and covered by mucous substances.[96]

Type IV cells = basal cells — These cells are not especially mentioned in all papers which use the type I, II, III classification, possibly because they do not reach the taste pit and do not contribute to the sensory epithelium of the TB. But these cells — one or two per TB — are regularly found in the basal region of mammalian TBs and therefore they must be included as TB cells, as done in earlier papers by Murray.[14,16] Basal cells are part of the base of the TB. They lie directly on the basal lamina which separates the TB epithelium from the subjacent connective tissue. These round or polygonal cells do not have elongated processes which reach the taste pore. The cytoplasm of basal cells stains relatively dark and is rich in organelles such as mitochondria, rough endoplasmic reticulum, ribosomes, and intermediate filaments. Basal cells do not contain dense cored vesicles and no dark granules, as type I cells do. In rare cases, basal cells may be in mitosis. Type IV = basal cells resembling neighboring marginal cells are considered to be the stem or precursor cells to the other cells of the TB.[14,19,80] It should be emphasized that the mammalian type IV (basal) cells have nothing to do with the Merkel-like basal cells in the TBs of lower vertebrates (see above).[18,23,25]

Marginal cells — Marginal cells are less specialized epithelial cells which lie close to the peripheral cells of the sensory epithelium. They are rich in intermediate filaments and connected to each other and to the cells at the TB border by interdigitations and desmosomes. There is some evidence that basally situated marginal cells may divide and migrate into the center of the TB transforming there to type I to III cells and (possibly?) also to basal cells.[188,189]

Taste bud pore and pit — The apical endings of TB cells I to III do not project directly onto the epithelial surface. They reach into the taste pit which is filled with an electron-dense, intensely staining mucous substance. Only the type III cells' nondivided and bluntly ending structure may penetrate the TB pore and thus is in contact with the oral cavity (but not in TBs of fungiform papillae).[11,14,183,184] As freeze-fracture preparations show, the type III cell's process has only a few, but relatively large, intramembranous particles (IMP) and, additionally, vesicular protrusions which probably are associated with endocytosis. In contrast, the microvillar processes of type I and type II cells have large numbers of smaller IMPs. IMPs are regarded to serve chemoreception; whether they function as receptor sites (receptor proteins) or whether they are enzymes is not clear.[190]

By use of carbohydrate histochemical techniques, the content of the taste pit is characterized as a "neutral mucopolysaccharide" (see above; now glycoprotein).[93,174,191] As demonstrated by the use of a lectin-binding technique at the EM level, this mucous substance derives from the dark staining granules of type I cells.[192] The secretory product of type I cells has been shown to bind the synthetic sweet-tasting protein, thaumatin, but there is no direct evidence for gustatory functions.[216] Light microscopical lectin histochemistry reveals that the mucus contains *N*-acetyl-galactosamine, fucose, mannose, and *N*-acetyl-glucosamine.[175] Interestingly, the contents of the taste pores vary between the TBs from the papillae foliatae and p. vallatae (sialic acid, galactose, *N*-acetyl-galactosamine, *N*-acetyl-glucosamine) and the papillae fungiformes (*N*-acetyl-galactosamine, *N*-acetyl-glucosamine, fucose).[193]

Type III cells appears to contribute to the taste pits mucous substances, too; they seem to secrete phospholipids (deriving from lamellar bodies) which lower the surface tension within the taste pit. This may allow fluids to penetrate the taste pore and pit easily, facilitating the spreading of sapid substances over the receptor villi (Reference 209; cf. References 151, 160).

The function of the mucous substances within the taste pit remains unclear yet. According to Bannister[194] the mucus has a function in sorting the arriving molecules, as a chromatography medium. Possibly the mucus serves as an ion reservoir.[194] Last but not least, the mucus possibly is involved in molecule-binding events, as lectin-binding studies may demonstrate.[175,192,193] In summary, as in other vertebrate groups, the functions of the mucus may be associated with perireceptor events.[141,159,172]

It is suggested that the mucus within the TB pore is functionally connected with the salivary secretions of the salivary glands, especially the von Ebner glands. There seems to exist a parallel between the von Ebner glands in the tongue and the Bowman glands in the olfactory organ.[195,197] This is likely since a glycoprotein was isolated[198] in von Ebner glands which resembles the "odorant-binding protein" (OBP). OBP may play a key role in olfactory chemoreception.[199]

Synaptic connections — Synapses and synapse-like connections between TB cells and nerve fibers, afferent as well as, to some extent, efferent ones, are described in many of the mammalian TB publications of the past 25 years. Due to the aforementioned confusion in TB cell nomenclature, it is not easy to compare the data concerned with TB synaptology and to report them in an unmistakable and satisfactory way. Refined methods have led to more accurate results in the last decade. High voltage electron microscopy (HVEM, allowing TEM of 0.25 μm thick semithin tissue sections) has been applied to mammalian TBs and their synapses, also.[78,81] In the TEM, synapse-like formations appear to be scarcely equipped with organelles and synaptic specializations.

In the HVEM, however, some of these structures have turned out to be synapses; others have not.

In the HVEM, all TB cells of the mouse vallate papilla, comprising dark cell (type I), intermediate cells, and light cells (type II), have synaptic contact to nerve fibers.[78,80,81] Synapses lie in the nuclear regions of the cells. At the cells (presynaptic) side, clear vesicles (40 to 70 nm in diameter) and membrane specializations occur. The synapses are either spot- (macular) or finger-like. The latter is characterized by a finger-like neuronal (postsynaptic) profile which protrudes into the TB cell (presynaptic). These synapses are found in all different types of cells. Both types of synapses may be located adjacent to one another, forming a synaptic complex onto a single neuronal process. All the named TB cells are regarded to be gustatory (sensory) cells.[78] Interestingly, up to five TB cells have synaptic contacts with one intragemmal nerve fiber. But dark (type I) and light (type II) cells do not synapse onto the same nerve fiber. Possibly this means that there is a correlation between taste cell morphology and sensory responsiveness.[81]

In contrast to TBs in the mouse circumvallate papilla, quite different synaptic relationships are found in foliate TBs of the rabbit. As Murray[96] postulated, type II and type III cells are afferently innervated and therefore both considered to be chemosensory cells. This point of view has been corrected by HVEM.[99] Only type III cells form one to six synapses with a nerve fiber, and there is evidence for both convergence and divergence of type III cell synaptic input to nerves. The active zones of type III cell- synapses are of different size and the presynaptic membrane specializations vary. In some cases there are considerably more dense-cored vesicles (80 to 140 μm in diameter) than clear vesicles (40 to 60 nm) on the presynaptic side. Possibly this means that functionally different types of synapses occur within a TB. In the HVEM subsurface cisternae of the endoplasmic reticulum are only seldom found.[99] They are located beneath the type III cell membrane at sites of apposition with nerve fibers; they are regarded as efferent. Efferent synapses are possibly more numerous during development. In macaque fetal TBs efferent synapses regularly occur beside afferent synapses, with and without subsurface cisternae.[200,201]

Type III cells are rich in dense-cored vesicles (see above). These are believed to be the storage sites of biogenic monoamines (e.g., Reference 99), and, consequently, the type III cell must be aminergic. There is indirect evidence to believe that type III cells do not have positive acetylcholinesterase reactions, as demonstrated in the TEM.[202] More direct evidence for aminergic transmission is given by fluorescence- and TEM-histochemical tests which proved serotonin (5HT) as a transmitter candidate in the mouse TB.[75,76,203,210,211] EM-immunohistochemistry, done on monkey and rabbit TBs, shows 5-HT antibody binding within the dense-cored vesicles, especially of type III cells.[204] However, there are regularly two populations of vesicles and therefore two different transmitters (an aminergic and a cholinergic one?) within a TB cell.

It is still unclear how TB cell-neurotransmission actually works: which transmitter may play the main role and which one might possibly serve the modulation of the main pathway (see Chapter 11)?

Besides aminergic (and cholinergic) systems within the TB, there also occur nerve fibers which contain a battery of immunohistochemically demonstrated neuropeptides. SP occurs in intra- and perigemmal nerve fibers of rat TBs.[205,212-215] Some of the fibers are exclusively immunoreactive to CGRP, some of them to both, SP and CGRP.[215] In the TB of the rat, synapsin I (a neuronal phosphoprotein) also is present. It is located throughout the cytoplasm of the nerve fiber and in association with vesicles (not synaptic vesicles).[206] However, TB cells are not immunoreactive for synapsin I, possibly because TB cells, as axonless cells, belong to the "short" receptor cells (photoreceptors, hair cells) which all seem to be free of synapsin I.[206]

In the guinea pig TB, there are the following positive immunohistochemical (LM) stainings: TB cells contain spot-35-protein (SP-35) and/or NSE; nerve fibers of the dermal papilla contain NFP and S-100-protein.[207] In the TEM, SP-35 is located within type III cells which are in synaptic contact with intragemmal nerve fibers; types I and II cells are SP-35 negative.[208]

These immunocytochemical data may help to resolve two important questions: (1) Which cells are the main gustatory receptor cells? The findings listed above favor the type III cells. (2) Which kind of cells are the type III cells? Since these cells have neurotransmitters and peptides similar to those found in neurons and paraneurons, type III cells belong to the group of paraneurons.[145,177]

IV. NERVES INNERVATING TASTE BUDS IN VERTEBRATES

Nerve fibers innervating the TBs of representatives of different vertebrate groups are derived from the facial (VII), glossopharyngeal (IX), and vagal (X) cranial nerves.

In **fish,** the facial nerve supplies TBs of the barbels, lips, and rostral palate. In many species, such as ictalurid catfish, a recurrent facial nerve leads from the facial lobe caudally to the buds on the fish's body surface.[217,218] Most of the orobranchial TBs are innervated by the vagal nerve, whereas the glossopharyngeal nerve plays a minor role for the gustatory supply in the oral cavity.[21,219,220]

In **amphibians,** the main gustatory nerve of the taste organs situated on the tongue and in the pharynx is the glossopharyngeal nerve (Frog[230]; *Necturus*[231]). According to Chernetzki,[232] sympathetic efferent fibers also enter the taste disks. The palatal mucosa is innervated by a palatine branch of the facial nerve which probably also supplies TDs situated in the soft palate.[232] Also the chorda tympani plays a role in TB innervation in amphibians.[234]

There are few data about nerves innervating TBs in **reptiles.** Fischer[221] and Schwenk (in lizards)[222] describe the chorda tympani as responsible for TB innervation. Jacobs[223] found similar results in *Lacerta viridis.*

The main gustatory nerves in **birds** are branches of the facial nerve (chorda tympani) in TBs of the anterior mandibule and also of the lower beak oral cavity in chicken;[224-226] medial branch of the facial nerve in duck[227]. The glossopharyngeal nerve carries fibers from posterior TBs[228] and from the ventral surface of the tongue.[229]

In **mammals,** TBs occur throughout the oral cavity, and gustatory information is thus transduced by different cranial nerves. On the anterior two thirds of the tongue, TBs are concentrated on fungiform papillae and innervated by axons of the chorda tympani, a branch of the N.intermedius.[235] This nerve is regarded as a part of the facial nerve. In some individuals, gustatory impulses of the fungiform TBs may travel through a shunt from the chorda tympani through the otic ganglion to the greater petrosal nerve.[236,237] Axons of this nerve, and a branch of the facial nerve,[238,239] carry gustatory information from most TBs (85%) of the soft palate. The remaining TBs are innervated by the deep petrosal branch of the glossopharyngeal nerve.[238]

Most TBs of (circum)vallate and foliate papillae are bilaterally innervated by the glossopharyngeal nerve.[240] Foliate papillae also receive axons of the chorda tympani.[241,242] The superior laryngeal branch of the vagal nerve innervates TBs of the larynx.[243,244]

Recent studies have shown that gustatory information may not be exclusively associated with the aforementioned cranial nerves, but also with axons of the trigeminal (V) nerve. In rat 75% of fibers entering the fungiform papillae are derived from the trigeminal nerve and only 25% from the chorda tympani.[245] Dissection experiments of both lingual and chorda tympani nerves in rat revealed that the lingual nerve may maintain the TBs when only the chorda tympani is destroyed.[246] However, direct gustatory pathways by way of the trigeminal nerve are at least doubtful (review: Reference 237).

V. CONCLUDING REMARKS

In summary, Figure 11 illustrates, more or less simplifying, the TB morphology of five different vertebrate groups. TBs consist of different types of cells, and these are termed partly by the same and partly by different names in different nomenclatures. There exist convergences in view of cell types, but also divergences, as in frog TDs where numerous cell types occur. Despite all the common characters of TBs from different animals of different systematic positions, it seems ineffective to use one obligatory classification that is valid for all vertebrate taste organs. Regarding comparative biology, it does not seem useful to regard all the light or all the basal cells as strictly established cell types; they also may have completely different functions. On the other hand, cells that appear different may perform the same function. These observations and experiences lead us to the opinion that TBs and TB cells should be considered within the context of their respective systematic group and characterized by using the terms of their group TB nomenclature. It does not matter, for instance, if one describes a "new" frog TD by using the fixed terms of amphibian nomenclature; but it is difficult to compare and to interpret data, such as histochemical work, between different species of different systematic positions by applying different nomenclatures. Consequently, the main vertebrate groups should have their own classifications and this chapter has attempted to provide such a classification. Under the phylogenetic aspect, it seems likely that taste organs were not consecutively and progressively developed from one vertebrate class to the next. There are good reasons to believe that taste organs in distinct vertebrate groups are highly specialized, both morphologically and functionally. This is demonstrated by frog TDs. There is too little information about taste organs from numerous animals belonging to different systematic groups which would allow us to postulate the "phylogenetic tree" of taste organs. It seems more fruitful to concentrate on animals of well-known systematic positions, to analyze morphology and function of their taste organs, and then, in a final step, to compare and discuss the results with data from neighbored systematic groups.

In evolution, the innervation of the taste organs by cranial nerves VII, IX, and X, seems to be constant. However, there is some variability of innervation within representatives of different systematic groups.

ACKNOWLEDGMENTS

The authors wish to express their appreciation for the assistance of Mrs. Karin Tiedemann (secretary), Mr. Gerd Geiger, Mr. Hermann Guckes (electron microscopy), Mr. Manfred Mauz (photography), and Mr. Mihnea Nicolescu (drawing).

ABBREVIATIONS

BL	basal lamina
Ca	capillary
Cb	basal cell
Cd	dark cell
Cl	light cell
Cm	marginal cell
Cs	sensory cell
D	desmosome
E	epidermis
ER	endoplasmic reticulum
Fi	intermediate filament
Fm	microfilament
G	Golgi-system
I	interdigitation
Ip	paracrystalline inclusion
Jt	tight junction
M	mitochondrium
MC	mucus-cell
Mt	microtubule
Nf	nerve fiber
NP	nerve fiber plexus of TB
Nu	nucleus
Pi	taste pit
Po	taste pore
Pr	rodlike process
R	ribosome
r	rough
s	smooth
TB	taste bud
TD	taste disc
Ve	synaptic vesicle
Vl	large receptor villus
Vs	small receptor villus
W	wing cell
I	Type I cell (dark)
II	Type II cell (light)
III	Type III cell (somewhat light)

REFERENCES

1. **Waller, A.,** Minute structure of the papillae and nerves of the tongue of the frog and toad, communicated by R. Owen, *Philos. Trans.* R. Soc London, Pt. I, 139, 1849.
2. **Leydig, F.,** Über die Haut einiger Süβwasserfische, *Z. Wiss. Zool.*, 3, 1, 1851.
3. **Lovén, Ch.,** Beiträge zur Kenntnis vom Bau der Geschmackswärzchen der Zunge, *Arch. Mikrosk. Anat.*, 4, 96, 1868.
4. **Schwalbe, G.,** Das Epithel der Papillae vallatae, *Arch. Mikrosk. Anat.*, 3, 504, 1867.
5. **Schwalbe, G.,** Über die Geschmacksorgane der Säugetiere und des Menschen, *Arch. Mikrosk. Anat.*, 4, 154, 1868.
6. **Oppel, A.,** Nerven und Sinnesorgane der Zunge und Mundhöhle, in *Lehrbuch der vergleichenden mikroskopischen Anatomie der Wirbeltiere*, 3. Teil: Mundhöhle, Bauchspeicheldrüse und Leber, Fischer, Jena, 1900, 425.
7. **Ebner, V. V.,** Von den Geschmacksknospen. Von den Nerven der Geschmacksknospen, in *Handbuch der Gewebelehre des Menschen*, Koelliker, A., Ed., 6th ed., Vol. 3, Engelmann, Leipzig, 1902, 18.
8. **Herrick, C. J.,** The organ and sense of taste in fishes, *Bull. U.S. Fish. Comm. For 1902*, 237, 1903.
9. **Kolmer, W. C.,** Geschmacksorgan, in *Handbuch der mikroskopischen Anatomie des Menschen*. Bd. 3, Teil 1: *Haut, Milchdrüse, Geruchsorgan, Geschmacksorgan, Sehorgan*, Möllendorf, W. von, Ed., Springer-Verlag, Berlin, 1927, 154.
10. **Boeke, J.,** Organe mit Endknospen und Endhügeln nebst eingesenkten Organen, in *Handbuch der vergleichenden Anatomie der Wirbeltiere*, Bolk, L., Göppert, E., Kallius, E., and Lubosch, W., Eds., Urban und Schwarzenberg, Berlin, 1934, 949.
11. **Graziadei, P. P. C.,** The ultrastructure of vertebrate taste receptors, in *Olfaction and Taste III. Proceedings of the 3rd International Symposium*, Rockefeller University Press, New York, 1969, 315.
12. **Jeppsson, P. H.,** Studies on the structure and innervation of taste buds. An experimental and clinical investigation, *Acta Oto-Laryngol. Suppl. (Stockholm)*, 259, 1, 1969.
13. **Murray, R. G.,** Ultrastructure of taste receptors, in *Handbook of Sensory Physiology*, Beidler, L. M., Ed., Vol. 4, *Chemical Senses*, Part 2, *Taste*, Springer-Verlag, Berlin, 1971, 31.
14. **Murray, R. G.,** The ultrastructure of taste buds, in *The Ultrastructure of Sensory Organs*, Friedmann, I., Ed., North-Holland, Amsterdam, 1973, 1.
15. **Murray, R. G.,** Gustatory receptor cells, in *Functional Morphology of Receptor Cells*, Rohen, J. W., Ed., Akademie der Wissenschaften und der Literatur Mainz, Steiner Verlag, Wiesbaden, 1978, 88.
16. **Murray, R. G. and Murray, A.,** The anatomy and ultrastructure of taste endings, in *Taste and Smell in Vertebrates*, Wolstenholme, G. E. W. and Knight, J., Eds., J. & A. Churchill, London, 1970, 3.
17. **Murray, R. G. and Murray, A.,** Relations and possible significance of taste bud cells, in *Contributions to Sensory Physiology*, Neff, W. D., Ed., Vol. 5, Academic Press, New York, 1971, 47.
18. **Reutter, K.,** Taste organ in the bullhead (Teleostei), *Adv. Anat. Embryol. Cell. Biol.*, 55, 1, 1978.
19. **Roper, S. D.,** The cell biology of vertebrate taste receptors, *Annu. Rev. Neurosci.*, 12, 329, 1989.
20. **Cordier, R.,** Sensory cells, in *The Cell: Biochemistry, Physiology, Morphology*, Brachet, J. and Mirsky, A. E., Eds., Academic Press, New York, 1964, 313.
21. **Atema, J.,** Structures and functions of the sense of taste in the catfish *(Ictalurus natalis)*, *Brain Behav. Evol.*, 4, 273, 1971.
22. **Kapoor, B. G., Evans, H. E., and Pevzner, R. A.,** The gustatory system in fish, *Adv. Mar. Biol.*, 13, 53, 1975.

23. **Reutter, K.,** Taste organ in the barbel of the bullhead, in *Chemoreception in Fishes*, Hara, T. J., Ed., Elsevier, Amsterdam, 1982, 77.
24. **Reutter, K.,** Chemoreceptors, in *Biology of the Integument*, Bereiter-Hahn, J., Matoltsy, A. G., and Richards, K. S., Eds., Vol. 2, Springer-Verlag, Berlin, 1986, 586.
25. **Reutter, K.,** Structure of the peripheral gustatory organ, represented by the silurid fish *Plotosus lineatus* (Thunberg), in *Fish Chemoreception*, Hara, T., Ed., Chapman & Hall, London, 1992, 60.
26. **Tucker, D.,** Fish chemoreception: peripheral anatomy and physiology, in *Fish Neurobiology: Brainstem and Sensory Organs*, Northcutt, R. G. and Davis, R. E., Eds., Vol. 1. University of Michigan Press, Ann Arbor, 311.
27. **Caprio, J. (1988),** Peripheral filters and chemoreception cells in fishes, in *Sensory Biology of Aquatic Animals*, Atema, J., Fay, R. R., Popper, A. N., and Tavolga, W. N., Eds., Springer-Verlag, New York, 1988, 313.
28. **Jakubowski, M. and Whitear, M.,** Comparative morphology and cytology of taste buds in teleosts, *Z. Mikrosk.-Anat. Forsch.*, 104, 529, 1990.
29. **Reutter, K., Breipohl, W., and Bijvank, G. J.,** Taste bud types in fishes. II. Scanning electron microscopical investigations on *Xiphophorus helleri* Heckel (Poeciliidae, Cyprinodontiformes, Teleostei), *Cell Tissue Res.*, 153, 151, 1974.
30. **Reutter, K.,** Ist rasterelektronenmikroskopisch eine "naturgetreue" Darstellung schleimbedeckter Oberflächen möglich? Untersuchungen an den Rezeptorfeldern der Geschmacksknospen des Zwergwelses, *Verh. Anat. Ges.*, 75, 575, 1981.
31. **Ovalle, W. K. and Shinn, S. L.,** Surface morphology of taste buds in catfish barbels, *Cell Tissue Res.*, 178, 375, 1977.
32. **Lane, E. B. and Whitear, M.,** Sensory structures at the surface of fish skin. I. Putative chemoreceptors, *Zool. J. Linn. Soc.*, 75, 141, 1982.
33. **Fujimoto, S. and Yamamoto, K.,** Electron microscopy of terminal buds on the barbels of the silurid fish, *Corydoras paleatus*, *Anat. Rec.*, 197, 133, 1980.
34. **Ono, R. D.,** Fine structures and distribution of epidermal projections associated with taste buds on the oral papillae in some loricariid catfishes (Siluroidei: Loricariidae), *J. Morphol.*, 164, 139, 1980.
35. **Rajbanshi, V. K.,** Scanning microscopical and histological studies of barbel epithelium of *Heteropneustes fossilis* (Bloch), *Zool. Beitr.*, 25, 73, 1979.
36. **Ezeasor, D. N.,** Distribution and ultrastructure of taste buds in the oropharyngeal cavity of the rainbow trout, *Salmo gairdneri* Richardson, *J. Fish Biol.*, 20, 53, 1982.
37. **Kawakita, K., Marui, T., and Funakoshi, M.,** Scanning electron microscope observations on the taste buds of the carp (*Cyprinus carpio* L.), *Jpn. J. Oral Biol.*, 20, 103, 1978.
38. **Walker, E. R., Fidler, S. F., and Hinton, D. E.,** Morphology of the buccopharyngeal portion of the gill in the fathead minnow *Pimephales promelas* (Rafinesque), *Anat. Rec.*, 200, 67, 1981.
38. **Hossler, F. E. and Merchant, L. H.,** Morphology of taste buds on the gill arches of the mullet *Mugil cephalus*, and the killifish *Fundulus heteroclitus*, *Am. J. Anat.*, 166, 299, 1983.
40. **Reutter, K.,** Typisierung der Geschmacksknospen von Fischen. I. Morphologische und histochemische Untersuchungen an *Xiphophorus helleri* Heckel (Poeciliidae, Cyprinodontiformes, Teleostei), *Z. Zellforsch.*, 143, 409, 1973.
41. **Jakubowski, M.,** New details of the ultrastructure (TEM, SEM) of taste buds in fishes, *Z. Mikrosk. Anat. Forsch.*, 97, 849, 1983.
42. **Farbman, A. I. and Yonkers, J. D.,** Fine structure of the taste bud in the mud puppy, *Necturus maculosus*, *Am. J. Anat.*, 131, 353, 1971.
43. **Cummings, T. A., Delay, R. J., and Roper, S. D.,** Ultrastructure of apical specializations of taste cells in the mudpuppy, *Necturus maculosus*, *J. Comp. Neurol.*, 261, 604, 1987.

44. **Delay, R. J. and Roper, S. D.,** Ultrastructure of taste cells and synapses in the mudpuppy *Necturus maculosus, J. Comp. Neurol.*, 277, 268, 1988.
45. **Roper, S. D.,** The microphysiology of peripheral taste organ, *J. Neurosci.*, 12, 1127, 1992.
46. **Fährmann, W.,** Licht- und elektronenmikroskopische Untersuchungen an der Geschmacksknospe des neotenen Axolotls (*Siredon mexicanum* SHAW), *Z. Mikrosk.-Anat. Forsch.*, 77, 117, 1967.
47. **Toyoshima, K., Miyamoto, K., and Shimamura, A.,** Fine structure of taste buds in the tongue, palatal mucosa and gill arch of the axolotl, *Ambystoma mexicanum, Okaj. Folia Anat. Jpn.*, 64, 99, 1987.
48. **Samanen, D. W. and Bernard, R. A.,** Scanning electron microscopy of the taste bud of the mudpuppy, *Necturus maculosus,* in *33rd Ann. Proc. Electron Microscopy Soc. Am.*, Bailey, G. W., Ed., Las Vegas, NV, 530.
49. **Bethe, A.,** Die Nervenendigungen im Gaumen und in der Zunge des Frosches, *Arch. Mikrosk.-Anat.*, 44, 185, 1895.
50. **Jaeger, C. B. and Hillman, D. E.,** Morphology of gustatory organs, in *Frog Neurobiology — A Handbook*, Llinas, R. and Precht, W., Eds., Springer-Verlag, Berlin, 1976, 588.
51. **Uga, S. and Hama, K.,** Electron microscopic studies on the synaptic region of the taste organ of carps and frogs, *J. Electron Microsc. (Tokyo)*, 16, 269, 1967.
52. **Graziadei, P. P. C.,** The ultrastructure of vertebrate taste receptors, *Olfaction and Taste III, Proc. 3rd Int. Symp.*, Pfaffmann, C., Ed., Rockefeller University Press, New York, 1969, 315.
53. **Graziadei, P. P. C. and De Han, R. S.,** The ultrastructure of frog's taste organs, *Acta Anat. (Basel)*, 80, 563, 1971.
54. **Goosens, N. and Vandenberghe, M.-P.,** The basal cells in the papillae fungiformis of the tongue of the common frog, *Rana temporaria* L., *Arch. Histol. Jpn.*, 36, 173, 1974.
55. **Düring, M. von and Andres, K. H.,** The ultrastructure of taste and touch receptors of the frog's taste organ, *Cell Tissue Res.*, 165, 185, 1976.
56. **Röhlich, P. and Pevzner, R. A.,** The chemoreceptor surface of the taste organ in the frog *Rana esculenta*. A freeze-fracture analysis, *Cell Tissue Res.*, 224, 409, 1982.
57. **Richter, H.-P., Avenet, P., Mestres, P., and Lindemann, B.,** Gustatory receptors and neighbouring cells in the surface layer of an amphibian taste disc: in situ relationship and response to cell isolation, *Cell Tissue Res.*, 254, 83, 1988.
58. **Suzuki, Y. and Takeda, M.,** Filaments in the cells of frog taste, *Organ. Zool. Sci.*, 6, 487, 1989.
59. **Stensaas, L. J.,** The fine structure of fungiform papillae and epithelium of the tongue of a south american toad, *Calyptocephalella gayi, Am. J. Anat.*, 131, 443, 1971.
60. **Sagmeister, H., Gubo, G., Lametschwandtner, A., Simonsberger, P., and Adam, H.,** A new cell type in the taste buds of anurans. A scanning and transmission electron microscopic study, *Cell Tissue Res.*, 183, 553, 1977.
61. **Jasinski, A.,** Light and scanning microscopy of the tongue and its gustatory organs in the common toad, *Bufo bufo* (L.), *Z. Mikrosk.-Anat. Forsch., (Leipzig)*, 93, 465, 1979.
62. **Toyoshima, K. and Shimamura, A.,** Comparative study of ultrastructure of the lateral-line organs and the palatal taste organs in the african clawed toad, *Xenopus laevis, Anat. Rec.*, 204, 371, 1982.
63. **Witt, M.,** The ultrastructure of the taste disc in the red-bellied toad *Bombina orientalis* (Discoglossida, Salientia), *Cell Tissue Res.*, 271, 59, 1992.
64. **Korte, G. E.,** Ultrastructure of the taste buds of the red-eared turtle, *Chrysemys scripta elegans, J. Morphol.*, 163, 231, 1980.
65. **Uchida, T.,** Ultrastructural and histochemical studies on the taste buds in some reptiles, *Arch. Histol. Jpn.*, 43, 459, 1980.

66. **Pevzner, R. A. and Tikhonova, M. A.,** Ultrastructural organization of the taste buds of reptiles. I. Tortoises, *Tsitologiia,* 21, 132, 1979.
67. **Pevzner, R. A. and Tikhonova, M. A.,** Ultrastructural organization of the taste buds of reptiles. II. Lizards, *Tsitologiia,* 21, 1266, 1979.
68. **Vinnikov, Ya. and Pevzner, R. A.,** The structure and cytochemistry of the vertebrate taste buds, *Proc. XIIth ORL World Congr.,* Surján, L. and Bodo, Gy., Eds., Publ. House Hung. Acad. Sci., Budapest, 1981, 637.
69. **Dmitrieva, N. A.,** Fine structural pecularities of the taste buds of the pigeon *(Columba livia), Dokl. Akad. Nauk. SSSR,* 23, 874, 1981.
70. **Suzuki, Y. and Takeda, M.,** Ultrastructure of taste buds in birds, *Jpn. J. Oral. Biol.,* 26, 669, 1984.
71. **Ganchrow, D. and Ganchrow, J. R.,** Gustatory ontogenesis in the chicken: an avian-mammalian comparison, *Med. Sci. Res.,* 17, 223, 1989.
72. **Ganchrow, D., Ganchrow, J. R., and Goldstein, R. S.,** Ultrastructure of palatal taste buds in the perihatching chick, *Am. J. Anat.,* 192, 69, 1991.
73. **Sprissler, Ch.,** Licht- und elektronenmikroskopische Untersuchungen an den Geschmacksknospen der Japanischen Wachtel, *Coturnix coturnix japonica,* Thesis, Med. Fac. Univ., Tübingen, 1993.
74. **Sprissler, Ch. and Reutter, K.,** Ultrastructure of taste buds in the quail, *Coturnix coturnix japonica,* in *Proc. 20th Göttingen Neurobiol. Conf.,* Elsner, N. and Richter, D. W., Eds., Thieme, Stuttgart, 1992, 279.
75. **Takeda, M.,** An electron microscopic study on the innervation in the taste buds of the mouse circumvallate papillae, *Arch. Histol. Jpn.,* 39, 257, 1976.
76. **Takeda, M., Shishido, Y., Kitao, K., and Suzuki, Y.,** Monoamines of taste buds in the fungiform and foliate papillae of the mouse, *Arch. Histol. Jpn.,* 45, 239, 1982.
77. **Suzuki, Y. and Takeda, M.,** Scanning electron microscopic observation of the basement membrane in mouse taste buds, *J. Electron Microsc.,* 36, 98, 1987.
78. **Kinnamon, J. C., Taylor, B. J., Delay, R. J., and Roper, S. D.,** Ultrastructure of mouse vallate taste buds. I. Taste cells and their associated synapses, *J. Comp. Neurol.,* 235, 48, 1985.
79. **Royer, S. M. and Kinnamon, J. C.,** Ultrastructure of mouse foliate taste buds: synaptic and nonsynaptic interactions between taste cells and nerve fibers, *J. Comp. Neurol.,* 270, 11, 1988.
80. **Delay, R. J., Kinnamon, J. C., and Roper, S. D.,** Ultrastructure of mouse vallate taste buds. II. Cell types and cell lineage, *J. Comp. Neurol.,* 253, 242, 1986.
81. **Kinnamon, J. C., Sherman, T. A., and Roper, S. D.,** Ultrastructure of mouse vallate taste buds. III. Patterns of synaptic connectivity, *J. Comp. Neurol.,* 270, 1, 1988.
82. **Farbman, A. I.,** Fine structure of the taste bud, *J. Ultrastruct. Res.,* 12, 328, 1965.
83. **Takeda, M. and Hoshino, T.,** Fine structure of taste buds in the rat, *Arch. Histol. Jpn.,* 37, 395, 1975.
84. **Gomez-Ramos, P. and Rodriguez-Echandia, E. L.,** The fine structural effect of sialectomy on the taste bud cells in the rat, *Tissue Cell,* 11, 19, 1979.
85. **Endo, Y.,** Exocytotic release of neurotransmitter substances from nerve endings in the taste buds of rat circumvallate papillae, *Arch. Histol. Cytol.,* 51, 489, 1988.
86. **Miller, R. L. and Chaudhry, A. P.,** Comparative ultrastructure of vallate, foliate and fungiform taste buds of golden Syrian hamster, *Acta Anat.,* 95, 75, 1976.
87. **Miller, R. L. and Chaudhry, A. P.,** An ultrastructural study on the development of vallate taste buds of the golden Syrian hamster, *Acta Anat.,* 95, 190, 1976.
88. **Whitehead, M. C., Beeman, C. S., and Kinsella, B. A.,** Distribution of taste and general sensory nerve endings in fungiform papillae of the hamster, *Am. J. Anat.,* 173, 185, 1985.
89. **Yoshie, S., Wakasugi, C., Teraki, Y., and Fujita, T.,** Fine structure of the taste bud in guinea pigs. I. Cell characterization and innervation patterns, *Arch. Histol. Cytol.,* 53, 103, 1990.

90. **Engström, H. and Rytzner, C.,** The fine structure of taste buds and taste fibers, *Ann. Otol. (St. Louis)*, 65, 361, 1956.
91. **Trujillo-Cenóz, O.,** Electron microscope study of the rabbit gustatory bud, *Z. Zellforsch.*, 46, 272, 1957.
92. **Nemetschek-Gansler, H. and Ferner, H.,** Über die Ultrastruktur der Geschmacksknospen, *Z. Zellforsch.*, 63, 155, 1964.
93. **Scalzi, H. A.,** The cytoarchitecture of gustatory receptors from the rabbit foliate papillae, *Z. Zellforsch.*, 80, 413, 1967.
94. **Olivieri-Sangiacomo, C.,** Ultrastructural modifications of denervated taste buds, *Z. Zellforsch.*, 108, 397, 1970.
95. **Toyoshima, K. and Shimamura, A.,** A scanning electron microscopic study of taste buds in the rabbit, *Biomed. Res.*, 2(Suppl.), 459, 1981.
96. **Murray, R. G.,** The mammalian taste bud type III cell: a critical analysis, *J. Ultrastruct. Mol. Struct. Res.*, 95, 175, 1986.
97. **Toyoshima, K. and Tandler, B.,** Dividing type II cell in rabbit taste bud, *Anat. Rec.*, 214, 161, 1986.
98. **Toyoshima, K. and Tandler, B.,** Modified smooth endoplasmic reticulum in type II cells of rabbit taste buds, *J. Submicrosc. Cytol.*, 19, 85, 1987.
99. **Royer, S. M. and Kinnamon, J. C.,** HVEM serial-section analysis of rabbit foliate taste buds. I. Type III cells and their synapses, *J. Comp. Neurol.*, 306, 49, 1991.
100. **Murray, R. G. and Murray, A.,** The fine structure of taste buds of Rhesus and Cynomalgus monkeys, *Anat. Rec.*, 138, 211, 1960.
101. **Ide, C. and Munger, B. L.,** The cytologic composition of primate laryngeal chemosensory corpuscles, *Am. J. Anat.*, 158, 193, 1980.
102. **Arvidson, K., Cottler-Fox, M., and Friberg, U.,** Taste buds of the fungiform papillae in Cynomolgus monkey, *J. Anat.*, 133, 271, 1981.
103. **Farbman, A. I., Hellekant, G., and Nelson, A.,** Structure of taste buds in foliate papillae of the Rhesus monkey, *Macaca mulatta*, *Am. J. Anat.*, 172, 41, 1985.
104. **Toyoshima, K. and Tandler, B.,** Modified endoplasmic reticulum in taste bud cells of the Japanese monkey, *J. Submicrosc. Cytol. Pathol.*, 21, 205, 1989.
105. **Paran, N., Mattern, C. F. T., and Henkin, R. I.,** Ultrastructure of the taste bud of the human fungiform papilla, *Cell. Tissue Res.*, 161, 1, 1975.
106. **Arvidson, K., Cottler-Fox, M., and Friberg, U.,** Fine structure of taste buds in the human fungiform papilla, *Scand. J. Dent. Res.*, 89, 297, 1981.
107. **Ferrell, F.,** Taste bud morphology in the fetal and neonatal dog, *Neurosci. Biobehav. Rev.*, 8, 175, 1984.
108. **Dougbag, A. E.-S.,** Scanning electron microscopic studies on the morphogenesis of the lingual gustatory papillae of camel *(Camel dromedarius)*. I. Morphogenesis of the fungiform papillae, *Z. Mikrosk.-Anat. Forsch. (Leipzig)*, 101, 881, 1987.
109. **Mistretta, C. M.,** Developmental neurobiology of the taste system, in *Smell and Taste in Health and Disease*, Getchell, T. V., et al., Eds., Raven Press, New York, 1991, 35.
110. **Murray, R. G.,** Cell types in rabbit taste buds, in *Olfaction and Taste III, Proc. of the 3rd Int. Symp.*, Pfaffman, C., Ed., Rockefeller University Press, New York, 1969, 331.
111. **Beidler, L. M.,** Innervation of rat fungiform papilla, in *Olfaction and Taste III, Proc. of the 3rd Int. Symp.*, Pfaffman, C., Ed., Rockefeller University Press, New York, 1969, 352.
112. **Jasiński, A. and Miodoński, A.,** Blood vessels in the tongue of the kitten: scanning electron microscopy of microcorrosion casts, *Anat. Embryol.*, 155, 347, 1979.
113. **Kishi, Y., Takahashi, K., and Trowbridge, H.,** Vascular network in papillae of dog oral mucosa using corrosive resin casts with scanning electron microscopy, *Anat. Rec.*, 226, 447, 1990.

114. **Ohshima, H., Yoshida, S., and Kobayashi, S.,** Blood vascular architecture of the rat lingual papillae with special reference to their relations to the connective tissue papillae and surface structures: a light and scanning electron microscope study, *Acta Anat.*, 137, 213, 1990.
115. **Mattern, C. F. T., Daniel, W. A., and Henkin, R. I.,** The ultrastructure of the human circumvallate papilla. I. Cilia of the papillary crypt, *Anat. Rec.*, 167, 175, 1970.
116. **Toyoshima, K. and Shimamura, A.,** The occurrence of ciliated and mucous cells in the peripapillary trench of the rat tongue, *Anat. Rec.*, 195, 301, 1979.
117. **Reutter, K.,** Die Geschmacksknospen des Zwergwelses *Amiurus nebulosus* (Lesueur). Morphologische und histochemische Untersuchungen, *Z. Zellforsch.*, 120, 280, 1971.
118. **Whitear, M.,** Cell specialization and sensory function in fish epidermis, *J. Zool. (London)*, 163, 237, 1971.
119. **Jakubowski, M. and Whitear, M.,** Ultrastructure of taste buds in fishes, *Fol. Histochem. Cytobiol.*, 24, 310, 1986.
120. **Trujillo-Cenóz, O.,** Electron microscope observations on chemo- and mechanoreceptor cells of fishes, *Z. Zellforsch.*, 54, 654, 1961.
121. **Desgranges, J. C.,** Sur l'existence de plusieurs types de cellules sensorielles dans les bourgeons du goût des barbillons du Poisson-chat, *C. R. Acad. Sci. Ser. D (Paris)*, 261, 1095, 1965.
122. **Desgranges, J. C.,** Sur la double innervation des cellules sensorielles des bourgeons du goût des barbillons du Poisson-chat, *C.R. Acad. Sci. Ser. D (Paris)*, 263, 1103, 1966.
123. **Desgranges, J. C.,** Sur les bourgeons du goût du Poisson-chat **Ictalurus melas**: ultrastructure des cellules basales, *C.R. Acad. Sci. Ser. D (Paris)*, 274, 1814, 1972.
124. **Hirata, Y.,** Fine structure of terminal buds on the barbels of some fishes, *Arch. Histol. Jpn.*, 26, 507, 1966.
125. **Uga, S. and Hama, K.,** Electron microscopic studies on the synaptic region of the taste organ of carps and frogs, *J. Electron Microsc. (Tokyo)*, 16, 269, 1967.
126. **Welsch, U. and Storch, V.,** Die Feinstruktur der Geschmacksknospen von Welsen (*Clarias batrachus* (L.) und *Kryptopterus bicirrhis* (Cuvier et Valenciennes)), *Z. Zellforsch.*, 100, 552, 1969.
127. **Storch, V. N. and Welsch, U. N.,** Electron microscopic observations on the taste buds of some bony fishes, *Arch. Histol. Jpn.*, 32, 145, 1970.
128. **Schulte, E. and Holl, A.,** Untersuchungen an den Geschmacksknospen der Barteln von *Corydoras paleatus* Jenyns. I. Feinstruktur der Geschmacksknospen, *Z. Zellforsch.*, 120, 450, 1971.
129. **Schulte, E. and Holl, A.,** Feinbau der Kopftentakel und ihrer Sinnesorgane bei *Blennius tentacularis* (Pisces, Blenniiformes), *Mar. Biol.*, 12, 67, 1972.
130. **Crisp, M., Lowe, G. A., and Laverack, M. S.,** On the ultrastructure and permeability of taste buds of the marine teleost *Ciliata mustela*, *Tissue Cell*, 7, 191, 1975.
131. **Grover-Johnson, N. and Farbman, A. I.,** Fine structure of taste buds in the barbel of the catfish, *Ictalurus punctatus*, *Cell Tissue Res.*, 169, 395, 1976.
132. **Joyce, E. C. and Chapman, G. B.,** Fine structure of the nasal barbel of the channel catfish, *Ictalurus punctatus*, *J. Morphol.*, 158, 109, 1978.
133. **Pevzner, R. A.,** Electron microscopic study of the taste buds of the eel, *Anguilla anguilla*, *Tsitologiia*, 20, 1112, 1978.
134. **Pevzner, R. A.,** The fine structure of taste buds of the ganoid fishes. I. Adult acipenseridae, *Tsitologiia*, 23, 760, 1981.
135. **Pevzner, R. A.,** The fine structure of taste buds of the ganoid fishes. II. Fries during endo-exogenous feeding, *Tsitologiia*, 23, 867, 1981.
136. **Connes, R., Granie-Prie, M., Diaz, J. P., and Paris, J.,** Ultrastructure des bourgeons du goût du téléostéen marin *Dicentrarchus labrax* L., *Can. J. Zool.*, 66, 2133, 1988.
137. **Reutter, K.,** SEM-study of the mucus layer on the receptor field of fish taste buds, in *Olfaction and Taste VII*, Starre, H. van der, Ed., IRL Press Limited, London, 1980, 107.

138. **Reutter, K.,** Specialized receptor villi and basal cells within the taste bud of the european silurid fish, *Silurus glanis* (Teleostei), in *Olfaction and Taste IX*, Roper, D. and Atema, J., Eds., Ann. N.Y. Acad. Sci., New York, 1987, 570.
139. **Witt, M. and Reutter, K.,** Electron microscopical demonstration of lectin binding sites at taste buds and adjacent epithelia of the catfish *Silurus glanis* (Teleostei), in *Lectins-Biology, Biochemistry, Clinical Biochemistry*, Kocourek, J. and Freed, D. L. J., Eds., Vol. 7, Sigma Chemical Company, St. Louis, MO, 1990, 359.
140. **Witt, M. and Reutter, K.,** Electron microscopic demonstration of lectin binding sites in the taste buds of the European catfish *Silurus glanis* (Teleostei), *Histochemistry*, 94, 617, 1990.
141. **Getchell, M. L., Zieliński, B., and Getchell, T. V.,** Odorant and autonomic regulation of secretion in the olfactory mucosa, *Molecular Neurobiology of the Olfactory System*, Margolis, F. and Getchell, T. V., Eds., Plenum Press, New York, 1988, 71.
142. **Toyoshima, K.,** Ultrastructure and immunohistochemistry of the basal cells in the taste buds of the loach, *Misgurnus anguillicaudatus*, *J. Submicrosc. Cytol. Pathol.*, 21, 771, 1989.
143. **Nada, O. and Hirata, K.,** The monoamine-containing cell in the gustatory epithelium of some vertebrates, *Arch. Histol. Jpn.*, 40, 197, 1977.
144. **Toyoshima, K., Nada, O., and Shimamura, A.,** Fine structure of monoamine-containing basal cells in the taste buds on the barbels of three species of teleosts, *Cell Tissue Res.*, 235, 479, 1984.
145. **Fujita, T., Kanno, T., and Kobayashi, S.,** *The Paraneuron*, Springer-Verlag, Tokyo, 1988.
146. **Raderman-Little, R.,** The effect of temperature on the turnover of taste bud cells in catfish, *Cell. Tissue Kinet.*, 12, 269, 1979.
147. **Emmerling, M. R., Sobkowicz, H. M., Levenick, C. V., Scott, G. L., Slapnick, S. M., and Rose, J. E.,** Biochemical and morphological differentiation of acetylcholinesterase-positive efferent fibers in the mouse cochlea, *J. Electron Microsc. Technol.*, 15, 123, 1990.
148. **Pappas, G. D. and Waxman, S. G.,** Synaptic fine structure-morphological correlates of chemical and electronic transmission, *Structure and Function of Synapses*, Pappas, G. D. and Purpura, D. P., Eds., North-Holland, Amsterdam, 1972, 1.
149. **Reutter, K. and Bardele, C.,** Ultrastrukturelle Untersuchung der Geschmacksorgane des Zwergwelses anhand von Gefrierätz-Präparaten, *Verh. Anat. Ges.*, 77, 747, 1983.
150. **Kinnamon, S. C., Cummings, T. A., and Roper, S. D.,** Isolation of single taste cells from lingual epithelium, *Chem. Senses*, 13, 355, 1988.
151. **Whitear, M.,** Apical secretion from taste bud and other epithelial cells in amphibians, *Cell Tissue Res.*, 172, 389, 1976.
152. **Jain, S. and Roper, S. D.,** Immunocytochemistry of gamma-aminobutyric acid, glutamate, serotonin, and histamine in *Necturus* taste buds, *J. Comp. Neurol.*, 307, 675, 1991.
153. **Toyoshima, K. and Shimamura, A.,** Monoamine-containing basal cells in the taste buds of the newt *Triturus pyrrhogaster*, *Arch. Oral Biol.*, 32, 619, 1987.
154. **Żuwała, K.,** Developmental changes in the structure of mucous membrane in the oral cavity and taste organs in tadpoles of the frog, *Rana temporaria* (SEM), *Acta Biol. Cracov.*, 33, 59, 1991.
155. **Żuwała, K. and Jakubowski, M.,** Development of taste organs in *Rana temporaria*. Transmission and scanning electron microscopy study, *Anat. Embryol.*, 184, 363, 1991.
156. **Carmignani, M. P. A., Zaccone, G., and Cannata, F.,** Histochemical studies on the tongue of anuran amphibian. I. Mucopolysaccharide Histochemistry of the papillae and the lingual glands in *Hyla arborea* L., *Rana esculenta* L. and *Bufo vulgaris* Laur., *Ann. Histochim.*, 20, 47, 1975.

157. **Raviola, E. and Osculati, F.,** La fine struttura dei recettori gustativi della lingua di rana, *Inst. Lomb. (Rend. Sci),* B 101, 599, 1967.
158. **Pevzner, R. A.,** Electron microscopic investigation of receptor and supporting cells in taste buds of *R. temporaria, Tsitologiia,* 12, 971, 1970.
159. **Sbarbati, A., Zancanaro, C., Franceschini, F., Balercia, G., Morroni, M., and Osculati, F.,** Characterization of different microenvironments at the surface of the frog's taste organ, *Am. J. Anat.,* 188, 199, 1990.
160. **Sbarbati, A., Ceresi, E., and Accordini, C.,** Surfactant-like material on the chemoreceptorial surface of the frog's taste organ: an ultrastructural and electron spectroscopic imaging study, *J. Struct. Biol.,* 107, 128, 1991.
161. **Sbarbati, A., Franceschini, F., Zancanaro, C., Cecchini, T., Ciaroni, S., and Osculati, F.,** The fine morphology of the basal cell in the frog's taste organ, *J. Submicrosc. Cytol. Pathol.,* 20, 73, 1988.
162. **Toyoshima, K., Honda, E., Nakahara, S., and Shimamura, A.,** Ultrastructural and histochemical changes in the frog taste organ following denervation, *Arch. Histol. Jpn.,* 47, 31, 1984.
163. **Hartschuh, W., Weihe, E., and Reinecke, M.,** The Merkel cell, in *Biology of the Integument,* Bereiter-Hahn, et al., Eds., Vol. 2, *Vertebrates,* Springer-Verlag, Berlin, 1986, 605.
164. **De Han, R. S. and Graziadei, P.,** The innervation of frog's taste organ. "A histochemical study", *Life Sci.,* 13, 1435, 1973.
165. **Krokhina, E. M., Esakov, A. I., and Savushkina, M. A.,** Catecholamine-containing cells of the taste buds of the tongue of the frog *(Rana temporaria), Arch. Anat. Microsc. Morphol. Exp.,* 64, 67, 1975.
166. **Hirata, K. and Nada, O.,** A monoamine in the gustatory cell of the frog's taste organ. A. fluorescence histochemical and electron microscopic study, *Cell Tissue Res.,* 159, 101, 1975.
167. **Nada, O. and Hirata, K.,** Ultrastructural evidence for the uptake of 5,6-dihydroxytryptamine in the frog's gustatory cell, *Am. J. Anat.,* 144, 393, 1975.
168. **Hirata, K. and Nada, O.,** Cytoarchitecture of monoamine-containing cells in the frog's gustatory epithelium, *Experientia,* 33, 1223, 1977.
169. **Sbarbati, A., Zancanaro, C., Franceschini, F., and Osculati, F.,** Basal cells of the frog's taste organ: fluorescence histochemistry with the serotonin analogue 5,7-dihydroxytryptamine in supravital conditions, *Bas. Appl. Histochem.,* 33, 289, 1989.
170. **Kuramoto, H.,** An immunohistochemical study of cellular and nervous elements in the taste organs of the bullfrog, *Rana catesbeiana, Arch. Histol. Cytol.,* 51, 205, 1988.
171. **Toyoshima, K. and Shimamura, A.,** An immunohistochemical demonstration of neuron-specific enolase inthe Merkel cells of the frog taste organ, *Arch. Histol. Cytol.,* 51, 295, 1988.
172. **Zancanaro, C., Sbarbati, A., Franceschini, F., Balercia, G., and Osculati, F.,** The chemoreceptor surface of the taste disc in the frog, *Rana esculenta.* An ultrastructural study with lanthanum nitrate, *Histochem. J.,* 22, 480, 1990.
173. **Carmignani, M. P. A. and Zaccone, G.,** Histochemical studies on the tongue of anuran amphibians. II. Comparative morphochemical study of the taste buds and the lingual glands in *Bufo viridis* Laurenti and *Rana graeca* Boulenger with particular reference to the mucosaccharide histochemistry, *Cell Molec. Biol.,* 22, 203, 1977.
174. **Reutter, K. and Klessen, C.,** Vergleichend-kohlenhydrathistochechemische Untersuchung der Geschmacksknospen niederer und höherer Wirbeltiere, *Verh. Anat. Ges.,* 73, 1019, 1979.
175. **Witt, M. and Reutter, K.,** Lectin histochemistry on mucous substances of the taste buds and adjacent epithelia of different vertebrates, *Histochemistry,* 88, 453, 1988.
176. **Desgranges, J. C.,** Ultrastructure de la cellule gustative d'*Alytes obstetricans* Laurenti (Anoura): Terminasion réceptrice, relations intercellulaires et double innervation, *C. R. Acad. Sci. Ser. D (Paris),* 279, 319, 1974.

177. **Fujita, T.,** Present status of paraneuron concept, *Arch. Histol. Cytol.*, 52(Suppl.), 1, 1989.
178. **Hirata, K. and Kanaseki, T.,** Substance P-like immunoreactive fibers in the frog taste organs, *Experientia*, 43, 386, 1987.
179. **Iwasaki, S.-I., Miyata, K., and Kobayashi, K.,** Studies on the fine structure of the lingual dorsal surface in the frog, *Rana nigromaculata*, *Zool. Sci.*, 3, 265, 1986.
180. **Iwasaki, S.-I. and Kobayashi, K.,** Fine structure of the dorsal tongue surface in the japanese toad, *Bufo japonicus* (Anura, Bufonidae), *Zool. Sci.*, 5, 331, 1988.
181. **Iwasaki, S. and Wanichanon, C.,** Fine structure of the dorsal lingual epithelium of the frog, *Rana rugosa*, *Tissue Cell*, 23, 385, 1991.
182. **Cottler-Fox, M., Arvidson, K., Hammarlund, E., and Friberg, U.,** Fixation and occurrence of dark and light cells in taste buds of fungiform papillae, *Scand. J. Dent. Res.*, 95, 417, 1987.
183. **Arvidson, K.,** Scanning electron microscopy of fungiform papillae on the tongue of man and monkey, *Acta Oto-Laryngol.*, 81, 496, 1976.
184. **Kullaa-Mikkonen, A. and Sorvari, T. E.,** A scanning electron microscopic study of the dorsal surface of the human tongue, *Acta Anat.*, 123, 114, 1985.
185. **Imfeld, T. N. and Schroeder, H. E.,** Palatal taste buds in man: topographical arrangement in islands of keratinized epithelium, *Anat. Embryol.*, 185, 259, 1992.
186. **Settembrini, B. P.,** Papilla palatina, nasopalatine duct and taste buds of young and adult rats, *Acta Anat.*, 128, 250, 1987.
187. **Murray, R. G., Murray, A., and Fujimoto, S.,** Fine structure of gustatory cells in rabbit taste buds, *J. Ultrastruct. Res.*, 27, 444, 1969.
188. **Beidler, L. M. and Smallman, R. L.,** Renewal of cells within taste buds, *J. Cell Biol.*, 27, 263, 1965.
189. **Farbman, A. I.,** Renewal of taste bud cells in rat circumvallate papillae, *Cell Tissue Kinet.*, 13, 349, 1980.
190. **Jahnke, K. and Baur, P.,** Freeze-fracture study of taste bud pores in the foliate papillae of the rabbit, *Cell Tissue Res.*, 200, 245, 1979.
191. **Brouwer, J. N. and Wiersma, A.,** Location of taste buds in intact taste papillae by a selective staining method, *Histochemistry*, 58, 145, 1978.
192. **Ohmura, S., Horimoto, S., and Fujita, K.,** Lectin cytochemistry of the dark granules in the type 1 cells of syrian hamster circumvallate taste buds, *Arch. Oral Biol.*, 34, 161, 1989.
193. **Witt, M. and Miller, I. J., Jr.,** Comparative lectin histochemistry on taste buds of foliate, circumvallate and fungiform papillae of the rabbit tongue, *Histochemistry*, 98, 173, 1992.
194. **Bannister, L. H.,** Possible functions of mucus at gustatory and olfactory systems, in *Transduction Mechanism in Chemoreception*, Poynder, T. M., Ed., Information Retrieval, London, 1974, 39.
195. **Gurkan, S. and Bradley, R. M.,** Secretion of von Ebner's glands influence responses from taste buds in rat circumvallate papilla, *Chem. Senses*, 13, 655, 1988.
196. **Ohsawa, I., Yoshida, M., Ishikawa, S., and Gotoh, M.,** A study about the relation between human taste and saliva, 22nd. Jpn. Symp. on Taste and Smell (Abstr.), *Chem. Senses*, 14, 303, 1988.
197. **Getchell, M. L. and Getchell, T. V.,** β-adrenergic regulation of the secretory granule content of acinar cells in olfactory glands of the salamander, *J. Comp. Physiol.*, 155, 435, 1984.
198. **Schmale, H., Holtgreve-Grez, H., and Christiansen, H.,** Possible role for salivary gland protein in taste reception indicated by homology to lipophilic-ligand carrier protein, *Nature*, 343, 366, 1990.
199. **Pevsner, J., Reed, R. R., Feinstein, P. G., and Snyder, S. H.,** Molecular cloning of odorant-binding protein: member of a ligand carrier family, *Science*, 241, 336, 1988.

200. **Zahm, D. S. and Munger, B. L.,** Fetal development of primate chemosensory corpuscles. II. Synaptic relationships in early gestation, *J. Comp. Neurol.*, 219, 36, 1983.
201. **Zahm, D. S. and Munger, B. L.,** Fetal development of primate chemosensory corpuscles. I. Synaptic relationships in late gestation, *J. Comp. Neurol.*, 213, 146, 1983.
202. **Jahnke, K.,** Der feinstrukturelle Nachweis der Acetylcholinesterase-Aktivität in den Geschmacksknospen der Papillae foliatae des Kaninchens, *Arch. Klin. Exp. Ohren-Nas. Kehlkopfheilkd.*, 203, 125, 1972.
203. **Takeda, M., Shishido, Y., Kitao, K., and Suzuki, Y.,** Biogenic monoamines in developing taste buds of mouse circumvallate papillae, *Arch. Histol. Jpn.*, 44, 485, 1981.
204. **Fujimoto, S., Ueda, H., and Kagawa, H.,** Immunocytochemistry on the localization of 5-hydroxytryptamine in monkey and rabbit taste buds, *Acta Anat.*, 128, 80, 1987.
205. **Nishimoto, T., Akai, M., Inagaki, S., Shiosaka, S., Shimizu, Y., Yamamoto, K., Senba, E., Sakanaka, M., Takatsuki, K., Hara, Y., Takagi, H., Matsuzaki, T., Kawai, Y., and Tohyama, M.,** On the distribution and origins of substance P in the papillae of the rat tongue: an experimental and immunhistochemical study, *J. Comp. Neurol.*, 207, 85, 1982.
206. **Finger, T. E., Womble, M., Kinnamon, J. C., and Ueda, T.,** Synapsin I-like immunoreactivity in nerve fibers associated with lingual taste buds of the rat, *J. Comp. Neurol.*, 292, 283, 1990.
207. **Yoshie, S., Wakasugi, C., Teraki, Y., Iwanaga, T., and Fujita, T.,** Immunocytochemical localizations of neuron-specific proteins in the taste bud of the guinea pig, *Arch. Histol. Cytol.*, 51, 379, 1988.
208. **Yoshie, S., Wakasugi, C., Teraki, Y., Iwanaga, T., and Fujita, T.,** Fine structure of the taste bud in guinea pigs. II. Localization of spot 35 protein, a cerebellar Purkinje cell-specific protein, as revealed by electron-microscopic immunocytochemistry, *Arch. Histol. Cytol.*, 54, 113, 1991.
209. **Toyoshima, K. and Tandler, B.,** Dense-cored vesicles and unusual lamellar bodies in type III gustatory cells in taste buds of rabbit foliate papillae, *Acta Anat.*, 135, 365, 1989.
210. **Nada, O. and Hirata, K.,** The occurrence of the cell type containing a specific monoamine in the taste bud of the rabbit's foliate papilla, *Histochemistry*, 43, 237, 1975.
211. **Takeda, M. and Kitao, K.,** Effect of monoamines on the taste buds in the mouse, *Cell Tissue Res.*, 210, 71, 1980.
212. **Lundberg, J. M., Hökfelt, T., Änggard, A., Pernow, B., and Emson, P.,** Immunohistochemical evidence for substance P immunoreactive nerve fibers in the taste buds of the cat, *Acta Physiol. Scand.*, 107, 389, 1979.
213. **Yamasaki, H., Kubota, Y., Takagi, H., and Tohyama, M.,** Immunoelectron-microscopic study on the fine structure of substance-P-containing fibers in the taste buds of the rat, *J. Comp. Neurol.*, 227, 380, 1984.
214. **Yamasaki, H. and Tohyama, M.,** Ontogeny of substance P-like immuno-reactive fibers in the taste buds and their surrounding epithelium of the circumvallate papillae of the rat. II. Electron microscopic analysis, *J. Comp. Neurol.*, 241, 493, 1985.
215. **Finger, T. E.,** Peptide immunohistochemistry demonstrates multiple classes of perigemmal nerve fibers in the circumvallate papilla of the rat, *Chem. Senses*, 11, 135, 1986.
216. **Farbman, A. I., Ogden-Ogle, C. K., Hellekant, G., Simmons, S. R., Albrecht, R. M., and Van Der Wel, H.,** Labeling of sweet taste binding sites using a colloidal gold-labeled sweet protein, thaumatin, *Scanning Microscopy*, 1, 351, 1987.
217. **Davenport, C. J. and Caprio, J.,** Taste and tactile recordings from the ramus recurrens facialis innervating flank taste buds in the fish, *J. Comp. Physiol.*, 147, 217, 1982.
218. **Finger, T. E., Drake, S. K., Kotrschal, K., Womble, M., and Dockstader, L.,** Postlarval growth of the peripheral gustatory system in the channel catfish, *Ictalurus punctatus*, *J. Comp. Neurol.*, 314, 55, 1991.

219. **Herrick, C. J.,** The cranial nerves and cutaneous sense organs of the north american siluroid fishes, *J. Comp. Neurol.*, 11, 177, 1901.
220. **Finger, T. E.,** Gustatory pathway in the bullhead catfish. I. Connections of the anterior ganglion, *J. Comp. Neurol.*, 165, 513, 1976.
221. **Fischer, J. H.,** Die Gehirnnerven der Saurier. Abhandlungen aus dem Gebiet der Naturwissenschaften, Hamburg (cited in Cords, 1904; Ref 229).
222. **Schwenk, K.,** Occurrence, distribution and functional significance of taste buds in lizards, *Copeia*, 1, 91, 1985.
223. **Jacobs, V. L.,** The sensory component of the facial nerve of a reptile *(Lacerta viridis)*, *J. Comp. Neurol.*, 184, 537, 1979.
224. **Gentle, M. J.,** The chorda tympani nerve and taste in the chicken, *Experientia*, 39, 1002, 1983.
225. **Gentle, M. J.,** Sensory functions of the chorda tympani nerve in the chicken, *Experientia*, 40, 1253, 1984.
226. **Ganchrow, J. R., Ganchrow, D., and Oppenheimer, M.,** Chorda tympani innervation of anterior mandibular taste buds in the chicken *(Gallus domesticus)*, *Anat. Rec.*, 216, 434, 1986.
227. **Krol, C. P. M. and Dubbeldam, J. L.,** On the innervation of taste buds by the N. facialis in the mallard *Anas platyrhynchos* L., *Neth. J. Zool.*, 29, 267, 1979.
228. **Gentle, M. J.,** The lingual taste buds of *Gallus domesticus*, *Br. Poult. Sci.*, 12, 245, 1971.
229. **Cords, E.,** Beiträge zur Lehre vom Kopfnervensystem der Vögel, *Anat. Hefte*, 26, 49, 1904.
230. **Rapuzzi, G. and Casella, C.,** Innervation of the fungiform papillae in the frog tongue, *J. Neurophysiol.*, 28, 154, 1965.
231. **Samanen, D. W. and Bernard, R. A.,** Response properties of the glossopharyngeal taste system of the mudpuppy *(Necturus maculosus)*, *J. Comp. Physiol.*, 143, 143, 1981.
232. **Chernetzki, K. E.,** Cephalic sympathetic fibers in the frog, *J. Comp. Neurol.*, 122, 173, 1964.
233. **Gaupp, E.,** A. *Ecker's und R. Wiederheim's Anatomie des Frosches. 3. Abth. Lehre von den Eingeweiden, dem Integument und den Sinnesorganen*, Vieweg & Sohn, Braunschweig, 1904.
234. **Ballard, W. W.,** Comparative anatomy and embryology, Ronald Press, New York, 1964, 295.
235. **Whitehead, M. C., Beeman, C. S., and Kinsella, B. A.,** Distribution of taste and general sensory nerve endings in fungiform papillae of the hamster, *Am. J. Anat.*, 173, 185, 1985.
236. **Schwartz, H. G. and Weddell, G.,** Observations on the pathways transmitting the sensation of taste, *Brain*, 61, 99, 1938.
237. **Pritchard, T. C.,** The primate gustatory system, in *Smell and Taste in Health and Disease*, Getchell, T. V., et al., Eds., Raven Press, New York, 1991, 109.
238. **Miller, I. J., Jr. and Spangler, K. M.,** Taste bud distribution and innervation on the palate of the rat, *Chem. Senses*, 7, 99, 1982.
239. **Barry, M. A. and Frank, M. E.,** Response of the gustatory system to peripheral nerve injury, *Exper. Neurol.*, 115, 60, 1992.
240. **Oakley, B.,** On the specification of taste neurons in the rat tongue, *Brain Res.*, 75, 85, 1974.
241. **Oakley, B.,** Reformation of taste buds by crossed sensory nerves in the rat's tongue, *Acta Physiol. Scand.*, 79, 88, 1970.
242. **Oakley, B.,** Taste bud development in the rat vallate and foliate papillae, in *Mechanoreceptors: Development, Structure and Function*, Hnik, P., Soukup, T., Vejsada, R., and Zelena, J., Eds., Plenum Press, New York, 1988, 17.

243. **Bradley, R. M., Stedman, H. M., and Mistretta, C. M.,** Superior laryngeal nerve response patterns to chemical stimulation of sheep epiglottis, *Brain Res.*, 276, 81, 1983.
244. **Stedman, H. M., Bradley, R. M., Mistretta, C. M., and Bradley, B. E.,** Chemosensitive responses from the cat epiglottis, *Chem. Senses*, 5, 233, 1980.
245. **Farbman, A. I. and Hellekant, G.,** Quantitative analyses of the fiber population in rat chorda tympani nerves and fungiform papillae, *Am. J. Anat.*, 153, 509, 1978.
246. **Hard Af Segerstad, C., Hellekant, G., and Farbman, A. I.,** Changes in number and morphology of fungiform taste buds in rat after transection of the chorda tympani or chorda-lingual nerve, *Chem. Senses*, 14, 335, 1989.

Chapter 3

THE GENERAL SOMATIC AFFERENT TERMINALS IN ORAL MUCOSAE

Bryce L. Munger

TABLE OF CONTENTS

0-8493-5341-6/93/$0.00 + $.50

The oral cavity, as defined in texts of gross anatomy,[1,2] begins at the lips and extends to the pharynx. The oral cavity is further subdivided into a vestibule that extends externally from the lips and cheeks to the teeth and gums internally and the oral cavity proper inside the teeth. The lining of the oral cavity is remarkably diverse and is, for all intents and purposes, identical to that of the pharynx and portions of the larynx.

I. THE NATURE OF ORAL MUCOSAE

A working concept regarding the nature of the lining of the oral cavity is defined succinctly by Ross et al.[3] based on the functional demands of the mucous membrane as (1) a masticatory mucosa, (2) a general or nonmasticatory mucosa, and (3) specialized mucosa of the dorsal surface of the tongue. Each of these mucosae are in fact specialized and unique in their own way.

The masticatory mucosa covers the gingiva or gums (bony tissue surrounding the roots of the teeth) and hard palate and consists of a stratified squamous epithelium and underlying lamina propria equivalent to the dermis of the skin.[3] As in the case of the skin, the superficial lamina propria is less dense in terms of the connective tissue components and contains more abundant ground substance. This superficial lamina propria furthermore undulates forming projections analogous to the dermal papillae of the skin. The papillary projections of the lamina propria are prominent in the gingiva and are relatively inconspicuous in masticatory mucosa elsewhere overlying bone, referred to as alveolar mucosa. Unlike the skin, however, there is no submucosa in the entire oral cavity except in the lips and cheeks and selected areas of the hard palate.[3] The dense portion of the lamina propria over the masticatory mucosa is continuous with the periosteum of the bones of the maxilla, mandible, and palate. The superficial layers of the stratified squamous epithelium are densely cornified, though not consistently keratinized as reviewed by Ross et al.[3]

The general, or nonmasticatory or lining, mucosa covers the lips, cheeks, undersurface of the tongue, and the floor of the mouth. This lining mucosa does have a lamina propria with shallow papillae projecting into the epithelium as well as a submucosa, except for the undersurface of the tongue.[3] The epithelium is a stratified squamous epithelium that is typically nonkeratinizing, but the vermilion borders of the lips do have a keratinizing stratified squamous epithelium. As is the case of the masticatory mucosa the lamina propria resembles that of skin with a superficial, less compact, connective tissue that undulates between the valleys of the overlying epithelium that in fact resemble the rete ridges and rete pegs of hairy skin or, to some extent glabrous skin.

The specialized mucosa of the dorsal surface of the tongue is most unique in having a diverse cellular composition and no submucosa. The epithelium is specialized into various projections referred to as papillae and is in part a true keratinizing epithelium. The most densely keratinized epithelium is found in the filiform papillae that are conical projections of epithelium and underlying lamina propria. The lamina propria is unique, consisting of dense connective tissue firmly adherent to the underlying musculature. The fungiform papillae are mushroom-shaped projections that typically have taste buds at their apex. The large circumvallate papillae have taste buds in the deep moat surrounding the projections that also serve as the site of termination of von Ebner's glands. Taste buds are usually absent on the oral surface of circumvallate papillae. A fourth type of papilla, the foliate papillae, are linear depressions of the epithelium and typically bear taste buds facing the moat or cavity and thus resemble circumvallate papillae.

II. AFFERENT INNERVATION

The sensory innervation of the oral cavity is unique in having many distinctly different functional types of axons coexisting in a common mucosa. The oral cavity and its posterior extension, the oropharynx, is in turn innervated by two different cranial nerves mediating general afferent (sensory) functions. The fifth or trigeminal nerve (V) provides general somatic afferent (GSA) or sensory innervation from the tip of the tongue to the level of the circumvallate papillae, and the glossopharyngeal or ninth cranial nerve (IX) provides general visceral afferent (sensory) innervation to the posterior 1/3 of the tongue and the remainder of the pharynx. The oral cavity thus has both GSA (V) as well as GVA (IX) nerves subserving the functional modalities of touch, pain, and temperature. The oral cavity in addition has special visceral afferent (SVA) axons mediating taste. The functional wiring of taste is further complicated by the fact that the anterior two thirds of the tongue receives taste fibers from the geniculate ganglion via the chorda tympani from the 7th cranial nerve and thus reaching the lingual surface via branches of the lingual nerve (V), whereas the posterior one third of the tongue receives taste fibers from the glossopharyngeal nerve from the inferior ganglion of that nerve.[1,2,4] The double anatomical source of the sensory innervation is thus analogous to the apposition of adjacent spinal segments or dermatomes. Since the somata of neurons subserving SVA (taste) as compared to GSA and GVA fibers to the tongue are separated, lesions can differentially affect sensibility of taste and touch in the tongue and pharynx. This indeed does happen with unilateral loss of the gag reflex, a visceral reflex, indicating a lesion of IX and insensitivity to touch over the anterior 2/3 of the tongue indicating a lesion of V. What is even more intriguing is the fact that IX as well as V are subject to idiopathic neuralgia (paroxysmal attacks of pain with specific trigger points) that can be very disabling.[5] The present author[4] has postulated that the pathogenesis of trigeminal neuralgia might be related to competition of axons between successive divisions of V since the trigger point typically is at the angle of the mouth at the junction of maxillary and mandibular divisions of V. The same could hold for glossopharyngeal neuralgia based on competition between the peripheral fields of IX and V.

The present review will emphasize aspects of the sensory innervation of the oral cavity that are similar to general body skin, despite the disparate source of the original innervation. The present author's original review of cutaneous sensory receptors[6] and most recent reviews[7-9] span a period of some 20 years, but what was stated as the major problems in 1971 remain the major problems in 1992; specifically, what are the anatomical correlates for functional specificity of all sensory nerve terminals whether they be in the trunk, digit, or oral cavity? Stated another way, can we identify functionally specific sensory receptors based on histological and/or ultrastructural characteristics? The answer is a qualified yes as noted in the material that follows.

The present analysis will rely on the past experience of the present author in dealing with cutaneous sensory receptors in the skin of the digits[9-11] and the face with studies on the lip[12,13] providing a transition to the oral cavity.[14,15] The author has extensive unpublished developmental material dealing with the epiglottis, pharynx, and larynx that will be used to compare oral with extraoral sensory innervation. As we shall see, the oral cavity is not that unique when compared to classical somatic truncal cutaneous derivatives. Wherever possible, we will compare the author's personal experience that relies heavily on tissue from monkeys with published human equivalent material, especially the works of Gairns.[16] Plates I and III from Gairns' paper have been reproduced and will be referred to as Figures 1 to 12 illustrating intraepithelial Merkel terminals (Figures 1 and 2), corpuscular endings (Figures 3 to 8 and 10 and 11) and free nerve endings (FNEs), some of which are also intraepithelial (Figures 9 and 12).

III. CORPUSCULAR RECEPTORS

The sensory receptors of the skin and oral cavity that are the easiest to understand conceptually are corpuscular receptors. Corpuscular sensory receptors are defined on the basis of their cellular demarcation from the general connective tissue compartment of the dermis or lamina propria. Such corpuscles have been often called encapsulated receptors, but many examples have been described where a capsule per se cannot be identified. Thus, the present author suggested that this class of sensory terminals should be referred to as corpuscular receptors.[6] This concept as developed in this previous review is valid today as well.[7] These sensory nerve terminals have specialized cells associated with the afferent axons that are thought to be specialized Schwann cells. These cells produce variably prominent thin cytoplasmic lamellae and hence the suggestion that they be called lamellar cells.[6] Lamellar cells can be easily identified by light and electron microscopy. In the former the nuclei associated with the convolutions of the terminal axon (Figures 3 to 8) are the nuclei of lamellar cells.

Several different types of corpuscular receptors can be recognized by light and electron microscopy in the oral cavity as well as the skin based on the configuration of the terminal axon and the relationship to the lamellae and lamellar cells. The largest corpuscular receptor is the well-studied Pacinian corpuscle that is prominent in the subcutaneous tissues of glabrous palmar and plantar skin and in the cat mesentery.[6,7] Since Pacinian corpuscles have not been identified in oral mucosa, we will not describe these complex sensory endings. The reviews cited all have detailed accounts of the organization of these receptors. Pacinian corpuscles are innervated by large A beta fibers.

The corpuscular receptors that can be found in oral mucosa are depicted in Figures 3 to 8. Meissner corpuscles, as illustrated in Figures 3 to 5, are also present in glabrous palmar and plantar skin. The configuration of the terminal axon in Meissner corpuscles resembles that of ribbon candy in that the axon winds back and forth in a relatively regular manner. Groups of Meissner corpuscles also can be found as illustrated in Figure 4. Figure 5 was not identified as a Meissner corpuscle by Gairns, but would be consistent with small Meissner corpuscles characteristic of rodent and marsupial glabrous skin.[7] In contrast to Meissner corpuscles, other corpuscles lack a regular pattern of the terminal axon and the present author has referred to these receptors as glomerular or simple corpuscles.[13-15] Figures 6 to 8 are examples, although not identified explicitly by Gairns.

An example of a developing Meissner corpuscle is illustrated in Figure 13 and a portion of a mature corpuscle in Fig. 14 from the hard palate and lip of a monkey.[17-20] Meissner corpuscles are defined ultrastructurally on the basis of the presence of many thin cytoplasmic lamellae separating the successive convolutions of the axon. The basal lamina of the lamellar cell is the point of separation of neural from the connective tissue compartment, and no basal lamina is present between the sensory axon and the lamellae. The lamellar cell body is located at the margin of the sensory corpuscle and is best visualized in a developing Meissner corpuscles as in Figure 13. The number of lamellae may be considerable with literally dozens of thin lamellae separating successive turns of the axon. Lamellae of sensory corpuscles typically contain many pinocytotic vesicles and have a thin basal lamina.

In addition to Meissner corpuscles identical to those of the digital glabrous skin, the oral mucosa also has another form of a corpuscular receptor that is characterized by more simple coiling of the terminal axon and only scant lamellae are present. The term the present author has used is a "glomerular terminal",[14,15] and similar corpuscles are present in the genitalia as genital end bulbs.[21] Gairns[16] refers to these loosely wound corpuscles as simple endings or in some cases Krause end bulbs, (Figures 6, 7, 8). These figures from Gairns' human material are indistinguishable from the monkey material in the publications of the present author. Thus, all primates have a similar repertoire of corpuscular receptors in all oral mucosae including lips, palate, and dorsum of the tongue.

Many, but not all, corpuscular receptors in the oral mucosae (and skin) have an intimate relationship to the overlying epithelium. The relationship

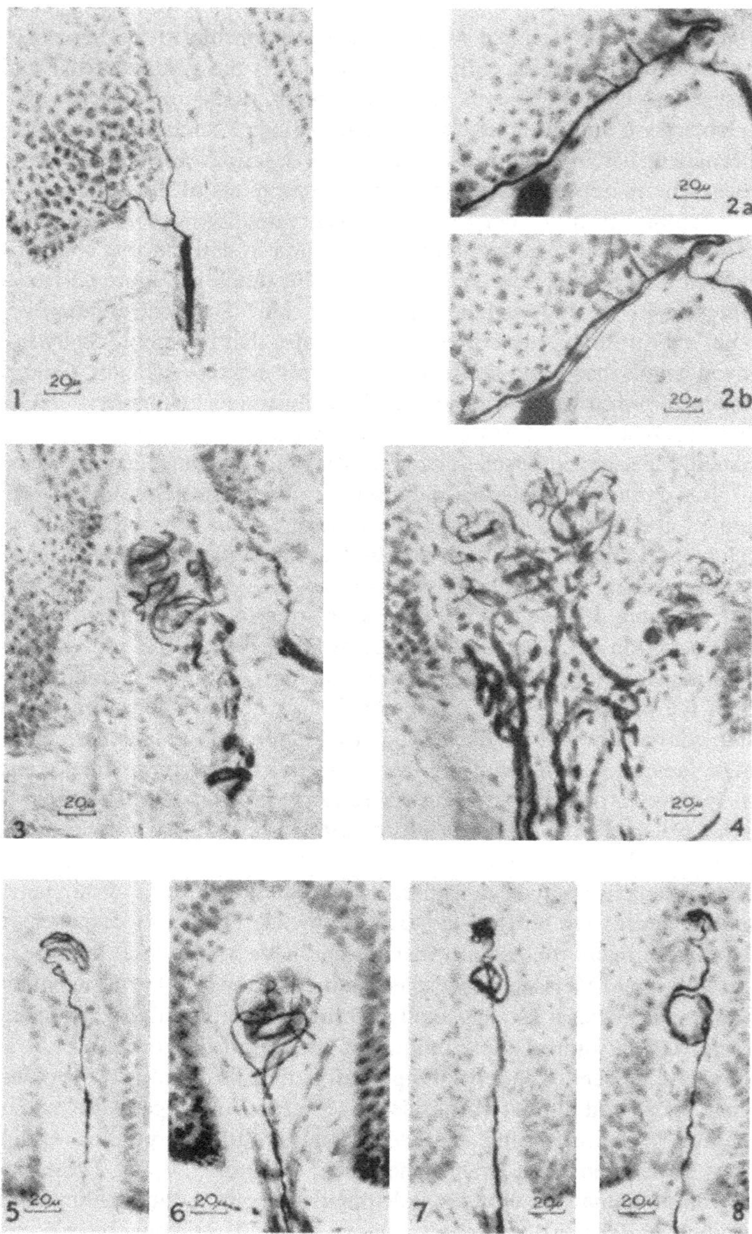

FIGURES 1–8. Figures 1 and 2. These plates are all examples of silver-impregnated 15- to 20-micron thick frozen sections using a modification of the Bielschowsky method. Figures 1 and 2 illustrate intraepithelial axons derived from a parent myelinated axon that repeatedly bifurcates in the epithelium presumably innervating a cluster of Merkel cells. Based on the magnification bar the parent axon would be in the A beta range typical of axons innervating

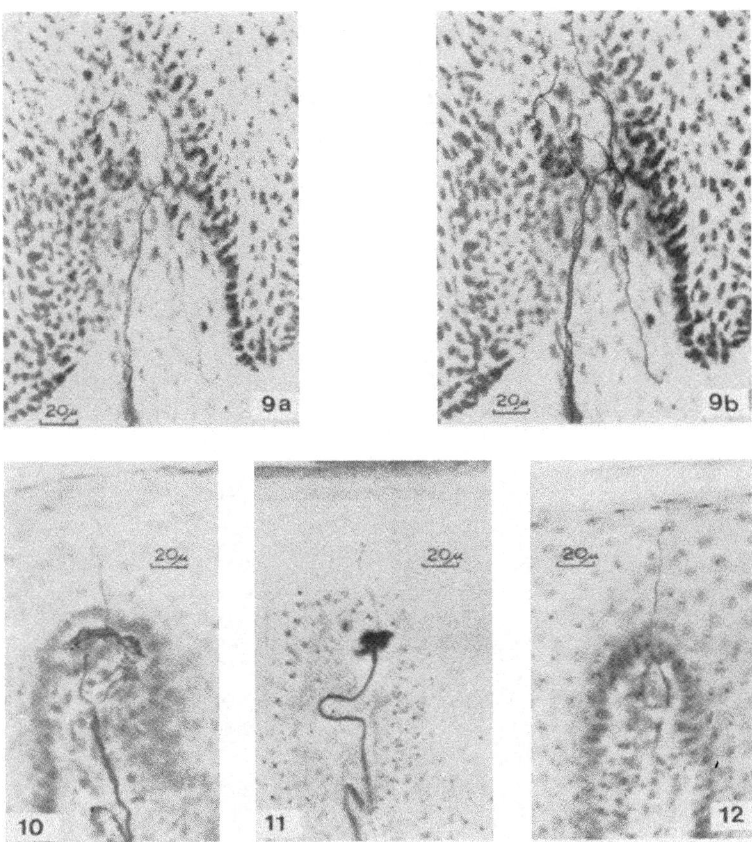

FIGURES 9–12. Figures 9 and 12. In the opinion of the present author these micrographs illustrate typical FNEs with extensions into the overlying epithelium. Gairns did consider Figure 9 to be unmyelinated axons, and these two serial photographic plates are convincing evidence of the terminals typical of C fibers. **Figures 10 and 11.** These axons could be part of a small corpuscle, but more likely they are FNEs as well as derived from myelinated A delta fibers. (From Gairns, F. W., *Q. J. Exp. Physiol.*, 40, 40, 1955. With permission.)

Merkel cells. **Figures 3, 4, and 5.** Figures 3 and 4 are typical Meissner corpuscles with a single corpuscle in Figure 3 and multiple corpuscles in adjacent papillae of the lamina propria. The corpuscle in Figure 3 is somewhat deeper in the lamina propria than is typical and the corpuscle to the right in Figure 4 tightly abuts the overlying epithelium as does the corpuscle in Figure 5. **Figures 6, 7, and 8.** These corpuscles are all examples of glomerular or simple corpuscles with the axon forming a skein with little pattern. The corpuscle in Figure 6 is deep in the papilla of the lamina propria and the other two tightly abut the overlying epithelium. (From Gairns, F. W., *Q. J. Exp. Physiol.*, 40, 40, 1955. With permission.)

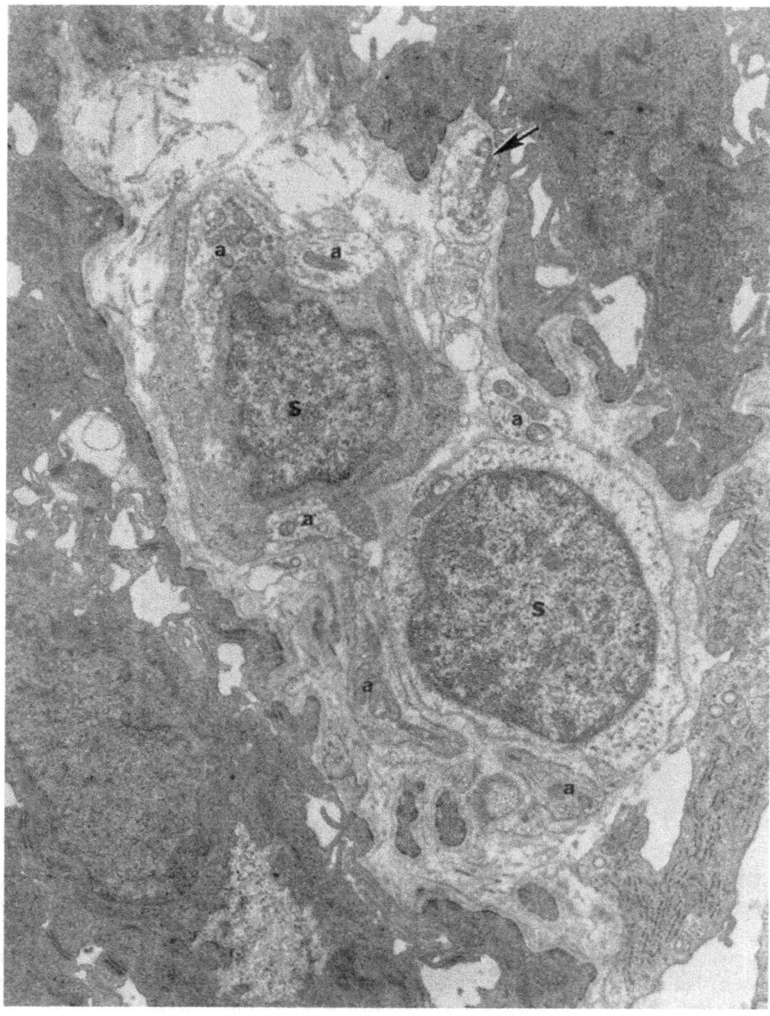

FIGURE 13. This electron micrograph is taken from high in a papilla of the lamina propria from the hard palate of a 4.5-month-old monkey embryo. This site typically has Meissner corpuscles and this presumptive Meissner corpuscle is identical to that illustrated in Renehan and Munger's study[22] of developing primate glabrous digital skin. Two Schwann cells envelope numerous axons (a) that in many cases have no Schwann cell investment. At the arrow, an axonal process crosses the basal lamina of the epithelium and directly abuts a process from a basal epithelial cell, again similar to the situation described in the digit. Magnification × 12,000.

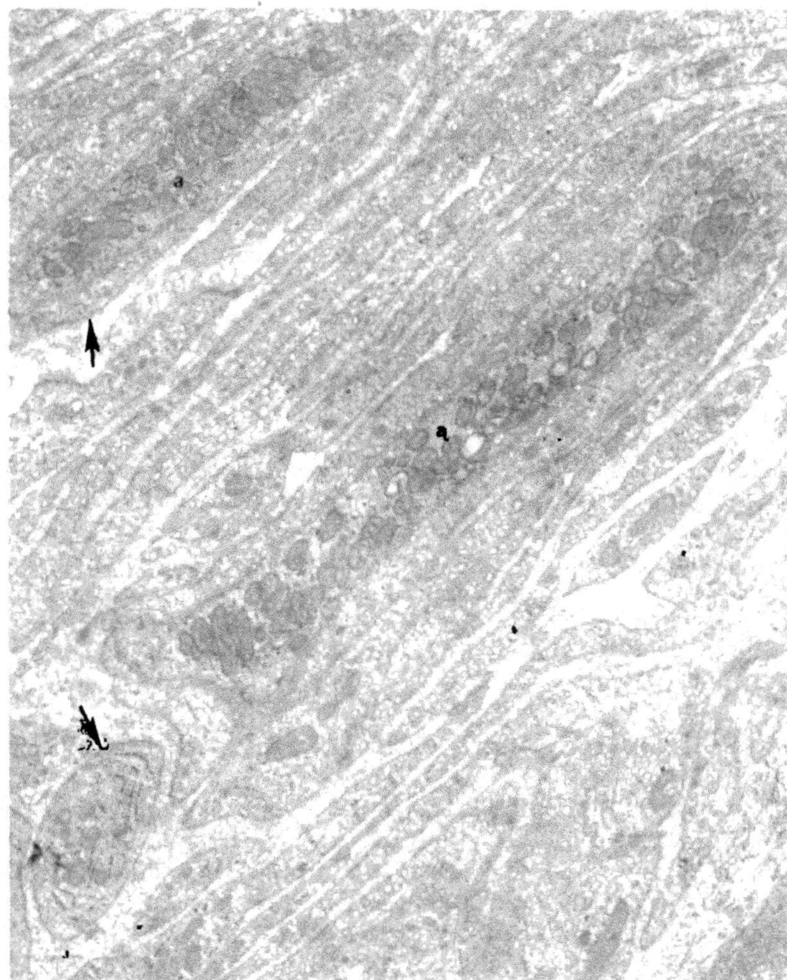

FIGURE 14. This micrograph depicts a portion of a relatively mature Meissner corpuscle from a 4-d-old rhesus monkey and is taken from the vermillion border of the lip. Two axonal profiles (a) containing numerous mitochondria are separated from one another by numerous lamellae. The cell bodies of lamellar cells are not in the field illustrated. The delicate basal lamina of the lamellae are indicated by the arrows. The lamellae are characterized by the numerous pinocytotic vesicles and scant mitochondria. Magnification × 20,000.

during development as described by Renehan and Munger[22] includes the penetration of the sensory axon into the epidermis, and the example illustrated in Figure 13 also has an axon directly abutting a process from a basal epithelial cell. Figures 4, 5, 7, and 8 all illustrate this intimate relationship of corpuscular receptors and the epithelium with axons in Figures 10 to 12 actually entering the epithelium from a presumptive simple corpuscle in Figures 10 and 11. As noted below, Figures 9 and 12 are considered to be FNEs. Furthermore, Halata and Munger[12] noted examples in the adult monkey lip where no basal lamina separated lamellae and axons from the overlying stratified squamous epithelium. This intimate relationship implies the existence of tropic and trophic interactions between sensory nerves, lamellar cells, and stratified squamous epithelia. As noted below, when axons cross the epidermal-dermal junction to reach Merkel cells the basal lamina of the epidermis and Schwann cells becomes continuous, and the neuroepithelium becomes continuous with the stratified squamous epithelium.[23]

Meissner corpuscles are thought to represent sensitive, rapidly adapting (RA) mechanoreceptors transducing light mechanical deformation or low frequency vibrations.[6] In the oral cavity Meissner corpuscles have been identified in the monkey lip,[12,13] tongue,[14] and hard palate.[15] We do not have any physiological data on isolated glomerular corpuscles to indicate their best stimulus for transduction or their adaptive properties. As reviewed by Halata and Munger[21] there are no data to suggest that genital end bulbs in any way respond functionally as do Meissner corpuscles. In fact the converse is true in that genital end bulbs in fact are presumably high threshold mechanoreceptors. It would be presumptuous to assume that glomerular endings of the oral mucosa are functionally identical to genital end bulbs, but the best information available would suggest that they are certainly not functionally a modified Meissner corpuscle.

IV. MERKEL CELL NEURITE COMPLEXES

The biology of Merkel cells has been recently reviewed by the present author,[8] and numerous controversies still surround our understanding of the nature and function of these ubiquitous cells. When they were first visualized in electron micrographs,[23] most neurobiologists were very skeptical as to the importance of these cells. We were surprised at the number of Merkel cells present in monkey palatal epithelium,[14,15] as well as being surprised at their total absence in lingual epithelium.[17-20] Figures 1 and 2 are considered to be innervated Merkel cells based on the studies by the present author. In oral mucosa, clusters of Merkel cells are typically present at the base of epithelial pegs analogous to papillary ridges in glabrous skin.

Several studies have appeared claiming that some cells in taste buds are Merkel cells based on the presence of membrane-limited granules of 100 to 150 nm in diameter that do resemble the granules of Merkel cells (see Chapters 2 and 11). In the opinion of the present author, subtle differences exist indicating that cells in taste buds[18-20,24] and Merkel cells[68,24] are different. Merkel cells are defined on the basis of their content of cytoplasmic granules resembling those present in protein secreting endocrine cells (Figures 15 and 16). The granules are polarized in the cytoplasm towards the abutting axon

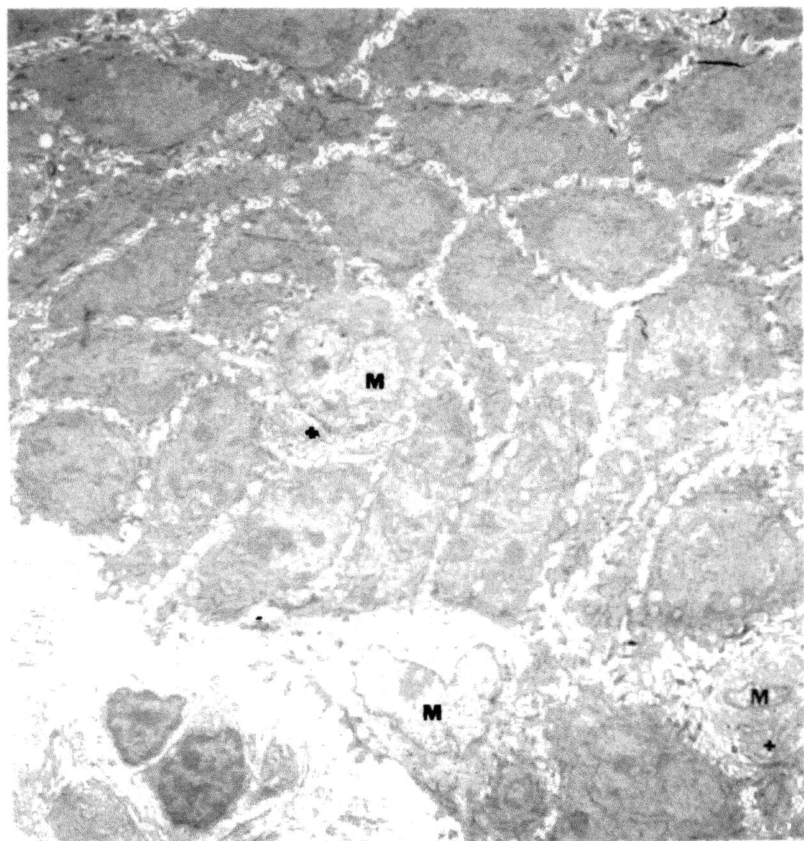

FIGURE 15. This low magnification electron micrograph depicts a cluster of Merkel cells (M) at the base of a rete peg from the lip of a 4-d-old rhesus monkey. Two of the Merkel cells have associated axons (a), while the central Merkel cell at the bottom does not. Noninnervated Merkel cells are commonly seen in juvenile tissue. The typical secretory granules of Merkel cells are polarized towards the axons and are depicted at higher magnification in Figure 16. Magnification × 5300.

that is shaped like a spoon, hence the name Merkel disk, tactile disk, or meniscus, etc. The Golgi apparatus is on the side of the cell away from the axon. Desmosomes are present between Merkel cells and contiguous keratinocytes, and the presence of desmosomes is necessary for absolute identification of Merkel cells.[8,23] Studies of the present author have consistently concluded that Merkel cells are derived from keratinocytes and are not of neural crest origin. They do have many neuronal properties including a content of numerous biologically active peptides. These neuronal properties have led to the suggestion of Fujita[25] that Merkel cells should be considered to be

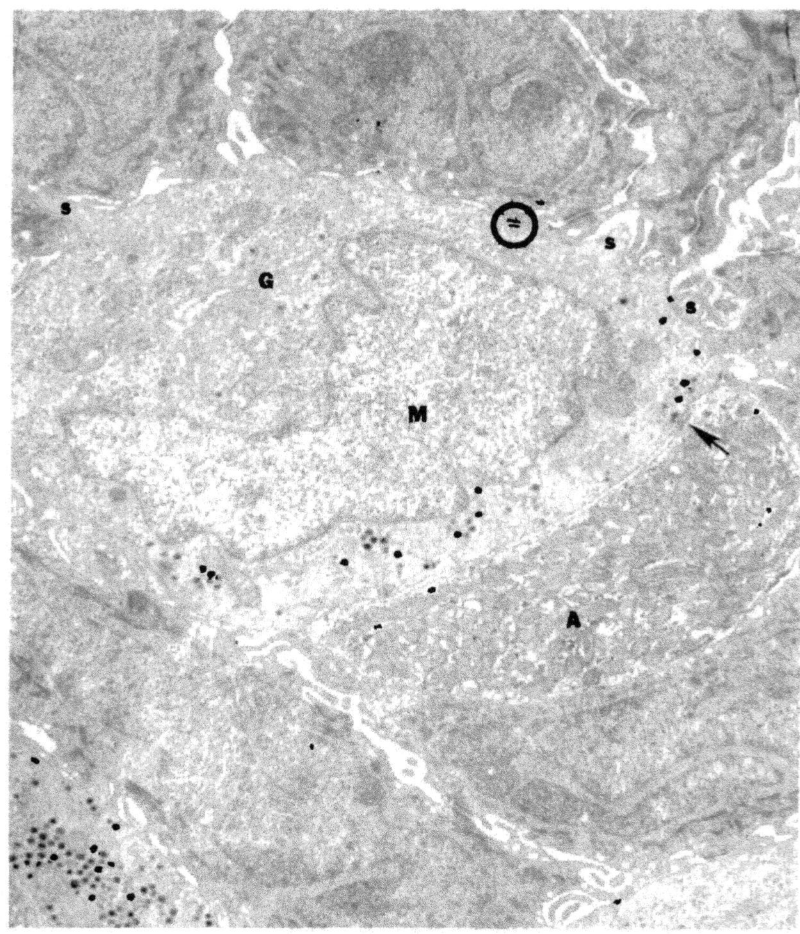

FIGURE 16. A Merkel cell (M) and associated axon (A) are taken from the same specimen as depicted in Figure 15. The Merkel cell secretory granules are polarized towards the axon and are relatively sparse, typical of developing tissues. The Golgi apparatus (G), where secretory granules are formed, is on the opposite side of the nucleus from the axon. A desmosome (circle) is present between the Merkel cell and abutting epithelial cell. The arrow depicts a small membrane density in the axon resembling a synapse. A portion of a second Merkel cell is present below the figure number. Magnification × 18,000.

paraneurons. Since Merkel cells can only be identified on the basis of their secretory granules, presence of desmosomes, and characteristic cytoplasmic spikes,[7] transmission electron microscopy (TEM) is required to establish a firm identification.

In contrast, the cells present in taste buds that contain endocrine-appearing secretory granules have only scant such granules that are not tightly polarized towards the axon and have no cytoplasmic spikes. Instead, they have abundant presumptive synaptic vesicles and subsurface cisternae that are polarized towards the abutting axons.[18-20,24]

While Gairns[16] knew of the existence of Merkel disks and cited the observation of Botezat from 1901 that they were present in human palatal mucosa, Gairns did not recognize Merkel terminals in his work on oral mucosa. However, his Figures 1 and 2 are clearly clusters of Merkel cells and associated intraepidermal axons. The presence of clusters of Merkel cells in rete pegs in the skin of the lip led Halata and Munger[12] to refer to these specialized rete pegs as *Tastscheibe* or touch spots analogous to *Haarscheiben* of hairy skin. We can conclude that Merkel terminals are a ubiquitous component of the mucosa of palate and lips and have no explanation for their absence in lingual mucosa.

The axons associated with Merkel cells are A beta fibers subserving the function of slowly adapting (SA) sensitive mechanoelectric transducers. The fact that Merkel cells differentiate prior to innervation has led the present author to conclude, as originally suggested,[23] that these cells are most likely trophic or tropic for sensory neurons. The recent finding by Vos et al.[26] that nerve growth factor (NGF) is present in Merkel cells would provide convincing proof of a trophic function. The fact that sensory axons grow directly towards Merkel cells during development[9-11] suggests a tropic function as well. In collaborative studies with Dykes laboratory,[27,28] we have found the persistence of SA receptors in allografted monkey skin immunosuppressed with cyclosporine, and we have failed to find Merkel cells in electron micrographs studied to date. This finding would negate any putative role for Merkel cells in mechanoelectric transduction and place the burden on the associated sensory axon.[7]

V. FREE NERVE ENDINGS

The term "free nerve ending" (FNE) embodies a concept more than a verifiable anatomical entity. In the earlier review of the present author[6] the concept was not considered as possibly verifiable, as the difficulty was knowing when one is at the end of an axon in the connective tissue compartments of the body. Yet axons ending blindly in the midst of the connective tissues were thought to be responsible for pain and thermal sensibility and carried over A delta as well as C fibers or unmyelinated axons. Kruger et al.[29] provided convincing physiological and ultrastructural evidence identifying thinly myelinated nociceptors (the terminals of A delta fibers) in cat hairy skin. They noted that FNE axon terminals (1) frequently had a 1:1 relationship with the investing Schwann cell, (2) areas of the axon lacked a Schwann cell ensheathment with the axon directly abutting the basal lamina, and (3) the axon frequently was intimately associated with an epithelium, i.e., the epidermis. These findings in many ways confirmed Cauna's exhaustive serial section

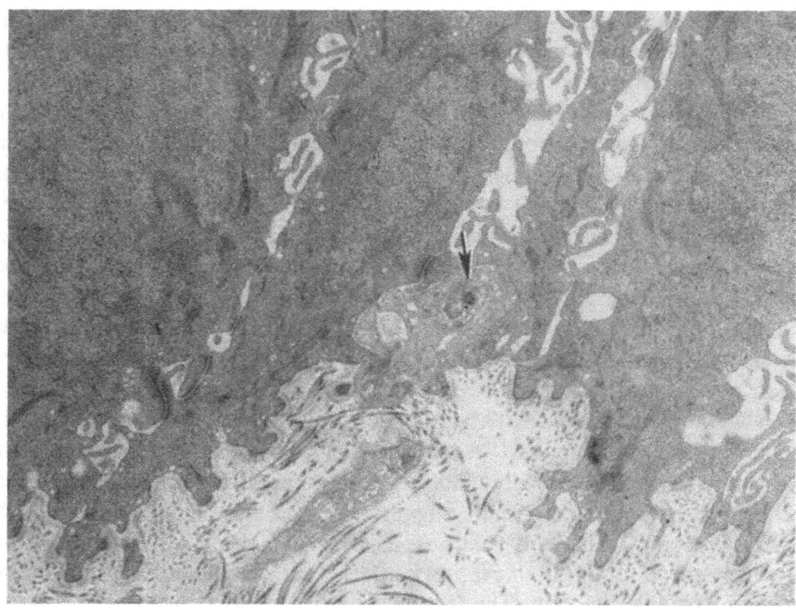

FIGURE 17. This figure and Figure 18 should be compared with one another as both depict presumptive FNEs. The FNE in Figure 17 is very small and probably is a terminal from a C fiber. The axon is within the epithelium sandwiched between basal cells of the epithelium. Fibers such as this often can be found in the superficial layers of the epithelium. Magnification × 15,000.

TEM study of the plexus of C fibers that is present in hairy skin. He also observed areas of axons without a Schwann cell investment and an intimate relationship to the epidermis. Munger and Halata[12] used similar criteria in their analysis of vibrissal innervation and tentatively identified FNEs in both vibrissae and guard hairs. The developing axons in Figure 13 thus resemble the putative FNEs of Figures 17 and 18.

Throughout the oral mucosa, FNEs can be identified by both light and electron microscopy. Axons terminating high in the papillae of the lamina propria are tentatively identified as a FNE. Figures 9 and 12 are in the opinion of this author FNEs of C fibers, and Figures 10 and 11 are probably FNEs derived from larger diameter myelinated A delta fibers. By TEM and applying the criteria developed by Kruger et al.,[29] axons intimately associated with an epithelium can be identified as being terminals of either C fibers or A delta fibers such as Figures 17 and 18.

FNEs also constitute the fibers considered as the accessory innervation of corpuscular receptors or the accessory fibers of Timofeew[31] and can be derived from either A delta or C fibers. Halata and Munger[12] provided convincing evidence of a small thinly myelinated A delta fiber intimately asso-

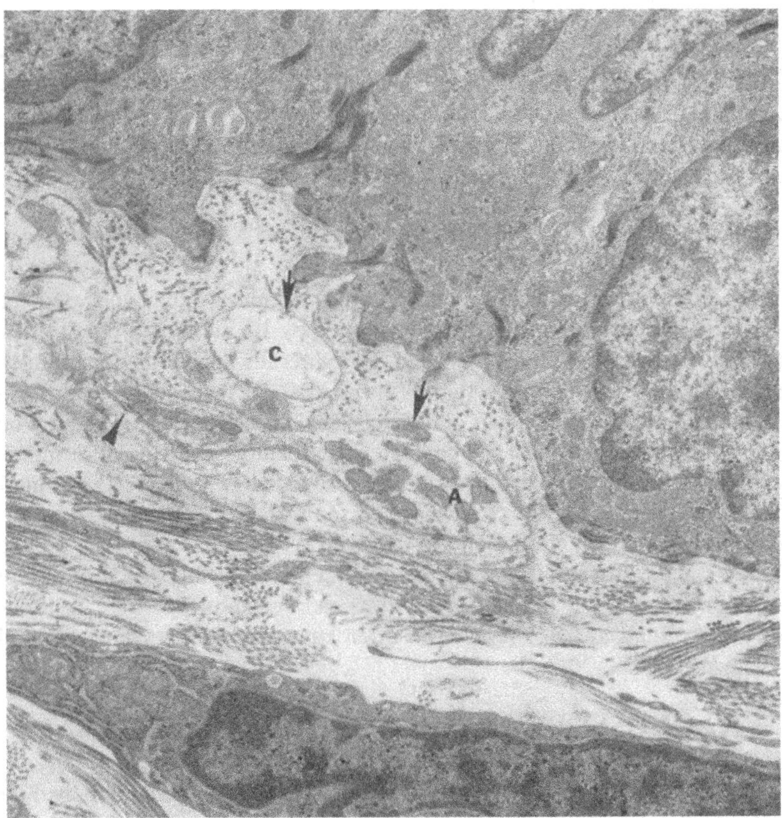

FIGURE 18. The FNE is this micrograph contains portions of an axon (A) and a growth cone (C). Several areas of axonal membrane lack an investing Schwann cell (arrows) and the growth cone approximates the basal lamina of the epithelium. The arrowhead indicates a junctional complex between the axon and the investing Schwann cell. Magnification × 17,000.

ciated with a Meissner corpuscle in the monkey lip. Gairns[16] also recognized the existence of accessory innervation in the corpuscles of human oral mucosa, although he did not cite Timofeew. Some of the C fibers as accessory fibers could be autonomic fibers that could possibly modulate the function of the corpuscular mechanoreceptors.

The FNEs of A delta and C fibers are also the axons containing biologically active peptides such as substance P (SP) as reviewed by Nishimoto et al.[32] and calcitonin gene-related protein (CGRP) as reviewed by Silverman and Kruger.[33] These two peptides have an important role in the pathogenesis of vasogenic edema and could be candidates for the modulation of other mechanosensitive axons in oral mucosa and skin. Silverman and Kruger[33] further argue persuasively that these axons form a subset of sensory neurons that would be the putative transducers of a primitive chemical sensibility or common chemical sense due to their intraepithelial position in the mucosae of the oral (and nasal) cavities. Thus, this concept could form a logical explanation for chemoresponsive trigeminal neurons as described by Simon and Sostman.[34] The role of other biologically active peptides that are present in A delta and C fibers is speculative at the present time.

VI. RUFFINI TERMINALS

Since the first documentation of slowly adapting hairs and associated ultrastructural definition of Ruffini terminals by Biemesderfer et al.,[35] these complex endings have been defined in numerous locations including the periodontal ligament by Byers[36] and Maeda et al.[37] We have not encountered Ruffini terminals elsewhere in oral mucosa in our studies to date, but would expect them to be present. The characteristic feature of a Ruffini terminal is the relationship of the sensory axon and associated Schwann cell to bundles of collagen. The axon and Schwann cell, as well as connective tissue cells called septal cells, encircle bundles of axons as a hand would grip a bundle of soda straws. Ruffini terminals are also identical in ultrastructure to Golgi tendon organs (GTOs) and both are SA mechanoreceptors sensitive to stretch.[7] Ruffini terminals are also reputed to often have an accessory innervation by A delta or C fibers.

VII. GLOSSOPHARYNGEAL NERVE TERMINALS

The sensory receptors associated with IX have not been extensively explored and the author's personal experience has only been published in a monograph[17] and one paper devoted to taste buds in the larynx.[24] However, in our unpublished material we have data that would conclude that highly specialized corpuscular receptors are present in the larynx and pharynx, and the sensory innervation of the posterior tongue is more complex than previously reported. The nature of the sensory innervation of the posterior tongue and the remainder of the pharynx is similar to that of the anterior portion of the tongue with numerous corpuscular receptors and FNEs. The terminals of IX thus seem to have the same basic repertoire as the terminals of V with the exception that some extremely large tangled skeins of sensory axons can be found in laryngeal lamina propria.

VIII. CENTRAL PROJECTIONS OF ORAL AFFERENTS

An authoritative account on the form and function of the cranial nerves is afforded by Brodal.[38] Brodal reviews the literature thoroughly on the course and distribution of both V, VII, and IX including the rationale for having the modalities of touch, pain, and temperature of V considered to be GSA and the same modalities of IX considered to be GVA. The basis for this separation of function is based on the evolutionary biology of the cranial nerves as described in many other publications of Brodal, but basically the gag reflex is a visceral reflex with IX being the afferent arc and vagal fibers (X) being the efferent arc. The taste fibers from VII (via V) and IX (and X from the lower airway and digestive system) are processed in part in the nucleus of the solitary tract, but collaterals also project to the sensory nuclei of V including the spinal trigeminal nucleus as well as the reticular formation. This projection to the nucleus of V from fibers of IX provides for the anatomical basis for integration of the contiguous peripheral sensory fields of the respective cranial nerves for the similar modalities of sensation.

The somatic and visceral general afferent sensibility is processed in the trigeminal complex of the brain stem and upper cervical spinal cord. The central projections are widely distributed from the level of entry of the nerve in the pons down into the upper cervical spinal cord as the spinal tract and nucleus of V. Controversy still exists as to the functional subdivisions within the nucleus, and a detailed review of these disputed connections and functional implications are beyond the scope of this review. Brodal[38] can be consulted for a detailed review of the complexities of trigeminal connections. He also notes that taste per se may be conveyed over fibers of V and would be processed as noted above in the solitary complex.

IX. UNIQUE FEATURES OF TASTE RECEPTORS

The chemical senses are unique in that the overwhelming consensus is that the transduction is performed by a cell, albeit in the case of nasal mucosa the cells are derived from an epithelial placode and a central axon invades the central nervous system to form the first synapse. In the case of taste, the epithelial cells themselves are thought to be the receptive element due to the fact that the cells have typical synaptic vesicles abutting synapses. Furthermore the synapses (and the majority of axonal profiles) are present towards the base of the taste bud. However, the most compelling evidence is the studies from DeSimone's laboratory as reviewed by Qing et al.[39]

We have referred previously to the superficial similarity of Merkel cells and those cells identified as chemosensory on the basis of the serial section studied by conventional TEM[24] as well as high voltage TEM by Kinnamon et al.[40] If the apical portions of taste bud cells are indeed linked by tight junctions as is the case in conventional lingual epithelium as described by Holland et al.,[41] then the chemically active substances should not diffuse to the level of the axons and the epithelial cells would have to be the site of chemoelectric transduction. The question is where are the receptors for gustatory substances, and the answer to the question will require technically very difficult methods.

REFERENCE

1. **Williams, P. L., Warwick, R., Dyson, M., and Banister, L. H.,** *Gray's Anatomy,* 36th Br. ed., W. B. Saunders, Philadelphia, 1989.
2. **Clemente, C. D.,** *Gray's Anatomy,* 30th Am. Ed., Lea and Febiger, Philadelphia, 1985.
3. **Ross, M. H., Reith, E. T., and Romrell, L. J.,** *Histology. A Text and Atlas,* 2nd ed., Williams and Wilkins, Baltimore, 1989.
4. **Munger, B. L.,** Trigeminal system, somatic sensory receptors, in *Encyclopedia of Human Biology,* Dulbecco, R., Ed. Academic Press, San Diego, CA, 1991, 671.
5. **Rowland, L. P., Ed.,** *Merritt's Textbook of Neurology,* 8th ed., Lea and Febiger, Philadelphia, 1989.
6. **Munger, B. L.,** Patterns of organization of peripheral sensory receptors, in *Handbook of Sensory Physiology,* Vol I, *Principles of Receptor Physiology,* Loewenstein, W. R., Ed., Springer-Verlag, 1971, 523.
7. **Munger, B. L. and Ide, C.,** The structure and function of cutaneous sensory receptors: a review, *Arch. Hist. Cytol.,* 51, 1, 1988.
8. **Munger, B. L.,** The biology of Merkel cells, in *Biochemistry, Physiology and Molecular Biology of the Skin,* Goldsmith, L. A., Ed., Oxford University Press, Oxford, 1991, 836.
9. **Dell, D. A. and Munger, B. L.,** The early embryogenesis of papillary (sweat duct) ridges in primate glabrous skin: the dermatotopic map of cutaneous mechanoreceptors and dermatoglyphics, *J. Comp. Neurol.,* 244, 511, 1986.
10. **Moore, S. J. and Munger, B. L.,** The early ontogeny of the afferent nerves and papillary ridges in human digital glabrous skin, *Dev. Brain Res.,* 48, 119, 1989.
11. **Morohunfola, K., Jones, T. E., and Munger, B. L.,** The differentiation of the skin and its appendages. I. Normal development of papillary ridges, *Anat. Rec.,* 232, 587, 1992.
12. **Munger, B. L. and Halata, Z.,** The sensory innervation of the primate facial skin. I. Hairy skin, *Brain Res. Rev.,* 5, 45, 1983.
13. **Halata, Z. and Munger, B. L.,** The sensory innervation of the primate facial skin. II. Vermilion border and mucosa of lip, *Brain Res. Rev.,* 5, 81, 1983.
14. **Munger, B. L.,** Cytology and ultrastructure of sensory receptors in the adult and newborn primate tongue, in *Development of the Fetus and Infant, 4th Symposium on Oral Sensation and Perception,* Bosma, J. F., Ed., DHEW Publications (NIH) #73-946, Bethesda, MD, USDHEW, 1973, 75.
15. **Munger, B. L.,** Specificity in the development of sensory receptors in primate oral mucosa, in *Development of Upper Respiratory Anatomy and Function: Implications for Sudden and Unexpected Infant Death,* Bosma, J. F., Ed., DHEW Publications (NIH), Bethesda, MD, 1975, 96 (or chap. 7).
16. **Gairns, F. W.,** The sensory nerve endings of the human palate, *Q. J. Exp. Physiol.,* 40, 40, 1955.
17. **Munger, B. L.,** Sensorineural status of the perinatal primate larynx, in *Sudden Infant Death Syndrome,* Tildon, J. T., Roeder, L. M., and Steinschneider, A., Eds., Academic Press, New York, 1983, 491.
18. **Zahm, D. S. and Munger, B. L.,** Fetal development of primate chemosensory corpuscles. I. Synaptic relationships in late gestation, *J. Comp. Neurol.,* 213, 146, 1983a.
19. **Zahm, D. S. and Munger, B. L.,** Fetal development of primate chemosensory corpuscles. II. Synaptic relationships in early gestation, *J. Comp. Neurol.,* 219, 36, 1983b.
20. **Zahm, D. S. and Munger, B. L.,** The innervation of the primate fungiform papilla — development, distribution, and changes following selective ablation, *Brain Res. Rev.,* 9, 147, 1985.
21. **Halata, Z. and Munger, B. L.,** The neuroanatomical basis for the protopathic sensibility of the human glans penis, *Brain Res.,* 371, 205, 1986.

22. **Renehan, W. and Munger, B. L.,** The development of Meissner corpuscles in primate digital skin, *Dev. Brain Res.*, 51, 35, 1990.
23. **Munger, B. L.,** The intraepidermal innervation of the snout skin of the opossum. A light and electron microscopic study, with observations on the nature of Merkel's "Tastzellen", *J. Cell Biol.*, 26, 79, 1965.
24. **Ide, C. and Munger, B. L.,** The cytologic composition of primate laryngeal chemosensory corpuscles, *Am. J. Anat.*, 158, 193, 1980.
25. **Fujita, T.,** Concept of paraneurons, *Arch. Histol. Jpn.*, 40(Suppl.), 1, 1977.
26. **Vos, P., Start, F., and Pittman, R. N.,** Merkel cells in vitro: production of nerve growth factor and selective interactions with sensory neurons, *Dev. Biol.*, 144, 281, 1991.
27. **Samulack, D. D., Munger, B. L., Dykes, R. W., and Daniel, R. K.,** Neuroanatomical evidence of reinnervation in primate allografted (transplanted) skin during cyclosporine immunosuppression, *Neurosci. Lett.*, 72, 1, 1986.
28. **Samulack, D. D., Dykes, R. W., and Munger, B. L.,** Neurophysiological aspects of allogeneic skin and upper extremity composite tissue transplantation in primates, *Transplant. Proc.*, 20, 279, 1988.
29. **Kruger, L., Perl, E. R., and Sedivec, M. J.,** Fine structure of myelinated mechanical nociceptor endings in cat hairy skin, *J. Comp. Neurol.*, 198, 137, 1981.
30. **Cauna, N.,** The free penicillate nerve endings of the human hairy skin, *J. Anat.*, 115, 277, 1973.
31. **Timofeew, T.,** Ueber eine besonders Art von eingekapselten Nervenendigungen in der manlichen Geschlechtsorganen von Saugetieren, *Anat. Anz.*, 40, 1896.
32. **Nishimoto, T., Akai, M., Inagaki, S., Shiosaka, S., Shimizu, Y., Yamamoto, K., Senba, E., Sakanaka., M., Takatsuki, K., Hara, Y., Takagi, H., Matsuzaki, T., Kawai, Y., and Tohyama, M.,** On the distribution and origins of substance P in the papillae of the rat tongue: an experimental and immunohistochemical study, *J. Comp. Neurol.*, 207, 85, 1982.
33. **Silverman, J. D. and Kruger, L.,** Calcitonin-gene-related-peptide immunoreactive innervation of the rat head with emphasis on specialized sensory structures, *J. Comp. Neurol.*, 280, 303, 1989.
34. **Simon, S. A. and Sostman, A. L.,** Electrophysiological responses to nonelectrolytes in lingual nerve of rat and in lingual epithelia of dog, *Arch. Oral Biol.*,
35. **Beimesderfer, D., Munger, B. L., Binck, J., and Dubner, R.,** The pilo-Ruffini complex: a non-sinus hair and associated slowly-adapting mechanoreceptor in primate facial skin, *Brain Res.*, 142, 197, 1978.
36. **Byers, M. R.,** Sensory innervation of the periodontal ligament of rat molars consists of unencapsulated Ruffini-like mechanoreceptor nerve endings, *J. Comp. Neurol.*, 231, 500, 1985.
37. **Maeda, T., Soto, O., Kobayoski, S., Iwanaga, T., and Fujita, T.,** The 7 ultrastructure of Ruffini endings in the periodontal ligament of rat incisors with special reference to the terminal Schwann cells (K-cells), *J. Anat.*, 167, 117, 1989.
38. **Brodal, A.,** *Neurological Anagomy in Relation to Clinical Medicine*, 3rd ed., Oxford University Press, New York, 1981.
39. **Qing, Y., Heck, G. L., and DeSimone, J. A.,** The anion paradox in sodium taste reception: resolution by voltage-clamp studies, *Science*, 254, 724, 1991.
40. **Kinnamon, J. C., Taylor, B. T., Elay, R. J., and Roper, S. J.,** Ultrastructure of mouse vallate taste buds. I. Taste cells and their associated synapses, *J. Comp. Neurol.*, 235, 48, 1985.
41. **Holland, V. F., Zampighi, G. A., and Simon, S. A.,** Morphology of fungiform papillae in canine lingual epithelium: location of intercellular junctions in the epithelium, *J. Comp. Neurol.*, 279, 13, 1989.

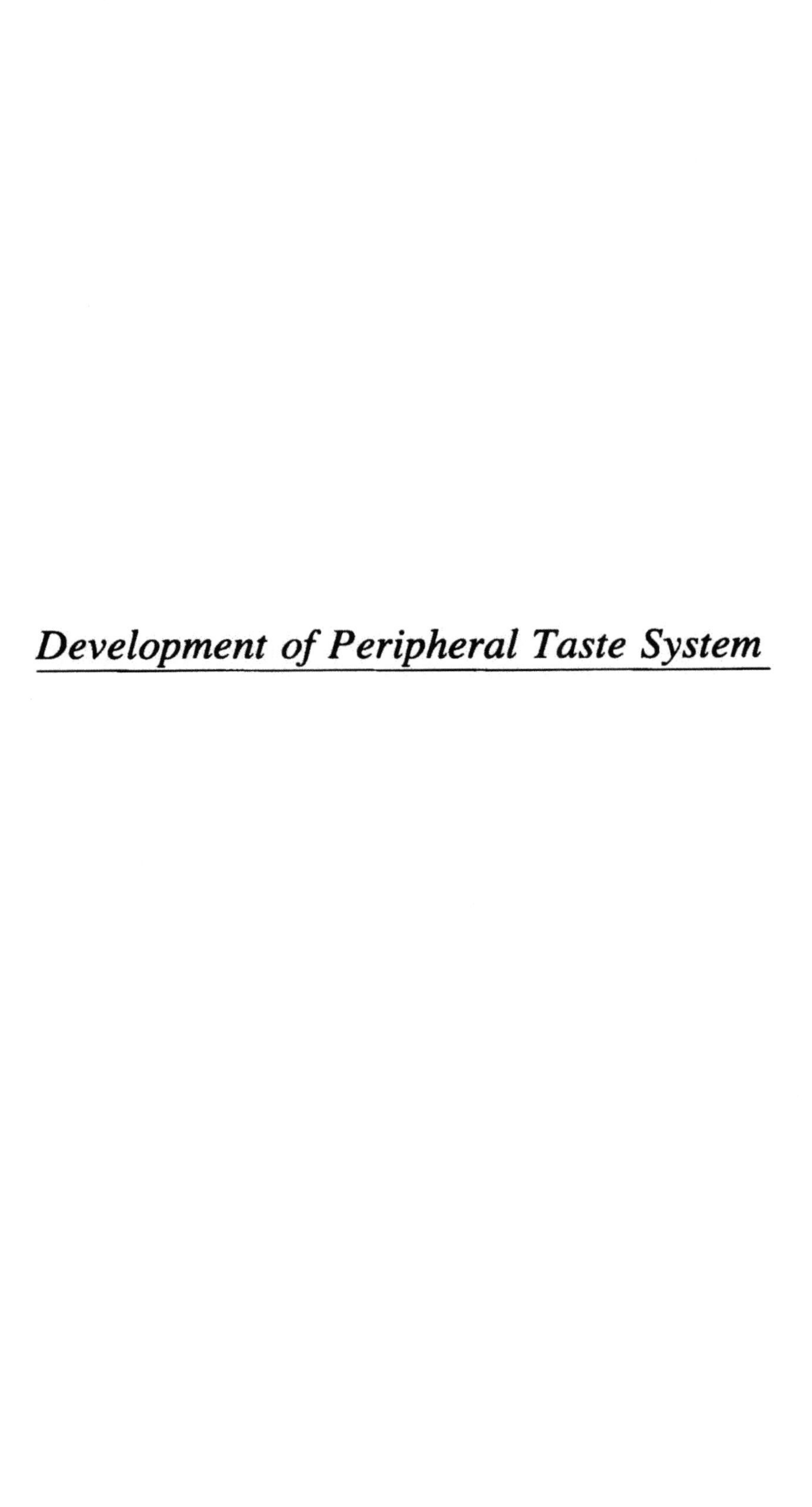

Development of Peripheral Taste System

Chapter 4

CONTROL MECHANISMS IN TASTE BUD DEVELOPMENT

Bruce Oakley

TABLE OF CONTENTS

0-8493-5341-6/93/$0.00 + $.50

I. INTRODUCTION TO TASTE BUD DEVELOPMENT

The present review of taste bud development focuses on cellular control mechanisms. The classical embryology of tongue and palate and the development of electrophysiological responses of gustatory neurons in rat and sheep have been reviewed elsewhere.[1,2] The development of electrophysiological taste responses is considered in Chapter 5. An important objective of developmental studies is to describe the sequence of cell and tissue changes that lead to the adult form. This constitutes a search for developmental mechanisms that generalize widely both across species and across taste buds found at different locations in the same species. This is not to imply that variations across species and taste bud populations are uninteresting. It is quite remarkable, for example, that the maturation of chorda tympani nerve responses to NaCl is profoundly delayed in sheep,[3] rat,[4] and mouse,[5] but not in the hamster chorda tympani[6] nor in the IXth nerve of mouse and sheep.[5,7] The molecular and behavioral implications of these nerve and species differences should prove to be interesting.

While our understanding of mechanisms controlling taste bud development is limited, recent efforts have provided some answers and several questions to prompt future research. The present assessment of developmental control mechanisms begins with a summary of the developmental anatomy of taste buds and ends with a brief consideration of the likelihood of gustatory mediation of suckling behavior in newborn rats.

The characterization and evaluation of any developing system depends upon a clear understanding of the adult condition. It is difficult to interpret development if there is no consensus on the mature structure. Taste buds pose a special challenge because adult taste buds are not static structures. The cells in taste buds turn over; the average lifespan of a taste cell is about 10 days in rat fungiform[8,9] and vallate papillae.[10] This means that both the birth and maturation of receptor cells and the formation of synaptic contacts are not restricted to development. Flux associated with turnover is an integral feature of adult taste buds. The neurotrophic dependence of adult taste buds provides further opportunities for morphogenetic flux. Taste bud nerve-dependence is present at the earliest developmental periods tested.[11,12] Nerve interruption causes taste bud degeneration.[13,14] Taste buds typically reform after nerve regeneration.[14] While damage to taste axons might be uncommon in nature, taste cell turnover necessarily remodels axonal endings and may alter the local level of neurotrophic support of taste receptor cells. If developing taste buds emulate the development of nerve dependent cutaneous receptors and muscle spindles, they may be especially vulnerable to impaired innervation during development.[15]

Taste buds are characteristically found in the epithelium of the epiglottis, pharynx, palate, and tongue of mammals. Such taste buds have much the same appearance. Many are associated with papillae visible as bumps or as

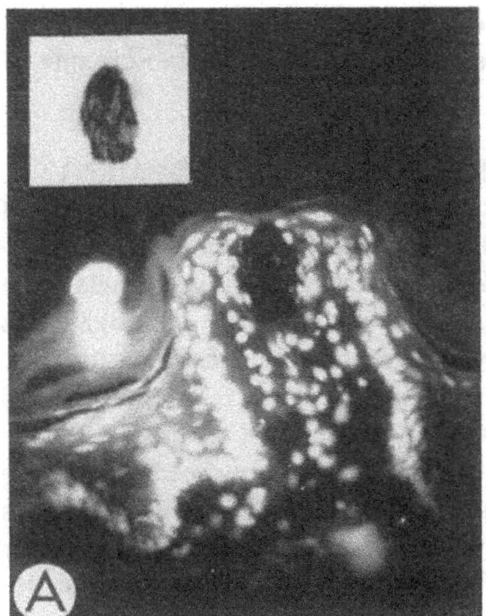

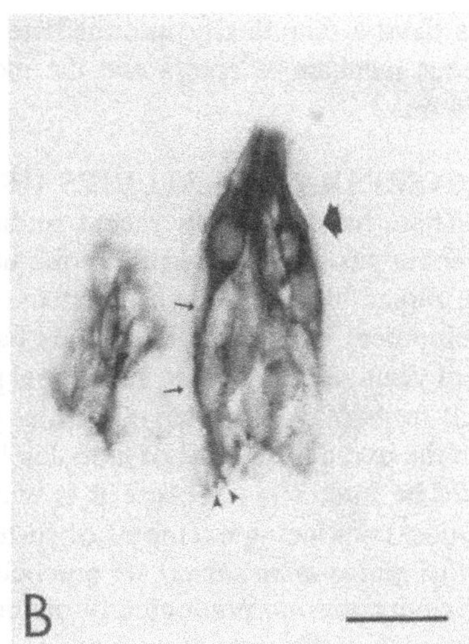

FIGURE 1. The unique keratin-like immunoreactivity of taste buds is a useful marker of taste cell differentiation. (A) Ths 20-μm frozen section of a rat fungiform papilla was double stained to reveal the nuclei of most cells in the papilla (Hoechst stain, fluorescence optics) and the keratin-like immunoreactivity of the receptor cells (inset). The position of the taste bud in the papilla is clearly evident, since the peroxidase reaction product in the taste cells obscured the fluorescent nuclei of the taste cells. Inset: In a bright field micrograph of this papilla the only cells with keratin 19-like immunoreactivity were differentiated cells of the taste bud (monoclonal LP2K supplied by Dr. E. B. Lane; Vectstain ABC, Vector Laboratories). (B) Keratin 8-like immunoreactivity of taste buds in the rat palate (monoclonal antibody LE41, Sigma; Vectstain ABC). The fusiform shape (small arrows) and endfeet (arrowheads) are evident. Some taste cells have apical nuclei (heavy arrow). The scale line is 21 μm for (B) and 50 μm for (A).

grooves in the tongue. The most notable morphological difference among mammalian taste buds is the variation in the number of cells per bud. The spatial density of taste buds also varies; on the tongue it ranges from one taste bud per fungiform papilla in many rodents (Figure 1A) to hundreds in rodent foliate and vallate papillae to thousands in these papillae in ungulates and primates.[16] An individual taste bud is a cluster of slender receptor cells arranged rather like the segments of an orange (Figure 1B). Chemical solutions stimulate the apices of receptor cells via a 2 to 4 μm diameter pore through the epithelial surface. The first afferent synapse lies in the taste bud itself — between receptor cell and axon. Taste solutions arouse receptor cells which in turn synaptically activate taste axons within the taste bud. Inhibitory synaptic interactions in the taste bud may account for the suppression of sensory messages observed in rat[17] and frog.[18] Taste buds can be considered sensory processing modules with nerve-dependent cells and continually realigning synapses (see also Roper in Chapter 11). As signal processing modules, taste buds have a complexity intermediate between the relative simplicity of cutaneous mechanoreceptors and the more sophisticated and stable circuits of the retina.

A. COMPARATIVE STUDIES OF TASTE BUD DEVELOPMENT

There have been few recent studies on the comparative development of vertebrate taste buds. In spite of the utility of fish for experimental embryology, rather little is known about the development of their taste buds. The development of urodele taste buds has not been intensively investigated in recent years, although their functional properties are being closely scrutinized. Adult frogs (*Rana* sp.) have peculiar taste disks that arise in metamorphosis from the ovoid taste buds of tadpoles.[19] Cell lineage studies of frog taste buds would be interesting because it is widely believed that frog taste buds have the most convincing examples of sustentacular or supporting cells. Chickens *(Gallus gallus domesticus)* are precocial animals; nearly all taste buds in their oral cavity are morphologically mature at hatching. The morphological development of chicken taste buds has many similarities with that of mammals.[20,21]

There is substantial variation in the timing of taste bud development. In precocial species, taste buds mature *in utero* or *in ovo*, whereas in altricial animals the development of taste buds can be primarily a postnatal process. While the timing of taste bud development may reflect life history patterns, the underlying developmental mechanisms are probably similar across mammals. Since the taste buds in rat, sheep, and monkey have been the focus of most research on development, they will be emphasized here. Primates and sheep are born with many morphologically mature taste buds, in contrast to the immaturity of the taste system of newborn rats. In both precocial and altricial species, taste buds begin development as somewhat disorderly clusters of polygonal cells lying beneath the surface of the epithelium. As the cells elongate, they stretch from the basal lamina to the epithelial surface they later penetrate. Taste cells develop apical microvilli before penetrating the keratinized surface squames.[22] A similar penetration or reopening of a sealed pore occurs in taste bud regeneration. Inhibition of squame-generating epidermal proliferative units (EPU)[23] directly beneath the taste pore site may assist in pore formation. The regular margins of the pore and the similarity of development and regeneration suggest the taste cells may actively contribute to the taste pore.

B. TYPES OF TASTE CELLS AND THEIR DEVELOPMENT

Most authors recognize three or four types of cells in *mature* mammalian taste buds (see also Ruetter, Chapter 2). Dark cells (type I) have dense cytoplasm, abundant membrane-bound granules in the cell apex, and indented nuclei. Light cells (type II) have a less electron dense cytoplasm and a large oval nucleus with a smooth nuclear membrane. From the more frequent occurrence of synaptic contacts and numerous dense-core vesicles, a type III cell has been considered by some investigators to be the true taste receptor cell.[24-26] Basal cells (type IV) lie on the basolateral margin of the taste bud, are rather undifferentiated, may be polygonal or somewhat elongated, and may be contacted by a nerve fiber. The dark cells are the most abundant and the basal cells the least abundant. Recent studies show that all taste cell types can make synaptic connections with taste axons.[27] In view of these findings it is questionable whether only type III cells arouse taste axons. Some investigators believe that mammalian taste buds also have supporting cells.[28,29]

As the term is used in biological studies of epithelia, a "basal cell" is one of the cuboidal or polygonal cells in the germinative or basal cell layer which contacts the basement membrane. A cycling subset of these cells comprises the active stem cells, whose suprabasal daughters migrate superficially to maintain the continually renewing epithelium. If one applies the distinction between basal and suprabasal cells to taste buds, some undifferentiated cells on the flanks of taste buds may be migrating suprabasal daughters rather than true stem cells for the taste bud. Although there is agreement that basal cells are probable forerunners of other cells, there is no consensus on the number of morphological or functional lineages that arise from basal cells. Some assessments of cell turnover with tritiated thymidine favor a single lineage with cells progressively changing appearance as they age, e.g., through basal, dark, intermediate, and light (BDIL).[8,30] Others workers favor two lineages leading to dark and light cells, respectively,[10] or three lineages leading to dark cells, to light cells, or to type III cells.[25]

The difficulty in appraising whether there are one or several morphological types of receptor cells is rivaled by the vexing problem of classifying taste receptor cells into functional types. Thus, even if the weight of evidence favors one line of cells whose appearance changes with age, it leaves unsettled the problem of substantial cell-to-cell differences in functional responses. Intracellular recordings from taste cells reveal considerable functional variation among receptor cells.[31,32] Individual taste buds probably have several functional cell types.[31,33] Moreover, the maturational state of receptor cells may not be reflected in the functional responses of the axons. Stewart and Hill (Chapter 5) have shown immunologically identified amiloride channels, the presumptive mediators of salt taste, are present in neonatal rat taste cells well before the primary axons develop robust NaCl responses.

The loss of an aged cell and its replacement somewhere else in the bud, perhaps by a cell of different sensitivity, is a threat to the stability of taste quality coding. A possible basis for stable signals in spite of cell turnover is that taste specificity is governed by cell age; a given taste axon restricts its contacts to cells of the same age. This interesting model is not without its problems. First, if a cell's taste function changes with age, an axon would have to shift its contacts to other cells or propagate aberrant signals. And to eliminate transitional periods of confusing blended signals, multiple downtimes might be required as axons disengage, receptor cells transform from one transduction mode to another quite different mode,[34] and axons re-engage. Second, electrophysiological studies have shown that taste buds composed primarily of young (newly regenerated) receptor cells did not have a restricted range of effective taste chemicals; young receptor cells generated primary axon response profiles closely resembling those mediated by adult taste fibers.[35]

As an alternative to functional changes with cell aging, one might postulate stable connections engendered by affinities between an axonal ending and its appropriate receptor cell. Several lines of evidence indicate that axons make preferential connections with taste cells. An axon does not dictate its receptor cell's chemical specificity.[36,37] Yet its branches are believed to innervate receptor cells of similar chemical specificity.[38] Most taste axons branch and innervate at least two taste buds. When such taste buds are individually and independently chemically stimulated they produce the same relative taste responses in the parent axon.[38] This is consistent with the view that taste axons make synaptic contact with matching types of taste receptor cells.[34] The number of cells innervated by one axon has been reported to be less than six in mouse.[39] Two groups have found that an axon's branches tend to be restricted to one morphological type of taste cell.[25,39]

In *immature* taste buds, the cells begin as relatively undifferentiated polygonal cells and later elongate. In the first ultrastructural examination of rat fungiform taste bud development, Farbman[22] identified a progressive elaboration of cell types associated with taste buds. Dark cells preceded the appearance of light cells. Zahm and Munger[28,40] have examined the development of fungiform taste buds in *Macaca mulatta* monkeys. Young taste buds were composed of cells intermediate in position and cytology between the surrounding extragemmal cells and the single type of chemosensory cell recognized in the core of the adult bud. Afferent synapses onto axons were present in young chemosensory cells in *Macaca;* putative efferent synapses developed later. Tight junctions and desmosomes were observed between chemosensory cells, but gap junctions were only seen between extragemmal cells. Because of their late appearance in monkey taste bud development, sustentacular cells, defined by apical secretory granules and a lack of synaptic contacts, may play a role beyond support. In ultrastructural examinations of developing vallate taste buds in hamsters, Miller and Chaudhry[41] evaluated the relationship between light and dark cells. Mature light cells were frequently present before mature dark cells. The absence of transitional cells between light and dark led the authors to conclude that dark and light cells had separate lineages. Takeda et al.[42] noted that type III cells appeared first, followed by type I and II cells in developing mouse vallate taste buds. Thus, developmental studies are divided on the issue of taste cell types and their lineage. Diverse opinions on cell types in development are to be expected, given the less differentiated nature of immature cells and the varied views on the cellular make-up of mature taste buds.

C. NERVE FIBERS: THEIR CONNECTIONS AND DEVELOPMENT

Acetylcholine and bioamines are venerable putative transmitter molecules in adult taste buds (see Reutter in Chapter 2 and Roper in Chapter 11). The availability of antibodies to identify some of the burgeoning number of neuropeptides has prompted immunocytochemical studies of taste buds.[43-46] Although calcitonin gene-related peptide (CGRP) and substance P axons are abundant in many tissues, relatively few neuronal somata in the ganglia containing taste neurons (geniculate, petrosal, and vagal ganglia) were CGRP or substance P positive.[44] Similarly, while CGRP and substance P positive axons are common in the perigemmal areas, they were infrequent within taste buds. Electron microscopy suggests that the substance P axons pass through the taste bud without synapsing.[46] Nonsubstance P axons arrive in developing vallate taste buds 2 days (E19) before substance P axons (E21).[47] No substance P fibers were observed to synaptically contact cells in or around taste buds. In summary, there is no compelling evidence that CGRP or substance P axons play a role in taste reception or in the trophic support of taste buds.[46-48] The presence of other endogenous neuroactive agents needs to be evaluated.

Afferent synapses between some taste cells and axons are evident in developing fungiform taste buds (Figure 2A). Some cell types are slow to form synaptic connections; synapses with developing palatal light cells were not detected.[49] Developing nerve endings with clear vesicles make numerous axo-axonal synapses[25,28,47] (Figure 2B). The regions of increased membrane density tend to be symmetric and vesicles are present in both axonal processes. The paucity of axo-axonal contacts in mature taste buds suggests that axonal contacts are characteristic of axons which are in the process of establishing connections with receptor cells. Like the transient connections to developing or regenerating skeletal muscle cells, taste buds might also be expected to have an abundance of axo-axonal contacts during regeneration as well as development. This has been reported.[50] This suggests that axo-axonal contacts reflect an interim phase of impermanent contacts that are not sustained once an axon has effective receptor cell connections.

II. DEVELOPMENTAL CONTROL MECHANISMS

The focus of this review of taste bud development is to emphasize experimental analyses of mechanisms controlling taste bud innervation and development. In this context, it is convenient to apportion taste bud development into the following 12 events, beginning with the formation of taste papillae and ending with programmed cell death (apoptosis).

A. THE FORMATION OF TASTE PAPILLAE

Although some taste buds on the epiglottis and palate lack specialized papillae,[51,52] most taste buds are associated with papillae evident as protrusions

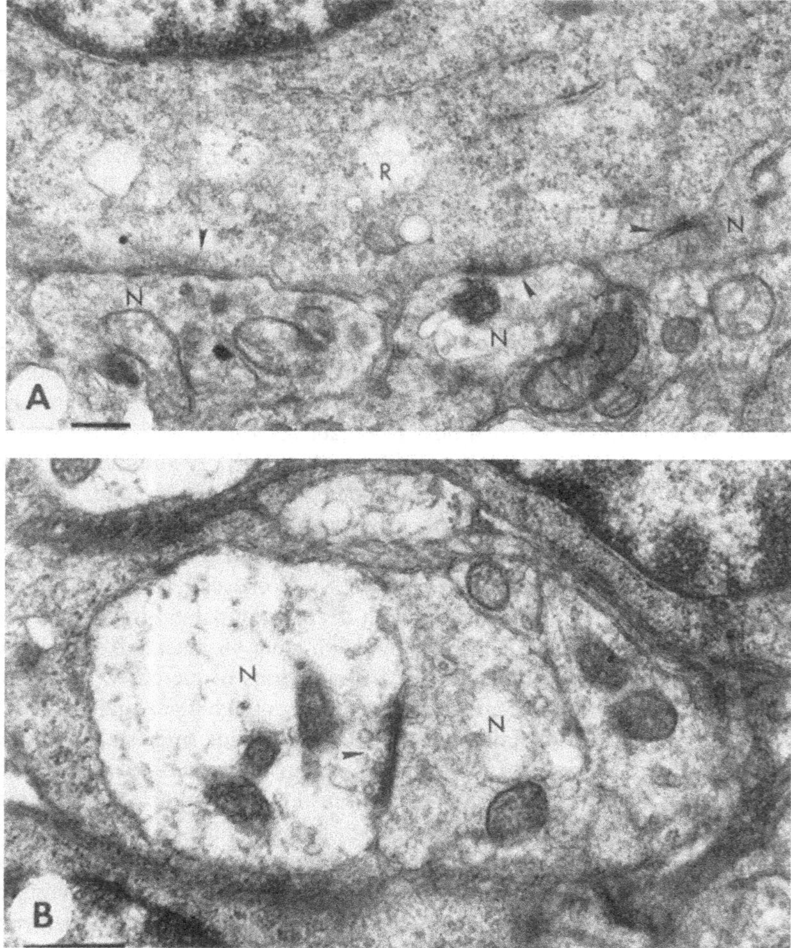

FIGURE 2. Transmission electron micrographs of fungiform taste buds in a 4-day-old rat. (A) A taste receptor cell (R) makes afferent synapses (arrowheads) with at least three nerve processes (N). HVEM. (B) An axo-axonal synapse (arrowhead) between two nerve endings (N) is relatively symmetric; vesicles and synaptic densification occur on both sides. TEM. Scale lines in A and B are 0.5 μm. (Courtesy of John C. Kinnamon and Cynthia Church.)

or furrows. Fungiform papillae and the circumvallate papilla on rat and mouse tongue are representative examples. There are about 90 fungiform papillae on each side of the anterior portion of rat and mouse tongue. They are most abundant at the tip and extend posteriorly in loosely defined rows.[53] On the posterior portion of the rat and mouse tongue is a single midline vallate papilla, less than 1 mm in diameter. The chorda tympani and lingual nerves innervate the fungiform papillae, whereas the vallate papilla receives bilateral innervation from the IXth nerve. The beginnings of vallate trenches appear on embryonic day 13 (E13) in mouse, and E14 to 15 in hamster and rat[41,54-56] vs. E16 for rat foliate.[57] Nerve fibers may contribute to development of the vallate papilla, given that it is innervated from its inception.[56] Fungiform papillae and underlying axons are first evident in rats at E15.[58] It has been possible to show the underlying nerve fibers are not required for the initial formation of fungiform papillae. Nascent fungiform papillae emerged after 2 d of organ culture of E14 rat tongues whose epithelium had never been innervated. However, sustained maturation may require innervation since the denervated fungiform papillae fail to mature further *in vitro*.[59,60] Recent experiments have clarified the role of nerves. In the absence of innervation the morphogenetic program for filiform spines governs the adult fungiform papilla,[61] as it seems to during *in vitro* maturation in a nerve-free environment.[62] The classic transplantation experiments of Billingham and Silvers[63] showed that even after the lingual epithelium and its basal cells had been transplanted to the underlying mesenchyme of hairy skin, the lingual basal cells typically sustained the character of the lingual epithelium and its filiform spines.

B. THE DISTRIBUTION OF TASTE AXONS AND THEIR CAPACITY TO FIND NASCENT GUSTATORY PAPILLAE

Counts of axonal terminations in the epithelium of the adult gerbil tongue indicate that well over 95% of regenerated chorda tympani axons terminate within, rather than between, the fungiform papillae.[64] Similarly, in rat embryos the tendency of developing chorda tympani fibers to be localized to the fungiform papilla was evident from the earliest times (E15).[58] Hence, while there is scant data on taste axon pathfinding in development, it would not be surprising if taste axons followed substrate pathways to their corresponding papillae. Virtually no taste buds reformed after the IXth nerve was reconnected to the distal stump of the hypoglossal nerve, even though regenerating IXth nerve axons were less than 1 mm away from gustatory papillae.[65] Thus, there appears to be little capacity for the adult gustatory epithelium to attract taste axons from a distance or for putative neurotrophic agents to diffuse more than 1 mm to the gustatory epithelium.

In adults, lingual nerve fibers surround the fungiform taste buds, yet infrequently invade them.[66] In spite of the marginal (perigemmal) location of lingual nerve axons, an equal and substantial number of fungiform taste buds reformed whether the transected chorda tympani was sutured to its own distal stump or to the stump of the lingual nerve proper.[67] Evidently regenerating chorda tympani taste axons that follow the lingual nerve pathways make a sufficiently close approach to the cells critical for reforming taste buds.

C. DETERMINANTS OF THE SITE OF TASTE BUD FORMATION

Studies of taste bud regeneration in adults suggest that taste buds will only form in specific gustatory epithelium.[14] In development, it is not known whether taste buds are also limited to such characteristic sites by the location of competent tissue or alternatively by the location of taste axons. Even in the case of fungiform papillae that seem to be laid down without benefit of nerve fiber influence, the gustatory competence of their cells may be nerve dependent.

D. THE INDUCTION OF TASTE BUDS BY NERVE FIBERS

Immature taste buds arise shortly after axons arrive in the embryonic vallate papilla. Early denervation results in an absence of taste buds in adult vallate papillae. However, a shortfall of taste buds after such denervation fails as a test of induction since one would predict the subsequent absence of taste buds in adults merely because adult taste buds are themselves nerve-dependent. The induction of taste buds by nerves can be tested, however, by exploiting the bilateral innervation of the rat vallate papilla. For example, when the left IXth nerve was removed at 3 d of age and the right IXth nerve simultaneously crushed, only 30 vallate taste buds formed after the right crushed nerve regenerated. In contrast with both IXth nerves crushed on day 3, a mean of 144, not 60, taste buds was formed. The excess of 84 taste buds must represent a synergistic interaction between the 2 nerves. That is, the two IXth nerves were responsible for an initial induction of numerous taste buds.[11,12]

E. A SENSITIVE PERIOD FOR TASTE BUD INDUCTION

The deleterious effects of early removal of one IXth nerve were most pronounced during a sensitive period that was maximal from 0 to 10 d postpartum.[12] Since the measured sensitive period is a composite of sensitive periods for several taste buds, the sensitive period for an individual taste bud may be much shorter. Similarly, perturbed developmental environments may alter taste buds by mechanisms whose true timing may be difficult to specify. For example, maternal salt deprivation must begin before E9 to prevent the normal development of NaCl responses in rat chorda tympani nerve.[68] But because plasma sodium levels are defended from sharp reductions in sodium excretion,[69] it is difficult to specify at what time sodium levels fall enough to create the cellular changes that later retard taste responses to NaCl.

F. THE GUSTATORY EPITHELIUM IS NERVE DEPENDENT

A cross-innervating chorda tympani nerve will support hundreds of taste buds in a normal vallate papilla, but supports only a few taste buds in a vallate papilla subjected to an early postnatal period of denervation. Early denervation causes critical vallate cells to die or to adopt a nongustatory fate.[70]

G. THE TIMING AND CONTROL OF TASTE BUD MATURATION

It is convenient to define taste bud maturity as the moment when a taste pore is formed, as thereafter the receptor cells can be stimulated by taste solutions. Taste pore formation is a timed occurrence; it is not regulated by taste bud volume. During the 15 to 60 d postnatal period, about 10 d were required for a given taste bud to mature (10.5 ± 0.9 d, mean ± 1 SD). In younger rats, taste buds matured more rapidly.[71]

H. COMPETITIVE AND SYNERGISTIC INTERACTIONS OF TASTE AXONS FOR GUSTATORY TARGETS

Several experiments indicate that some taste axons, rather than competing for targets, synergistically develop additional taste buds, perhaps by increasing the number of stem cells committed to taste.[11,12,72]

I. TARGET DEPENDENCE OF DEVELOPING GUSTATORY AXONS

While synergism among axons raises the prospect that taste axon survival is independent of target derived trophic factors, it does not preclude that possibility. As with the known target dependence of trigeminal axons,[73] the target-dependence of taste axons would be most likely manifested early on when the tongue epithelium is first innervated, which in rats occurs during the third embryonic week.[58] However, in view of the plasticity necessitated by receptor cell turnover, it is possible that, unlike trigeminal axons, taste axons never engage in a competitive struggle for target derived trophic factors. The issue needs to be experimentally evaluated during embryogenesis.

J. THE TIME COURSE OF TASTE BUD PROLIFERATION

All populations of rat taste buds so far investigated mature after birth. For example, mature taste buds in the palate[52] and in the nasopalatine papilla and canals[49] are first evident during the second postnatal week. Few mature fungiform taste buds were detected by scanning electron microscopy of the tongues of newborn rats.[74] Similarly, at birth there are virtually no mature rat foliate or vallate taste buds. The full complement of taste buds is completed after 1.5 months (foliate) and 3 months (vallate).[15,55] A similar postnatal proliferation period for vallate and foliate taste buds has been described in hamsters.[41,75] Investigation of the continually enlarging vallate papilla in 10- to 60-d-old neonatal rats demonstrated that new taste buds are added primarily in growth zones at the rostro-caudal extremes of the elongating vallate trench.[71] The constant taste bud density after day 40 suggests that the rate of addition of vallate taste buds is limited by expansion of gustatory tissue.[55] In contrast to taste buds increasing in number up to a fixed density, the progressive increase in surface area of the lamb epiglottis leads to a decline in taste bud density before stabilizing at adult numbers.[51]

K. TASTE CELL DIFFERENTIATION

During their final phase of maturation, mouse vallate taste buds will sequester biogenic monoamines,[42] as do selected cells in developing taste buds of frogs.[76] This developmental evidence is consistent with the view that biogenic amines play a role in signal processing in the taste bud (see Reutter, Chapter 2 and Roper, Chapter 11).

Reliable markers of differentiated taste receptor cells would be a valuable contribution to the examination of developing gustatory epithelium. If present in developing taste receptor cells, keratins might serve as useful markers, particularly if they differ from the keratins expressed in ordinary keratinocytes that comprise the interpapillary tissue. Keratins 8, 18, and 19 are present in all taste buds thus far examined in rats, gerbils, and rabbits.[65,77] While keratin 19 is not found in the laterally adjoining epithelium, it is present in the luminal cells of the ducts of von Ebner's glands. Regardless of embryological origin of the host tissue (e.g., branchial arches 1 through 4 and the lateral palatal projections), the elongated cells of all taste buds in diverse regions of the oral cavity contain keratins 8, 18, and 19. Consequently, monoclonal antibodies against these keratins will be useful markers to identify differentiated taste cells and their slender extensions under various conditions (Figures 1 and 3).

L. APOPTOSIS (PROGRAMMED CELL DEATH)

The programmed death of developing cells is a common mode of sculpting tissue and pruning neural circuits of weak or improper connections. In gustatory systems a developmental reduction in the number of taste cells would be most evident as a loss of entire taste buds. The number of taste buds in chickens rises to a maximum at E19 and remains constant thereafter.[20] Nor is there evidence for a significant decline in numbers of rodent fungiform taste buds which seem to be maintained indefinitely at one taste bud per fungiform papilla. However, fungiform taste buds appear to be lost in sheep.[78] While Liem et al.[79] reported a decline from an average of 25 to 20 taste buds in rat soft palate, Srivastava and Vyas[52] found no decline from a stable maximum of 115 soft palate taste buds during the same 3- to 8-week postnatal period. Both studies used Wistar rats. In the hamster soft palate, mature taste buds reach a peak of 115 that is constant from 3 weeks to at least 14 weeks.[80] In some locations taste buds are only present in perinatal animals. For example, the few taste buds (<10) present on the dorsal surface of the neonatal vallate papilla disappear from adult rats[55] (Figure 3A). A most dramatic example of taste bud loss occurs in dolphins *(Stenella coeruleoalba)* from which all oral taste buds disappear.[81] Thus, in some instances there appears to be a developmental overproduction of taste buds. The causes or functional significance of taste bud losses have not been determined.

In addition to the 12 events that have been described, it is useful to compare taste bud regeneration with development and to summarize morphogenetic programs in a schematic diagram.

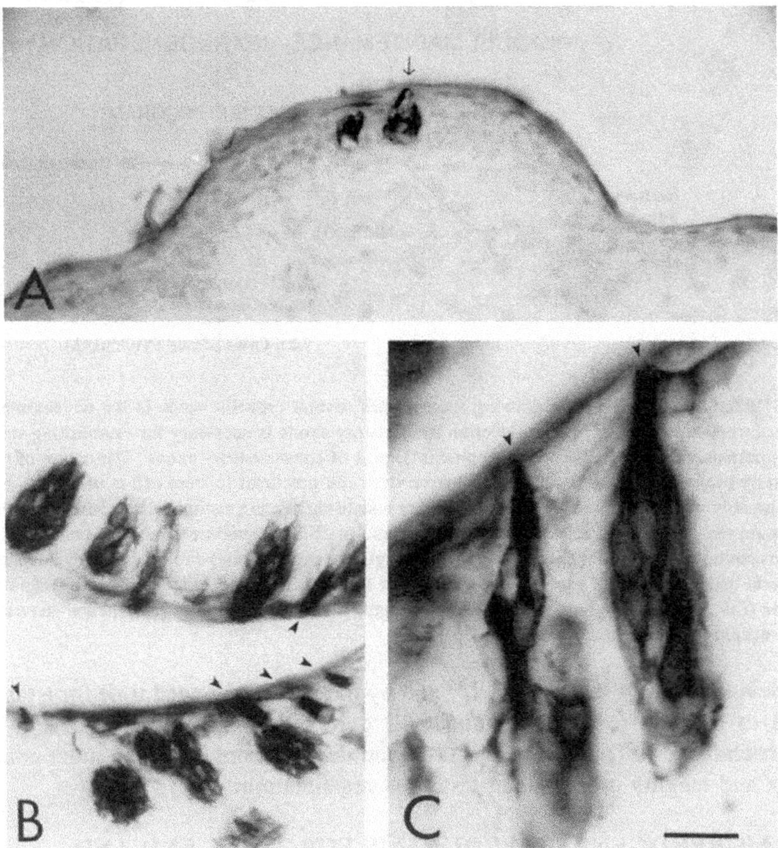

FIGURE 3. Keratin 8-like immunoreactivity reveals the developmental differentiation of taste buds. (A) Immunoreactive taste buds (arrow) are evident on the dorsal surface of the newborn rat vallate papilla. These buds later disappear. (B) and (C): By 10 days after birth there are numerous taste buds in various stages of differentiation in the trenches of vallate papilla. The density of staining with antibodies against keratin is particularly useful in revealing fine cellular extensions, such as sites where the palisade array of taste cell apical processes extend to the epithelial surface (arrowheads). (Monoclonal antibody LE41, Sigma; Vectstain ABC). The scale line in C is 28 μm for (A), 30 μm for (B), and 12 μm for (C).

M. TASTE BUD REGENERATION DOES NOT RECAPITULATE TASTE BUD DEVELOPMENT

Although the regeneration of taste buds may replicate the late developmental steps, regeneration does not recapitulate the initial developmental steps. This is demonstrated by the permanent disabling of gustatory epithelium with denervation during the sensitive period,[70] by the capacity of one gustatory nerve to maintain more than twice as many adult taste buds as it will induce

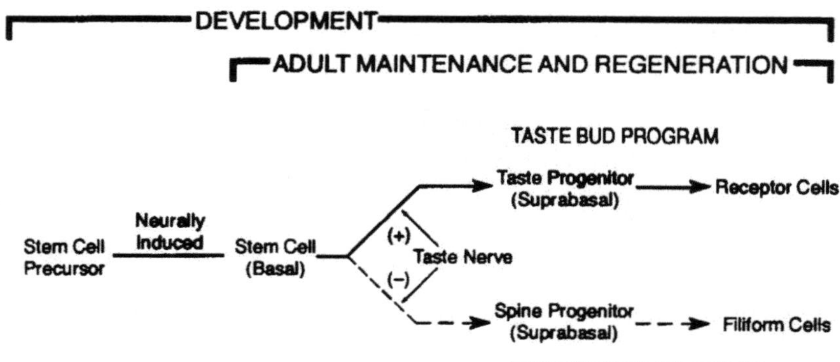

FIGURE 4. This schematic drawing incorporates several probable steps in the development of mammalian taste buds. Early induction by gustatory axons is necessary for establishing stem cells permanently responsive to the trophic influence of chemosensory axons. The nature of the neurally triggered cellular transformation from stem cell precursor to stem cell is unknown, but presumably would involve either cell division or a state change in existing cells. (Gustatory stem cells are the germinative cells whose daughters lead to differentiated taste receptor cells.) Taste axons control two morphogenetic programs in fungiform papillae: they promote the development of taste buds and prevent expression of a filiform spine. Denervated fungiform papillae default to the filiform spine program.[61] Brackets above signify that some initial developmental steps are not recapitulated in regeneration.

in development (496 vs. 225),[11,12] and by the more prolonged time for a taste bud to develop to maturity (about 10 d in development vs. 2 to 3 d in regeneration[11,35,71]). Apparently, taste bud development may be a more complex and lengthy process than taste bud regeneration.

N. MORPHOGENETIC PROGRAMS FOR TASTE BUD AND FILIFORM SPINE DEVELOPMENT

A flow diagram is a convenient way to summarize our limited understanding of lineages and control mechanisms in taste bud development. Figure 4 is a working model that outlines some interesting features of fungiform taste bud development. Since denervated taste buds can reform even after disappearing completely, they must reform from residual basal/stem cells which survive denervation for weeks, perhaps months.[82] Because early denervation makes the gustatory epithelium permanently unresponsive to taste axons, innervation during the period of induction must trigger a conversion step that establishes the stem cells. That is, innervation leads to an altered cell state or perhaps to cell division that converts a stem cell precursor to a stem cell capable of reforming taste buds after denervation in adults. If we assume that taste buds arise from a basal to suprabasal lineage, as is well established for the renewal of other epidermal cells,[23] the gustatory stem cells may be depicted as giving rise to suprabasal progeny that migrate and differentiate into mature cells. For simplicity the model assumes a taste axon controls not two separate lineages but a binary switch for a single stem cell lineage. The anterior dorsal surface of mammalian tongues is normally covered by a field of protective filiform spines[61,83] and scattered spine-free fungiform papillae. In the absence of taste axons, fungiform papillae develop an ectopic filiform spine closely resembling a normal filiform spine.[61] When taste axons are present, they inhibit ectopic filiform spine formation from keratinocytes and promote the differentiation of the fungiform taste bud. The early inductive steps are probably unique to development. The working model of morphogenetic programs in Figure 4 suggests the potential for analysis both of the positive regulatory role of taste axons in promoting the formation of taste buds and of the negative regulatory role in inhibiting the formation of filiform spines.

III. ON THE BEHAVIORAL SIGNIFICANCE OF TASTE BUD DEVELOPMENT

A rat's sensory capacities at birth should reflect the functional demands posed during its early life history. Presumably there is a state of receptor responsiveness that will optimize the establishment of suckling and milk ingestion and minimize nonnutritive behavioral diversions like licking the salt derived from urine or maternal saliva. A newborn altricial rodent survives only if it promptly and consistently suckles. Olfactory cues seem to be essential for suckling by rats.[84] If taste were also essential, one would expect to find numerous taste buds that were adequately developed and responsive to lactose. There is a virtual absence of mature taste buds in newborn rats (e.g., 0 mature and ca. 30 immature vallate taste buds[71]). Regenerating fungiform taste buds too immature to have a taste pore do not mediate nerve discharges.[35] These observations suggest a minimal role for gustation in suckling by newborn rats. This inference is supported by most behavioral assessments of taste. One- to five-d-old rats are either unresponsive to mild taste solutions or fail to discriminate among them.[85-87] However, it has been reported that 1-d-old rats react to a droplet of saccharin or sucrose by licking and appear to reject quinine by mouth opening while reacting only slightly to water.[88]

Species like rats and sheep that prefer NaCl as adults may employ the sluggish development of NaCl taste reception to minimize early NaCl ingestion that would compete with suckling. The slow development of rat chorda tympani nerve responses to NaCl is consistent with a strategy of excluding NaCl-triggered licking that competes with suckling. In contrast, hamsters, which show no preference to NaCl, develop their taste receptor apparatus for NaCl promptly. It is tempting to infer that the lack of a hamster NaCl preference eliminates the selective advantage of the delayed maturation of nerve sensitivity to NaCl. This hypothesis would predict that neonatal hamsters have no NaCl preference. This should be tested since adult taste preferences are suspect predictors of neonatal preferences. Who would have anticipated that 18-d-old rats would strongly prefer 0.2 *M* NH_4Cl — a chemical that is never preferred at any concentration by adult rats[86,87]?

In rats, taste pores do not form in the palate until several days after birth.[52,79] In contrast, suckling in hamsters might be assisted by palatal taste sensitivity, since 39% of hamster palatal taste buds are mature at birth.[80] The position of palatal taste buds in the oral cavity and their known sugar sensitivity in adults[89] makes palatal taste buds putative candidates for monitoring the gustatory components of milk in hamster. At about 0.1 *M*, lactose is the dominant carbohydrate in milk and the principal obligatory gustatory stimulant for young mammals, yet it has been the exceptional behavioral or electrophysiological investigation of mammalian neonates that has incorporated lactose as a standard stimulus.

IV. CONCLUDING REMARKS

Taste buds provide a model to examine the biology of cellular mechanisms that control epithelial morphogenesis. Increased understanding of such mechanisms has shown that taste buds are rather more complicated sensory processing modules than previously appreciated. Consequently, syntheses of morphological and functional information from several laboratories will be valuable in formulating working hypotheses of cellular interactions. For example, it is known that: (1) there is an upper limit in size of taste sensory units (the number of receptor cells innervated by a taste axon[39]), (2) some developing taste cells lack ultrastructurally identifiable synaptic connections,[39] (3) robust NaCl-best axonal discharges arise days, perhaps weeks, after taste cells express immunocytochemically identified amiloride channels (Stewart and Hill, Chapter 5), (4) the size of the receptive field of sheep NaCl-best fibers diminishes as responses to NaCl improve,[91] and (5) receptor cells with unusual response profiles are associated with some of the largest receptive fields.[38] Collectively, these observations raise the possibility that the receptive fields of NaCl-best axons become smaller because axons both preferentially support short branches and seek their preferred receptor cell type. One may speculate that an initial local shortage of the preferred receptor cell type prompts a wide search by axonal branches; as cells nearby become NaCl-best and receptive to synaptogenesis, NaCl receptive fields constrict.

One of the most important issues confronting taste biologists is the basis for apparently stable sensory signals for taste quality in the presence of continual receptor cell turnover. Developmental models of axon-receptor affiliations that attempt to account for stable taste quality signals[15,34,91] should consider that taste responses are tissue-specific[36,37] and that branches of taste axons innervate similar kinds of structural and functional receptor cells.[25,38,39] Clearly many fascinating problems remain to be explored in the dynamic control of chemoreceptor cell development and renewal.

ACKNOWLEDGMENT

I thank Lianna Wong and Rose Wagner for their assistance. Supported in part by NIH Grant DC00083 and NSF Dir-9014275.

REFERENCES

1. **Mistretta, C. M.,** Taste development, in *Development of Sensory Systems in Mammals,* Coleman, J. R., Ed., John Wiley & Sons, New York, 1990, 567.
2. **Mistretta, C. M.,** Developmental neurobiology of the taste system, in *Smell and Taste in Health and Disease,* Getchell, T. V., Doty, R. L., Bartoshuk, L. M., and Snow, J. B., Eds., Raven Press, New York, 1991, chap. 3.
3. **Mistretta, C. M. and Bradley, R. M.,** Neural basis of developing salt taste sensation: response changes in fetal, postnatal, and adult sheep, *J. Comp. Neurol.,* 215, 199, 1983.
4. **Hill, D. L., Mistretta, C. M., and Bradley, R. M.,** Developmental changes in taste response characteristics of rat single chorda tympani fibers, *J. Neurosci.,* 2, 782, 1982.
5. **Ninomiya, Y. T., Tanimukai, T., Yoshida, S., and Funakoshi, M.,** Gustatory neural responses in preweanling mice, *Physiol. Behav.,* 49, 913, 1991.
6. **Hill, D. L.,** Development of chorda tympani nerve taste responses in the hamster, *J. Comp. Neurol.,* 268, 346, 1988.
7. **Mistretta, C. M. and Bradley, R. M.,** Developmental changes in taste responses from glossopharyngeal nerve in sheep and comparisons with chorda tympani responses, *Dev. Brain Research,* 11, 107, 1983.
8. **Beidler, L. M. and Smallman, R. L.,** Renewal of cells with taste buds, *J. Cell Biol.,* 27, 263, 1965.
9. **Conger, A. D. and Wells, M. A.,** Radiation and aging effect on taste structure and function, *Radiat. Res.,* 37, 31, 1969.
10. **Farbman, A. I.,** Renewal of taste bud cells in rat circumvallate papillae, *Cell Tissue Kinet.,* 13, 349, 1980.
11. **Hosley, M., Hughes, S. E., and Oakley, B.,** Neural induction of taste buds, *J. Comp. Neurol.,* 260, 224, 1987.
12. **Hosley, M. A., Hughes, S. E., Morton, L. L., and Oakley, B.,** A sensitive period for the neural induction of taste buds, *J. Neurosci.,* 7, 2075, 1987.
13. **Ganchrow, J. R. and Ganchrow, D.,** Long-term effects of gustatory neurectomy on fungiform papillae in the young rat, *Anat. Rec.,* 225, 224, 1989.
14. **Oakley, B.,** Trophic competence in mammalian gustation, in *Taste, Olfaction and the Central Nervous System,* Pfaff, D., Ed., Rockefeller University Press, New York, 1985, 92.
15. **Oakley, B.,** Taste bud development in rat vallate and foliate papillae, in *Mechanoreceptors,* Hnik, P., Soukup, T., Vejsada, R., and Zelena, J., Eds., Plenum Press, New York, 1988, 17.
16. **Bradley, R. M.,** Tongue topography, in *Handbook of Sensory Physiology IV, Chemical Senses, Taste,* Part 2, Beidler, L. M., Ed., Springer-Verlag, New York, chap. 1.
17. **Riddle, D. R., Hughes, S. E., Belczynski, C. R., DeSibour, C. L., and Oakley, B.,** Inhibitory interactions among rodent taste axons, *Brain Res.,* 533, 113, 1990.
18. **Murayama, N.,** Interactions among different sensory units within a single fungiform papilla in the frog tongue, *J. Gen. Physiol.,* 91, 685, 1988.

19. **Zuwala, K. and Jakubowski, M.**, Development of taste organs in Rana temporaria, *Anat. Embryol.*, 184, 363, 1991.
20. **Ganchrow, J. R. and Ganchrow, D.**, Taste bud development in chickens, *Anat. Rec.*, 218, 88, 1987.
21. **Ganchrow, D. and Ganchrow, J. R.**, Gustatory ontogenesis in the chicken: an avian-mammalian comparison, *Med. Sci. Res.*, 17, 223, 1989.
22. **Farbman, A. I.**, Electron microscope study of the developing taste bud in rat fungiform papilla, *Dev. Biol.*, 11, 110, 1965.
23. **Potten, C. S. and Allen, T. D.**, Control of epidermal proliferative units (EPUs). An hypothesis based on the arrangement of neighboring differentiated cells, *Differentiation*, 3, 117, 1975.
24. **Murray, R. G.**, The ultrastructure of taste buds, in *The Ultrastructure of Sensory Organs*, Friedman, I., Ed., Elsevier, New York, 1974, 3.
25. **Murray, R. G. and Murray, A.**, Relations and possible significance of taste bud cells, *Contrib. Sens. Physiol.*, 5, 47, 1971.
26. **Murray, R. G., Murray, A., and Fujimoto, S.**, Fine structure of gustatory cells in rabbit taste buds, *J. Ultrastruct. Res.*, 27, 444, 1969.
27. **Kinnamon, J. C., Taylor, B. J., Delay, R. J., and Roper, S. D.**, Ultrastructure of mouse vallate taste buds. I. Taste cells and their associated synapses, *J. Comp. Neurol.*, 253, 48, 1985.
28. **Zahm, D. S. and Munger, B. L.**, Fetal development of primate chemosensory corpuscles. II. Synaptic relationships in early gestation, *J. Comp. Neurol.*, 219, 36, 1983.
29. **Chan, K. Y. and Byers, M. R.**, Anterograde axonal transport and intercellular transfer of WGA-HRP in trigeminal-innervated sensory receptors of rat incisive papilla, *J. Comp. Neurol.*, 234, 201, 1985.
30. **Delay, R. J., Kinnamon, J. C., and Roper, S. D.**, Ultrastructure of mouse vallate taste buds. II. Cell types and cell lineage, *J. Comp. Neurol.*, 253, 242, 1986.
31. **Tonosaki, K. and Funakoshi, M.**, Intracellular taste responses of mouse, *Comp. Biochem. Physiol.*, 78A, 651, 1984.
32. **Tonosaki, K. and Funakoshi, M.**, The mouse taste cell response to five sugar stimuli, *Comp. Biochem. Physiol.*, 79, 625, 1984.
33. **Ozeki, M. and Sato, M.**, Responses of gustatory cells in the tongue of rat to stimuli representing four taste qualities, *Comp. Biochem. Physiol.*, 41A, 391, 1972.
34. **Oakley, B.**, On the neurotrophic support of sensory receptor cells, in *Olfaction and Taste X*, Doving, K. B., Ed., GCS Press, Oslo, 1990, 186.
35. **Cheal, M. L., Dickey, W. P., Jones, L. B., and Oakley, B.**, Taste fiber responses during reinnervation of fungiform papillae, *J. Comp. Neurol.*, 172, 627, 1977.
36. **Oakley, B.**, Altered temperature and taste responses from cross-regenerated sensory nerves in the rat's tongue, *J. Physiol.*, 188, 353, 1967.
37. **Nejad, M. S. and Beidler, L. M.**, Taste responses of the cross-regenerated greater superficial petrosal and chorda tympani nerves of the rat, in Olfaction *Annals New York Acad. Sci.*, Vol. 510, Roper, S. D. and Atema, J., Eds., N.Y. Acad. Sci., New York, 1987, 523.
38. **Oakley, B.**, Receptive fields of cat taste fibers, *Chem. Senses Flavor*, 2, 52, 1975.
39. **Kinnamon, J. C., Sherman, T. A., and Roper, S. D.**, Ultrastructure of mouse vallate taste buds. III. Patterns of synaptic connectivity, *J. Comp. Neurol.*, 270, 1, 1988.
40. **Zahm, D. S. and Munger, B. L.**, Fetal development of primate chemosensory corpuscles. I. Synaptic relationships in late gestation, *J. Comp. Neurol.*, 213, 146, 1983.
41. **Miller, R. L. and Chaudhry, A. P.**, An ultrastructural study on the development of vallate taste buds of the golden Syrian hamster, *Acta Anat.*, 95, 190, 1976.
42. **Takeda, M., Shishido, Y., Kitao, K., and Suzuki, Y.**, Biogenic monoamines in developing taste buds of mouse circumvallate papillae, *Arch. Histol. Jpn.*, 44, 485, 1981.

43. **Finger, T. E.,** Peptide immunohistochemistry demonstrates multiple classes of perigemmal nerve fibers in the circumvallate papilla of the rat, *Chem. Senses*, 11, 135, 1986.
44. **Silverman, J. D. and Kruger, L.,** Calcitonin-gene-related-peptide-immunoreactive innervation of the rat head with emphasis on specialized sensory structures, *J. Comp. Neurol.*, 280, 303, 1989.
45. **Luts, A., Montavon, P., Lindstrand, K., and Sundler, F.,** Peptide containing nerve fibers in the circumvallate papillae, *Regulatory Peptides*, 27, 209, 1990.
46. **Yamasaki, H. and Tohyama, H.,** Ontogeny of substance P-like immunoreactive fibers in the taste buds and their surrounding epithelium of the circumvallate papilla of the rat. II. Electron microscopic analysis, *J. Comp. Neurol.*, 241, 4934, 1985.
47. **Yamasaki, H., Kubota, Y., Takagi, H., and Tohyama, M.,** Immunoelectron-microscopic study of the fine structure of substance-P-containing fibers in taste buds of the rat, *J. Comp. Neurol.*, 227, 380, 1984.
48. **Kinnman, E. and Aldskogius, H.,** The role of substance P and calcitonin gene-related peptide containing nerve fibers in maintaining fungiform taste buds in the rat after a chronic chorda tympani nerve injury, *Exp. Neurol.*, 113, 85, 1991.
49. **Settembrini, B. P.,** Papilla palatina, nasopalatine duct and taste buds of young and adult rats, *Acta Anat.*, 128, 250, 1987.
50. **Fujimoto, S. and Murray, R. G.,** Fine structure of degeneration and regeneration in denervated rabbit vallate taste buds, *Anat. Rec.*, 168, 393, 1970.
51. **Bradley, R. M., Cheal, M. L., and Kim, Y. H.,** Quantitative analysis of developing epiglottal taste buds in sheep, *J. Anat.*, 130, 25, 1980.
52. **Srivastava, H. C. and Vyas, D. C.,** Postnatal development of rat soft palate, *J. Anat.*, 128, 97, 1979.
53. **Miller, I. J., Jr. and Preslar, A. J.,** Spatial Distribution of rat fungiform papillae, *Anat. Rec.*, 181, 679, 1975.
54. **Rehmer, H.,** Die Entwicklung der Papilla vallata und der Anzahl ihrer Geschmacksknospen beim Goldhamster, (Mesocricetus auratus Waterhouse, 1939), *Anat. Anz.*, 125, 274, 1969.
55. **Hosley, M. A. and Oakley, B.,** Postnatal development of the vallate papillae and taste buds in rats, *Anat. Rec.*, 218, 216, 1987.
56. **Ahpin, P., Ellis, S., Arnott, C., and Kaufman, M. H.,** Prenatal development and innervation of the circumvallate papilla in the mouse, *J. Anat.*, 162, 33, 1989.
57. **State, F. A., El-Eishi, H. I., and Naga, I. A.,** The development of taste buds in the foliate papillae of the albino rat, *Acta Anat.*, 89, 452, 1974.
58. **Farbman, A. I. and Mbiene, J. P.,** Early development and innervation of taste bud-bearing papillae on the rat tongue, *J. Comp. Neurol.*, 304, 172, 1991.
59. **Farbman, A. I.,** Differentiation of foetal rat tongue homografts in the anterior chamber of the eye, *Arch. Oral Biol.*, 16, 51, 1971.
60. **Farbman, A. I.,** Differentiation of lingual filiform papillae of the rat in organ culture, *Arch. Oral Biol.*, 18, 197, 1973.
61. **Oakley, B., Wu, L. H., Lawton, A., and DeSibour, C.,** Neural control of ectopic filiform spines in adult tongue, *Neuroscience*, 36, 831, 1990.
62. **Baratz, R. S. and Farbman, A. I.,** Morphogenesis of rat lingual filiform papillae, *Am. J. Anat.*, 143, 283, 1975.
63. **Billingham, R. E. and Silvers, W. K.,** Studies on the conservation of epidermal specificities of skin and certain mucosas in adult mammals, *J. Exp. Med.*, 125, 429, 1967.
64. **Cheal, M. L. and Oakley, B.,** Regeneration of fungiform taste buds: temporal and spatial characteristics, *J. Comp. Neurol.*, 172, 609, 1977.
65. **Oakley, B.,** Karatin 19-like immunoreactivity in fusiform cells of mammalian taste buds, submitted.
66. **Whitehead, M. C., Beeman, C. S., and Kinsella, B. A.,** Distribution of taste and general sensory nerve endings in fungiform papillae of the hamster, *Am. J. Anat.*, 173, 185, 1985.

67. **Oakley, B.,** Reformation of taste buds by crossed sensory nerves in the rat's tongue, *Acta Physiol. Scand.*, 79, 88, 1970.
68. **Hill, D. L. and Przekop, P. R., Jr.,** Influences of dietary sodium on functional taste receptor development: a sensitive period, *Science*, 241, 1826, 1988.
69. **Thiels, E., Verbalis, J. G., and Stricker, E. M.,** Sodium appetite in lactating rats, *Behav. Neurosci.*, 104, 742, 1990.
70. **Oakley, B.,** The gustatory competence of the lingual epithelium requires neonatal innervation, *Dev. Brain Res.*, in press.
71. **Oakley, B., LaBelle, D. E., Riley, R. A., Wilson, K., and Wu, L. H.,** The rate and locus of development of rat vallate taste buds, *Dev. Brain Res.*, 58, 215, 1991.
72. **Oakley, B.,** Axons do not necessarily compete for targets, in *Olfaction and Taste IX, Annals New York Acad. Sci.*, Vol. 510, Roper, S. D. and Atema, J., Eds., N.Y. Acad. Sci., New York, 1987, 574.
73. **Davies, A. M. and Vogel, K. S.,** Developmental programmes of growth and survival in early sensory neurons, *Philos. Trans. R. Society London*, 331, 259, 1991.
74. **Mistretta, C. M.,** Topographical and histological study of the developing rat tongue palate and taste buds, in *Third Symposium on Oral Sensation and Perception: The Mouth of the Infant*, Bosma, J. F., Ed., Charles C. Thomas, Springfield, IL, 1972, 163.
75. **Miller, I. J., Jr. and Smith, D. V.,** Proliferation of taste buds in the foliate and vallate papillae of postnatal hamsters, *Growth Dev. Aging*, 52, 123, 1988.
76. **Hirata, K. and Nada, O.,** A fluorescence histochemical study of the monoamine-containing cell in the developing frog taste organ, *Histochemistry*, 67, 65, 1980.
77. **Takeda, M., Obara, N., and Suzuki, Y.,** Keratin filaments of epithelial and taste-bud cells in the circumvallate papillae of adult and developing mice, *Cell Tissue Res.*, 260, 41, 1990
78. **Mistretta, C. M., Gurkan, S., and Bradley, R. M.,** Morphology of Chorda tympani fiber receptive fields and proposed neural rearrangements during development, *J. Neurosci.*, 8, 73, 1988.
79. **Liem, R. S. B., van Willigen, J. D., Copray, J. C. V. M., and Ter Horst, G. J.,** Corpuscular bodies in the palate of the rat. I. Morphology and distribution, *Acta Anat.*, 138, 56, 1990.
80. **Belecky, T. L. and Smith, D. V.,** Postnatal development of palatal and laryngeal taste buds in the hamster, *J. Comp. Neurol.*, 293, 646, 1990.
81. **Komatsu, S. and Yamasaki, F.,** Formation of the pits with taste buds at the lingual root in the striped dolphin, *Stenella coeruleoalba, J. Morphol.*, 164, 107, 1980.
82. **Jeppsson, P. H.,** Studies on the structure and innervation of taste buds. An experimental and clinical investigation, *Acta Otolaryngol. Suppl. (Stockholm)*, 259, 1, 1969.
83. **Hume, W. S.,** Stem cells in oral epithelia, in *Stem cells: Their Identification and Characterisation*, Potten, C. S., Ed., Churchill Livingstone, New York, 1973, 233.
84. **Teicher, M. H. and Blass, E. M.,** First suckling response of the newborn albino rat: the roles of olfaction and amniotic fluid, *Science*, 198, 635, 1977.
85. **Hall, W. G. and Bryan, T. E.,** The ontogeny of feeding in fats. IV. Taste development as measured by intake and behavioral responses to oral infusions of sucrose and quinine, *J. Comp. Physiol. Psychol.*, 95, 240, 1981.
86. **Moe, K. E.,** The ontogeny of salt preference in rats, *Dev. Psychobiol.*, 19, 185, 1986.
87. **Bernstein, I. L. and Courtney, L.,** Salt preference in the preweanling rat, *Dev. Psychobiol.*, 20, 443, 1987.
88. **Ganchrow, J. R., Steiner, J. E., and Canetto, S.,** Behavioral displays to gustatory stimuli in newborn rat pups, *Dev. Psychobiol.*, 19, 163, 1986.
89. **Harada, S. and Smith, D. V.,** Gustatory sensitivities of the hamster's soft palate, *Chem. Senses*, 17, 37, 1992.

90. **Nagai, T., Mistretta, C. M., and Bradley, R. M.,** Developmental decrease in size of peripheral receptive fields of single chorda tympani nerve fibers and relation to increasing NaCl taste sensitivity, *J. Neurosci.*, 8, 64, 1988.
91. **Oakley, B.,** Neuronal-epithelial interaction in mammalian gustatory epithelium, in *Regeneration of Vertebrate Sensory Receptor Cells,* Vol. 160, Whelan, J., Ed., J. Wiley & Sons, Chichester (Ciba Foundation Symposium), 1991, 277.

Chapter 5

THE DEVELOPING GUSTATORY SYSTEM: FUNCTIONAL, MORPHOLOGICAL AND BEHAVIORAL PERSPECTIVES

Robert E. Stewart and David L. Hill

TABLE OF CONTENTS

0-8493-5341-6/93/$0.00 + $.50

I. INTRODUCTION

During the past decade, the development of the taste system has become the focus of increasing scientific attention. Numerous investigations, primarily in sheep and rat, have described the functional and anatomical maturation of the gustatory system at several levels of the neuraxis. For example, detailed neurophysiological characterization of single primary taste axon properties during development has revealed that major age-related changes in taste function occur in the peripheral taste system. Recordings of central taste neuron activity indicate that developmental modifications take place in brainstem structures as well. More recently, anatomical studies of central gustatory pathway development have become available. The findings from these studies extend existing knowledge from studies of functional development. The goal of this chapter is to summarize our current understanding of normal peripheral and central taste development. In addition, we will briefly describe what is known about the development of taste-guided behaviors. For convenience, we present these findings in separate sections for "sweet", "sour", "bitter", and "salty" stimuli. For the relevance of these stimulus classes to coding, the reader is referred to Chapters 12 and 13.

Although there have been efforts to characterize gustatory responses to "sweet", "sour", and "bitter" stimuli, the majority of work on taste system development has focused upon salt taste. Detailed studies of changes in taste system responses to salt taste stimuli show that several factors may combine to foster maturation of taste function, both peripherally and centrally. The gradual *morphological* maturation of taste buds and their afferent neuronal elements (see Chapter 4) might account for changes in taste system function during development. However, as we will describe in this chapter, alterations in receptor cell membrane components and cellular biochemical characteristics also may occur during development. These changes may reflect the action of humoral agents upon the nascent taste system. Similarly, we will show that development of central taste system response properties, especially with regard to salts, cannot be predicted solely by the maturation of peripheral taste function. Instead, at each level of the developing gustatory system, unique and apparently independent processes occur which result in the expression of mature taste function.

II. DEVELOPMENT OF PERIPHERAL TASTE FUNCTION

The functional development of the mammalian peripheral taste system, evaluated by examination of taste fiber neurophysiological response characteristics, has been the subject of increasing scientific interest for more than a decade. The goal of this section is to present the findings of such neurophysiological investigations. In comparison to the literature which exists regarding the morphological development of the peripheral taste system (see Chapter 4), information regarding the *functional* development of the peripheral taste system is relatively limited. Nonetheless, studies that have appeared reveal significant age-related alterations in peripheral taste system response properties. While a detailed review of peripheral morphological development is beyond the scope of this chapter, the reader is urged to bear in mind the numerous, temporally intertwined events which characterize the appearance and maturation of taste buds and their afferent neural elements. These events must have important consequences for changing peripheral taste function. In addition, alterations in peripheral taste function may reflect age-related changes in the expression and regulation of taste stimulus transduction (see Chapter 6). Finally, as will be seen in a later section, changes in peripheral taste function during development can be used to predict changes in central taste system neurophysiology and in emerging taste-related behaviors. Indeed, the development of peripheral taste function serves as a valuable conceptual reference for understanding developmental changes in taste system function at higher levels of the neuraxis.

A. BITTER TASTE

To date, no reports have appeared in the literature which rigorously examine the development of peripheral taste responses to bitter, typically alkaloid, taste stimuli in mammals. In addition, existing findings are conflicting. In rat, whole-nerve chorda tympani responses to quinine hydrochloride exhibit an apparent age-related decrease in magnitude.[1] At 3 weeks of age, rat pups show significantly greater relative responses to concentrations of quinine above 1 m*M* when compared with responses in 12-week-old adults. In addition, the slope of the response-concentration function for quinine in 3-week-old rats appears to be considerably steeper at stimulus concentrations above 0.1 m*M*. The meaning of these findings, however, is not entirely clear due to the unusual method used to express the neural data in this study. Specifically, summated whole nerve responses to quinine in both preweanling and adult rats were expressed as a ratio of *adult responses* to a 0.1 *M* NaCl *standard stimulus solution*. Because these workers also reported an age-dependent rise in chorda tympani sodium sensitivity, the apparent decrease in quinine sensitivity might represent an unchanging quinine response relative to an increasing NaCl reference response. Finally, the adult chorda tympani nerve in rats shows a relatively low sensitivity to quinine;[2,3] so discriminating age-related changes in chorda tympani sensitivity may be difficult against an already low signal to noise ratio. Data from other species also tend to refute the finding of decreasing chorda tympani sensitivity to quinine.

While the results above suggest that the rat chorda tympani nerve becomes less sensitive to quinine with increasing age, others[4] have recently shown that mouse chorda tympani nerve responses to quinine do not alter during development. When expressed relative to a 0.1 *M* NH_4Cl reference response, mouse chorda tympani responses to quinine are comparable between 7- to 10-d-old and 8 to 16-week-old animals. Similar findings have been reported for quinine-elicited responses in the chorda tympani nerve of hamster at 14 to 20, 25 to 35, and 55 to 73 d of age.[5] When expressed relative to the response to 0.1 *M* NH_4Cl, hamster chorda tympani responses to quinine remain stable with advancing age. Thus, the lack of consonance between these results and those of Yamada[1] most probably relates to the nature of the reference stimulus during development. That is, a changing reference stimulus (i.e., *increasing* NaCl sensitivity) can invoke an apparent fluctuation in peripheral taste system sensitivity to other stimuli. Although the developmental nature of NH_4Cl responses in mouse and hamster is not established, the rat chorda tympani does not alter in sensitivity to this taste stimulus.[6]

In contrast to these findings, the mouse *glossopharyngeal nerve* becomes more sensitive to quinine during postnatal development.[4] The magnitude of responses to quinine (expressed relative to a 0.1 *M* NH_4Cl reference response) in the glossopharyngeal nerve of adult mice is significantly greater than that in the chorda tympani. In preweanling mice, however, relative response magnitudes for quinine are similar in the chorda tympani and the glossopharyngeal nerves. Therefore, there appears to be an age-related increase in glossopharyngeal nerve sensitivity to quinine. Such an increase could be due to several factors.

An obvious explanation for a developmental increase in mouse glossopharyngeal nerve sensitivity to quinine is the postnatal appearance of circumvallate taste buds.[7] That is, the sensitivity of the glossopharyngeal nerve to quinine may reflect the number of taste buds which contribute to the whole nerve response. Finally, changes in glossopharyngeal nerve sensitivity to quinine might be due to changes in the composition of taste cell membranes and/or their properties, possibly including the membrane insertion or activation of quinine-sensitive transduction components. Determination of this last possibility might be best accomplished with more detailed studies of glossopharyngeal responses to quinine during development. The high sensitivity to quinine of the adult glossopharyngeal nerve in several species, such as rat and hamster,[2,3,8] makes it an attractive system for studying developmental trends in bitter taste function.

B. SWEET TASTE

Although there is a dearth of knowledge about the ontogeny of peripheral sweet taste function in mammals, one report has appeared in which developmental changes in sweet taste were investigated. Hill[5] examined the development of chorda tympani nerve taste responses to several mono- and disaccharide sugars in preweanling (14 to 20-d-old), recently weaned (25- to 35-d-old), and adult (55- to 73-d-old) hamsters. The hamster provides an especially sensitive model system for the examination of taste responses to sugars, as hamster chorda tympani nerve responses to sugars are considerably more robust than are those in the rat chorda tympani nerve.[9,10] Remarkable age- and stimulus-dependent increases in whole chorda tympani nerve sensitivity to sugars occur in the hamster.[5]

When expressed relative to a 0.1 *M* NH_4Cl reference response, hamster whole chorda tympani nerve responses to the **monosaccharide** sugars, glucose and fructose, increase with advancing age. Specifically, steady-state responses to these sugars at all but the lowest stimulus concentrations are significantly larger in adult hamsters as compared with preweanling animals. Hamsters between 25 and 35 d of age, however, exhibit chorda tympani responses to monosaccharide sugars which are intermediate to, though not significantly different from, preweanling and adult responses (Figures 1A and 1B). A significantly different pattern of response development to disaccharide sugars occurs in hamster.

While hamster chorda tympani responses to glucose and fructose are similar to each other, yet increase in a graded fashion between suckling and adulthood, steady-state responses to disaccharide sugars appear in a more abrupt manner later in postnatal development. Relative chorda tympani responses to the **disaccharide** sugars, maltose and sucrose, are much greater in adult hamsters than in preweanling or recently weaned animals. In preweanling and recently-weaned hamsters, chorda tympani responses to maltose and sucrose are similar in magnitude (Figure 1C and 1D). Therefore, while recently-weaned hamsters express responses to monosaccharides which are intermediate to those of preweanling and adult hamsters, their responses to disaccharides are similar in magnitude to those of preweanling animals. A developmental pattern of chorda tympani responses to saccharin similar to that described for responses to disaccharide sugars is also observed (Figure 1E). These results suggest strongly that mono- and disaccharide sugar response properties in hamster chorda tympani mature at different rates. Specifically, responses to monosaccharide sugars are adult-like earlier than are responses to disaccharide sugars. Given evidence for the existence of specific membrane receptor sites[11-13] and receptor-linked second messenger transduction processes[14,15] for "sweet" taste stimuli, these developmental neurophysiological changes may reflect alterations in receptor molecule affinity, increases in receptor density, enhancement of intracellular messenger systems, and/or age-dependent induction of totally separate receptor types.

It is notable that by 25 d of age, hamster fungiform papilla taste buds have essentially achieved adult form.[16] Therefore, it is unlikely that morphological immaturity (e.g., lack of afferent innervation, lack of taste pore, incomplete taste cell complement) can account for the late onset of adult-like, disaccharide-stimulated taste responses. Instead, these changes imply that processes involving the maturation of transduction elements in taste cells are important in the emergence of peripheral taste system sensitivity to "sweet" stimuli. These processes apparently occur *subsequent* to taste cell morphological maturation.

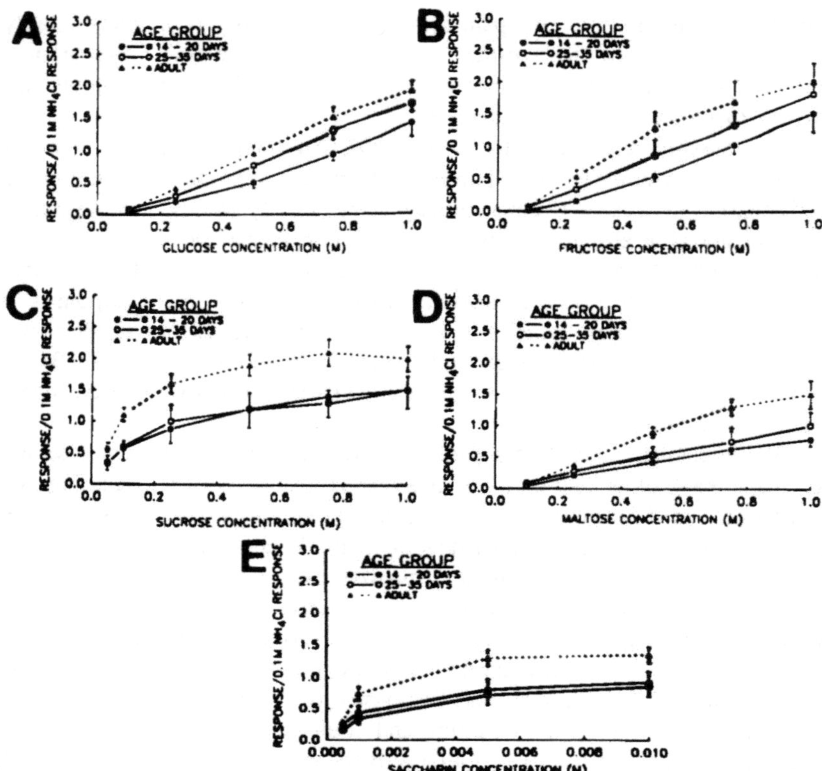

FIGURE 1. Sugar response-concentration functions of hamster chorda tympani nerve during development. Relative response magnitudes for monosaccharide sugars glucose (A) and fructose (B) in preweanling hamsters (14 to 20 d old) are significantly lower vs. adult responses. Although not significantly different in magnitude from either preweanling or adult responses, monosaccharide responses observed in 25- to 35-day-old hamster chorda tympani are segregated from preweanling and adult values. In contrast, chorda tympani responses to disaccharide sugars, sucrose (C), and maltose (D) are similar in magnitude in preweanling and recently weaned hamsters. These responses are significantly smaller than responses seen in adults. The pattern of response development to the nonsugar sweetener, saccharin (E), is similar to that observed for disaccharides. Together, these results suggest that a differential rate of development occurs for "sweet" taste responses, depending upon the nature of the stimulus. (From Hill, D. L., *J. Comp. Neurol.*, 268, 346, 1988. With permission.)

Others have examined, in considerably less detail, the development of peripheral taste responses to sweet stimuli in rats and mice. Yamada[1] failed to find any significant difference in relative magnitudes of sucrose-stimulated chorda tympani nerve responses in 3- vs. 12-week-old rats. Again, these findings are unclear given the method of data expression. Age-related alterations in chorda tympani or glossopharyngeal nerve responses to sucrose are not seen in mice aged 7 to 10 d vs. 12 to 16 weeks of age.[4] The apparent lack of a significant developmental change in rat chorda tympani nerve sensitivity to sucrose could relate to the low amplitude of multifiber responses to sweet taste stimuli observed in that species.[5] Emergent information suggests that a large proportion of sweet taste information may be transmitted via the greater superficial petrosal branch of the facial nerve, which innervates taste buds located on the palate and in the nasoincisor ducts.[17-19] It may be that developmental changes in sweet taste sensitivity could be detected in the greater superficial petrosal nerve simply by virtue of its greater sensitivity in to sweet stimuli. Indeed, further information about processes involved in the regulation of sweet taste development may become available only when appropriate model taste systems (e.g., hamster or gerbil chorda tympani nerve or rat greater superficial petrosal nerve) are used in developmental investigations.

It is abundantly clear that investigations of bitter and sweet taste development have been largely neglected by taste physiologists, perhaps because of the lack of robust sweet taste responses in the rat chorda tympani. However, this exciting area of taste research might be successfully pursued in other species. Insight about the mechanisms which underlie "sweet taste" in the mature, adult gustatory system could be forthcoming from such investigations. Significant strides in this regard have been made in studies of salt, specifically sodium, taste development.

C. SALT TASTE

Major understanding of systems-level alterations in salt taste function which occur during development has been derived from studies conducted in sheep and in rat. In an initial report nearly 20 years ago, Bradley and Mistretta[20] showed that the fetal sheep chorda tympani was responsive to a variety of taste stimuli. Subsequently, it has been shown that dramatic, progressive increases in the magnitude of sheep whole chorda tympani responses to LiCl and NaCl occur during development.[21] Specifically, when expressed relative to a 0.5 *M* NH_4Cl reference response, chorda tympani responses to NaCl and LiCl in young fetal sheep (at 110 and 130 d gestation) are much lower in magnitude compared with responses to NH_4Cl and KCl. In addition, relative response magnitudes to NaCl and LiCl in the young fetus chorda tympani are significantly smaller vs. NaCl and LiCl chorda tympani responses in perinatal lambs (term, about 147 d), in 30- to 60-d-old lambs, and in adult sheep. Chorda tympani responses to NaCl and LiCl recorded from young lambs (30 to 60 d of age) are significantly lower in relative magnitude as compared with responses from adult sheep. In contrast, with advancing age, KCl becomes a less effective stimulus relative to NaCl and LiCl, but remains stable relative to NH_4Cl (Figure 2). Together, these results indicate a dramatic shift in rank order effectiveness of salt taste stimuli in the developing ovine chorda tympani, due primarily to an age-dependent increase in chorda tympani sensitivity to NaCl and LiCl (see legend to Figure 2).

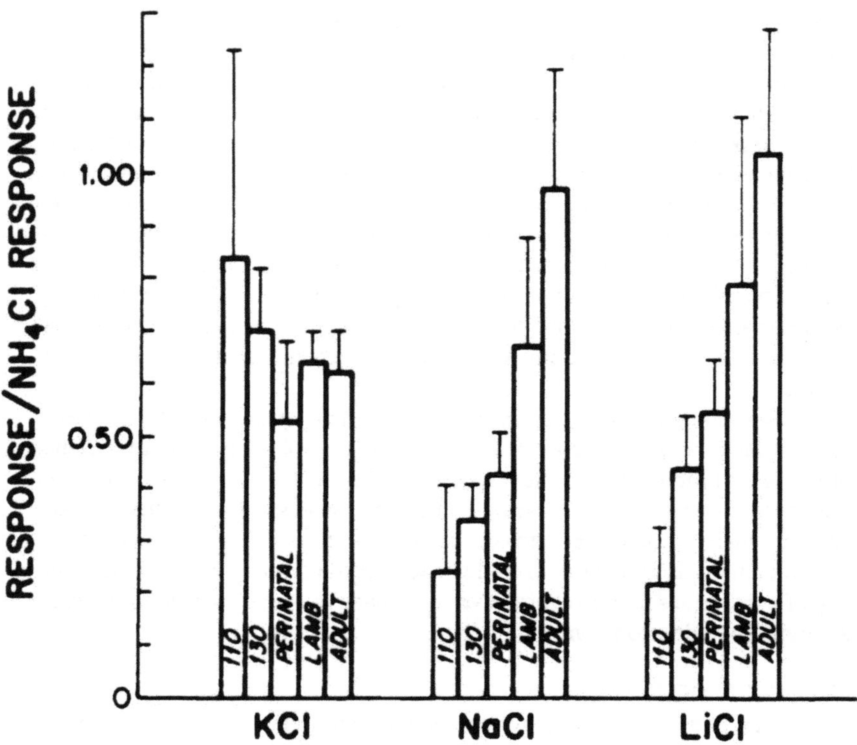

FIGURE 2. Multiunit sheep chorda tympani response ratios (mean ± SD) to several salt stimuli during development. It is apparent that relative responses to KCl do not change appreciably between gestation and adulthood. In contrast, an orderly and progressive increase in relative response magnitude to NaCl and LiCl is observed between gestation and adulthood. These changes represent a striking rearrangement in the rank order effectiveness of salt stimuli across these ages: in fetuses at about 110 d of gestation, the rank order is $NH_4Cl > KCl >> NaCl = LiCl$; in adult sheep the rank order is $LiCl = NaCl = NH_4Cl > KCl$. (From Mistretta, C. M. and Bradley, R. M., *J. Comp. Neurol.*, 215, 199, 1983. With permission.)

Single chorda tympani fiber analysis supports and clarifies the results described above. The age-related increase in NaCl sensitivity can be accounted for by an increase in the proportion of chorda tympani fibers which respond maximally to NaCl.[22] At the same time, the mean response frequency to NaCl and NH_4Cl among groups of fibers most sensitive to these stimuli does not change significantly. That is, the mean response frequencies of "NaCl-best" and "NH_4Cl-best" fibers does not change with age, but the number of "NaCl-best" fibers in the chorda tympani increases significantly between late gestation and adulthood. (We term these fibers as "NaCl-best" and "NH_4Cl-best", without attributing any coding connotation to this categorization; see Chapter 12 for further discussion on this issue.) Notably, mean response frequencies to KCl decrease significantly with advancing age. Mistretta and Bradley[21] attribute increases in peripheral chorda tympani sensitivity to NaCl and LiCl to age-related changes in the proportion of taste cells which possess appropriate transduction elements in their apical membranes, namely amiloride-sensitive Na^+ channels.[22] In an attempt to establish the generalizability of developmental alterations in salt taste function in other taste nerves, Mistretta and Bradley[23] recorded taste responses from the glossopharyngeal nerve of fetal, early postnatal, and adult sheep. Unlike the chorda tympani, whole glossopharyngeal nerve responses to salt stimuli, expressed relative to the 0.5 *M* NH_4Cl response, are similar across ages. A similar finding in the mouse glossopharyngeal nerve was described by Ninomiya and co-workers.[4] Therefore, processes which regulate the development of salt responses in the glossopharyngeal nerve are different from those which occur during chorda tympani development. This disparity might be predicted, since recent data shows that the glossopharyngeal nerve does not exhibit amiloride-sensitivity.[24] Therefore, an essential difference in transduction mechanisms in taste receptor cells innervated by the glossopharyngeal and chorda tympani nerves could explain differences in sodium response development observed in these nerves.

Remarkably, highly similar developmental changes in chorda tympani responsiveness to salts have been observed in the rat and mouse. Yamada,[1] Hill and Almli,[25] and Ferrell and co-workers[26] have demonstrated that rat chorda tympani responses to NaCl and LiCl increase relative to NH_4Cl during development. A significant, gradual increase in chorda tympani sensitivity to NaCl and LiCl occurs roughly between 15 and 45 d of age, when responses to NaCl and LiCl exhibit adult-like magnitudes (Figure 3). Potassium chloride stimuli also appear to increase slightly, though not significantly, in effectiveness relative to NH_4Cl.[26] Subsequent single chorda tympani fiber analyses established the neural basis for these developmental changes in chorda tympani salt responses.

The prominent age-related changes in chorda tympani responses to NaCl and LiCl relate primarily to differential changes in individual fiber sensitivity to these salts.[6] Specifically, in 14 to 20-d-old rats, nearly 90% of all single chorda tympani fibers respond to 0.1 *M* and 0.5 *M* solutions of NaCl, LiCl, KCl, and NH_4Cl. However, mean response frequencies attending stimulation with NaCl and LiCl increase nearly twofold between 14 to 20 d of age and adulthood (Figure 4A). No significant age-dependent change in NH_4Cl-elicited mean response frequencies is observed (Figure 4B). Furthermore, mean response frequencies elicited by 0.5 *M* but not 0.1 *M* KCl increase significantly during development. Finally, when chorda tympani fibers are classified according to best-stimulus characteristics, the number of fibers which best-respond to NaCl and LiCl increases greatly between 14 to 20 days of age and adulthood. Just the opposite is true for NH_4Cl-best fibers; their number decreases significantly between 14 to 20 d of age and adulthood. Taken together, these findings suggest that an orderly change in taste cell properties must

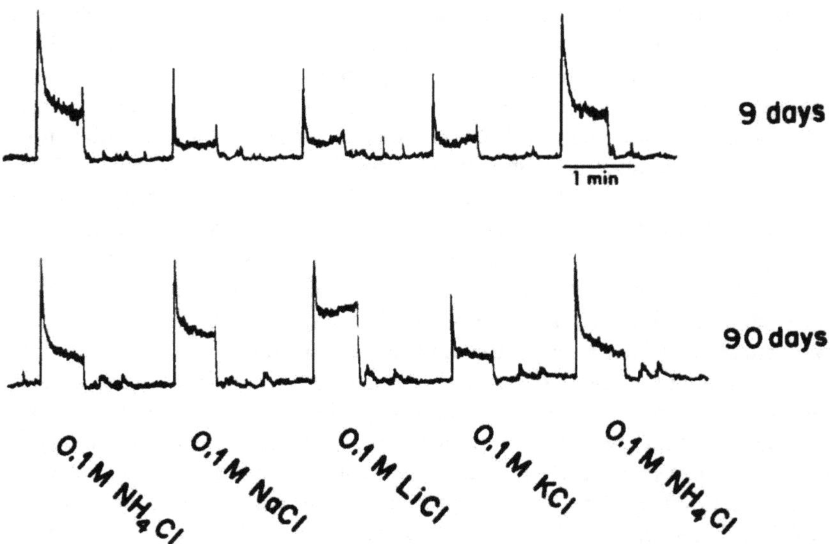

FIGURE 3. Integrated whole chorda tympani nerve response records from rats aged 9 and 90 d. Relative to a 0.1 *M* NH_4Cl reference response, NaCl and LiCl elicit small magnitude responses early in development. By adulthood, however, NaCl and LiCl elicit greater responses than NH_4Cl (From Hill, D. L., et al., *J. Neurosci.*, 2, 782, 1982. With permission.)

occur to some, but not to all, salt stimuli. While developmental processes which promote taste bud and/or afferent neuronal maturation could contribute in part for these observed changes, other evidence strongly suggests that specific alterations in receptor cell membrane components are responsible for changes in peripheral taste responses to NaCl and LiCl.

Hill and Bour[27] demonstrated that increases in rat chorda tympani sensitivity to NaCl and LiCl occur in parallel with the initial appearance of chorda tympani sensitivity to the epithelial Na^+ channel blocker, amiloride. Chorda tympani responses of rat pups aged 12 to 13 d are insensitive to the inhibitory influence of lingually applied, 100 μM amiloride. In contrast, rats aged 29 to 31 and 90 to 110 d show amiloride suppression of NaCl and LiCl responses, the extent of which is proportional to the degree of chorda tympani sensitivity to these stimuli at these ages (Figure 5). It is currently thought that amiloride-sensitive sodium channels are the major, if not only, sodium taste transduction pathway in the mammalian taste system.[28-31] Hill and Bour[27] concluded that the graded increase in NaCl and LiCl responses during development is due to an increase in the expression of functional, amiloride-sensitive sodium channels in taste cell membranes. An extraordinarily similar developmental progression in the ontogeny of amiloride-sensitive sodium responses was observed by Ninomiya and colleagues[4] in the chorda tympani of mice. In addition to these neurophysiological data, recent, unpublished immunohistochemical evidence from our laboratory suggests that amiloride-sensitive

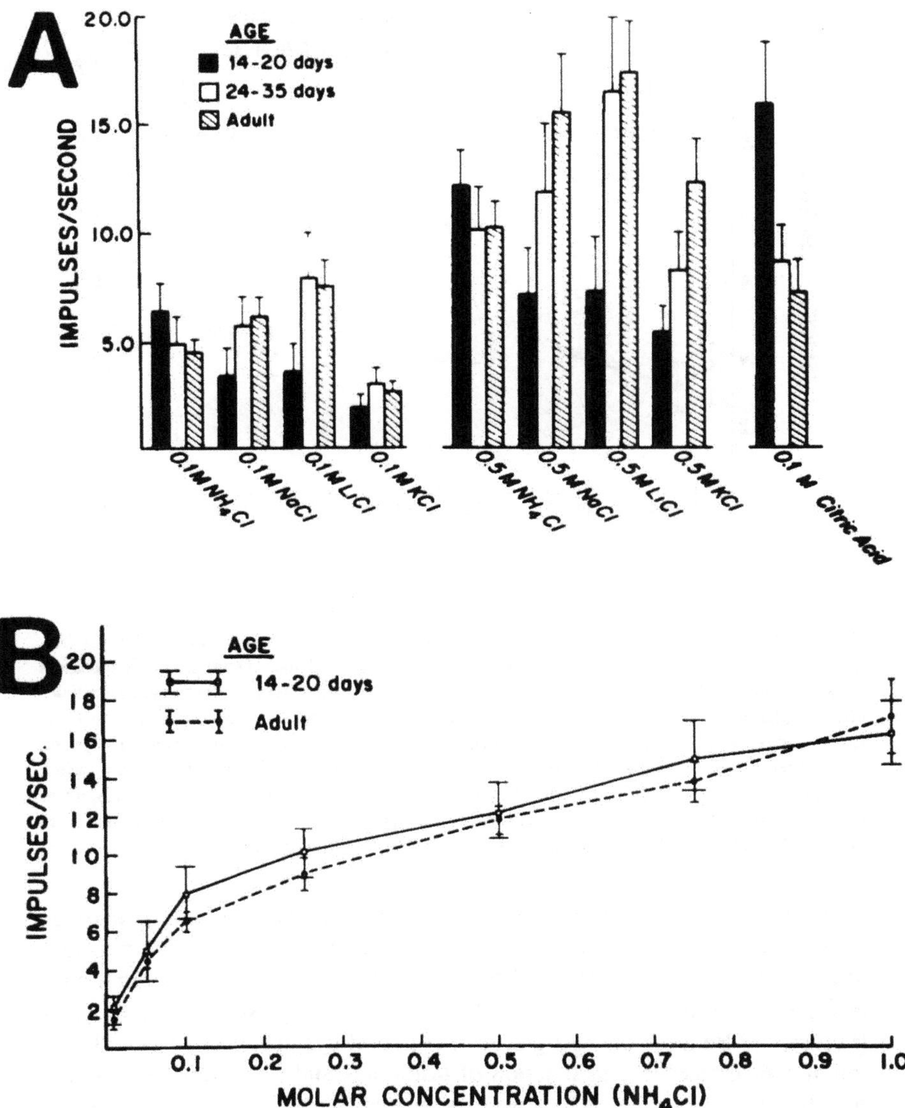

FIGURE 4. (A) Response frequencies (mean ± SEM) of individual chorda tympani fibers in rat during development. Mean response frequencies attending stimulation with NaCl and LiCl increase with advancing age, while no age-dependent change in NH_4Cl-elicited response frequency is observed (see also 4B). Mean response frequencies to 0.1 *M* KCl do not change during development, although 0.5 *M* KCl-stimulated response frequencies increase between 14 to 20 days of age and adulthood. (B) Ammonium chloride response-concentration functions for single chorda tympani fibers in young and adult rats. (From Hill, D. L., et al., *J. Neurosci.*, 2, 782, 1982. With permission.)

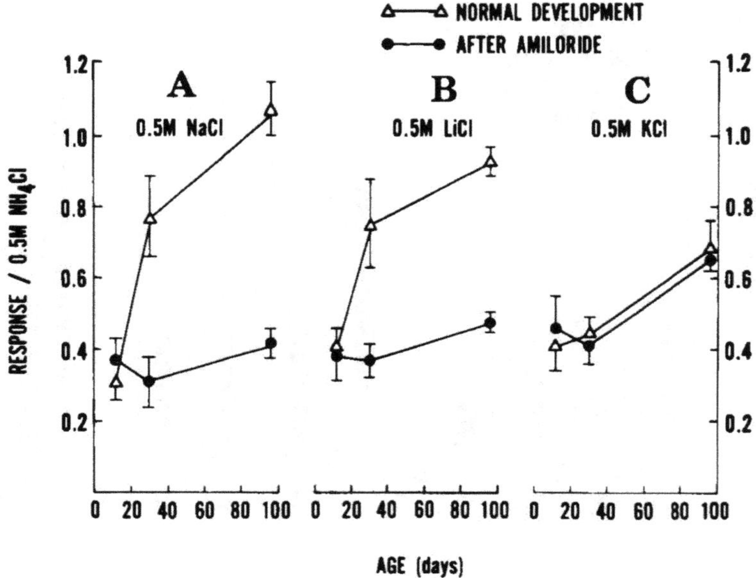

FIGURE 5. Mean chorda tympani nerve relative responses to 0.5 *M* solutions of NaCl (A), LiCl (B), and KCl (C) from rats between 12 and 100 days of age. Open triangles represent responses observed during development. Closed circles represent relative response magnitudes after lingual pretreatment with 500 μ*M* amiloride. It is apparent that as responses to NaCl and LiCl increase with age, there is a concomitant increase in the extent of amiloride suppression of these responses. In contrast, KCl responses are insensitive to the inhibitory effects of amiloride. This proportional increase in the inhibitory potency of amiloride on NaCl and LiCl responses reflects a progressive addition of amiloride-sensitive channels to receptor cell membranes. (From Hill, D. L. and Bour, T. C., *Dev. Brain Res.*, 20, 310, 1985. With permission.)

channels are present within taste cells of rat fungiform taste buds as early as the day after birth, long before the chorda tympani nerve exhibits amiloride-sensitivity (Figures 6A, 6B, and 6C). The pattern of channel-like immunoreactivity is similar to that described previously by Simon and his colleagues.[32,33] Because of the significant temporal dysjunction between (1) the initial appearance[34,36] and function of taste buds and (2) initial taste system sensitivity to amiloride, it has been speculated that some endo- or paracrine events might act postnatally to activate quiescent sodium transduction pathways in the developing rat peripheral taste system. This idea was first proposed by Hill[37,38] and later by Przekop et al.[39] based upon studies which examined the influence of maternal dietary Na^+ upon the development of the rat peripheral gustatory system.

Based upon the success of studies examining the role of environment in shaping developing sensory systems[40-43] a similar approach was used in studies of the developing gustatory system. Hill[37] manipulated dietary sodium levels

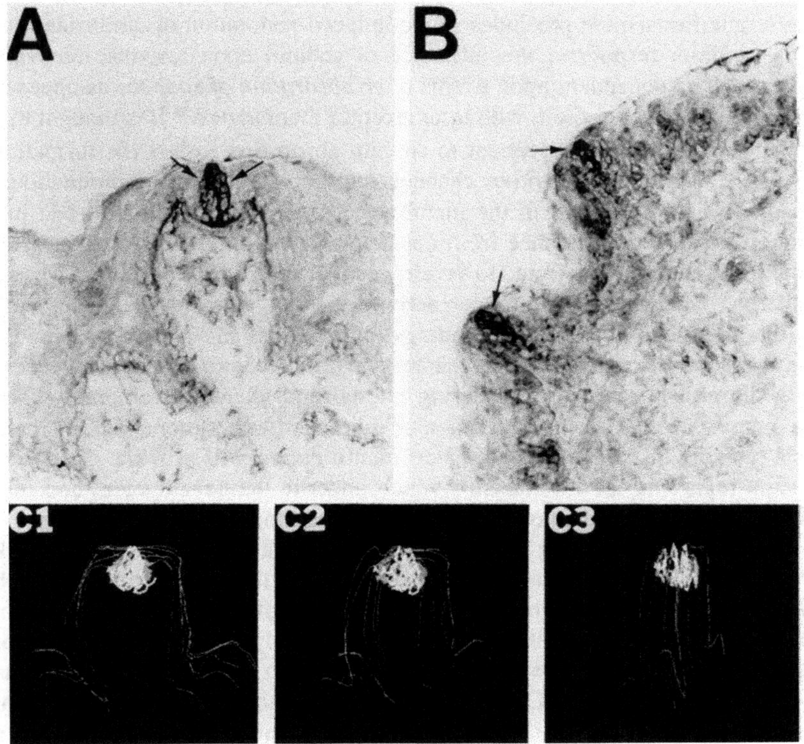

FIGURE 6. Visualization of amiloride-sensitive sodium channel-like immunoreactivity in taste buds (arrows) of adult (A) and 1-d-old (B) rat fungiform papillae. A polyclonal antibody (approx. 2 μg protein/ml) raised against purified bovine renal epithelial sodium channel protein was visualized indirectly with an avidin-biotin-peroxidase method modified from Reference 33. The pattern of staining indicates that amiloride-sensitive channels are located along apical *and* basolateral domains of receptor cells within the central core of taste buds. Less intense staining of cells within the lamina propria and along the superior aspect of the epidermis is also noted. (C1) to (C3): Three-dimensional reconstructions of the distribution of sodium channel immunoreactivity in an adult rat taste bud. This reconstruction shows extension by receptor cells of tapered processes toward the apex (shown in white) of the taste bud (shown in gray). In contrast to the spindle-like organization of labeled cells in the adult rat taste bud, labeled 1-d-old rat taste receptor cells are distributed in a more globular, less organized manner, reflective of the degree of morphological maturation present at this age.

starting early in gestation. Restriction of maternal dietary Na^+ (0.03% dietary NaCl) beginning on or before gestational day 8, then continued during suckling and after weaning, results in profoundly and specifically reduced chorda tympani sensitivity to sodium salts at adulthood. These reduced responses are due to a lack of functional amiloride-sensitive NA^+ channels.[38] Remarkably, these channels appear to be highly labile, since amiloride-sensitivity can be restored by repletion of dietary sodium[37] or by limited access to isotonic saline.[39] Because pretreatment of sodium-restricted animals with the diuretic/

natriuretic furosemide precludes saline-induced restoration of amiloride-sensitive sodium responses, the influence of sodium upon response recovery appears to be dependent upon events ***after absorption of sodium***, as opposed to direct contact of sodium with taste receptor membranes.[39] It is thought that some humoral event consequent to sodium absorption fosters the formation of nascent amiloride-sensitive channels or the activation of nonfunctional channels already present in the membrane. Again, recent, unpublished immunohistochemical evidence from our laboratory supports the latter idea, as young and adult rats subjected to developmental Na^+ restriction exhibit normal patterns of channel-like immunoreactivity in taste epithelia. Together with results described above, these findings indicate that some humoral event(s) may precede and/or direct the appearance of taste receptor cell responsiveness to sodium taste stimuli, both during normal and altered development. The factors responsible for the activation of quiescent taste system sodium-channels remains to be determined. Their identification will provide significant insight regarding the mechanisms which underlie the development and regulation of taste transduction pathways by circulating or paracrine factors.

While such activating events may be occurring during the second week of postnatal life in rats and mice, it appears that they have occurred much earlier in the hamster. In hamster, the magnitude of relative chorda tympani responses to NaCl and LiCl *decreases* significantly between 14 to 20 d of age and adulthood[5] (Figure 7). Interestingly, in 7- to 9-d-old hamsters, chorda tympani responses to NaCl and LiCl are suppressed up to 65% by lingual treatment with amiloride. Since these responses are expressed relative to a 0.1 *M* NH_4Cl reference response, it is not yet clear whether a down-regulation of amiloride-sensitive membrane components can account for this age-dependent change in sodium and lithium sensitivity, or if a developmental change in single unit responsiveness (e.g., an increase) to NH_4Cl might occur. Regardless of which process impacts on the changing hamster gustatory system, it is likely that changes in gustatory sensitivity to NaCl reflect important adaptive mechanisms for the maintenance of fluid and electrolyte homeostasis in this desert-inhabiting rodent.

The results discussed above relate primarily to changes in peripheral taste system neural sensitivity and to alterations in transduction processes which might account for them. In addition to changes in overall peripheral taste sensitivity to various taste stimuli, remarkable age-related modification of taste receptive fields also occurs. By applying discrete electrical stimulation to individual taste papillae in fetal and young lambs and in adult sheep, it has been demonstrated that a decrease in the number of fungiform papillae innervated by individual chorda tympani fibers occurs between 130 d of gestation and about 60 d postnatally.[22] Specifically, late-emerging, smaller receptive fields are correlated with high responsiveness to NaCl, whereas larger receptive fields common in fetuses respond most vigorously to NH_4Cl. These relationships between field size and responsiveness are maintained in

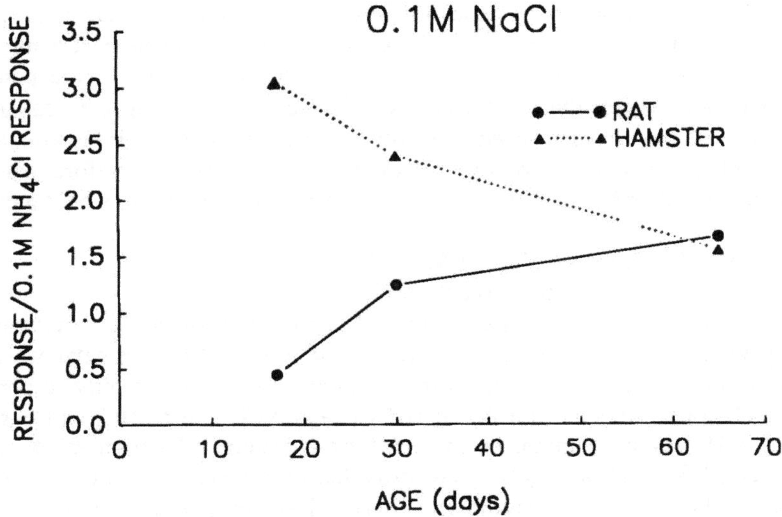

FIGURE 7. Responses to 0.1 *M* NaCl (expressed relative to a 0.1 *M* NH_4Cl reference response) from chorda tympani nerves of hamster (triangles) and rat (circles) during development. Remarkably divergent developmental changes in peripheral sodium sensitivity occur in these species. Hamsters demonstrate a profound age-related *decrease* in sodium sensitivity, while rats exhibit a striking age-related *increase* in sodium-sensitivity. (From Hill, D. L., *J. Comp. Neurol.*, 268, 346, 1988. With permission.)

fetuses, lambs, and adult sheep; however, smaller, highly NaCl-responsive receptive fields are more numerous in older animals. Functional changes in peripheral receptive fields may be related both to increases in the proportion of fibers that respond best to NaCl[22] and also to postnatal decreases in the average number of fungiform papilla taste buds (but not of fungiform papillae).[44] These results intimate that paring back of taste bud number *and/or* chorda tympani fiber peripheral terminal fields occurs during development of the peripheral taste system. Similar processes in other peripheral and central neuronal systems have been observed.[45-47]

III. DEVELOPMENT OF THE CENTRAL GUSTATORY PATHWAY

As is apparent from the preceding section, our understanding of peripheral gustatory development has progressed to where we soon may be able to understand the developmental regulation of specific receptor cell membrane components important in taste transduction. By comparison, our understanding of central gustatory development remains far behind what is known about the peripheral taste system, even though progress has proceeded rapidly during the past decade. The following description of the functional development of central gustatory structures relies on relatively few studies. Indeed, neural responses have been examined developmentally in only two central structures, the nucleus of the solitary tract and the parabrachial nuclei in the pons. Furthermore, these studies have been conducted in only two species, rat and sheep, and they have focused primarily on salt responses in order to complement earlier studies of peripheral gustatory development. Therefore, much like the peripheral taste system, little is known about other classes of stimuli.

A. FUNCTIONAL DEVELOPMENT

1. Nucleus of the Solitary Tract (NST)

As observed for responses in the rat chorda tympani nerve,[6] multiunit responses from the NST can be recorded within the first postnatal week. Nucleus of solitary tract neurons in rats aged 5 to 7 d consistently respond to high concentrations (0.5 *M*) of NH_4Cl, and KCl, and to HCl and citric acid.[48] However, as might be predicted from patterns of peripheral development, NST neurons in young rats often fail to respond to NaCl (Figure 8). That is, the lack of NST neuron sensitivity to NaCl may simply reflect a lack of functional amiloride-sensitive sodium channels in taste receptor cells. Thus, the neural message is not sufficiently strong to be transmitted across the first central synapse to *all* NST neurons. However, the observation that *many* central neurons in week-old rats are capable of responding to a variety of stimuli indicates that enough of the neural signal may be present to mediate early taste-guided behaviors.

In rats older than 14 d, single NST neurons generally respond to all stimuli, suggesting that peripheral responses are transmitted reliably to the CNS. While NST neurons receive neural information from the anterior tongue before weaning (21 d), their response frequencies to some stimuli continue to change significantly with age.[48] Specifically, average response frequencies to NaCl, LiCl, KCl, sucrose, and sodium saccharin increase between 35 and 50 d, whereas there are no developmental changes in average frequencies to NH_4Cl, quinine, and citric and hydrochloric acids[48] (Figure 8). Therefore, the neural code for many taste stimuli alters throughout postnatal development in the rat, while the code for other stimuli remains constant with age. It is interesting to note that NST neurons increase in their responsiveness with age to stimuli associated with nutritive value. In contrast, it appears as though the central neurons are "set" to respond in a mature way early in development to stimuli often associated with toxicity.[49,50] Such a scheme implies an adaptive correlate to central taste development.

Although many of the maturational processes in the peripheral gustatory system are mirrored in the NST, there must also be maturational processes unique to the first central relay. For example, developmental alterations in NST neural response frequencies parallel changes in chorda tympani fibers in that developmental changes occur to the same stimuli for both neural levels. However, the increases in salt response frequencies of NST neurons occur at

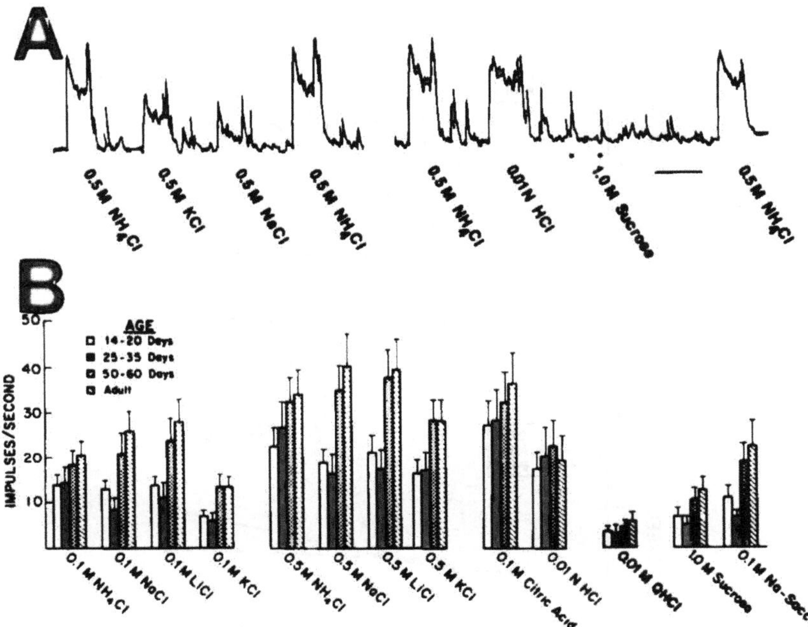

FIGURE 8. Neurophysiological responses from the rat nucleus of the solitary tract (NST). Many, but not all, stimuli elicit multiunit responses in the NST of a rat aged 7 days postnatal (A). Dots mark the time of application and rinse for sucrose. The time bar denotes 1 min. In rats 14 d postnatal and older, the majority of neurons respond to all stimuli, but response frequencies (mean ± SEM) increase to NaCl, LiCl, KCl, sucrose, and sodium saccharin (B). Response frequencies to NH_4Cl, acids, and quinine remain constant during development. (From Hill, D. L., et al., *J. Neurophysiol.*, 50, 879, 1983. With permission.)

later ages compared to chorda tympani fibers (Figure 9). Thus, developmental changes observed in NST responses reflect the developmental processes inherent to the periphery, yet also reflect processes *unique* to the central relay. It is also noteworthy that there is an overall increase in response frequencies in the central taste system compared to the periphery.[48] For each stimulus used in single neuron recordings from the chorda tympani and NST, there is an amplification in the evoked CNS signal at all ages (Figure 9). That is, there is an increase in the signal as it traverses the first synaptic relay. Related to this issue is the specificity of stimuli to which NST neurons respond. Compared to chorda tympani fibers, individual NST neurons respond to a broader array of stimuli at all ages.[48] As described previously, the majority of chorda tympani neurons in young and adult rats respond differentially to only one salt. In contrast, the majority of NST neurons in rats of all ages respond equally well to all salts. Therefore, the developmental change in stimulus specificity noted in the periphery is not apparent in the NST, further

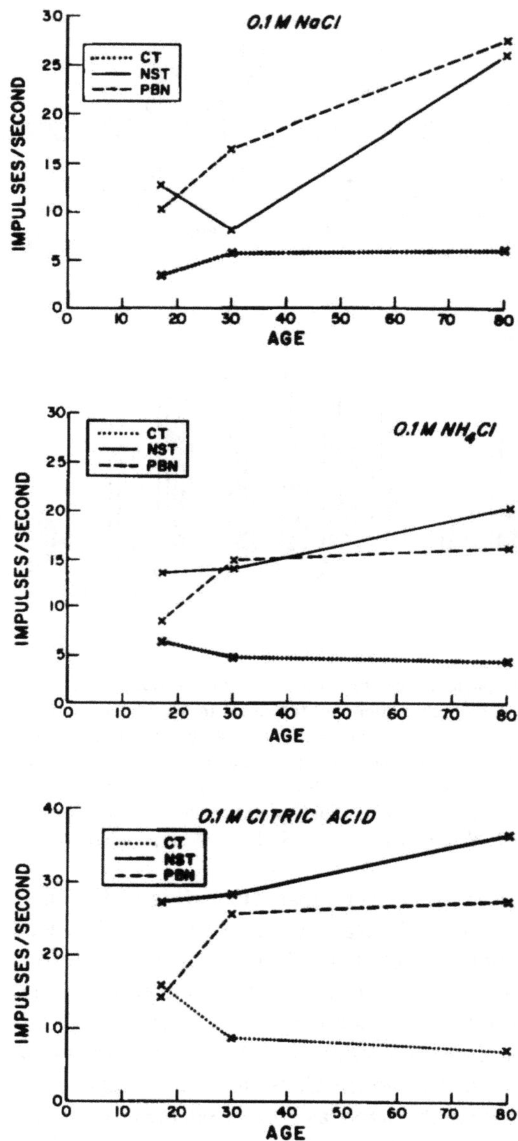

FIGURE 9. Unique developmental changes in responsiveness occur at each of three neural levels to 0.1 *M* NaCl (top), 0.1 *M* NH_4Cl (center), and 0.1 *M* citric acid (bottom). Average response frequencies (impulses/second) are shown for chorda tympani fibers (dotted line), nucleus of the solitary tract neurons (NST; dashed line), and parabrachial nuclei neurons (PBN; dashed lines). Data points are placed at the midpoint of the respective age group. (From Hill, D. L., *J. Neurophysiol.*, 57, 481, 1987. With permission.)

emphasizing distinct developmental processes in the peripheral and central gustatory systems.

Similar to the postnatal development of central taste responses in rat, NST responses in developing sheep alter dramatically to some, but not to all, stimuli.[51,52] For example, NH_4Cl, KCl, and citric acid are the only stimuli to elicit NST responses in sheep fetuses younger than 114 d gestation; no responses occur to NaCl or LiCl, despite the presence of peripheral responses. As fetuses age, responses to NaCl and LiCl emerge and later become some of the most effective stimuli.[51,53] Thus, as in the rat, NST neurons fail to respond to NaCl during periods when the periphery is functional, and they exhibit greater responsiveness to NaCl and LiCl with age.

Findings indicating the "uniqueness" of the NST compared to the peripheral taste system can also be used as evidence that chorda tympani neurons converge onto NST neurons progressively with age. Specifically, the relative lack of stimulus specificity in the NST, the presence of a response magnitude amplification in central neurons, and the relative increase in spontaneous activity levels for NST neurons in both rat and sheep suggests that there is convergence between the chorda tympani and the NST.[48,53] While functional correlates of convergence in the rat NST are not available, a perspective on these processes is available from receptive field studies of fetal and postnatal sheep.[53] Compared to the *decrease* in receptive field size for chorda tympani fibers that occurs with age,[54] receptive field size *increases* in the NST.[53] Moreover, the receptive field size of central neurons is greater than those in the chorda tympani at perinatal and lamb ages, but not in fetuses.[53] Thus, it appears that the greatest amount of functional convergence occurs between fetal and perinatal ages. It should be noted that the period during which central receptive fields change the most corresponds with the period during which there are increases in spontaneous frequencies and increases in responses to some, but not to all, salts. Such stimulus-specific processes have been proposed to amplify weak peripheral input elicited by NaCl in the central gustatory system.[36,53]

2. Parabrachial Nuclei (PBN)

Similar to first and second order taste neurons, third order gustatory neurons in the rat PBN exhibit major changes in response characteristics during development. Specifically, PBN neurons in rats 4 to 7 days postnatal respond to many, but not to all, chemical stimuli.[55] In rats aged 14 d and older, however, the majority of PBN neurons respond to *all* stimuli. Therefore, PBN neurons become responsive to more stimuli between the first and second week of the rat's life, even though many are functional well before the peripheral gustatory system is mature. Again, this demonstrates that neurophysiological taste responses reach central neurons (at least third order neurons) within the first postnatal week, yet the signal becomes refined during a prolonged period of postnatal development.

Comparison of PBN response development with chorda tympani and NST neuron response development indicates that developmental changes in the PBN are unique to this structure.[55] That is, response changes relate to maturational events specific to the PBN, in addition to developmental alterations in lower order taste neuron properties. Generally, the absolute response magnitudes, the pattern of changes in response frequencies to each stimulus, and the lack of response specificity in PBN neurons are similar to those noted in the NST. The major difference is that response frequencies to *all* stimuli increase from 14 to 20 d of age to 25 to 35 d, suggesting that major convergence of NST inputs onto PBN neurons occurs during this period (Figure 9).

B. ANATOMICAL DEVELOPMENT

As noted for the literature on the functional development of central gustatory nuclei, only seminal works on the anatomical development of central gustatory relays were published before this past decade. However, our understanding of the anatomical development of these areas has increased substantially with the recent work of Lasiter and his colleagues. We will focus primarily on their work that describes central developmental processes in the gustatory system of the postnatal rat. The reader should also refer to a recent review by Mistretta[36] concerning embryonic developmental neuroanatomy of central gustatory nuclei.

1. Nucleus of the Solitary Tract

Three cranial nerves innervate taste receptors located in the oropharyngeal cavity (see Oakley, Chapter 4), all of which project to the NST in a rostral to caudal gradient (see Travers, Chapter 13). Of these three recipient zones in the NST, the anterior pole of the nucleus, which receives input from the front of the tongue via axons of the chorda tympani nerve, has been most extensively studied developmentally. Although chorda tympani axons synapse in the NST as early as postnatal day 1, organization of the terminal field is not complete until about postnatal day 25.[56,57] During this period, axons appear to migrate caudally, with the terminal field ultimately occupying twice the volume observed in newborns (Figures 10 and 11). In contrast, axons of the glossopharyngeal nerve, which innervates the posterior tongue, initially enter the NST at about postnatal day 10 and expand within the intermediate NST until as late as postnatal day 45[56] (Figure 11). Interestingly, overlap between the chorda tympani and glossopharyngeal nerve terminal fields begins at about postnatal day 10 and is complete at about postnatal day 20[56] (Figure 11). Therefore, the terminal fields within the NST do not develop simultaneously, but exhibit a rostral-to-caudal axis of maturation.

Concomitant with the development of peripheral nerve terminal fields in the NST, there are dramatic alterations in second order neurons intrinsic to the nucleus.[57] Nucleus of solitary tract neurons in rat are present beginning

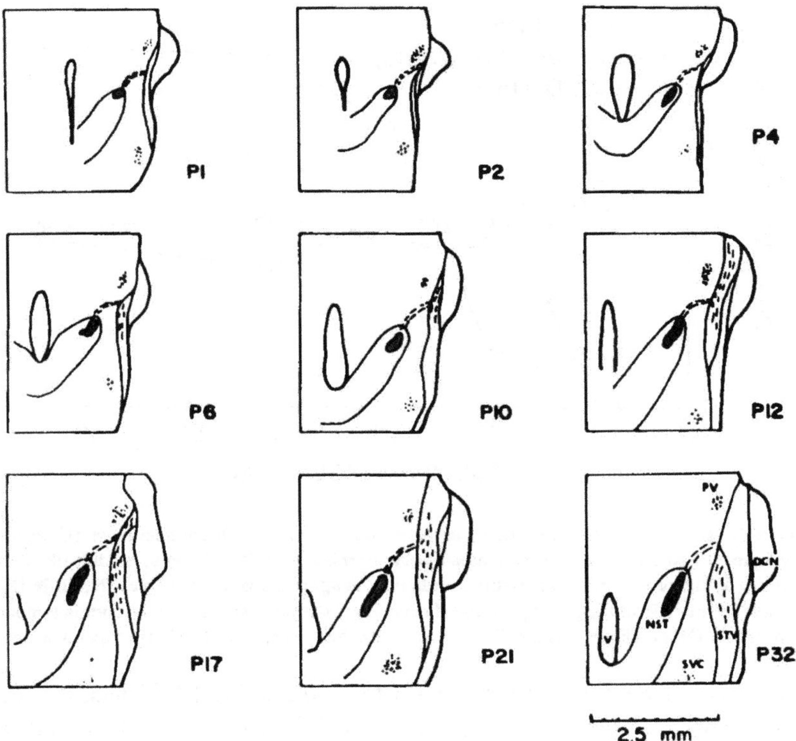

FIGURE 10. Location and extent of HRP-labeled terminal fields in the developing postnatal rat nucleus of the solitary tract (NST). Terminal fields are denoted by filled areas and drawings of the caudal brainstem are shown in horizontal section. The rostrocaudal extent of labeling increases during development, whereas the mediolateral extent of label is similar with age. Abbreviations are DCN, dorsal cochlear nucleus; NST, nucleus of the solitary tract; PN, principle trigeminal nucleus; STV, spinal trigeminal tract; SVC, caudal division of the spinal trigeminal nucleus; and V, fourth ventricle. (From Lasiter, P. S., et al., *Brain Res. Bull.*, 22, 313, 1989. With permission.)

at embryonic day 12, have their peak proliferation period at embryonic day 13, and are complete by embryonic day 16.[58] First order dendrites of fusiform, multipolar, and ovoid neurons, as well as second order fusiform and ovoid dendrites, increase threefold between postnatal day 4 and day 20. Adult-like dendritic lengths for all first order dendrites occur by approximately postnatal day 25, whereas a later period, between postnatal day 30 and 70, is significant for increases in second order dendrites of multipolar neurons. As might be expected, increases in the activity of the metabolic markers cytochrome oxidase, succinate dehydrogenase, and NADH occur correspondingly with the period of first order dendritic growth.[60] Finally, the expression of synaptophysin, a membrane glycoprotein present in clear synaptic vesicles, increases

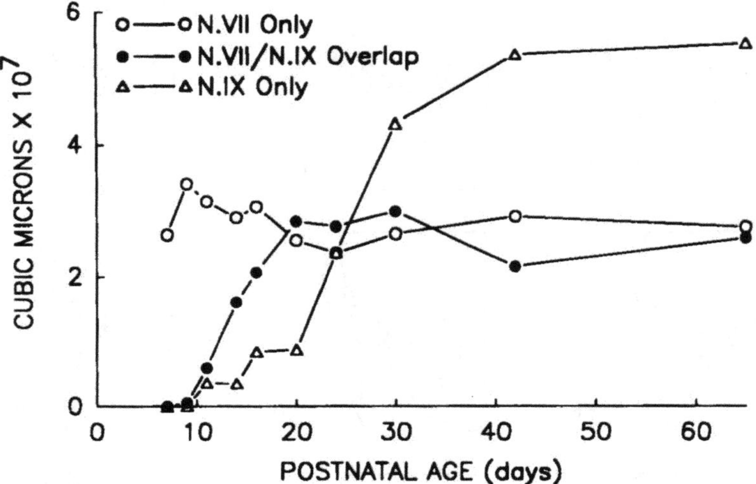

FIGURE 11. Areas of terminal fields in the rat nucleus of the solitary tract have differential developmental patterns. The chorda tympani nerve terminal field (N.VII Only) is adult-like first, followed by the area overlapped by the chorda tympani/glossopharyngeal nerves (N.VII/N.IX), and then by the field of the glossopharyngeal nerve that does not overlap with the chorda tympani nerve (N.IX Only). (From Lasiter, P. S., *Brain Res. Bull.*, 28, 667, 1992. With permission.)

in the rostral NST from postnatal day 1 to postnatal day 10.[59] Thus, the first 10 postnatal days may be an important period during which chorda tympani axons are making functional synaptic connections onto developing NST neurons. This period precedes the time of dendritic elaboration of rostral NST neurons.[57]

2. Development of Projection Neurons to Parabrachial Nuclei

The sequence of taste neuron projections to succeeding neural levels occurs *serially.*[56] That is, projections from the NST to PBN do not occur until after gustatory nerves make significant contacts within the NST. Specifically, there is a delay of approximately 3 weeks between primary terminal field development in the rostral NST (anterior tongue recipient zone) and development of NST projections to the PBN. In contrast, there is a delay of approximately 1 week between primary terminal field development in the intermediate NST (posterior tongue recipient zone) and the development of NST projections to the PBN.[56] Thus, based upon terminal field development, it might be predicted that the anatomical maturation of rat PBN neurons is later than NST neurons.

3. Parabrachial Nuclei Development

Although terminal field development occurs sequentially along the gustatory neuraxis, dendritic development of neurons in the NST and in the PBN occurs contemporaneously. Golgi studies reveal that dendritic growth in the rat PBN occurs between postnatal ages 16 and 35 d.[60] Activity of cytochrome oxidase, succinate dehydrogenase, and NADH also increase during this developmental period.[60] Thus, although the time period of central gustatory projections seems to occur serially, the developmental patterns of neuronal dendritic elaboration in the NST and PBN appear to occur in parallel.

IV. BEHAVIORAL DEVELOPMENT OF THE GUSTATORY SYSTEM

From the findings presented in this review, specific predictions can be made about age-related changes in taste-guided behaviors. Such information has indeed been useful in understanding gustatory behavioral development. However, perhaps due to the involvement of a variety of intero- and exteroceptive systems in the detection of and preferences for chemical stimuli, the correspondence between the biological substrates and behavioral outcomes is not always good. Also, different experimental procedures may contribute to the lack of correspondence. For example, investigators use different modes of stimulus delivery, different stimuli, different stimulus concentrations, and measure different variables to assess behavioral responses. Therefore, it is not entirely surprising that this variability, along with the limited stimulus arrays used in functional studies, results in behavioral outcomes that are often unexpected. Nonetheless, it is instructive to examine the findings available on taste-related behavioral development in order to compare them with findings about the development of functional and structural substrates. For convenience, we will discuss the behavioral findings obtained in animals by the categories of "sweet", "sour", "bitter", and "salty" stimuli. Additionally, since many of the results obtained from neurophysiological studies of salt responses in developing mammals have guided human psychophysical studies, we will provide a very brief description of such data. The reader is referred to a recent, extensive review of human taste development for more information on this topic.[61]

A. SWEET TASTE DEVELOPMENT

Using orofacial reactions as the dependent variable, Ganchrow et al.[62] demonstrated that rats aged 4 d postnatal display adult-like behavioral responses to sucrose and saccharin. Additionally, drops of sucrose, lactose, and saccharin put into the mouths of rats 4 to 10 d of age elicit mouthing behaviors characteristic of appetitive behaviors.[63,64] Results from these studies indicate that rats *detect* "sweet" solutions within the first postnatal week. Findings from Hall and Bryan[65] and from Johanson and Shapiro[66] showed that rats at about 6 d postnatal can *discriminate* the taste of sucrose from the taste of water by ingesting more of the former, and by increasing their intake accordingly with increasing stimulus concentration. While these rats exhibit some components of adult-like behavioral responses within the first postnatal week, mature preferences for sucrose do not appear until about 15 d postnatal.[65] Although these reports indicate that the early postnatal rat perceives "sweet" stimuli, it is not clear whether the *quality* of these stimuli are regarded similarly in young and adult rats. Specifically, does a "sweet" stimulus have the same taste quality for young compared to adult rats?

B. SOUR AND BITTER TASTE DEVELOPMENT

Ganchrow et al.[62] also showed that rats aged 4 d and younger are able to detect citric acid and quinine, and that the orofacial responses to these stimuli resemble responses noted in adults. Although the detection of quinine, as measured by ingestion, appears to occur within the first week, discrimination between it and water does not occur until 9 d of age.[65] In contrast, Johanson and Shapiro[66] demonstrated that 1-d-old rats suppress intake and display aversive behaviors to *high* concentrations of "sour" and "bitter" solutions. As mentioned earlier, the concentration of stimuli may contribute to the variability in the findings. While there is some disagreement about when young rats first perceive sour and bitter solutions, it is clear that the gustatory system is "set" to respond to these stimuli within the first 2 weeks postnatal. This is especially interesting since neurophysiological data show that peripheral and central neurons respond to these stimuli in an adultlike fashion within this period.

C. SALT TASTE DEVELOPMENT IN RAT

In part because of the abundance of information available on the functional development of salt taste, there has been considerable effort devoted to understanding the behavioral development of salt taste. In contrast to other stimulus classes, the major finding from such studies is that behavioral responsiveness to the prototypical salt, NaCl, changes dramatically during a prolonged period of development. For example, 5-d-old rats show no evidence of a preference for NaCl solutions over water, regardless of the NaCl concentration.[67] Similar conclusions are reported by Hall and Bryan[65] who show that 3-d-old pups reject NaCl. In contrast, rats aged 10 d actually *prefer* NaCl solutions that are aversive to adults[67,68] (Figure 12). Thus, behavioral responses to NaCl during early postnatal development in the rat are not similar to those in adults. In fact, mature preferences and aversions to NaCl do not occur until much later in development. Rats aged 25 days drink approximately 3 times more of hypertonic NaCl solutions than mature rats, and the magnitude of aversion seen in adults is not exhibited until approximately 48 d of age.[69] These results indicate that young rats fail to perceive the *quantitative* characteristics (i.e., concentration) of salt stimuli in a manner similar to adults.

The *qualitative* characteristics of NaCl also change with age.[70] Following pairing of illness with the taste of a monochloride salt, adult rats readily discriminate NaCl from NH_4Cl and from KCl. In contrast, early postweaning

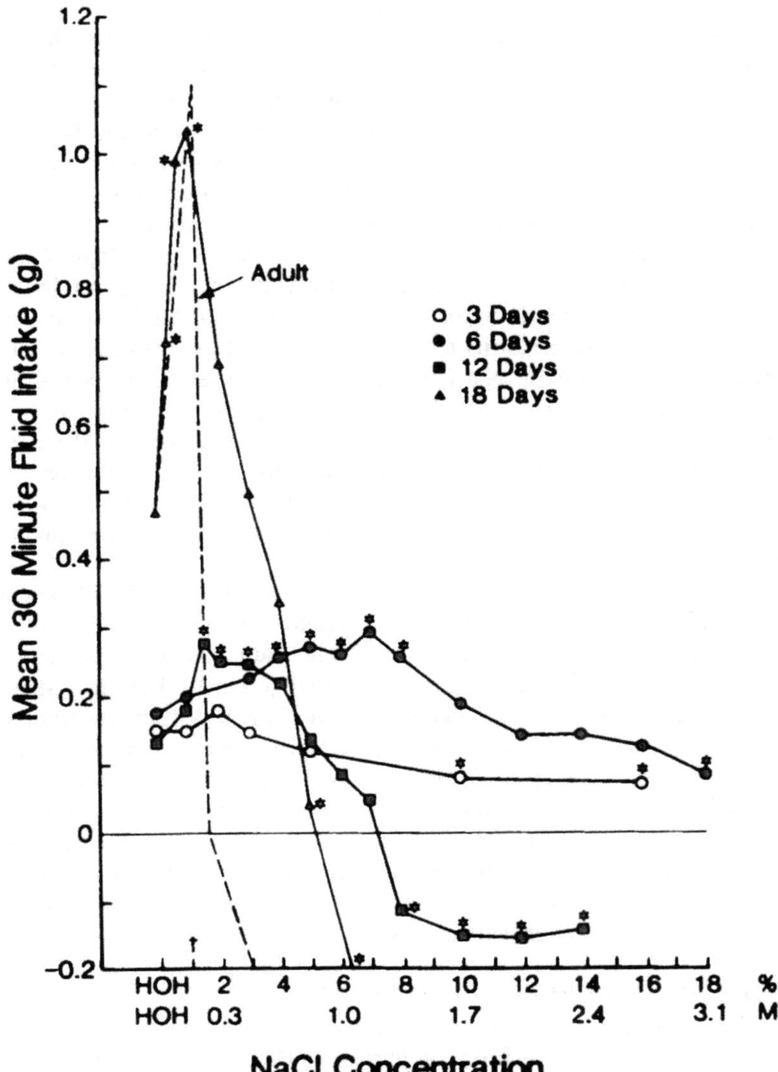

FIGURE 12. Age-related changes in mean intakes (g) occur to NaCl compared to water in the developing rat. Adult-like preferences are only observed in rats aged 18 days postnatal. All younger groups exhibit intakes similar to water (3 d) or exhibit preferences for NaCl at concentrations not preferred by adults (6 and 12 d). (From Moe, K. E., *Dev. Psychobiol.*, 19, 185, 1986. With permission.)

rats are unable to make the same discrimination, even though they learn the association between illness and the taste of the salt. This qualitative change in the taste of NaCl probably relates to the addition of functional amiloride-sensitive sodium channels to the taste receptor cell, as well as to central gustatory development. As stated earlier, young rats that have relatively small NaCl-elicited chorda tympani responses are correspondingly less sensitive to amiloride.[27] This amiloride-insensitive portion of the response may have a "sour/salty" taste that is generalizable to all halogenated salts but not to pure "salty" sodium salts. Evidence for this comes from electrophysiological and behavioral experiments where the amiloride-insensitive NaCl response in the chorda tympani of adult rats is due to the halogen anion, whereas the amiloride-sensitive portion relates primarily to sodium.[31,70,71] To this extent, the neural and behavioral responses to NaCl in immature rats resemble responses in amiloride-treated adult rats, indicating that the increase in the activity of this sodium transducer is the predominant component responsible for the age-dependent increase in behavioral NaCl taste sensitivity. However, since central taste system maturation also occurs during this period, processes in addition to taste receptor cell maturation also likely contribute to changing taste-guided behaviors.

The only other salt that has been used in behavioral taste development studies is NH_4Cl. Preference for high concentrations of NH_4Cl are evident in rats aged 10 d,[68] in striking contrast with the adult rats' aversion to the stimulus. Moe[67] also reported a preference for NH_4Cl in 12-d-old rats; however, Kehoe and Blass[72] showed that NH_4Cl (and NaCl) disrupts suckling at 10 d of age. The apparent discrepancy in results may relate to the use of the suckling conditions compared to other methods of stimulus delivery. Regardless, these findings are of importance because they represent conclusions drawn from a monochloride salt that elicits adult-like functional responses early in development[25,26] to a stimulus that is not preferred by adults. Therefore, the result that early postnatal rats prefer such a stimulus is not what one would predict from neurophysiological data.

D. SALT TASTE DEVELOPMENT IN HUMAN

Striking similarities exist in behavioral taste development between rats and humans. Premature and full-term human newborns discriminate among taste stimuli and exhibit affective responses to acids, sugars, and some bitter stimuli.[73] However, it has been difficult to demonstrate that the human newborn can detect and respond to NaCl. Observations from developmental studies indicate that mature behavioral responses to NaCl are not expressed until about 4 years of age.[74] Thus, there is a gradual shift in behavioral response to various concentrations of NaCl during the first few years of postnatal life, which, like the rat, may indicate changing structural and functional substrates.

V. SUMMARY AND FUTURE DIRECTIONS

Impressive morphological, physiological, and behavioral changes characterize the developing mammalian gustatory system. The most notable changes occur to sodium salts. In this chapter, we have described findings which show that the sensitivity of the gustatory system of sheep, rat, and hamster changes with age to some, but not to all, taste stimuli. It is possible that this trend reflects an adaptive role for the gustatory system. That is, stimuli generally regarded as "appetitive" become more effective in eliciting gustatory responses during development, while stimuli regarded as "aversive" are as effective in young animals as they are in adults. The former class of stimuli are normally supplied by the mother until the time of weaning, when the gustatory system becomes capable of responding to all stimuli in an adult-like fashion. The latter class of stimuli may relate to harmful materials in the environment. Thus, the preweanling animal exploring its environment is able to detect substances that are often associated with toxicity. Therefore, the sequence in which responses to nutritive and aversive stimuli become mature may reflect maximized survival potential. Perhaps the best examples to illustrate how the developing gustatory system is "set" to respond to potentially harmful, but not to preferred chemicals, are the patterns of NaCl response development among species. As noted in preceding sections, sodium salts become dramatically more effective in eliciting peripheral and central neural responses in developing sheep and rat. In contrast, NaCl elicits peripheral taste responses in early postnatal hamsters as effectively as it does in adults. The differences in functional developmental patterns correspond to differences in behavioral taste preferences. Adult rats and sheep prefer hypo- and isotonic concentrations of NaCl, whereas hamsters avoid NaCl at all concentrations.[75-77] Furthermore, ingestion of sodium without compensatory ingestion of water, as would be likely for desert animals, would seriously compromise extracellular fluid volume. Therefore, the widely divergent patterns of functional development to individual stimuli among species may reflect evolutionary and environmental pressures. A better understanding of these controls on transduction components will provide a link in the relationship between gustatory function and survival of a species.

Another major point that we have stressed relates to the sequential maturation of structural and functional events that occurs at successive neural levels. In most instances, especially with regard to functional development, the maturation of successively higher neural structures can be accounted for, only in part, by the prior development of lower order neurons. We have emphasized that *unique* events must also occur which are inherent to central gustatory relays. For example, the pattern of NaCl response development in NST is similar to the pattern noted in chorda tympani neurons. However, the match is not exact. Specifically, central functional development cannot be explained fully by the development of amiloride-sensitive sodium channel function in taste receptor cells. Indeed, additional processes inherent to the NST must operate to influence the time course of NaCl response development. While beyond the scope of this chapter, significant new work is underway which focuses on identifying the mechanisms involved in central gustatory development. Much of this work employs early environmental manipulations in the developing rat as an experimental strategy to reveal factors important during normal development. Emerging evidence suggests that while neural activity is influential in shaping the properties of central gustatory neurons,[78] humoral factors (e.g., hormones, growth factors) circulating in the mother and/or fetus must also be involved.[79] Thus, as noted for peripheral gustatory development, nonneuronal influences may have very specific roles in the development of specific responses. Such findings illustrate how the normative findings presented in this chapter can lead to a more mechanistic understanding of peripheral and central gustatory development.

Finally, as we learn more of the morphological and functional characteristics of the developing taste system, we will be better able to understand the development of taste-guided behaviors. As noted earlier, there often is poor correspondence between behavioral outcomes and structural and functional events. This disparity may be due to methodological differences. A more important consideration, however, is that taste-guided behaviors are shaped by the complex interactions of multiple regulatory systems. In fact, even if the gustatory system alone were responsible for these behaviors, our existing knowledge would still be insufficient to accurately predict behavioral outcomes. Studies aimed at understanding changing neural codes of brainstem structures during development as well as studies focused on more central gustatory structures (thalamus and cortex) may reveal important neural correlates of taste-guided behaviors.

ACKNOWLEDGMENTS

We thank Dr. Dale Benos who generously supplied anti-amiloride-sensitive sodium channel antibodies. The preparation of this chapter was supported by NIH Training Grant #HD07323 and by NIH Grant #DC00407.

REFERENCES

1. **Yamada, T.,** Chorda tympani responses to gustatory stimuli in developing rats, *Jpn. J. Physiol.*, 30, 631, 1980.
2. **Frank, M. and Pfaffmann, C.,** Taste nerve fibers: a random distribution of sensitivities to four tastes, *Science*, 164, 1183, 1969.
3. **Nowlis, G. H. and Frank, M. E.,** Quality coding in gustatory systems of rat and hamster, in *Perception of Behavioral Chemicals*, Norris, D. M., Ed., Elsevier/North-Holland, Amsterdam, 1981, 59.

4. **Ninomiya, Y., Tanimukai, T., Yoshida, S., and Funakoshi, M.,** Gustatory neural responses in preweanling mice, *Physiol. Behav.*, 49, 913, 1991.
5. **Hill, D. L.,** Development of chorda tympani nerve responses in the hamster, *J. Comp. Neurol.*, 268, 346, 1988.
6. **Hill, D. L., Mistretta, C. M., and Bradley, R. M.,** Developmental changes in taste response characteristics of rat single chorda tympani fibers, *J. Neurosci.*, 2, 782, 1982.
7. **State, F. A. and Bowden, R. E.,** Innervation and cholinesterase activity of the developing taste buds in the circumvallate papilla of the mouse, *J. Anat.*, 118, 211, 1974.
8. **Appleberg, B.,** Species differences in the taste qualities mediated through the glossopharyngeal nerve, *Acta Physiol. Scand.*, 44, 129, 1958.
9. **Frank, M.,** An analysis of hamster afferent taste nerve response functions, *J. Gen. Physiol.*, 61, 588, 1973.
10. **Hyman, A. M. and Frank, M. E.,** Effects of binary taste stimuli on the neural activity of the hamster chorda tympani, *J. Gen. Physiol.*, 76, 143, 1980.
11. **Dastoli, F. R. and Price S.,** Sweet-sensitive protein from bovine taste buds: isolation and assay, *Science*, 154, 905, 1966.
12. **Cagan, R. H.,** Biochemical studies of taste sensation. I. Binding of ^{14}C-labelled sugars to bovine taste papillae, *Biochim. Biophys. Acta*, 252, 199, 1971.
13. **Cagan, R. H. and Morris, R. W.,** Biochemical studies of taste transduction. VI. Binding to taste tissue of tritiated monellin, a sweet tasting protein, *Proc. Natl. Acad. Sci. U.S.A.*, 76, 1692, 1979.
14. **Avenet, P., Hoffman, F., and Lindemann, B.,** Transduction in taste receptor cell requires cAMP-dependent protein kinase, *Nature*, 331, 351, 1988.
15. **Striem, B. J., Pace, U., Zehavi, U., Naim, M., and Lancet, D.,** Sweet tastants stimulate adenylate cyclase coupled to GTP-binding protein in rat tongue membranes, *Biochem. J.*, 260, 121, 1989.
16. **Miller, I. J., Jr.,** Taste bud development in the hamster, *Soc. Neurosci. Abstr.*, 9, 464, 1983.
17. **Nejad, M. S.,** The neural activities of the greater superficial petrosal nerve of the rat in response to chemical stimulation of the palate, *Chem. Senses*, 11, 283, 1986.
18. **Krimm, R. K., Nejad, M. S., Smith, J. C., Miller, I. J., and Beidler, L. M.,** The effect of bilateral sectioning of the chorda tympani and the greater superficial petrosal nerves on the sweet taste in the rat, *Physiol. Behav.*, 41, 495, 1987.
19. **Travers, S. P. and Norgren, R.,** Coding the sweet taste in the nucleus of the solitary tract: differential roles for anterior tongue and nasoincisor duct gustatory receptors in the rat, *J. Neurophysiol.*, 65, 1372, 1991.
20. **Bradley, R. M. and Mistretta, C. M.,** The gustatory sense in foetal sheep during the last third of gestation, *J. Physiol.*, 231, 271, 1973.
21. **Mistretta, C. M. and Bradley, R. M.,** Neural basis of developing salt taste sensation: response changes in fetal, postnatal, and adult sheep, *J. Comp. Neurol.*, 215, 199, 1983.
22. **Nagai, T., Mistretta, C. M., and Bradley, R. M.,** Developmental decrease in size of peripheral receptive fields of single chorda tympani nerve fibers and relation to increasing NaCl taste sensitivity, *J. Neurosci.*, 8, 64, 1988.
23. **Mistretta, C. M. and Bradley, R. M.,** Developmental changes in taste responses from glossopharyngeal nerve in sheep and comparisons with chorda tympani responses, *Dev. Brain Res.*, 11, 107, 1983.
24. **Formaker, B. K. and Hill, D. L.,** Lack of amiloride sensitivity in SHR and WKY glossopharyngeal taste responses to NaCl, *Physiol. Behav.*, 50, 765, 1991.
25. **Hill, D. L. and Almli, C. R.,** Ontogeny of chorda tympani nerve responses to gustatory stimuli in the rat, *Brain Res.*, 197, 27, 1980.
26. **Ferrell, M. F., Mistretta, C. M., and Bradley, R. M.,** Development of chorda tympani taste responses in rat, *J. Comp. Neurol.*, 198, 37, 1981.

27. **Hill, D. L. and Bour, T. C.,** Addition of functional amiloride-sensitive components to the receptor membrane: a possible mechanism for altered taste responses during development, *Dev. Brain Res.*, 20, 310, 1985.
28. **Heck, G. L., Mierson, S., and DeSimone, J. A.,** Salt taste transduction occurs through an amiloride-sensitive sodium transport pathway, *Science*, 223, 403, 1984.
29. **DeSimone, J. A. and Ferrell, F.,** Analysis of amiloride inhibition of chorda tympani taste response of rat to NaCl, *Am. J. Physiol.*, 249, R52, 1985.
30. **Formaker, B. K. and Hill, D. L.,** An analysis of residual NaCl taste response after amiloride, *Am. J. Physiol.*, 255, R1002, 1988.
31. **Ye, Q., Heck, G. L., and DeSimone, J. A.,** The anion paradox in sodium taste reception: resolution by voltage clamp studies, *Science*, 254, 724, 1991.
32. **Simon, S. A., Holland, V. F., Benos, D. J., and Zamphighi, G. A.,** Transport properties and proteins in canine circumvallate papilla, *Chem. Senses*, 14, 748 (Abstr.), 1989.
33. **Simon, S. A., Holland, V. F., Benos, D. J., and Zamphighi, G. A.,** Transport pathways in lingual epithelia, *J. Electronmicro. Tech.*, 1992, in press.
34. **Farbman, A. I.,** Electron microscope study of the developing bud in rat fungiform papilla, *Dev. Biol.*, 11, 110, 1965.
35. **Mistretta, C. M.,** Topographical and histological study of the developing rat tongue, palate and taste buds, in *3rd Symp. Oral Sensation and Perception: The Mouth of the Infant*, Bosma, J. F., Ed., Charles C Thomas, Springfield, IL, 1972, 163.
36. **Mistretta, C. M.,** Developmental neurobiology of the taste system, in *Smell and Taste in Health and Disease*, Getchell, T. V., Doty, R. L., Bartoshuk, L. M., and Snow, J. B., Eds., Raven Press, New York, 1991, chap. 3.
37. **Hill, D. L.,** Susceptibility of the developing rat gustatory system to the physiological effects of dietary sodium deprivation, *J. Physiol.*, 393, 413, 1987.
38. **Hill, D. L. and Przekop, P. R.,** Influences of dietary sodium on functional taste receptor development: a sensitive period, *Science*, 241, 1826, 1988.
39. **Przekop, P. R., Mook, D. G., and Hill, D. L.,** Functional recovery of the gustatory system after sodium deprivation during development: how much sodium and where, *Am. J. Physiol.*, 259, R786, 1990.
40. **Blakemore, G. and Cooper, G. F.,** Development of the brain depends on the visual environment, *Nature*, 228, 477, 1970.
41. **Hubel, D. H. and Wiesel, T. N.,** The period of susceptibility to the physiological effects of unilateral eye closure in kittens, *J. Physiol.*, 206, 419, 1970.
42. **Knudsen, E. I., Knudsen, P. F., and Esterly, S. D.,** A critical period for the recovery of sound localization accuracy following monaural occlusion in the barn owl, *J. Neurosci.*, 4, 1012, 1984.
43. **Brunjes, P. C. and Frazier, L. L.,** Maturation and plasticity in the olfactory system of vertebrates, *Brain Res. Rev.*, 11, 1, 1986.
44. **Mistretta, C. M., Gurkan, S., and Bradley, R. M.,** Morphology of chorda tympani fiber receptive fields and proposed neural rearrangements during development, *J. Neurosci.*, 8, 73, 1988.
45. **Lichtman, J. W.,** The reorganization of synaptic connexions in the rat submandibular ganglion during postnatal development, *J. Physiol.*, 273, 155, 1977.
46. **Rakic, P. and Riley, K. P.,** Regulation of axon number in primate optic nerve by prenatal binocular competition, *Nature*, 305, 135, 1983.
47. **Thompson, W. J.,** Changes in the innervation of mammalian skeletal muscle fibers during postnatal development, *Trends Neurosci.*, 9, 25, 1986.
48. **Hill, D. L., Bradley, R. M., and Mistretta, C. M.,** Development of taste responses in rat nucleus of solitary tract, *J. Neurophysiol.*, 50, 879, 1983.
49. **Nowlis, G. H.,** From reflex to representation: taste-elicited tongue movements in the human newborn, in *Taste and Development: The Genesis of Sweet Preference*, Weiffenbach, J., Ed., DHEW Publications No. (NIH) 77-1068, Bethesda, MD, 1977, 190.

50. **Scott, T. R. and Mark, G. P.,** The taste system encodes stimulus toxicity, *Brain Res.*, 414, 197, 1987.
51. **Bradley, R. M. and Mistretta, C. M.,** Developmental changes in neurophysiological taste responses from the medulla in sheep, *Brain Res.*, 191, 21, 1980.
52. **Mistretta, C. M. and Bradley, R. M.,** Taste responses in sheep medulla: changes during development, *Science*, 202, 535, 1978.
53. **Vogt, M. B. and Mistretta, C. M.,** Convergence in mammalian nucleus of solitary tract during development and functional differentiation of salt taste circuits, *J. Neurosci.*, 10, 3148, 1990.
54. **Nagai, T., Mistretta, C. M., and Bradley, R. M.,** Developmental decrease in size of peripheral receptive fields of single chorda tympani nerve fibers and relation to increasing NaCl taste sensitivity, *J. Neurosci.*, 8, 64, 1988.
55. **Hill, D. L.,** Development of taste responses in the rat parabrachial nucleus, *J. Neurophysiol.*, 57, 481, 1987.
56. **Lasiter, P. S.,** Postnatal development of gustatory recipient zones within the nucleus of the solitary tract, *Brain Res. Bull.*, 28, 667, 1992.
57. **Lasiter, P. S., Wong, D. M., and Kachele, D. L.,** Postnatal development of the rostral solitary nucleus in rat: dendritic morphology and mitochondrial enzyme activity, *Brain Res. Bull.*, 22, 313, 1989.
58. **Altman, J. and Bayer, S. A.,** Development of the brainstem in the rat. II. A thymidine radiographic study of the time of origin of neurons in the lower medulla, *J. Comp. Neurol.*, 194, 1, 1980.
59. **Lasiter, P. S. and Kachele, D. L.,** Postnatal development of protein P-38 ("Synaptophysin") immunoreactivity in pontine and medullary gustatory zones of rat, *Dev. Brain Res.*, 55, 57, 1989.
60. **Lasiter, P. S. and Kachele, D. L.,** Postnatal development of the parabrachial gustatory zone in rat: dendritic morphology and mitochondrial enzyme activity, *Brain Res. Bull.*, 21, 79, 1988.
61. **Beauchamp, G. K., Cowart, B. J., and Schmidt, H. J.,** Development of chemosensory sensitivity and preference, in *Smell and Taste in Health and Disease*, Getchell, T. V., Doty, R. L., Bartoshuk, L. M., and Snow, J. B., Eds., Raven, Press, New York, 1991, chap. 22.
62. **Ganchrow, J. R., Steiner, J. E., and Canetto, S.,** Behavioral displays to gustatory stimuli in newborn rat pups, *Dev. Psychobiol.*, 19, 163, 1986.
63. **Jacobs, H. L.,** Observations on the ontogeny of saccharine preference in the neonate rat, *Psychol. Sci.*, 1, 105, 1964.
64. **Jacobs, H. L. and Sharma, K. N.,** Taste versus calories: sensory and metabolic signals in the control of food intake, in *Neural Regulation of Food and Water Intake*, Morgane, J. P., Ed., Annals N. Y. Acad. Sci., 157, 1084, 1969.
65. **Hall, W. G. and Bryan, T. E.,** The ontogeny of feeding in rats. IV. Taste development as measured by intake and behavioral responses to oral infusions of sucrose and quinine, *J. Comp. Physiol. Psychol.*, 95, 240, 1981.
66. **Johanson, I. B. and Shapiro, E. G.,** Intake and behavioral responsiveness to taste stimuli in infant rats from 1 to 15 days of age, *Dev. Psychobiol.*, 19, 59, 1986.
67. **Moe, K. E.,** The ontogeny of salt preference in rats, *Dev. Psychobiol.*, 19, 185, 1986.
68. **Bernstein, I. L. and Courtney, L.,** Salt preference in the preweaning rat, *Dev. Psychobiol.*, 20, 443, 1987.
69. **Midkiff, E. E. and Bernstein, I. L.,** The influence of age and experience on salt preference of the rat, *Dev. Psychobiol.*, 16, 385, 1983.
70. **Formaker, B. K. and Hill, D. L.,** Alterations of salt taste perception in the developing rat, *Behav. Neurosci.*, 104, 356, 1990.
71. **Hill, D. L., Formaker, B. K., and White, K. S.,** Perceptual characteristics of the amiloride-suppressed sodium chloride taste response in the rat, *Behav. Neurosci.*, 104, 734, 1990.

72. **Kehoe, P. and Blass, E. M.,** Gustatory determinants of suckling in albino rats 5–20 days of age, *Dev. Psychobiol.*, 18, 67, 1985.
73. **Steiner, J. E.,** Human facial expression in response to taste and smell stimulation, *Adv. Child Dev.*, 13, 257, 1979.
74. **Beauchamp, G. K., Cowart, B. J., and Moran, M.,** Developmental changes in salt acceptability in human infants, *Dev. Psychobiol.*, 19, 17, 1986.
75. **Carpenter, J. A.,** Species differences in taste preference, *J. Comp. Physiol. Psychol.*, 49, 139, 1956.
76. **Denton, D. A.,** *The Hunger for Salt*, Springer-Verlag, New York, 1982.
77. **Wong, R. and Jones, W.,** Saline intake in hamsters, *Behav. Biol.*, 24, 474, 1978.
78. **Lasiter, P. S.,** Effects of early postnatal receptor damage on dendritic development in gustatory recipient zones of the rostral nucleus of the solitary tract, *Dev. Brain Res.*, 61, 197, 1991.
79. **Krimm, R. F. and Hill, D. L.,** The critical period for chorda tympani terminal fields in the NTS of sodium chloride deprived rats, *Neurosci. Abstr.*, 17, 231, 1991.

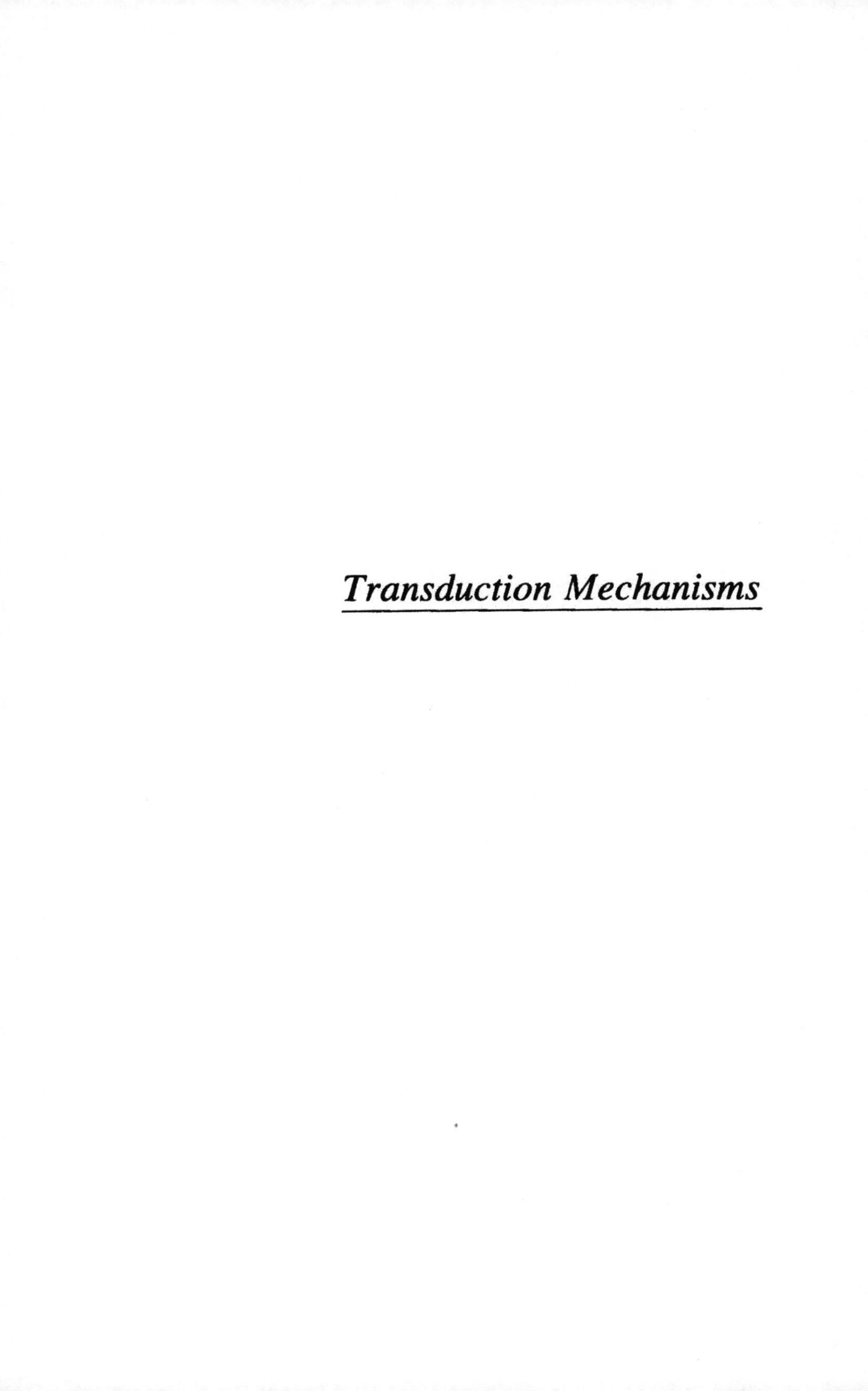

Transduction Mechanisms

Chapter 6

PERIPHERAL TRANSDUCTION MECHANISMS

P. Avenet, S. C. Kinnamon, and S. D. Roper

TABLE OF CONTENTS

0-8493-5341-6/93/$0.00 + $.50

I. INTRODUCTION

Taste cells inform the brain about the chemical environment of the oral cavity. Chemicals are typically classified into sweet, salty, sour, and bitter stimuli in relation to the sensations they evoke in humans. Taste stimuli are thought to interact only with the apical tips of receptor cells that protrude through the taste pore into the saliva. As a result of this interaction, taste cells depolarize, fire action potentials, and release transmitters at synapses with afferent nerve fibers.

The initial steps in taste transduction resemble those of synaptic transmission and hormone receptor interaction, suggesting that these mechanisms might share similar signal transduction mechanisms. There are, however, major differences between taste reception and signal transduction in other systems. First, taste stimuli are diverse, ranging from ions to large molecules. This diversity requires that there are a number of different transduction mechanisms. Second, the pH and osmolarity of the oral environment vary widely and are typically outside the range of normal physiological values for most tissues. Chemosensory mechanisms in taste must operate in such ranges.

A special feature of taste cells is their epithelial-like organization. Tight junctions separate the basolateral membrane, containing mainly voltage-gated channels, from the apical membrane in which one finds specialized chemosensitive receptors. The apical membrane is exposed to the highly variable oral environment and the basolateral membrane is bathed in a relatively constant *mileau interior*. Since channels exist in the apical membrane as well as the basolateral region, the membrane potential of taste cells will be directly affected by ionic changes in the oral cavity. Thus, to investigate the physiology of taste transduction, recording methods that preserve the polarity of the taste epithelium must be employed as well as methods that give direct information about the ion channels that are present.

We shall review here the different mechanisms that have been elucidated in recent years for transducing taste stimuli into receptor potentials. Figure 1 summarizes these mechanisms.

II. APICAL CONDUCTANCES: ION FLUX THROUGH OPEN CHANNELS

DeSimone and colleagues first presented evidence that taste stimuli can induce ionic fluxes through taste cells. When lingual epithelium from frogs, rats or dogs is mounted in an Ussing chamber, a transepithelial short circuit current develops when the apical side is exposed to NaCl (see also Chapter 8).[1] This current varies with the Na^+ concentration, is abolished by ouabain, a Na/K ATPase blocker, or by amiloride,[2] which blocks the common epithelial Na^+ channel of frog skin or kidney. An amiloride-blockable Na^+ current also develops when a recording pipette perfused with NaCl is placed directly over a single rat taste pore, suggesting that the current is generated by taste cells.[3]

The presence of such a channel in the apical membrane of taste cells implies that an increase in the sodium concentration of the oral cavity will depolarize the taste cells. This would provide a simple mechanism for the detection of salty (e.g., NaCl) stimuli.

In fact, there is now strong evidence that direct influx of Na^+ is the transduction mechanism for sodium salt taste. The transepithelial Na-dependent current measured in a small area of the tongue *in vivo* correlates well with the chorda tympani response measured simultaneously.[4] Furthermore, action potentials induced by NaCl can be measured directly from rat taste cells *in situ* with a suction electrode placed over the taste pore; these action potentials are blocked by amiloride.[3] Finally, amiloride blocks the chorda tympani response to sodium in rodents[2] and the perceived Na^+ taste intensity in humans.[5]

With the patch-clamp technique it is possible to study these channels directly. Whole-cell recordings have demonstrated the presence of an amiloride blockable current in isolated taste cells of the frog.[6] This current has been studied in outside out patches, where it is possible to analyze single channel events.[7] Because of the small conductance of each channel and the large number of channels in the patch membrane, it was necessary to use the technique of noise analysis. This method revealed a single channel conductance for the amiloride-sensitive Na^+ channel in the range of 1 to 2 pS and a low open probability.[8] This amiloride-sensitive channel in frog taste cells is different from the classical transporting epithelial channel with respect to open probability, selectivity, unitary conductance, and sensitivity to amiloride analogs.[7-9] In contrast, preliminary studies have shown that the rodent amiloride-sensitive channel in taste cells has properties similar to the amiloride-sensitive channel of transporting epithelia.[10,11]

Taste cells possessing apical Na^+ channels will be depolarized by any cation that permeates the channel. Protons are known to permeate the amiloride-sensitive sodium channel of transporting epithelia.[12] Recent evidence suggests that the influx of protons through Na^+ channels may be involved in the detection of acid (sour) stimuli. In hamsters, citric acid at pH 2.6 (3 m*M*) in the absence of Na^+ produces a strong excitation of taste cells that is blocked by amiloride.[13,14,14a]

The restricted selectivity of the amiloride-sensitive Na^+ channel in rodents for sodium (and protons) implies that other channels must exist to explain the taste of other salty stimuli. An amiloride-insensitive Na^+ response exists in young rats,[15] in mice,[16] and in *Necturus*.[17] The channel underlying amiloride-insensitive taste responses has not been studied directly, and its selectivity is unknown. Another channel, one that is voltage dependent, is localized in the apical membrane of *Necturus*.[18,19] This channel is potassium selective and TEA sensitive. This apical K^+ channel has been shown to be responsible for depolarization in response to K^+ salts.

While cation flux through an apical channel depolarizes taste cells, the magnitude of the effect will depend on the anion present. Indeed, it was

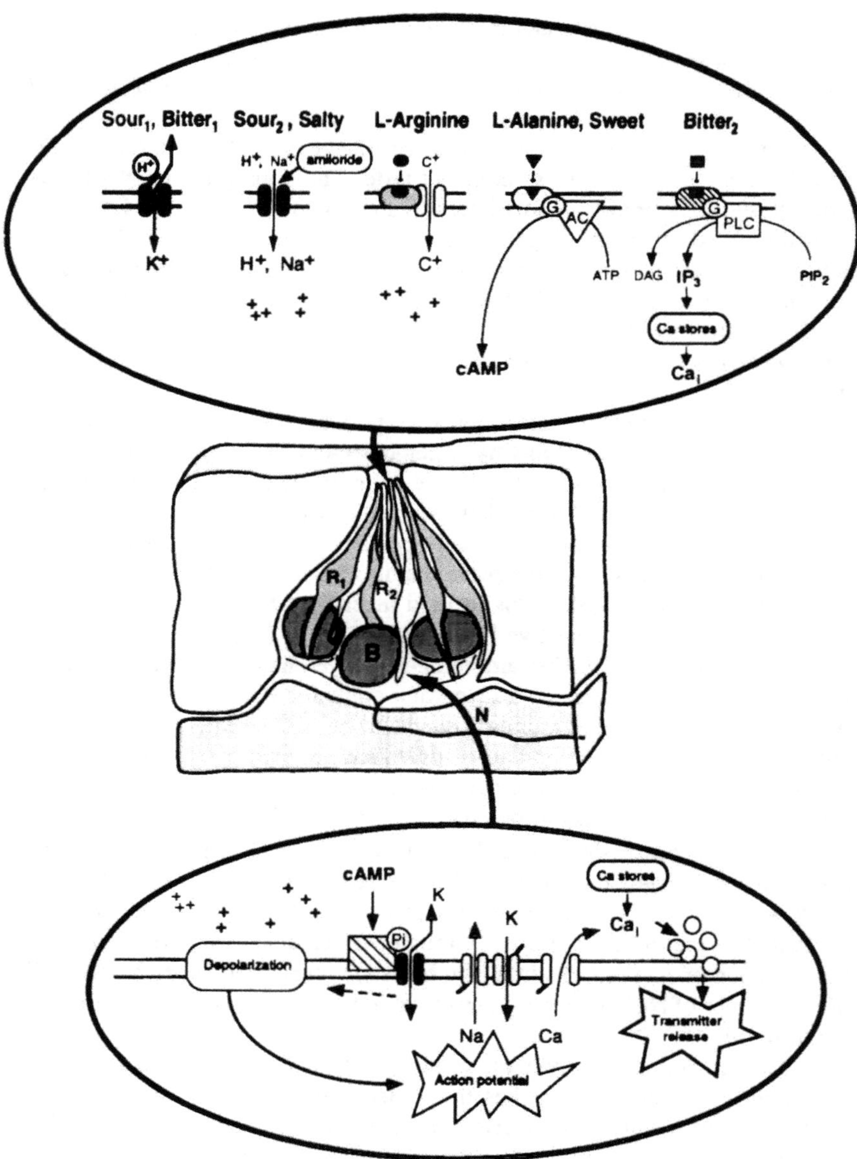

FIGURE 1. Summary of chemosensory transduction mechanisms occurring in taste buds. Events illustrated at the apical membrane of taste receptor cells include protons (sour) and bitter compounds blocking K^+ channels; protons and Na^+ passing through apical, amiloride-sensitive sodium channels; L-arginine binding to a ligand-gated, nonselective cation channel; L-alanine and sweet compounds activating a G protein by binding to appropriate membrane receptors and stimulating adenylyl cyclase; bitter compounds activating a G protein that is coupled to phospholipase C and resulting in an increase in inositol triphosphate.

recently shown that cations are less effective in depolarizing rat taste cells when the anion cannot permeate the tight junctions.[20,21] This is thought to be due to a current which hyperpolarizes taste cells when the anion cannot penetrate the tight junctions (see also Chapter 8). Additionally, McBride, Taylor, and Roper have shown that taste cells in *Necturus* possess a long-lasting Ca-dependent anion conductance that is selectively permeable for Cl^-.[22,55] Activation of this conductance by depolarizing receptor potentials would render the membrane susceptible to anion influx if a permeant anion species (e.g. Cl^-) was present in the extracellular spaces. These phenomena may help explain differences in the perception of salts of different anions. Ca-dependent anion fluxes may also explain adaption to a constant salt stimulus. Permeability change of the tight junctions in response to low pH has also been proposed to play a role in acid transduction.[23] Thus, taste transduction likely involves not only the taste cells but also their epithelial environment. Figure 2 illustrates these points.

III. BLOCK OF APICAL CONDUCTANCE

Apically located channels not only influence the cell membrane potential when they conduct ions but also when they are blocked by chemical stimuli. Low pH, for example, is known to block many channels. In *Necturus*, proton-block of apically localized K^+ channels is responsible for taste cell excitation. In agreement with this model, acid stimulation was shown to increase the membrane resistance and to prolong the action potential of taste receptor cells in *Necturus*.[24] In experiments performed in intact epithelium, K^+ channel blockers like TEA were effective when applied apically but not basolaterally in *Necturus*.[18] This suggests that most of the K^+ channels involved have an apical location. This high apical concentration of K^+ channels was demonstrated directly in isolated *Necturus* cells with patch clamp techniques.[19] Single channels were obtained in patches derived from the apical membrane of these taste cells. Several types of K^+ channels were found; these were voltage dependent and had conductances varying from 25 to 200 pS. Low pH reduced the open probability of these channels but only when applied to the external membrane surface.[25,26]

FIGURE 1 (continued). Events illustrated on the basolateral membranes of taste receptor cells include depolarization of the membrane consequent to an influx of cations at the apical membrane; reduction of potassium conductance due to cyclic AMP buildup which also depolarizes the membrane; generation of an action potential when the depolarization reaches threshold; influx of Ca^{2+} ions; release of neurotransmitter stores consequent to increased $[Ca^{2+}]_i$. Abbreviations: C^+, cations; G, G protein; AC, adenylyl cyclase; PLC, phospholipase C; DAG, diacylglycerol; IP_3, inositol 1,4,5-triphosphate; PIP_2, inositol 4,5-bisphosphate; R, taste receptor cell; B, basal taste cells; N, nerve bundle.

Several stimuli which taste bitter to humans, including quinine and $CaCl_2$, also depolarize *Necturus* taste cells by blocking the apical K^+ conductance.[26,27] Thus, K^+ channels at the apical tip may be important in transducing acids, K^+ salts, and bitter stimuli in the mud puppy (*Necturus*). Whether this animal can distinguish between these taste qualities is not known. Behavioral studies are needed to clarify this point.

IV. LIGAND-GATED APICAL CONDUCTANCES

Transduction mechanisms similar to ionotropic receptors (i.e., receptors directly coupled to a channel) of the central nervous system were demonstrated in taste cells of the catfish. Catfish taste bud membrane preparations bind amino acids with an extremely high affinity. This property, and the fact that numerous taste buds are found on the catfish epithelium, allowed Teeter and co-workers to isolate the putative receptor protein. When reconstituted into lipid bilayers, the protein displayed channel properties in response to L-arginine, an amino acid that produces a strong taste response in catfish.[28] Given the lack of other intracellular constituents, the presence of active channels in these experiments suggests that there is a direct coupling between the receptor and the channel.[29] A similar mechanism has been suggested in dogs, where an apical amiloride-sensitive cationic conductance is thought to be responsible for transepithelial current in response to sweet stimuli.[30]

It should be pointed out that any ionic flux through such a ligand-gated channel will be dependent on the concentration of the permeant ions present at the apical surface. Therefore, the detection of nonionic stimuli may be enhanced or reduced by the ionic composition of the medium in contact with the apical membrane of taste receptor cells. Channel gating can also be ion dependent as suggested by the proton-gated Ca^{2+} conductance in response to sour stimuli in the frog.[31]

V. BASOLATERAL CONDUCTANCES

An adenylyl cyclase whose activity was dependent on the concentration of saccharose has been demonstrated in membrane fractions derived from rat vallate papillae.[32,33] Thus, cAMP is a likely intracellular second messenger for sweet stimuli that can affect basolateral membrane channels. In frogs, patch-clamp studies have shown that millimolar extracellular cAMP concentrations, or micromolar intracellular concentrations, cause a depolarization linked to a decrease in the outward K^+ currents of the cell.[34,35] Forskolin, an adenylyl cyclase activator, mimics the effect of cAMP, and IBMX, a phosphodiesterase inhibitor, greatly prolongs the depolarization by inhibiting cAMP breakdown.[34] During whole cell patch clamp, cAMP depolarizes taste cells only if ATP is present in the pipette, suggesting that a phosphorylation process is involved. Indeed when ATP-γS, the nonhydrolyzable ATP analog of ATP, into the cell diffuses into the cell from the patch pipette, the cAMP depolarization is irreversible. Furthermore the depolarizing effect of cAMP is abolished if a protein kinase inhibitor is included in the patch pipette. Including the catalytic subunit of the kinase A causes the cell to depolarize irreversibly. Thus, it appears that basolateral K channels are closed by phosphorylation and this contributes to cell depolarization (see Figure 1).

K^+ channels subject to cAMP regulation have been obtained in inside out patches from frog and have a conductance of 44 pS.[35] The open probability of these channels is reduced by the catalytic subunit of the kinase A. A decrease in outward currents in response to sweet stimuli was also demonstrated in mammals and may operate according to the same mechanism.[36-38]

This intracellular-mediated response to sweet may be also used in other taste transduction mechanisms. In catfish, L-alanine also elicits a strong taste response but, in contrast to L-arginine, does not directly gate a channel. L-alanine has been shown to stimulate an increase of the second messengers, IP_3 and cAMP.[39] The targets of the intracellular messengers in this case are not known.

Channel blockage at the basolateral membrane has also been suggested to occur in response to highly lipophilic molecules like quinine, which have a well-known bitter taste in humans. Quinine has been shown to block Na^+ and K^+ currents in isolated amphibian taste cells.[34,40] It is not known if *in vivo* quinine can actually reach the basolateral membrane.

VI. ABSENCE OF MEMBRANE CONDUCTANCE CHANGES

The main effect of the taste cell depolarization is to open Ca^{2+} channels near synapses, allowing Ca^{2+} to enter the cell and trigger transmitter release. Alternative steps that lead to an increase in intracellular Ca^{2+} concentration have also been demonstrated in taste cells in response to bitter stimulation. Denatonium chloride is one of the most bitter stimuli known in humans. It has been shown with Fura-2 imaging techniques that denatonium elicits Ca^{2+} release from intracellular stores in isolated rat taste cells.[41] Denatonium interacts with an apical membrane receptor and causes the production of the second messenger IP_3. IP_3, in turn, causes release of intracellular Ca^{2+}.[42] A rise of intracellular Ca^{2+}, if it were to occur nearby active zones at synapses, would stimulate transmitter release.

VII. FURTHER STEPS

The taste cell depolarization leads to the generation of action potentials when the threshold of activation of the voltage-gated sodium channels, present in the basolateral membrane, is reached. The electrical excitability of taste cells was first demonstrated by Roper[43] and Kashiwayanagi[44] in intact taste buds in amphibians and later in isolated cells of a variety of species, including mammals.[34,45-48] Trains of action potentials have been measured in the intact epithelium of *Necturus* in response to $CaCl_2$.[49] Impulses have also been recorded extracellularly in intact rat and hamster tongues in response to salty, sweet, and acid stimuli.[14,50] Isolated rat taste cells produce action potentials in response to saccharin.[37] One effect of action potentials is to trigger Ca^{2+} currents. Voltage-dependent Ca^{2+} currents have been directly measured in amphibian[46] and mammalian[37] taste cells. Thus, one consequence of a train of impulses would be a sequence of pulsatile increases in intracellular Ca^{2+} and, hence, transmitter release.

Thus, it appears that the taste cell membrane potential plays a central function. Cell activation and transmitter release will take place whenever the threshold of voltage-dependent Ca^{2+} channels is reached (with the possible exception of some bitter stimuli, such as denatonium, as discussed above). The specific way by which the transduction mechanism operates does not, therefore, provide information about the taste quality (i.e., sweet, sour, salty, bitter) since the membrane depolarization and the transmitter release are integrative steps. This poses the problem of taste specificity if a single taste cell can be depolarized by different agents. This problem also exists at the level of single taste fibers which have been shown to be sensitive to several taste stimuli (see Chapter 12).

VIII. TASTE SPECIFICITY AND INTERACTION BETWEEN STIMULI

A complex picture of the chemospecifity of individual taste receptor cells is emerging. There are several factors that are involved when a taste cell responds to a mixture of chemical stimuli. Even if taste cells are specialized to detect one category of stimuli, the possibility exists that several different stimuli will effect a single cell. For example, we have commented above on the fact that the concentration and species of ions present at the mucosal surface will affect chemosensory transduction generally. This consideration will be especially important when mixtures of salts and other taste stimuli are present. Second, two or more substances are known to interact with the same receptor or channel at the apical membrane of taste cells. For example, acids affect salt responses due to the interaction between protons and Na^+ ions at amiloride-sensitive Na^+ channels in the apical membrane of taste cells.[14,51] Or, 5′ nucleotides are known to modulate the chemosensitivity to glutamate salts. Last, receptor potentials in taste cells are thought to be determined, in part, by the anion selectivity of the apical tight junctions and of the Ca^{2+}dependent anion conductance in the taste cell membrane (see Figure 2). These factors, and perhaps others that remain to be discovered, must be considered together to provide a complete picture of the specificity and intensity of responses in a single taste cell.

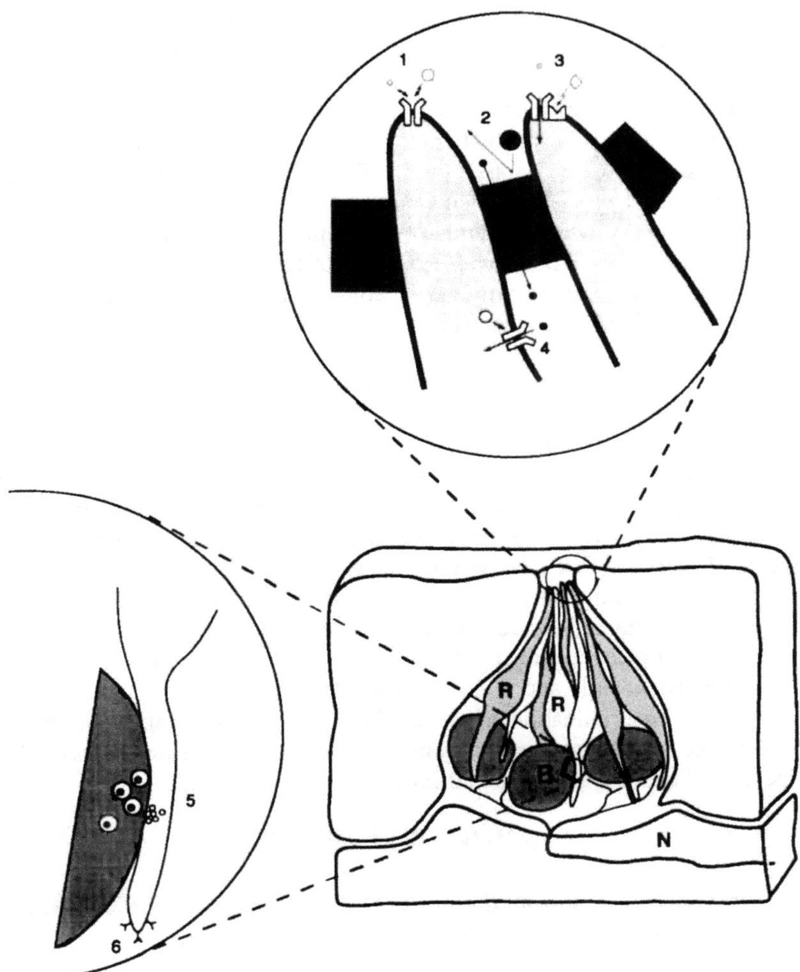

FIGURE 2. Schematic drawing showing factors that influence the chemosensitivity of individual taste cells, as discussed in the text. Two or more chemical stimuli may interact with one channel or receptor. For example, Na^+ (O) and H^+ (o) compete for the amiloride-sensitive channel, as shown at 1. The tight junctions that ring the apical tips of taste cells are permeable to anions such as Cl^- (•), but not others (●), as shown at 2. The ability of ligand-gated channels to generate receptor currents will depend upon the concentration and species of ions present in solution (o), as shown at 3. Anion influx (•) through selectively permeable Ca^{2+}-dependent anion channels, shown at 4, will modify receptor potentials. Electrical and synaptic interactions between taste cells may allow the activity in one cell to influence neighboring cells. For example, a bidirectional chemical synapse between a receptor cell (R) and a basal cell (B) is shown at 5 (see Chapter 11). Last, taste cells may possess hormonal receptors, as shown at 6, rendering them susceptible to circulating hormones.

In addition to the above factors, other mechanisms may operate to affect the selectivity and sensitivity of taste cells. It has been shown recently that taste cells make synapses not only with the afferent nerve fibers but also with neighboring taste cells, including basal cells.[52-54] Electrical and chemical synapses within the taste bud may provide a basis for lateral interactions among taste cells (see Chapter 11). The function of these synaptic connections is unknown. They may play an important role in modulating the taste cell signal, especially when mixtures of taste stimuli are present. One may also speculate on possible hormonal regulation of the taste cell sensitivity by way of basolaterally located hormonal receptors. Factors that influence the chemospecifity of individual taste cells are summarized in Figure 2.

IX. SUMMARY AND CONCLUSION

Several taste transduction mechanisms have been elucidated in the past few years due to the application of new methods such as patch-clamping, direct extracellular recording of the taste cell activity, or transepithelial current measurements, and new preparations such as isolated taste cells and lingual slices. It appears that a variety of transduction mechanisms are present in taste cells including cation flux through apical ion channels, stimulus block of apical ion channels, receptor-mediated release of second messengers, and direct ligand-gated channels. While these mechanisms have been demonstrated independently and in different species, many questions still exist concerning the discrimination of these stimuli. This is a complex problem for which one should take into account the taste cell environment and all the factors influencing the taste cell physiology.

ACKNOWLEDGMENT

This work was supported, in part, by NIH grants 2 Pol DC00244, 2 Rol DC00374 (SDR) and 2 Rol DC00766 (SCK).

REFERENCES

1. **DeSimone, J. A., Heck, G. L., and DeSimone, S. K.,** Active ion transport in dog tongue: a possible role in taste, *Science*, 214, 1039, 1981.
2. **Heck, G. L., Mierson, S., and DeSimone, J. A.,** Salt taste transduction occurs through an amiloride-sensitive sodium transport pathway, *Science*, 223, 403, 1984.
3. **Avenet, P. and Lindemann, B.,** Noninvasive recording of receptor cell action potentials and sustained currents from single taste buds maintained in the tongue: the response to mucosal NaCl and amiloride, *J. Membr. Biol.*, 124, 33, 1991.
4. **Heck, G. L., Persaud, K. C., and DeSimone, J. A.,** Direct measurement of translingual epithelial NaCl and KCl currents during the chorda tympani taste response, *Biophys. J.*, 55, 843, 1989.

5. **Schiffman, S. S., Lockhead, E., and Maes, F. W.,** Amiloride reduces the taste intensity of Na^+ and Li^+ salts and sweeteners, *Proc. Natl. Acad. Sci. U.S.A.*, 80, 6136, 1983.
6. **Avenet, P. and Lindemann, B.,** Amiloride-blockable sodium currents in isolated taste receptor cells, *J. Membr. Biol.*, 105, 245, 1988.
7. **Avenet, P. and Lindemann, B.,** Chemoreception of salt taste: the blockage of stationary sodium currents by amiloride in isolated receptor cells and excised membrane patches, in *Chemical Senses: Receptor Events and Transduction in Taste and Olfaction Reception*, Brand, J. G., et al., Vol. 1, Marcel Dekker, New York, 1989, 171.
8. **Avenet, P. and Lindemann, B.,** Fluctuation analysis of amiloride-blockable currents in membrane patches excised from salt-taste receptor cells, *J. Basic Clin. Physiol. Pharmacol.*, 1, 383, 1990.
9. **Avenet, P. and Lindemann, B.,** Salt-taste receptor currents inhibited by low concentrations of amiloride in membrane patches excised from chemo-sensory cells of the tongue, *Pflügers Arch.*, 413, R46, 1989.
10. **Schiffman, S. S., Suggs, M. S., Cragoe, E. J. Jr., and Erickson, R. P.,** Inhibition of taste responses to Na^+ salts by epithelial Na^+ channel blockers in gerbil, *Physiol. Behav.*, 47, 455, 1990.
11. **Schiffman, S. S., Frey, A. E., Suggs, M. S., Cragoe, E. J. Jr., and Erickson, R. P.,** The effect of amiloride analogs on taste responses in gerbil, *Physiol. Behav.*, 47, 435, 1990.
12. **Palmer, L. G.,** Ion selectivity of epithelial Na channels, *J. Membr. Biol.*, 96, 97, 1987.
13. **Gilbertson, T. A., Avenet, P., Kinnamon, S. C., and Roper, S. D.,** *In situ* recording from hamster fungiform taste cells: response to sour stimuli, *Soc. Neurosci.*, 17, Abstr., 1216, 1991.
14. **Gilbertson, T. A., Avenet, P., Kinnamon, S. C., and Roper, S. D.,** Proton currents through amiloride-sensitive Na channels in hamster taste cells: role in acid transduction, *J. Gen. Physiol.*, 100, 803, 1992.

14a. **Gilbertson, T. A., Roper, S. D., and Kinnamon, S. C.,** Proton currents through amiloride-sensitive Na channels in isolated hamster taste cells: enhancement by vasopressin and cAMP, *Neuron*, in press 1993.

15. **Hill, D. L.,** Development of chorda tympani nerve taste responses in the hamster, *J. Comp. Neurol.*, 268, 346, 1988.
16. **Tonosaki, K. and Funakoshi, M.,** Amiloride does not block taste transduction in the mouse (Slc:ICR), *Comp. Biochem. Physiol. A*, 94, 659, 1989.
17. **McPheeters, M. and Roper, S. D.,** Amiloride does not block taste transduction in the mudpuppy, *Necturus maculosus*, *Chem. Senses*, 10, 341, 1985.
18. **Roper, S. D. and McBride, D. W. Jr.,** Distribution of ion channels on taste cells and its relationship to chemosensory transduction, *J. Membr. Biol.*, 109, 29, 1989.
19. **Kinnamon, S. C., Dionne, V. E., and Beam, K. G.,** Apical localization of K^+ channels in taste cells provides the basis for sour taste transduction, *Proc. Natl. Acad. Sci. U.S.A.*, 85, 7023, 1988.
20. **Ye, Q., Heck, G. L., and DeSimone, J. A.,** The anion paradox in sodium taste reception: resolution by voltage-clamp studies, *Science*, 254, 724, 1991.
21. **Elliott, E. J. and Simon, S. A.,** The anion in salt taste: a possible role for paracellular pathways, *Brain Res.*, 535, 9, 1990.
22. **McBride, D. W. Jr. and Roper, S. D.,** Ca^{2+} dependent chloride conductance in *Necturus* taste cells, *J. Membr. Biol.*, 124, 85, 1991.
23. **Simon, S. A. and Garvin, J. L.,** Salt and acid studies on canine lingual epithelium, *Am. J. Physiol.*, 249, C398, 1985.
24. **Kinnamon, S. C. and Roper, S. D.,** Evidence for a role of voltage-sensitive apical K^+ channels in sour and salt taste transduction, *Chem. Senses*, 13, 115, 1988.
25. **Kinnamon, S. C. and Cummings, T. A.,** Properties of voltage-activated and inwardly-rectifying potassium channels in *Necturus* taste receptor cells, in Proc. Xth Int. Symp. Olfaction and Taste, 1990. Oslo.

26. **Cummings, T. A. and Kinnamon, S. C.,** Apical K^+ channels in *Necturus* taste cells: modulation by intracellular factors and taste stimuli, *J. Gen. Physiol.*, 99, 591, 1992.
27. **Bigiani, A. R. and Roper, S. D.,** Mediation of responses to calcium in taste cells by modulation of a potassium conductance, *Science* (Washington, D.C.), 252, 126, 1991.
28. **Kanwal, J. S., Hidaka, I., and Caprio, J.,** Taste responses to amino acids from facial nerve branches innervating oral and extra-oral taste buds in the channel catfish, *Ictalurus punctatus, Brain Res.*, 406, 105, 1987.
29. **Teeter, J. H., Brand, J. G., and Kumazawa, T.,** A stimulus-activated conductance in isolated taste epithelial membranes, *Biophys. J.*, 58, 253, 1990.
30. **Mierson, S., DeSimone, S. K., Heck, G. L., and DeSimone, J. A.,** Sugar-activated ion transport in canine lingual epithelium. Implications for sugar taste transduction, *J. Gen. Physiol.*, 92, 87, 1988.
31. **Miyamoto, T., Okada, Y., and Sato, T.,** Ionic basis of receptor potential of frog taste cells induced by acid stimuli, *J. Physiol. (London)*, 405, 699, 1988.
32. **Striem, B. J., Naim, M., and Lindemann, B.,** Generation of cyclic AMP in taste buds of the rat circumvallate papilla in response to sucrose, *Cell Physiol. Biochem.*, 1, 46, 1991.
33. **Striem, B. J., Pace, V., Zehari, U., Naim, M., and Lancet, D.,** Sweet tastants stimulate adenylate cyclase coupled to GTP-binding protein in rat tongue membranes, *Biochem. J.*, 260, 121, 1989.
34. **Avenet, P. and Lindemann, B.,** Patch-clamp study of isolated taste receptor cells of the frog, *J. Membr. Biol.*, 97, 223, 1987.
35. **Avenet, P., Hofmann, F., and Lindemann, B.,** Transduction in taste receptor cells requires cAMP-dependent protein kinase, *Nature*, 331, 351, 1988.
36. **Cummings, T. A., Avenet, P., Roper, S. D., and Kinnamon, S. C.,** Modulation of potassium currents by sweeteners in hamster taste cells, *Biophys. J.*, 59, 594A, 1991.
37. **Behe, P., DeSimone, J. A., Avenet, P., and Lindemann, B.,** Membrane currents in taste cells of the rat fungiform papilla. Evidence for two types of Ca currents and inhibition of K currents by saccharin, *J. Gen. Physiol.*, 96, 1061, 1990.
38. **Tonosaki, K. and Funakoshi, M.,** Cyclic nucleotides may mediate taste transduction, *Nature*, 331, 354, 1988.
39. **Teeter, J. H., Sugimoto, K., and Brand, J. G.,** Ionic currents in taste cells and reconstituted taste epithelial membrane, in *Chemical Senses: Receptor Events in Transduction in Taste and Olfaction Reception*, Brand, J. G., et al., Vol. 1, Marcel Dekker, New York, 1989, 151.
40. **Kinnamon, S. C. and Roper, S. D.,** Passive and active membrane properties of mudpuppy taste receptor cells, *J. Physiol. (London)*, 383, 601, 1987.
41. **Akabas, M., Dodd, J., and Al-Awqati, Q.,** Identification of electrophysiologically distinct subpopulations of rat taste cells, *J. Membr. Biol.*, 114, 71, 1990.
42. **Hwang, P. M., Verma, A., Bredt, D. S., and Snyder, S. H.,** Localization of phosphatidylinositol signaling components in rat taste cells: role in bitter taste transduction, *Proc. Natl. Acad. Sci. U.S.A.*, 87, 7395, 1990.
43. **Roper, S.,** Regenerative impulses in taste cells, *Science*, 220, 1311, 1983.
44. **Kashiwayanagi, M., Miyake, M., and Kurihara, K.,** Voltage-dependent Ca^{2+} channel and Na^+ channel in frog taste cells, *Am. J. Physiol.*, 244, C82, 1983.
45. **Sugimoto, K. and Teeter, J. H.,** Voltage-dependent ionic currents in taste receptor cells of the larval tiger salamander, *J. Gen. Physiol.*, 96, 809, 1990.
46. **Kinnamon, S. C. and Roper, S. D.,** Membrane properties of isolated mudpuppy taste cells, *J. Gen. Physiol.*, 91, 351, 1988.
47. **Miyamoto, T., Okada, Y., and Sato, T.,** Membrane properties of isolated frog taste cells: three types of responsivity to electrical stimulation, *Brain Res.*, 449, 369, 1988.
48. **Akabas, M. H., Dodd, J., and Al-Awqati, Q.,** A bitter substance induces a rise in intracellular calcium in a subpopulation of rat taste cells, *Science*, 242, 1047, 1988.

49. **Avenet, P. and Lindemann, B.,** Action potentials in epithelial taste receptor cells induced by mucosal calcium, *J. Membr. Biol.*, 95, 265, 1987.
50. **Avenet, P., Kinnamon, S., and Roper, S.,** *In situ* recording from hamster taste cells: responses to salt, sweet and sour, *Chem. Senses*, 16, 498, 1991.
51. **Gilbertson, T. A., Roper, S. D., and Kinnamon, S. C.,** Effects of acid stimuli on isolated hamster taste cells, *Soc. Neurosci. Abstr.*, 18, 844, 1992.
52. **Roper, S. D.,** The microphysiology of peripheral taste organs, *J. Neurosci.*, 12, 1127, 1992.
53. **Delay, R. J. and Roper, S. D.,** Ultrastructure of taste cells and synapses in the mudpuppy *Necturus maculosus*, *J. Comp. Neurol.*, 277, 268, 1988.
54. **Ewald, D. A. and Roper, S. D.,** Intercellular signalling in *Necturus* taste buds: chemical excitation of receptor cells elicits responses in basal cells, *J. Neurophysiol.*, 67, 1316, 1992.
55. **Taylor, R. S. and Roper, S. D.,** A calcium-dependant anion conductance in *Necturus* taste receptor cells, *Soc. for Neurosci-Abst.*, 18, 845, 1992.

Chapter 7

THE MOLECULAR BIOLOGY OF CHEMOTRANSDUCTION

Myles H. Akabas

TABLE OF CONTENTS

0-8493-5341-6/93/$0.00+$.50

I. INTRODUCTION

Over the past decade there have been major advances in our understanding of the biochemical and physiological bases of chemosensory transduction mechanisms. This has provided the foundation for the application of molecular biology to the study of chemosensory transduction. Molecular biology has already made major contributions to the study of olfaction where it has facilitated the study of the specific proteins involved in the olfactory signal transduction pathway. The goal of this chapter is to review the contributions of molecular biology to the study of chemosensory transduction.

The original goal of molecular biology was to elucidate the structure and expression of genes. In the pursuit of these studies, molecular biologists discovered a large number of enzymes whose function is to modify DNA. Through the use of these enzymes one can cut DNA at specific sites, ligate fragments together in new arrangements, synthesize large quantities of DNA, transcribe it into RNA, and express specific genes in new cells. As the techniques for manipulating DNA have developed, molecular biology has provided the capacity to study a wide range of biological problems, not limited to the structure and expression of genes. In particular, the techniques of molecular cloning, the identification of the DNA coding for specific proteins, have provided access to proteins that could not otherwise be purified and studied by more classical techniques of protein chemistry. This chapter will be divided into two parts. The first part will be an introduction to molecular biology, and the second a review of the recent applications of molecular biology to the study of chemosensory transduction. Because molecular biology has made a much greater contribution in the field of olfaction, that will be the major focus of this chapter.

II. APPROACHES TO MOLECULAR CLONING

This section will provide an overview of the processes of molecular cloning. For a more detailed discussion of molecular biology the reader can consult recent textbooks.[1] For technical details on the various procedures the reader should consult a laboratory manual.[2,3]

Molecular biology has developed a language of its own. Understanding the various terms is essential to be able to read the literature of molecular biology. The following section is a brief description of terms and procedures which can be skipped by those familiar with the terms.

A. DEFINITION OF COMMON TERMS USED IN MOLECULAR BIOLOGY

The basic enzymes used to manipulate DNA include restriction enzymes, ligase, polymerases, and reverse transcriptase.

1. *Restriction enzymes* are bacterial enzymes which cut DNA at precisely defined sequences; over 180 have been identified.
2. *Ligase* is the enzyme which pastes DNA fragments together, connecting the 3′ end of one fragment to the 5′ end of another.
3. *Polymerases* are enzymes which polymerize nucleoside triphosphates in a 5′ to 3′ direction in a template-dependent manner. Two major classes of polymerases exist; DNA polymerases, which make a DNA strand complementary to the DNA template, and RNA polymerases, which make a RNA strand complementary to the DNA template.
4. *Reverse transcriptase* is a viral enzyme which uses RNA as a template for the synthesis of DNA, unlike all other polymerases which use DNA as a template for synthesizing either DNA or RNA. It is essential for the construction of a complementary DNA (cDNA) library.
5. The *polymerase chain reaction (PCR)* is a process which amplifies a defined stretch of DNA. It requires a template (the DNA to be amplified), oligonucleotide primers (which define the limits of region to be amplified), and a thermostable DNA polymerase (Figure 1). The reaction mixture is first raised to 94°C to melt (separate) the DNA strands of the template. The temperature is lowered to about 37°C which allows the oligonucleotide primers to anneal to the single strands of DNA. The temperature is raised to 72°C, the temperature optimum for Taq polymerase, to permit the synthesis of the second strand starting at the oligonucleotide primer. This series of temperature steps is then repeated 20 to 30 times. PCR can easily amplify DNA a millionfold.
6. *Genomic DNA* is the genetic material of an organism. Within the genomic DNA there are segments of variable length called genes which code for specific proteins. In eukaryotic organisms, many genes are not continuous stretches of DNA. Rather, the regions which code for the protein, called exons, are interspersed with noncoding regions, called introns. The process of protein synthesis involves transcription of DNA into RNA in the nucleus. During the process of RNA synthesis the introns are spliced out of the mature *messenger RNA (mRNA)*. The mature mRNA is transported to the cytoplasm where it is translated into a protein by the ribosomes.

   The expression of individual genes is controlled by a series of regulatory elements called promoters and enhancers. *Promoters* are

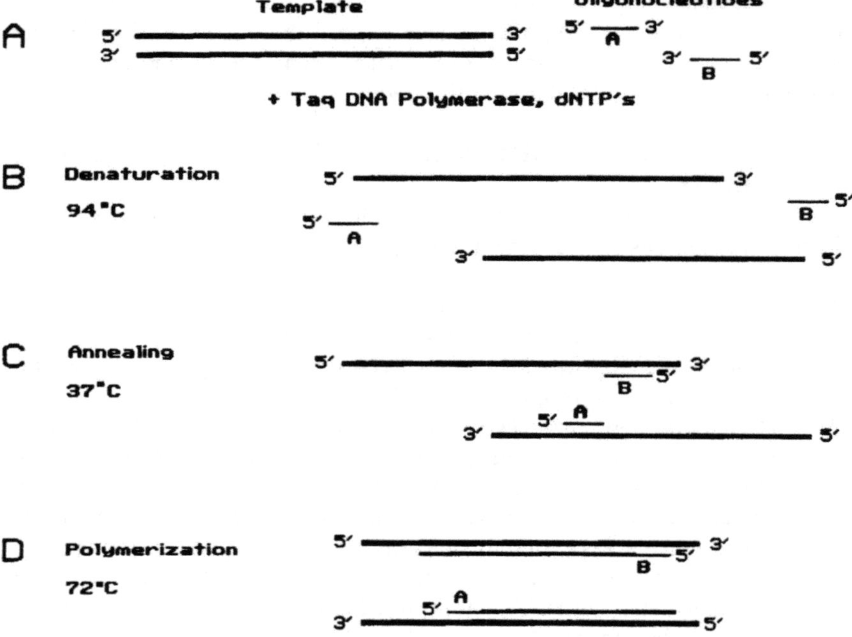

FIGURE 1. The polymerase chain reaction: (A) The ingredients for PCR, the template containing the DNA sequence to be amplified, the oligonucleotides, present in about 1000-fold molar excess compared to template, Taq DNA polymerase, a heat-stable DNA polymerase, and deoxyribonucleosides, dNTPs (dATP, dTTP, dCTP, dGTP). (B) In the first step the reaction mixture is heated to 94°C to separate the template DNA into single strands. (C) In the second step the mixture is cooled to between 37 to 55°C to allow the oligonucleotides to anneal to the single strand template DNA. (D) In the third step the temperature is raised to 72°C, the optimal temperature for extension from the oligonucleotide primers by Taq DNA polymerase. Steps (B), (C), and (D) are then repeated 20 to 30 times. In each subsequent cycle the product from previous cycles serves as template resulting in the tremendous amplification of the region between the oligonucleotide primers.

generally located close to the 5′ end of the gene; *enhancers* can be located anywhere, even thousands of base pairs away. *Transcription factors* are proteins which bind at regulatory elements. They may activate the expression of some genes and reduce the expression of other genes. Transcription factors may be tissue specific or they may be activated by hormones or environmental factors.[4-7]

7. *Complementary DNA (cDNA)* is a man-made form of DNA that is copied from mRNA using reverse transcriptase. cDNA lacks the introns (non-coding regions of DNA) that are present in eukaryotic genomic DNA. A cDNA coding for a protein generally consists of four regions: (1) a 5′ untranslated segment, (2) the coding region or open reading frame which codes for the specific protein, (3) a 3′ untranslated region and

(4) a poly A tail. In order to propagate a cDNA it is usually inserted into a *vector* which is either a bacteriophage (a virus which infects bacteria) or a plasmid. *Plasmids* are small circular pieces of DNA which can replicate in bacteria, often to the level of several hundred copies per cell, independently of the genomic DNA. Many different plasmids have been engineered for specific purposes, but in general they usually contain three major elements: (1) an origin of replication which permits the plasmid to replicate in the bacteria, (2) a selectable marker, usually an antibiotic resistance gene, which allows bacteria containing the plasmid to survive in the presence of the antibiotic, usually ampicillin, and (3) a multiple cloning site or polylinker which contains a series of unique restriction enzyme cleavage sites where the cDNA can be inserted. Plasmids may also contain initiation sites for RNA polymerases flanking the multiple cloning site which permits *in vitro* transcription of mRNA from the cDNA and eukaryotic promotors which facilitate expression of the cDNA in eukaryotic cells.

8. A *library* is a collection of recombinant vectors, containing either cDNAs or fragments of genomic DNA. In general, a cDNA library is constructed by purifying mRNA from the tissue of interest (Figure 2). The mRNA will code for all of the proteins that are being expressed in the tissue. cDNAs are synthesized from the mRNA and then ligated into an appropriate vector and transfected into bacteria.
9. *Molecular cloning* involves two distinct steps: (1) the construction of the cDNA library from mRNA purified from a specific tissue and (2) the development of suitable probes or assays to screen the cDNA library. The problem of screening the library originates in part from the large number of clones that must be screened, typically 10^5 to 10^6 clones. Therefore, the screening procedure must be both highly specific and highly sensitive. There are two basic approaches to screening a cDNA library: one based on structural information about the protein and the other based on the functional properties of the protein.

B. cDNA SCREENING USING STRUCTURAL INFORMATION

The classical method for screening a cDNA library is first to obtain structural information about the protein. This information is used to develop probes to identify the specific clones in the cDNA library. The first step in this process usually involves purifying the protein of interest. The purified protein is subjected to microsequencing to obtain partial amino acid sequence. Based on the amino acid sequence, oligonucleotide probes are synthesized. There is a complication in the synthesis of these oligonucleotides because many amino acids are coded for by more than one triplet codon. This is referred to as degeneracy. One must therefore synthesize an oligonucleotide probe which accounts for all or most of the potential codons for a given amino acid, that is, a degenerate oligonucleotide probe. These oligonucleotide probes are used to screen the cDNA library to identify the corresponding clones.[8,9]

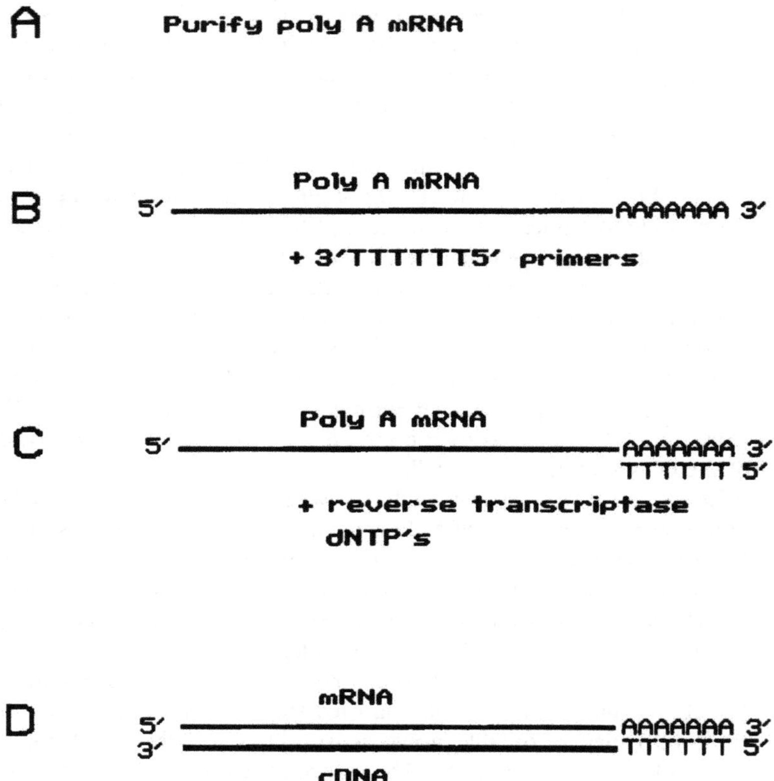

FIGURE 2. Synthesis of complementary DNA (cDNA). (A) Poly adenylated mRNA (poly A mRNA) is purified from the tissue of interest. (B) Oligonucleotide primers consisting of poly thymidine (5′ TTTTTT 3′) is added to the mRNA and allowed to anneal to the complementary poly A tail. This provides the initiation site necessary for reverse transcriptase. (C) Reverse transcriptase and deoxyribonucleosides, dNTPs (dATP, dTTP, dCTP, dGTP), are added. (D) Reverse transcriptase extends the poly T primer using the mRNA as template. The sequence of the DNA polymerized is complementary to the mRNA template. In subsequent steps the mRNA is eliminated, the second DNA strand is synthesized, and the resulting double-stranded cDNA is ligated into an appropriate vector and transfected into bacteria.

The screening process can be performed under conditions which require an exact or almost exact match between the probe and the clone, referred to as high stringency screening. Alternatively, screening conditions may be used which permit some degree of mismatch between the sequence of the probe and the clone, referred to as low stringency screening. This allows one to identify clones which are related to a specific protein.

If partial amino acid sequence information cannot be obtained, an alternative approach is to develop antibodies against the purified protein. The specific antibodies can be used to screen a cDNA library in an expression vector, such as lambda gt11.

A related approach, that has been widely used to clone tissue-specific members of a gene family, is to clone new members of the family based on homology to the members that have already been cloned.[10-15] This avoids the need to purify the protein from a specific tissue. The sequences of members of the gene family that have been cloned are compared. Highly conserved regions of the proteins are identified, and, based on the sequence of these regions, oligonucleotide probes are synthesized. These probes are used to screen a cDNA library derived from the tissue of interest at low stringency to identify related cDNAs.[11-13,16] Suitable oligonucleotides can also be devised to allow amplification of related cDNAs using the PCR. This is followed by cloning and sequencing of the amplified cDNAs.[10,14]

C. cDNA SCREENING USING FUNCTIONAL INFORMATION

The second approach to screening cDNA libraries is based on the functional properties of the protein to be cloned. In this approach, referred to as expression cloning, a method must be developed to express the protein of interest in a functional state. Screening is based on the ability to either express the function or knockout expression of the function. The major advantage of this approach is that it is not necessary to purify the protein to be cloned. This is particularly important in the case of proteins that are scarce, such as taste receptors, etc., or for other reasons are difficult to purify. The major disadvantages are (1) the need to develop a system to express the protein of interest and (2) the difficulty screening large numbers of clones in expression systems. A variety of functional properties can be used for screening including ligand binding, interaction with second messenger pathways, transport properties, enzyme activity, etc. One of the most extensively used expression systems for cloning, particularly for integral membrane proteins, is the Xenopus oocyte system.[17] In general, a cDNA library is constructed in an RNA expression vector which permits synthesis of mRNA by *in vitro* transcription. The library is divided into large pools of clones. mRNA is synthesized from each pool and injected into oocytes. The large size of the oocytes facilitates this process. Following injection of exogenous mRNA into Xenopus oocytes, the oocytes translate the injected mRNA and synthesize the corresponding proteins. If they are integral membrane proteins, they are expressed in the plasma membrane of the oocyte. Several days later the oocytes are assayed for the expression of the desired function. cDNA pools which express the desired function are subdivided and the process is repeated. Through an iterative process one ultimately obtains a single clone. There are several potential complications for this technique. First, the protein must be coded for by a single subunit. Second, in the case of G protein-linked receptors, the oocyte must express an appropriate G protein and second messenger pathways. Third, one must obtain a full length cDNA clone for the specific protein. Due to the inefficiencies of reverse transcriptase it is often difficult to obtain full length cDNA clones for long mRNAs. Despite these limitations, this process has been successfully used to clone the serotonin HT1c receptor (a G protein-linked receptor), the sodium-coupled glucose cotransporter and a variety of other proteins.[18,19]

A variation of this kind of expression cloning that has been used, is to knock out the expressed function.[20] In this case, one must have a large source of poly A mRNA that when it is injected into Xenopus oocytes induces the expression of the desired function. A cDNA library is subdivided into pools, and single-stranded complementary DNA is synthesized. This single-stranded cDNA is allowed to anneal with the poly A mRNA before it is injected into oocytes. Following injection into the oocytes, DNA:RNA hybrids are rapidly degraded by RNase H. cDNA pools which knock out the desired function are subdivided as above and through an iterative process a single clone, or several clones in the case of multi-subunit proteins, is/are obtained. The clones need not be full length, but can then be used to rescreen suitable libraries to obtain the entire cDNA. This variation of expression cloning has the advantage that it is not limited by the problems outlined above for expression cloning, but it does require a large source of poly A mRNA which may limit its usefulness in some situations. This technique has been successfully used to clone a chloride channel, from the electric organ of the ray, *Torpedo marmorata*.[20] This clone was then used to clone the homologous mammalian skeletal muscle chloride channel by the techniques described above.[15]

D. CLONING BY DNA-MEDIATED GENE TRANSFER

The cloning techniques described above all involve construction of a cDNA library. In order to accomplish this, one must first be able to isolate tissue-specific poly A mRNA. In some cases this is not possible. An alternative technique is to utilize DNA-mediated gene transfer to establish a cell line expressing the protein or function of interest.[21] In this technique, high molecular weight genomic DNA is transfected into a appropriate cell line, generally from a different species, that does not express the desired protein. During the transfection, recipient cells take up the exogenous DNA and integrate it into their own genome. On average each cell integrates about 1×10^6 base pairs of exogenous DNA, meaning that the entire human genome of 3×10^9 bp can be expressed in about 3000 independently transfected cells.[19,21] Following the transfection the cells are allowed to grow. The exogenous genes are expressed by the transfected cells. The transfected cells are screened to find cells expressing the desired protein, usually with a specific antibody or ligand. Cells expressing the desired protein are isolated and grown. High molecular weight genomic DNA is isolated from the expressing cell line and used to generate secondary transformants, which are rescreened for expression of the desired protein or function. The secondary transformants will contain only one or a few foreign genes. The gene of interest can be cloned from this highly enriched source. This technique has been used to clone a variety of proteins including aprt, lymphocyte CD4 and CD8 proteins, nerve growth factor receptor, etc.[19,23-25]

III. OLFACTION AND MOLECULAR BIOLOGY

Molecular biology has been applied more extensively to the study of the olfactory system than any other chemosensory modality. This is probably due to several factors. Of all of the chemosensory organs, the olfactory epithelium is the largest. The greater availability of olfactory tissue has led to a more extensive biochemical and physiological understanding of the process of olfactory transduction. This has provided the necessary foundation for the application of molecular cloning. I will first briefly review the biochemical and physiological information that has guided the molecular cloning of various components of the olfactory transduction system.[26,27] I will then examine the results of the cloning experiments and the insights that they have provided.

A. PHYSIOLOGY AND BIOCHEMISTRY OF OLFACTION

The first step in the olfactory transduction process involves receptors for the odorant molecules. Psychophysical experiments in humans strongly suggest that there are a large number of highly specific receptors for odorant compounds. First, the olfactory system displays stereospecificity with respect to certain odorants; that is, stereoisomers of certain compounds smell differently. This strongly supports the existence of specific receptors and argues against a nonspecific interaction with the cell membrane. Second, the compound androstenone is odorless to 1/3 of the population, smells urinous to 1/3, and perfume-like to the other third. There is a 100% concordance in the ability of monozygotic twins to smell androstenone, but only a 60% concordance in dizygotic twins.[28] This suggests that there is a genetic basis for the inheritance of the ability to smell androstenone and supports the hypothesis that there are specific receptors coded for by DNA. Third, over 30 specific anosmias have been documented in the human population suggesting the existence of a large family of specific receptors.[26,27] Based on these results and many others, it was hypothesized that a large family of specific receptor proteins are involved in the initial events in olfactory transduction and that these receptors are located in the olfactory cilia.

The seminal experiments of Lancet and co-workers and Snyder and co-workers suggested that the interaction of odorants with their receptors stimulated an increase in adenylyl cyclase activity.[29,30] They demonstrated that olfactory cilia possessed a very low basal level of adenylyl cyclase activity, but a very high level of activity, about 100 times higher than brain, following stimulation with nonodorant activators of adenylyl cyclase, such as forskolin. Odorants increased adenylyl cyclase activity and different odorants activated adenylyl cyclase to different levels. The activation by odorants was dependent on the presence of GTP, suggesting that G proteins were involved in the transduction process. Heterotrimeric G proteins were assumed to be coupling the olfactory receptors to adenylyl cyclase.[31] Biochemical studies demonstrated the presence of heterotrimeric GTP-binding proteins in olfactory cilia.

These G proteins were sensitive to ADP-ribosylation by cholera toxin suggesting that they were from the stimulatory G protein class, G_s.[31,32] Based on differences in sensitivity to various nonhydrolyzable GTP analogs, the authors suggested that the olfactory G protein might be a variant of the G_s protein.[32] Immunocytochemical studies also demonstrated the presence of G proteins in the olfactory epithelium.[33] These studies suggested that the olfactory receptors were members of the gene superfamily of seven transmembrane domain, G protein-linked receptors. They also suggested that a stimulatory G protein coupled the odorant receptors to adenylyl cyclase activation. This suggested that cAMP might act as an intracellular second messenger in the olfactory transduction process.

Biochemical purification of adenylyl cyclase from olfactory cilia suggested that the olfactory enzyme was distinct from other forms of the enzyme that had been purified from brain and myocardium.[34] The olfactory adenylyl cyclase was 180 kDa, 30 kDa larger than the brain enzyme. A monoclonal antibody against the brain form of adenylyl cyclase only slightly recognized the olfactory form. The turnover number of olfactory adenylyl cyclase was only 3.5 times higher than myocardial adenylyl cyclase, suggesting that most of the higher specific activity observed in olfactory cilia was due to higher content of adenylyl cyclase.[34] Thus, odorant binding to receptors stimulates an increase in activity of an olfactory-specific adenylyl cyclase. This results in the generation of cAMP.

The next step in the transduction process was suggested by the discovery of Nakamura and Gold that olfactory cilia contained cAMP-gated cation-selective channels, similar to the cGMP-gated channels in the visual transduction pathway.[8,35-37] The rise in cAMP induced by odorant binding was hypothesized to open the cAMP-gated channels resulting in depolarization of the olfactory neuron and firing of action potentials.[38] The discovery of this channel suggested that there were significant parallels between the olfactory and the visual transduction systems.[39,40]

The final step in the transduction process is termination of the signaling process. Termination must involve two distinct processes: One, the elimination of the odorant and, two, the metabolism of cAMP to turn off the transduction system. Using antibodies against liver cytochrome P450, several groups had suggested that this enzyme was present in the olfactory epithelium.[41] Lancet and co-workers purified two integral membrane proteins from bovine olfactory epithelium. Amino acid microsequencing of CNBr cleavage products of these proteins indicated that one was a cytochrome P450 and the other UDP-glucuronosyl transferase.[42] Using polyclonal antibodies against these two proteins, they demonstrated that they were localized in a microsomal membrane fraction derived from olfactory epithelium, but were not present in adjacent nasal respiratory epithelium. Most of these enzymes were located in the nonneuronal cells in the olfactory epithelium. They suggested that these enzymes may be involved in metabolism of odorants to nonodorous derivatives.[42]

There has been relatively little study of the enzymes involved in the degradation of cAMP in the olfactory neurons. By analogy to the visual system this is likely to be an important and highly regulated process probably involving olfactory-specific cAMP-phosphodiesterase.[43] It is also possible that receptor-kinases analogous to the β-adrenergic receptor-kinase or rhodopsin kinase may be involved in receptor-desensitization during prolonged odorant stimulation; however, further work will be necessary to demonstrate such activity.[44,45]

In summary, these biochemical and physiological studies suggested that the olfactory transduction pathway included (1) receptors that were members of the seven transmembrane domain, G protein-linked gene superfamily, (2) stimulatory heterotrimeric G proteins, (3) an olfactory-specific form of adenylyl cyclase, (4) an olfactory-specific cAMP-gated cation channel and (5) odorant metabolism pathways involving cytochrome P450 and UDP-glucuronosyl transferase. The analogy with the visual transduction system suggests that olfactory-specific cAMP-phosphodiesterase cascades and receptor-kinases are yet to be discovered which will control signal termination and adaptation processes.

B. MOLECULAR CLONING OF THE PROTEINS OF THE OLFACTORY TRANSDUCTION PATHWAY

1. Olfactory Receptors

Buck and Axel[10] devised a strategy to clone the olfactory receptors based on the hypothesis that the olfactory receptors would constitute a large family of genes which would be members of the seven transmembrane domain, G protein-linked gene superfamily.[46] Based on sequence comparison of known members of this gene superfamily, they identified conserved regions in transmembrane domains 2 and 7 and designed a series of degenerate PCR oligonucleotide primers. They used these primers with a rat olfactory cDNA library as the template for PCR amplification of presumptive olfactory receptor clones. Appropriately sized bands were digested with restriction enzymes to identify bands that contained multiple DNA sequences. Probes derived from such a PCR band were then used to screen a cDNA library to obtain multiple full length cDNA clones. Eighteen distinct cDNA clones were sequenced. The deduced protein sequences indicated that the clones clearly constituted a new multigene family within the superfamily of the seven transmembrane domain receptors. To support the contention that this gene family constituted the olfactory receptors, Northern blot analysis was used to show that mRNAs coding for this new gene family were expressed in olfactory epithelium, but not in a variety of other tissues including brain, heart, kidney, liver, lung, ovary, retina, or spleen. In addition, they demonstrated that expression was enriched in olfactory neurons compared to olfactory epithelium.

Analysis of the deduced protein sequences revealed several important features of the new family of receptors. By hydrophobicity analysis, each contained seven putative transmembrane domains, with a presumed extracellular N terminus and a cytoplasmic C terminus. There was one potential N-linked glycosylation site in the short N terminal domain. There were several potential serine and threonine phosphorylation sites in the third cytoplasmic loop which might be involved in receptor desensitization, as was shown for the adrenergic receptors.[46] Studies of other members of this gene superfamily have suggested that the third, fourth, and fifth membrane spanning domains formed part of the ligand-binding site. In those gene families which recognize a single ligand, such as the adrenergic or the muscarinic acetylcholine receptor families, these regions were highly conserved.[46] In the olfactory receptor family there was considerable sequence divergence in the third, fourth, and fifth putative membrane spanning domains. Since the ligands for the olfactory receptors are structurally diverse, one would expect that the ligand-binding site would not be conserved among members of the olfactory receptor gene family.

The analysis of the deduced protein sequences also suggested that there were at least three distinct subfamilies within the olfactory receptor gene family. Using genomic Southern blotting at high stringency, at least 70 different genes were identified. This sets a minimum for the number of olfactory receptor genes. Under the high stringency conditions used in this screening, more divergent members of a given subfamily were probably not detected. Lower stringency analysis suggested that there were 100 to 200 genes per haploid genome, but this too may be a lower limit for the total number of olfactory receptor genes. It is likely that other subfamilies of the olfactory receptor family exist, but were not identified by the original PCR screening procedure due to divergence from the sequences of the PCR primers. Additional experiments will be necessary to identify other olfactory receptor gene subfamilies.

The diversity of olfactory receptor genes does not appear to be due to DNA rearrangement and somatic mutation which is the mechanism used in the generation of immunoglobulin diversity.[47] It appears that each olfactory receptor is coded by a distinct gene. This raises the question of what is the mechanism of regulation of gene expression in individual olfactory neurons, that is, how an individual olfactory neuron determines which olfactory receptor gene(s) it will express.

The identification of a large family of olfactory receptors suggests that a significant portion of olfactory discrimination occurs at the level of the primary sensory neurons. This suggests that single olfactory neurons will express one or a limited number of olfactory receptor types. Now, it will be possible to examine whether single olfactory neurons express a single receptor gene, or whether they express multiple receptor genes. The importance of peripheral discrimination in the olfactory system would represent a significant difference from the visual system. In the visual system, only three types of photoreceptors are needed to discriminate the entire color spectrum. A given photoreceptor cell expresses only a single type of photoreceptor pigment. The ability to discriminate between colors is due to central nervous system processing of the relative intensity of stimulation of the three peripheral photoreceptors.[40]

Further work will be necessary to elucidate the physiology of the cloned olfactory receptors. At present it has not been demonstrated that these cloned receptors bind odorants or interact with the G protein-coupled second messenger systems. Nevertheless, the properties of this gene family that have already been elucidated strongly support the contention that these represent a portion of the odorant receptors.

Subsequent to the cloning of the odorant receptors, another group of investigators noted that a family of G protein-linked receptors, which they had cloned from dog and human testis by low stringency PCR but for which the ligands were unknown, belonged to the family of odorant receptors.[48] This family of G protein-linked receptors were expressed in sperm cells and were between 30 and 90% identical with the clones identified by Buck and Axel.[10,48] This raises the possibility that chemotaxis by sperm cells is mediated by receptors from the same family of receptors as the odorant receptors found in the olfactory neurons. Further work will be necessary to elucidate the role of odorant receptors in sperm physiology.

2. G Proteins in the Olfactory Epithelium

Biochemical evidence summarized above suggested the presence of heterotrimeric GTP-binding proteins in the olfactory epithelium and the involvement of a stimulatory G protein, G_s-like, in olfactory transduction. Reed and co-workers have cloned several G proteins from olfactory epithelium. To accomplish this they used screening strategies based on sequence conservation with G proteins which had been cloned from other tissues. Initially, using a partially degenerate oligonucleotide probe to a highly conserved 18 amino acid stretch in all G_α subunits, they cloned five different G_α subunits, G_s, G_o, and three G_is.[16] The G_s subunit was not enriched in olfactory neurons which suggested that it was not involved in odorant signal transduction. Using a fully degenerate oligonucleotide probe at a lower stringency, they identified a new G protein of the G_s class which they named G_{olf}.[49] The cDNA for G_{olf} had an open reading frame coding for a protein of 381 amino acids. The protein was 88% identical with $G_{s\alpha}$. mRNA for G_{olf} was only expressed in olfactory epithelium. The level of mRNA expression for G_{olf} declined after ipsilateral bulbectomy, which causes ipsilateral degeneration of the olfactory neurons. This strongly supported the localization of G_{olf} expression in the olfactory sensory neurons. Antipeptide antibodies to a unique sequence in G_{olf} were also generated. Using this antibody, G_{olf} protein was shown to be present only in the sensory neurons, particularly in the region of the olfactory cilia. Finally they demonstrated that expression of G_{olf} in the S49 cyc^- kin^- cell line, which lacks $G_{s\alpha}$, reconstituted stimulatory G protein activity. Thus, demonstrating that G_{olf} behaved functionally as a stimulatory G protein.[49,50]

It thus appears that a unique stimulatory $G_{s\alpha}$ protein, G_{olf}, has evolved in the olfactory system to couple the odorant receptors to adenylyl cyclase. It is interesting that in the visual transduction system a unique G protein, transducin, has also evolved to couple the photoreceptor proteins to their effector target, cyclic GMP-phosphodiesterase.[31,43]

3. Olfactory-Specific Adenylyl Cyclase

Biochemical studies had suggested the existence of an olfactory-specific adenylyl cyclase.[34] Oligonucleotide probes derived from the sequence of brain adenylyl cyclase (type I), which had been previously cloned, were used to screen a rat olfactory cDNA library at low stringency. A class of cDNA clones was identified that coded for a novel adenylyl cyclase (type III).[11] An open reading frame was identified which coded for a protein 1144 amino acids long, similar in length to the type I enzyme. Both proteins had similar hydrophobicity profiles, with two hydrophobic regions, one at the N terminus and one in the middle of the protein. Each hydrophobic region contained six putative membrane-spanning regions. There are three potential N-glycosylation sites in the type III sequence, but only one in the type I sequence, which may explain the difference in molecular weight observed in the biochemical study of these enzymes.[34] The greatest amount of sequence conservation between the type I and III enzymes was in the putative cytoplasmic domains. These are presumably the enzymatically active regions of the proteins.

By Northern blot analysis, type III adenylyl cyclase mRNA was only expressed in olfactory epithelium. In addition, the level of type III mRNA disappeared following bulbectomy. This implies that the type III mRNA is expressed in the sensory neurons and not in the supporting cells. The expression of the type III protein was studied by immunocytochemistry using antipeptide antibodies specific for the type III form of adenylyl cyclase. This demonstrated that type III adenylyl cyclase was located in the olfactory cilia, the presumed site of odorant detection and transduction.[11]

The cloned cDNAs for the type I and III enzymes were transfected into a clonal cell line to allow the functional properties of the two isoforms of adenylyl cyclase to be studied. The major distinction between the two forms of the enzyme was that the level of basal activity of the olfactory-specific form was about 50-fold lower than the brain form of the enzyme. This low level of basal activity confirmed previous observations of adenylyl cyclase activity in olfactory cilia.[29,30]

4. Olfactory-Specific cAMP-Gated Cation-Selective Channel

Electrophysiological studies of olfactory cilia and cells had indicated the existence of a cAMP-gated cation-selective channel with properties similar to those that had been reported for the cGMP-gated channel in rod cells of the visual transduction system.[35-37] Shortly after the cloning of the cGMP-gated channel from the photoreceptor cells, a closely related channel was cloned from an olfactory epithelial cDNA library by low stringency screening using the photoreceptor channel cDNA as a probe.[8,12,13,39,51] The olfactory and photoreceptor channels are very closely related, displaying 57% amino acid sequence identity. The olfactory channel cDNA had an open reading frame encoding a protein of 664 amino acids. Hydrophobicity analysis revealed six putative hydrophobic membrane spanning domains. It suggested that the N terminus is in the cytoplasm, as is the C terminus which contains the cyclic nucleotide binding domain. There are two potential N-linked glycosylation sites, but based on the proposed topology only one of them is on the extracellular domain of the protein. Northern blot analysis showed that the mRNA for the olfactory channel was only expressed in olfactory epithelium. The level of mRNA expression was markedly reduced by bulbectomy. This suggests that the channel was expressed in the olfactory sensory neurons and not in the supporting cells whose number is not altered by bulbectomy.

The electrophysiologic properties of the cloned olfactory cAMP-gated channel were studied following transient expression of the cDNA.[12,51] Patch-clamp recordings demonstrated the presence of a cyclic nucleotide-gated conductance in the transfected cells. The $K_{1/2}$ was about 40 μM for cAMP, 2 μM for cGMP, and 88 μM for cCMP. In the studies by Nakamura and Gold of the cAMP-gated channel in olfactory cilia, a bimodal distribution of $K_{1/2}$s for cAMP was observed with peaks at 2 and 40 μM.[35] The origin of this difference between the native channel and the cloned channel is unclear. The cloned channel was cation-selective, essentially equally permeant to monovalent cations, but blocked in a voltage-dependent manner by calcium and other divalent cations as was previously described for the channel in olfactory cilia.[12,35,37,51]

5. Enzymes Involved in Signal Termination

Signal termination is an important aspect of sensory transduction. Odorant molecules, which in general are volatile and lipophilic, need to be cleared from the olfactory epithelium. Recent work by Lancet and co-workers suggests that a combination of enzymes are involved in this process. They have cloned three olfactory-specific proteins, two members of the cytochrome P450 family and UDP-glucuronosyl transferase (UGT_{olf}).[52,53] These olfactory-specific enzymes were highly similar to homologous enzymes that have been cloned from liver. Immunolocalization studies have shown that these enzymes are mainly expressed in the subepithelial Bowman's glands and also in the supporting cells of the olfactory epithelium.[52-54] They do not appear to be highly expressed in the olfactory sensory neurons. Following functional expression of UGT_{olf} in transfected COS cells, it was shown that the substrate specificity matched that observed in microsomes derived from olfactory epithelium, but was different from the liver UGT enzyme. Thus, as with the proteins of the transduction pathway, the termination pathway has evolved olfactory-specific counterparts of enzymes found in other organs.

6. Other Olfactory-Epithelium Related Proteins

Several proteins have been purified and/or cloned from olfactory tissue that are not directly involved in the processes of olfactory transduction or signal termination. They include olfactory marker protein (OMP), odorant-binding protein (OBP), and a platelet-derived growth factor receptor.[55-58] OMP was originally purified from olfactory epithelium and identified as a protein specifically expressed by olfactory neurons. It is a soluble cytoplasmic protein with a molecular weight of 18,500 daltons. It constitutes about 1% of the cytoplasmic protein in the olfactory bulb. The actual function of this protein remains uncertain, but it has been cloned and the amino acid sequence derived. The sequence shares no homology with any other known proteins.[58]

Odorant-binding protein was initially identified by its ability to bind radioactively labeled odorant molecules. It has been purified to homogeneity by several groups.[27] It was a soluble protein with a molecular weight of about 19,000 daltons. Immunohistochemical studies showed that it was synthesized by the cells of the nasal glands, not by the olfactory neurons. It is secreted into the mucus layer overlying the olfactory cilia. The protein has been cloned from both frog and rat.[55,56] From the derived amino acid sequence it was apparent that this protein is a member of a family of proteins which serve as carriers for small lipophilic molecules such as α_2-microglobulin, retinol-binding protein, apolipoprotein D, etc. It has been suggested that OBP helps to transport lipophilic odorants across the mucus layer to the cell membrane of the olfactory cilia where the odorant receptors are presumably located.[27,55,56]

Reed and co-workers have identified an α-platelet-derived growth factor receptor (α-PDGF receptor) cDNA while screening an olfactory epithelium-derived cDNA library.[57] The receptor is similar to other PDGF receptors that have been cloned. It codes for a protein of about 120 kDa, with 8 potential N-glycosylation sites in the N terminal domain, a single membrane spanning segment, and a tyrosine kinase in the C terminal domain. Following expression of a full length cDNA in COS cells, the transfected cells expressed a PDGF receptor. In the olfactory epithelium and bulb, the α-PDGF receptor was expressed in nonneuronal cells, but the same PDGF receptor was also expressed in a variety of nonolfactory organs. It was hypothesized that the function of the PDGF receptor in olfactory tissue is to regulate the growth of glial cells, but this remains to be demonstrated.[57]

IV. TASTE AND MOLECULAR BIOLOGY

Over the past few years physiological experiments have begun to elucidate the mechanisms of transduction involved in various taste modalities.[59] These advances have come largely through the application of techniques capable of studying the responses of individual cells, such as patch-clamp recording and single cell microfluorimetry with calcium-sensitive fluorescent dyes. Biochemical studies of taste transduction have been limited by the lack of a preparation comparable to the olfactory cilia preparation, which is highly enriched in olfactory neuronal proteins. This difficulty in obtaining a preparation of taste receptor cells that is not contaminated by surrounding nonsensory lingual epithelial cells has limited biochemical studies of taste transduction.

In the absence of biochemical information on the proteins involved in taste transduction, one must use information based on homology to proteins cloned from other tissues in order to apply molecular biology to identify taste-specific proteins. Based on considerable physiologic information (reviewed below for specific taste modalities) which suggested that G-proteins were involved in some modalities of taste transduction, Margolskee and co-workers synthesized a series of degenerate oligonucleotide probes to conserved regions in the α-subunit of G-proteins. These oligonucleotides were used to amplify cDNA's encoding G-proteins from a taste-tissue-specific cDNA library. Six different α-subunits were identified. Five were identical to previously cloned α-subunits, but the sixth was a previously uncloned α-subunit whose expression was taste cell specific.[60] This taste cell specific G-protein α-subunit was named gustducin. It is homologous to transducin, a G-protein involved in visual transduction. Gustducin is a member of the α_i family and therefore is not involved in coupling taste receptors to stimulation of adenylate cyclase. The receptor and target for gustducin are unknown at present.

Sufficient information is now available to provide a foundation for using a similar approach, i.e., based on homology to previously cloned proteins, to apply molecular biology to the cloning of other components of the taste transduction pathways. In the remainder of this chapter I will review the mechanisms of transduction of various taste modalities highlighting the information which provides a basis for the application of molecular biology to the study of taste transduction mechanisms. This will not be a detailed review of taste transduction mechanisms which is beyond the scope of this chapter.[59]

A. BITTER TASTE

Over 50 years ago, genetic studies of human bitter taste perception identified a dimorphism in the human population in the ability to taste phenylthiocarbamide (PTC) as bitter. The inability to taste PTC was inherited in an autosomally recessive manner, whereas the detection of other bitter substances was unrelated to the ability to taste PTC. This suggests that there are at least two receptors involved in bitter taste and shows that proteins, coded for by DNA, are likely to be the receptors for bitter substances. The genetics of bitter taste perception has been studied more extensively in mice. The ability of different strains of mice to taste a variety of bitter substances has been shown to be genetically determined by a series of independent, autosomal, monogenic loci.[61-66] These loci were shown to be closely linked.[65] This suggests that there is a family of bitter taste receptors.

Insights into the nature of the bitter taste receptors and the transduction process has come from both physiological and biochemical experiments. Akabas et al.[67] showed that the bitter substance denatonium induced release of calcium from internal stores in a subpopulation of taste cells. Release of calcium from internal stores generally involves the second messenger inositol triphosphate (IP_3), which is generated by a G protein-coupled phospholipase C. In addition, since denatonium is membrane-impermeant this suggested that the receptor was a cell surface protein, most likely a member of the seven transmembrane domain, G protein-linked receptor superfamily. Subsequent biochemical studies confirmed that denatonium induced an increase in the turnover of IP_3 in a membrane preparation of rat lingual epithelium.[68]

Taken together these results suggest that bitter taste transduction involves (1) a family of receptors which are members of the seven transmembrane domain, G protein-linked gene superfamily, (2) G proteins, which by analogy to the visual and olfactory systems, may be gustatory-specific, and (3) a phospholipase C. Since the denatonium receptor appears to activate the IP_3 pathway much the same as the serotonin HT1c receptor, perhaps it too could be cloned by expression cloning in Xenopus oocytes in a manner similar to that used to clone the serotonin HT1c receptor.[18]

B. SWEET TASTE

The transduction of sweet taste has been the focus of considerable study over the past few years. Genetic studies from mice suggest that sucrose, saccharin, dulcin, and acesulfame have a common receptor, but physiological studies suggest that there are multiple receptors.[69,70] Lancet and co-workers showed that sweet substances activated adenylyl cyclase in a GTP-dependent fashion in a lingual epithelial membrane preparation.[71] Using patch-clamp recording, other groups have shown that activation of cAMP-dependent protein kinase (A-kinase) stimulated the closing of a class of potassium channels in taste cells leading to depolarization of the cells and firing of action potentials.[72,73] These results suggest that the sweet taste receptor(s) are probably also members of the seven transmembrane domain, G protein-linked receptor superfamily. They also suggest that a stimulatory G protein, adenylyl cyclase, and A-kinase are involved in the transduction process. Any or all of these proteins might be gustatory specific; further experiments will be needed to determine this issue. There is a caveat to these results. In studies of ion transport across lingual epithelium, several groups have shown that sugars increased the short circuit current through the epithelium, but various stimulators of second messenger pathways including 8-Br-cAMP, 8-Br-cGMP, forskolin, adenosine, and A23187 had no effect on short circuit current.[74] Further work will be necessary to resolve these apparently contradictory results.

C. SALT TASTE

The mechanism of salt taste transduction is more clearly defined than the other taste modalities. Salt-sensitive taste cells possess an amiloride-sensitive Na^+ selective ion channel.[75-79] Sodium ions on the surface of the tongue presumably pass through this channel depolarizing the taste cell, resulting in firing of action potentials and presumably the secretion of neurotransmitter.[75-78] Several groups are cloning the component proteins of amiloride-sensitive sodium channels from various other epithelia; perhaps probes from these groups could be used to clone the gustatory channel whose properties are distinct from those observed in similar channels from other tissues.[59,77,79]

Several studies have indicated that salt deprivation or volume depletion influences the perceived intensity or the magnitude of the integrated chorda tympani nerve response to a given salt stimulus.[59,80-83] This suggests that the salt taste cells may be regulated by hormonal systems which regulate body volume status (also see Chapter 6). One of the hormonal systems which is central to the body's response to volume depletion is the renin-angiotensin-aldosterone system.[83] It is quite possible that salt taste cells express receptors for angiotensin II and aldosterone. Receptors for both of these hormones have been cloned.[84-88] It may be possible using probes derived from these cDNA clones to demonstrate the expression of these receptors in a subpopulation of taste cells by *in situ* hybridization. Also, since angiotensin II stimulates an increase in intracellular calcium, it might be possible to demonstrate the functional presence of these receptors using calcium-sensitive fluorescent dyes and single cell microfluorimetry.[59,67]

D. SOUR TASTE

Several mechanisms have been proposed for the transduction of sour taste. At this time the details have not been clarified sufficiently to form the basis for molecular biological studies of sour taste.

E. UMAMI TASTE

The unique taste of monosodium glutamate has been called the umami taste. The molecular basis of umami taste transduction is unknown, but glutamate receptors in the central nervous system (CNS) have been extensively studied. Glutamate is a major excitatory neurotransmitter in the mammalian CNS. Upon binding glutamate, the ligand-gated ion channel receptors open a cation-selective channel leading to depolarization of the cell. Several distinctive classes of glutamate receptors have been defined based on differences in response to a variety of agonists and antagonists. Over the past few years a number of these glutamate receptors have been cloned and sequenced.[89-91] The question arises as to whether one of these CNS glutamate receptors or a close relative is expressed in a subpopulation of taste cells. Alternatively, glutamate might be the neurotransmitter between the taste cells and the gustatory nerves. Glutamate applied to the surface of the tongue may have limited access to some post-synaptic glutamate receptors. Using a panel of agonists and antagonists the pharmacology of the gustatory glutamate receptors was studied. However, the gustatory receptors did not fit into one of the classes of CNS glutamate receptors.[92] Nevertheless, using probes derived from the cloned CNS glutamate receptors at low stringency, it may be possible to identify gustatory-specific glutamate receptors.

F. AMINO ACID TASTE IN CATFISH

Catfish offer several unique aspects for the study of taste transduction. Foremost is the large number of gustatory cells which has facilitated extensive biochemical studies of taste transduction in catfish. The major drawback in studying taste transduction in catfish is that they taste amino acids rather than the more traditional bitter, sweet, salty, and sour modalities found in other animals. Biochemical and neurophysiological studies have identified at least two major taste receptor systems in catfish: one, responsive to L-alanine and other short chain neutral amino acids, which appears to be a G protein-linked receptor, and, two, responsive to L-arginine, which appears to be a ligand-gated cation channel.[93-95]

Since one can obtain relatively large amounts of taste tissue from catfish, it should certainly be possible to construct gustatory-specific cDNA libraries in RNA expression vectors which could be used for expression cloning studies of the catfish gustatory receptors.

G. FUTURE PROSPECTS FOR THE APPLICATION OF MOLECULAR BIOLOGY TO THE STUDY OF TASTE TRANSDUCTION MECHANISMS

The application of molecular biology to the study of taste transduction is just beginning, but the possibilities for future studies are plentiful. Considerable evidence has accumulated to suggest that the bitter and sweet taste receptors are members of the seven transmembrane domain, G protein-linked receptor superfamily.[39,40,71] Many procedures have been developed for cloning new members of this gene superfamily.[10,14,48] Likewise, similar techniques are available to clone new members of the G protein gene family, and preliminary results suggesting the existence of a gustatory-specific G protein have been reported.[60] Other possible targets for cloning in the gustatory system include the glutamate receptors and amino acid receptors from catfish.

V. INSECT PHEROMONE AND ODORANT RECEPTION AND MOLECULAR BIOLOGY

A class of low molecular weight, water soluble proteins has been purified from insect sensillum lymph, the fluid which surrounds the sensory cells. These proteins are hypothesized to bind pheromones or odorants and carry them through the sensilum lymph to the receptors on the surface of the sensory neurons. The purified proteins were subjected to N-terminal microsequencing to obtain a partial amino acid sequence. Degenerate oligonucleotide probes were synthesized based on the amino acid sequence and were used to screen cDNA libraries. cDNA's coding for pheromone-binding and odorant-binding proteins has been identified from several species of insects.[96-100] All of these cDNAs code for proteins of 140 to 170 amino acids in length. They all have an N-terminal signal sequence, consistent with the extracellular localization of the mature proteins. All of these proteins contain six conserved cysteine residues, suggesting that disulfide bonds are important in maintaining the structure of the mature proteins. Among various species of moths the pheromone-binding proteins are 60 to 90% similar in amino acid sequence.[99] However, the odorant-binding proteins are only about 30% similar to the pheromone-binding proteins.[96,97,99,100] The insect odorant-binding proteins are not homologous to the odorant-binding proteins cloned from vertebrates.[97]

VI. FUTURE DIRECTIONS

Molecular biology has already made substantial contributions to the study of olfactory transduction mechanisms. With the cloning of a number of olfactory-specific genes it should now be possible to identify gene regulatory elements that control the expression of olfactory-specific genes. This may permit the identification of master control genes which regulate the differentiation of basal cells into mature olfactory neurons. This will enable us to understand the process of development of olfactory neurons at a molecular level.

Molecular studies of the gustatory system are still in their infancy, but advances in our understanding of gustatory physiology and biochemistry are providing the foundation for the application of molecular biology to the study of taste transduction and to the study of development and differentiation of taste cells.

ACKNOWLEDGMENT

This work was supported in part by grants from the National Institute of Health, DC01019, and the National Science Foundation, BNS-8808028.

REFERENCES

1. **Darnell, J. E., Lodish, H. F., and Baltimore, D.,** *Molecular Cell Biology,* W. H. Freeman, New York, 1986.
2. **Sambrook, J., Fritsch, E. F., and Maniatis, T.,** *Molecular Cloning: A Laboratory Manual,* 2nd ed., 3 vols., Cold Spring Harbor Laboratory Press, Cold Spring Harbor, New York, 1989.
3. **Berger, S. L. and Kimmel, A. R.,** *Guide to Molecular Cloning Techniques,* Academic Press, New York, 1987.
4. **Evans, R. M.,** The steroid and thyroid hormone receptor superfamily, *Science,* 240, 889, 1988.

5. **Evans, T., Felsenfeld, G., and Reitman, M.,** Control of globin gene transcription, *Annu. Rev. Cell Biol.*, 6, 95, 1990.
6. **Davis, R. L., Cheng, P. F., Lassar, A. B., and Weintraub, H.,** The MyoD DNA binding domain contains a recognition code for muscle-specific gene activation, *Cell*, 60, 733, 1990.
7. **Weintraub, H., Dwarki, V. J., Verma, I., Davis, R., Hollenberg, S., Snider, L., Lassar, A., and Tapscott, S. J.,** Muscle-specific transcriptional activation by MyoD, *Genes Dev.*, 5, 1377, 1991.
8. **Kaupp, U. B., Niidome, T., Tanabe, T., Terada, S., Bonigk, W., Stuhmer, W., Cook, N. J., Kangawa, K., Matsuo, H., Hirose, T., Miyata, T., and Numa, S.,** Primary structure and functional expression from complementary DNA of the rod photoreceptor cyclic GMP-gated channel, *Nature*, 342, 762, 1989.
9. **Numa, S.,** A molecular view of neurotransmitter receptors and ionic channels, *Harvey Lect.*, 83, 121, 1987–88.
10. **Buck, L. and Axel, R.,** A novel multigene family may encode odorant receptors: a molecular basis for odor recognition, *Cell*, 65, 175, 1991.
11. **Bakalyar, H. A. and Reed, R. R.,** Identification of a specialized adenylyl cyclase that may mediate odorant detection, *Science*, 250, 1403, 1990.
12. **Dhallan, R. S., Yau, K.-W., Schrader, K. A., and Reed, R. R.,** Primary structure and functional expression of a cyclic nucleotide-activated channel from olfactory neurons, *Nature*, 347, 184, 1990.
13. **Ludwig, J., Margalit, T., Eismann, E., Lancet, D., and Kaupp, U. B.,** Primary structure of cAMP-gated channel from bovine olfactory epithelium, *FEBS Lett.*, 270, 24, 1990.
14. **Libert, F., Parmentier, M., Lefort, A., Dinsart, C., van Sande, J., Maenhaut, C., Simons, M.-J., Dumont, J. E., and Vassart, G.,** Selective amplification and cloning of four new members of the G protein-coupled receptor family, *Science*, 244, 569, 1989.
15. **Steinmeyer, K., Ortland, C., and Jentsch, T. J.,** Primary structure and functional expression of a developmentally regulated skeletal muscle chloride channel, *Nature*, 354, 301, 1991.
16. **Jones, D. T. and Reed, R. R.,** Molecular cloning of five GTP-binding protein cDNA species from rat olfactory neuroepithelium, *J. Biol. Chem.*, 262, 14241, 1987.
17. **Sigel, E.,** Use of Xenopus oocytes for the functional expression of plasma membrane proteins, *J. Membr. Biol.*, 117, 201, 1990.
18. **Julius, D., MacDermott, A. B., Axel, R., and Jessell, T. M.,** Molecular characterization of a functional cDNA encoding the serotonin 1c receptor, *Science*, 241, 558, 1988.
19. **Hediger, M. A., Coady, M. J., Ikeda, T. S., and Wright, E. M.,** Expression cloning and cDNA sequencing of the Na^+/glucose co-transporter, *Nature*, 330, 379, 1987.
20. **Jentsch, T. J., Steinmeyer, K., and Schwarz, G.,** Primary structure of Torpedo marmorata chloride channel isolated by expression cloning in Xenopus oocytes, *Nature*, 348, 510, 1990.
21. **Pellicer, A., Robins, D., Wold, B., Sweet, R., Jackson, J., Lowy, I., Roberts, J. M., Sim, G.-K., Silverstein, S., and Axel, R.,** Altering genotype and phenotype by DNA-mediated gene transfer, *Science*, 209, 1414, 1980.
22. **Lowy, I., Pellicer, A., Jackson, J. F., Sim, G.-K., Silverstein, S., and Axel, R.,** Isolation of transforming DNA: cloning the hamster part gene, *Cell*, 22, 817, 1980.
23. **Maddon, P. J., Littman, D. R., Godfrey, M., Maddon, D. E., Chess, L., and Axel, R.,** The isolation and nucleotide sequence of a cDNA encoding the T cell surface protein T4: a new member of the immunoglobulin gene family, *Cell*, 42, 93, 1985.
24. **Littman, D. R., Thomas, Y., Maddon, P. J., Chess, L., and Axel, R.,** The isolation and sequence of the gene encoding T8: a molecule defining functional classes of T lymphocytes, *Cell*, 40, 237, 1985.

25. **Chao, M. V., Bothwell, M. A., Ross, A. H., Koprowski, H., Lanahan, A. A., Buck, C. R., and Sehgal, A.,** Gene transfer and molecular cloning of the human NGF receptor, *Science*, 232, 518, 1986.
26. **Lancet, D.,** Vertebrate olfactory reception, *Annu. Rev. Neurosci.*, 9, 329, 1986.
27. **Snyder, S. H., Sklar, P. B., and Pevsner, J.,** Molecular mechanisms of olfaction, *J. Biol. Chem.*, 263, 13971, 1988.
28. **Wysocki, C., Jr. and Beauchamp, G.,** Ability to smell androstenone is genetically determined, *Proc. Natl. Acad. Sci. U.S.A.*, 81, 4899, 1984.
29. **Pace, U., Hanski, E., Salomon, Y., and Lancet, D.,** Odorant-sensitive adenylate cyclase may mediate olfactory reception, *Nature*, 316, 255, 1985.
30. **Sklar, P. B., Anholt, R. R., and Snyder, S. H.,** The odorant-sensitive adenylate cyclase of olfactory receptor cells. Differential stimulation by distinct classes of odorants, *J. Biol. Chem.*, 261, 15538, 1986.
31. **Gilman, A.,** G proteins: transducers of receptor-generated signals, *Annu. Rev. Biochem.*, 56, 615, 1987.
32. **Pace, U. and Lancet, D.,** Olfactory GTP-bonding protein: signal-transducing polypeptide of vertebrate chemosensory neurons, *Proc. Natl. Acad. Sci. U.S.A.*, 83, 4947, 1986.
33. **Anholt, R. R. H., Mumby, S. M., Stoffers, D. A., Girard, P. R., Kuo, J. F., and Snyder, S. H.,** Transduction proteins of olfactory receptor cells: identification of guanine nucleotide binding proteins and protein kinase C, *Biochemistry*, 26, 788, 1987.
34. **Pfeuffer, E., Mollner, S., Lancet, D., and Pfeuffer, T.,** Olfactory adenylyl cyclase: identification and purification of a novel enzyme form, *J. Biol. Chem.*, 264, 18803, 1989.
35. **Nakamura, T. and Gold, G. H.,** A cyclic nucleotide-gated conductance in olfactory receptor cilia, *Nature*, 325, 442, 1987.
36. **Yau, K.-W. and Baylor, D. A.,** Cyclic GMP-activated conductance of retinal photoreceptor cells, *Annu. Rev. Neurosci.*, 12, 289, 1989.
37. **Zufall, F., Firestein, S., and Shepherd, G. M.,** Analysis of single cyclic nucleotide-gated channels in olfactory receptor cells, *J. Neurosci.*, 11, 3573, 1991.
38. **Firestein, S. and Werblin, F.,** Odor-induced membrane currents in vertebrate olfactory receptor neurons, *Science*, 244, 79, 1989.
39. **Kaupp, U. B.,** The cyclic nucleotide-gated channels of vertebrate photoreceptors and olfactory epithelium, *Trends Neurosci.*, 14, 150, 1991.
40. **Nathans, J., Thomas, D., and Hogness, D. S.,** Molecular genetics of human color vision: the genes encoding blue, green and red pigments, *Science*, 232, 193, 1986.
41. **Dahl, A. R.,** The effect of cytochrome P-450 dependent metabolism and other enzyme activities on olfaction, in *Molecular Neurobiology of the Olfactory System*, Margolis, F. L. and Getchell, T. V., Eds., Plenum Press, New York, 1988, 51.
42. **Lazard, D., Tal, N., Rubinstein, M., Khen, M., Lancet, D., and Zupko, K.,** Identification and biochemical analysis of novel olfactory-specific cytochrome P-450IIA and UDP-glucuronosyl transferase, *Biochemistry*, 29, 7433, 1990.
43. **Stryer, L.,** Visual excitation and recovery, *J. Biol. Chem.*, 266, 10711, 1991.
44. **Lorenz, W., Inglese, J., Palczewski, K., Onorato, J. J., Caron, M. G., and Lefkowitz, R. J.,** The receptor kinase family: primary structure of rhodopsin kinase reveals similarities to the beta-adrenergic receptor kinase, *Proc. Natl. Acad. Sci. U.S.A.*, 88, 8715, 1991.
45. **Boekhoff, I. and Breer, H.,** Termination of second messenger signalling in olfaction, *Proc. Natl. Acad. Sci. U.S.A.*, 89, 471, 1992.
46. **Dohlman, H. G., Thorner, J., Caron, M. G., and Lefkowitz, R. J.,** Model systems for the study of seven transmembrane-segment receptors, *Annu. Rev. Biochem.*, 60, 653, 1991.
47. **Tonegawa, S.,** Somatic generation of immune diversity, *Biosci. Rep.*, 8, 3, 1988.

48. **Parmentier, M., Libert, F., Schurmans, S., Schiffmann, S., Lefort, A., Eggerickx, D., Ledent, C., Mollereau, C., Gerard, C., Perret, J., Groodegoed, A., and Vassart, G.,** Expression of members of the putative olfactory receptor gene family in mammalian germ cells, *Nature*, 355, 453, 1992.
49. **Jones, D. T. and Reed, R. R.,** G_{olf}: an olfactory neuron specific-G protein involved in odorant dignal transduction, *Science*, 244, 790, 1989.
50. **Jones, D. T., Masters, S. B., Bourne, H. R., and Reed, R. R.,** Biochemical characterization of three stimulatory GTP binding proteins, *J. Biol. Chem.*, 265, 2671, 1990.
51. **Altenhofen, W., Ludwig, J., Eismann, E., Kraus, W., Bonigk, W., and Kaupp, U. B.,** Control of ligand specificity in cyclic nucleotide-gated channels from rod photoreceptors and olfactory epithelium, *Proc. Natl. Acad. Sci. U.S.A.*, 88, 9868, 1991.
52. **Nef, P., Heldman, J., Lazard, D., Margalit, T., Jaye, M., Hanukoglu, I., and Lancet, D.,** Olfactory-specific cytochrome P-450: cDNA cloning of a novel neuroepithelial enzyme possibly involved in chemoreception, *J. Biol. Chem.*, 264, 6780, 1989.
53. **Lazard, D., Zupko, K., Poria, Y., Nef, P., Lazarovits, J., Horn, S., Khen, M., and Lancet, D.,** Odorant signal termination by olfactory UDP gluronosyl transferase, *Nature*, 349, 790, 1991.
54. **Zupko, K., Poria, Y., and Lancet, D.,** Immunolocalization of cytochromes P-450olf1 and P-450olf2 in rat olfactory mucosa, *Eur. J. Biochem.*, 196, 51, 1991.
55. **Lee, K.-H., Wells, R. G., and Reed, R. R.,** Isolation of an olfactory cDNA: similarity to retinol-binding protein suggests a role in olfaction, *Science*, 235. 1053, 1987.
56. **Pevsner, J., Reed, R. R., Feinstein, P. G., and Snyder, S. H.,** Molecular cloning of odorant-binding protein: member of a ligand carrier family, *Science*, 241, 336, 1988.
57. **Lee, K.-H., Bowen-Pope, D. F., and Reed, R. R.,** Isolation and characterization of the α platelet-derived growth factor receptor from rat olfactory epithelium, *Mol. Cell. Biol.*, 10, 2237, 1990.
58. **Rogers, K. E., Dasgupta, P., Gubler, U., Grillo, M., Khew-Goodall, Y. S., and Margolis, F. L.,** Molecular cloning and sequencing of a cDNA for olfactory marker protein, *Proc. Natl. Acad. Sci. U.S.A.*, 84, 1704, 1987.
59. **Akabas, M. H.,** Mechanisms of chemosensory transduction in taste cells, *Int. Rev. Neurobiol.*, 32, 241, 1990.
60. **McLaughlin, S. K., McKinnon, P. J., and Margolskee, R. F.,** Gustducin is a taste-cell-specific G protein closely related to the transducins, *Nature*, 357, 563, 1992.
61. **Lush, I. E.,** The genetics of tasting in mice. I. Sucrose octaacetate, *Genet. Res. Camb.*, 38, 93, 1981.
62. **Lush, I. E.,** The genetics of tasting in mice. II. Strychnine, *Chem. Sense*, 7, 93, 1982.
63. **Lush, I. E.,** The genetics of tasting in mice. III. Quinine, *Genet. Res. Camb.*, 44, 151, 1984.
64. **Lush, I. E.,** The genetics of tasting in mice. IV. The acetates of raffinose, galactose and β-lactose, *Genet. Res. Camb.*, 47, 117, 1986.
65. **Lush, I. E. and Holland, G.,** The genetics of tasting in mice. V. Glycine and cycloheximide, *Genet. Res. Camb.*, 52, 207, 1988.
66. **Whitney, G. and Harder, D. B.,** Single-locus control of sucrose octaacetate tasting among mice, *Behav. Genet.*, 16, 559, 1986.
67. **Akabas, M. H., Dodd, J., and Al-Awqati, Q.,** A bitter substance induces a rise in intracellular calcium in a subpopulation of rat taste cells, *Science*, 242, 1047, 1988.
68. **Hwang, P. M., Verma, A., Bredt, D. S., and Snyder, S. H.,** Localization of phosphatidylinositol signaling components in rat taste cells: role in bitter taste transduction, *Proc. Natl. Acad. Sci. U.S.A.*, 87, 7395, 1990.
69. **Lush, I. E.,** The genetics of tasting in mice. VI. Saccharin, acesulfame, dulcin and sucrose, *Genet. Res. Camb.*, 53, 95, 1989.
70. **Jakinovich, W., Jr. and Sugarman, D.,** Sugar taste reception in mammals, *Chem. Senses*, 13, 13, 1988.

71. **Striem, B. J., Pace, U., Zehavi, U., Naim, M., and Lancet, D.,** Sweet tastants stimulate adenylate cyclase coupled to GTP-binding protein in rat tongue membranes, *Biochem. J.*, 260, 121, 1989.
72. **Avenet, P., Hofmann, F., and Lindemann, B.,** Transduction in taste receptor cells requires cAMP-dependent protein kinase, *Nature*, 331, 351, 1988.
73. **Kinnamon, S. C.,** Patch clamp studies of sweet taste transduction in mammalian taste cells, *J. Gen. Physiol.*, 98, 8a, 1991.
74. **Simon, S. A., Labarca, P., and Robb, R.,** Activation by saccharides of a cation-selective pathway on canine lingual epithelium, *Am. J. Physiol.*, 256, R394, 1989.
75. **Heck, G. L., Mierson, S., and DeSimone, J. A.,** Salt taste transduction occurs through an amiloride-sensitive sodium transport pathway, *Science*, 223, 403, 1984.
76. **Schiffman, S. S., Lockhead, E., and Maes, F. W.,** Amiloride reduces the taste intensity of Na^+ and Li^+ salts and sweeteners, *Proc. Natl. Acad. Sci. U.S.A.*, 80, 6136, 1983.
77. **Avenet, P. and Lindemann, B.,** Amiloride-blockable sodium currents in isolated taste receptor cells, *J. Membr. Biol.*, 105, 245, 1988.
78. **Avenet, P. and Lindemann, B.,** Noninvasive recording of receptor cell action potentials and sustained currents from single taste buds maintained in the tongue: the response to mucosal NaCl and amiloride, *J. Membr. Biol.*, 124, 33, 1991.
79. **Smith, P. R. and Benos, D. J.,** Epithelial Na^+ channels, *Annu. Rev. Physiol.*, 53, 509, 1991.
80. **Contreras, R. J.,** Changes in gustatory nerve discharges with sodium deficiency, a single unit analysis, *Brain Res.*, 121, 373, 1977.
81. **Contreras, R. J. and Frank, M.,** Sodium deprivation alters neural responses to gustatory stimuli, *J. Gen. Physiol.*, 73, 569, 1979.
82. **Kosten, T. and Contreras, R. J.,** Adrenalectomy reduces peripheral neural responses to gustatory stimuli in the rat, *Behav. Neurosci.*, 99, 734, 1985.
83. **Ballermann, B. J., Zeidel, M. L., Gunning, M. E., and Brenner, B. M.,** Vasoactive peptides and the kidney, in *The Kidney*, Vol. 1, 4th ed., Brenner, B. M. and Rector, F. C., Jr., Eds., W. B. Saunders, Philadelphia, PA, 281, 1991.
84. **Monnot, C., Weber, V., Stinnakre, J., Bihoreau, C., Teutsch, B., Corvol, P., and Clauser, E.,** Cloning and functional characterization of a novel mas-related gene, modulating intracellular angiotensin II actions, *Mol. Endocrinol.*, 5, 1477, 1991.
85. **Jackson, T. R., Blair, L. A., Marshall, J., Goedert, M., and Hanley, M. R.,** The mas oncogene encodes an angiotensin receptor, *Nature*, 335, 437, 1988.
86. **Patel, P. D., Sherman, T. G., Goldman, D. J., and Watson, S. J.,** Molecular cloning of a mineralocorticoid (type I) receptor complementary DNA from rat hippocampus, *Mol. Endocrinol.*, 3, 1877, 1989.
87. **Tilley, W. D., Marcelli, M., Wilson, J. D., and McPhaul, M. J.,** Characterization and expression of a cDNA encoding the human androgen receptor, *Proc. Natl. Acad. Sci. U.S.A.*, 86, 327, 1989.
88. **Arriza, J. L., Weinberger, C., Cerelli, G., Glaser, T. M., Handelin, B. L., Housman, D. E., and Evans, R. M.,** Cloning of human mineralocorticoid receptor complementary DNA: structural and functional kinship with the glucocorticoid receptor, *Science*, 237, 268, 1987.
89. **Hollmann, M., O'Shea-Greenfield, A., Rogers, S. W., and Heinemann, S.,** Cloning by functional expression of a member of the glutamate receptor family, *Nature*, 342, 643, 1989.
90. **Nakanishi, N., Shneider, N. A., and Axel, R.,** A family of glutamate receptor genes: Evidence for the formation of heteromultimeric receptors with distinct channel properties, *Neuron*, 5, 569, 1990.
91. **Moriyoshi, K., Masu, M., Ishii, T., Shigemoto, R., Mizuno, N., and Nakanishi, S.,** Molecular cloning and characterization of the rat NMDA receptor, *Nature*, 354, 31, 1991.
92. **Faurion, A.,** Are umami taste receptor sites structurally related to glutamate CNS receptor sites, *Physiol. Behav.*, 49, 905, 1991.

93. **Brand, J. G., Bryant, B. P., Cagan, R. H., and Kalinoski, D. L.,** Biochemical studies of taste sensation. XIII. Enantiomeric specificity of the alanine taste receptor sites in catfish, *I. punctatus*, *Brain Res.*, 416, 119, 1987.
94. **Brand, J. G., Teeter, J. H., Kumazawa, T., Huque, T., and Bayley, D. L.,** Transduction mechanisms for the taste of amino acids, *Physiol. Behav.*, 49, 899, 1991.
95. **Teeter, J. H., Brand, J. G., and Kumazawa, T.,** A stimulus-activated conductance in isolated taste epithelial membranes, *Biophys. J.*, 58, 253, 1990.
96. **Raming, K., Krieger, J., and Breer, H.,** Molecular cloning of an insect pheromone-binding protein, *FEBS Lett.*, 256, 215, 1989.
97. **Gyorgyi, T. K., Roby-Shemkovitz, A. J., and Lerner, M. R.,** Characterization and cDNA cloning of the pheromone-binding protein from the tobacco hornworm, Manduca sexta: a tissue-specific developmentally regulated protein, *Proc. Natl. Acad. Sci. U.S.A.*, 85, 9851, 1988.
98. **Vogt, R. G., Prestwich, G. D., and Lerner, M. R.,** Odorant-binding-protein subfamilies associate with distinct classes of olfactory receptor neurons in insects, *J. Neurobiol.*, 22, 74, 1990.
99. **Krieger, J., Raming, K., and Breer, H.,** Cloning of genomic and complementary DNA encoding insect pheromone binding proteins: evidence for microdiversity, *Biochim. Biophys. Acta*, 1088, 277, 1991.
100. **Vogt, R. G., Rybczynski, R., and Lerner, M. R.,** Molecular cloning and sequencing of general odorant-binding proteins GOBP1 and GOBP2 from the Tobacco Hawk moth *Manduca sexta*: comparisons with other insect OBPs and their signal peptides, *J. Neurosci.*, 11, 2972, 1991.

Chapter 8

CELL BIOLOGY OF THE LINGUAL EPITHELIUM

John A. DeSimone and Gerard L. Heck

TABLE OF CONTENTS

0-8493-5341-6/93/$0.00 + $.50

I. ORAL CAVITY MILIEU

A. SALIVARY COMPOSITION

1. Major Salivary Glands

The alimentary tract contains single cells in the mucosal layer and collections of cells organized as acinar glands dedicated to ion and water secretion or absorption. The balance between secretion and absorption produces local environments that differ widely in ion composition, osmolarity, and acidity. The material composition of the oral cavity environment varies in ways that reflect ongoing physiological activity in a manner analogous to stimulated activity in lower regions of the gastrointestinal tract. The salivary glands (parotid, submaxillary, and sublingual) provide the major part of the aqueous milieu of the oral cavity. These are acinar glands with end pieces that form a primary secretion that is conducted through a system of convergent ducts that modify the secretion before emptying into the oral cavity. Typically, salivary composition varies considerably with flow rate, which depends on the extent of parasympathetic and sympathetic neural stimulation. At basal flow rates (<0.5 ml/min) saliva, relative to plasma, is hypotonic (ca. 90 mosmol), rather low in sodium (3 to 25 m*M*), chloride (ca. 25 m*M*), and bicarbonate (<25 m*M*), but notably high in potassium (25 to 30 m*M*).[1] In the stimulated state salivary tonicity more closely approaches that of plasma; sodium, chloride, and bicarbonate are at higher than basal levels, and, while potassium levels may fall somewhat, they still exceed plasma levels by a wide margin (ca. 20 m*M*).

a. End Piece Secretion

The end pieces (acinar cells) of the gland form the primary salivary secretion by coupling fluid movement to sodium chloride secretion. The key process involves transcellular chloride transport[2] and appears to be similar mechanistically to the transport processes that govern airway and gut secretions.[3] The chief features are an electroneutral $Na^+/K^+/2Cl^-$ transporter on the basolateral membranes of the end piece cells that utilizes the plasma-to-cell sodium gradient to accumulate chloride ion in the cells, and an apical membrane-passive chloride ion channel to release Cl^- ion to the saliva. Net Na^+ ion secretion is thought to occur mainly by a paracellular route. End piece secretion can be stimulated by β-adrenergic agonists that act through cAMP. There is evidence to suggest that salivary apical Cl^- ion channels belong to the channel class affected in cystic fibrosis.[4,5]

b. Duct-Epithelial Modification

The duct cells alter the primary secretion through a variety of ion transporters in the apical cell membranes. These include Na^+ and Cl^- channels, and HCO_3^--Cl^-, Na^+-H^+, and K^+-H^+ exchangers. These transporters are especially active during interstimulus periods, dramatically altering the primary secretion as it courses through the ductal system. This accounts for the dilute, hypotonic basal saliva. During stimulation end piece output of primary secretion increases, whereas the ductal Na^+ channel conductance decreases as does the efficiency of the Na^+-H^+ exchanger. The result is a saliva higher in Na^+, Cl^-, and HCO_3^-.

End piece and duct cells contain α- and β-adrenergic receptors as well as cholinergic receptors. Among the most potent of the sensory input to effect copious salivation are chemical tastants and irritants, especially acids. The taste-mediated reflex appears to involve synaptic interaction between second order sensory neurons in the nucleus tractus solitarius and parasympathetic secretomotor neurons in the salivatory nuclei.[6] The especially effective activation of salivation by acids through the taste system or through the trigeminal sensory system has an end result similar to the stimulation of the duodenal acid receptors by stomach acid. In both instances the stimulus evokes a secretion containing high levels of bicarbonate which, of course, neutralizes the stimulus. This is effected in the mouth through salivation, whereas in the lower alimentary tract, through secretin-stimulated pancreatic acinar glands.

2. Lingual Salivary Glands

The large circumvallate papillae on the posterior dorsal surface of the tongue and the foliate papillae along the posterior lateral margin of the tongue contain a salivary gland system that supplies its own peculiar secretory microenvironment. These are the submucosal von Ebner's glands (VEG). The circumvallate and foliate papillae contain large concentrations of taste buds in the papillary clefts. Because of their location, access to these taste buds is more restricted than in the case of the anterior lingual fungiform papillae. Recently, a protein with remarkable structural homology to the superfamily of hydrophobic molecule transporters has been cloned from VEG saliva.[7] The VEG protein may play a role in solubilizing hydrophobic taste stimuli, thereby facilitating their transport to taste buds. A binding capacity for small hydrophobic molecules has been shown in the structurally analogous olfactory binding protein. Alternatively, VEG protein may act as a buffer binding potentially toxic stimuli. The circumvallate and foliate taste-receptive fields are especially sensitive to bitter stimuli, many of which are small hydrophobic molecules that could readily partition into receptor cell membranes. The elaboration of binding proteins by VEG may be essential for the maintenance of taste sensitivity in these receptive fields.

II. ION TRANSPORT ACROSS THE LINGUAL EPITHELIUM

A. *IN VITRO* STUDIES

1. Symmetrical Bathing Media

The canine lingual epithelium was the first to be investigated extensively for ion transport.[8-11] The initial studies were aimed at determining whether these are pathways for transepithelial ion transport. Because the lingual epithelium can be dissected from the underlying muscle as a continuous sheet of tissue, the short circuit method was used.[12] This technique distinguishes active electrogenic processes from passive ion transport, and, when used with radioisotopes for the candidate transported ions, gives an accurate account of the individual ions that compose transcellular, metabolically linked ion flow. The results show conclusively that under conditions where passive driving forces for transepithelial ion fluxes have been eliminated (symmetrical bathing solutions of Krebs-Henseleit [KH] buffer and short-circuited transepithelial potential difference), there is a net sodium ion absorption and chloride ion secretion. Table 1 shows typical symmetrical control parameters. Sodium absorption accounts for about 56% of the short circuit current (SCC) and chloride secretion for about 24%. The remaining 20% has not been identified.

The electrical resistance reported for canine lingual epithelium ranges from 396 Ω-cm^2 (Reference 11) to 782 Ω-cm^2 (Reference 9) which places it most nearly in the category of "leaky" epithelia, i.e., ion transporting epithelia in which the transepithelial resistance is dominated by paracellular shunts.[13-15] However, the papillary structure and the heterogeneous cell types in the lingual epithelium make classification difficult. Ultrastructural studies affirm that the taste pore region constitutes a low resistance shunt whereas the nontaste related epithelium appears to be structurally tighter[16] (see Section V). This mosaic of a relatively tight epithelium punctuated by low resistance pathways through the taste sensory regions has important consequences for taste function (see below). Ouabain, when added to the submucosal bath under symmetrical conditions, eliminates the entire short circuit current.[8-11] This demonstrates that the active "injection" of metabolic energy is at the level of the ATP-driven Na-K pump. Simon et al.[17] have used immunocytochemical methods to localize the Na,K-ATPase to the basolateral membranes of canine taste cells using monoclonal antibodies against the α-subunit of the enzyme.

2. Hyperosmotic Salt Response

One of the more unusual features of the lingual epithelium is the two-component time course of the change in short circuit current when the mucosal bathing solution is changed from the null current NaCl concentration (usually 20 to 30 m*M*) to some hyperosmotic value. This change in bathing medium is especially interesting because it mimics a typical stimulation maneuver

TABLE 1
Ion Fluxes Across the Canine Lingual Epithelium

	Control[a]	3-*O*-MG[b]
I_{sc}	0.72 ± 0.06	1.44 ± 0.12
R (Ω cm²)	724 ± 51	682 ± 62
PD (mV)	13.2 ± 0.4	24.2 ± 1.1
J^{Na}(ms)	0.83 ± 0.08	1.63 ± 0.19
J^{Na}(sm)	0.43 ± 0.06	0.32 ± 0.04
J^{Na}(net)	0.40 ± 0.08	1.31 ± 0.16
J^{Cl}(ms)	0.49 ± 0.05	0.70 ± 0.14
J^{Cl}(sm)	0.65 ± 0.09	0.68 ± 0.09
J^{Cl}(net)	−0.17 ± 0.08	0.02 ± 0.07
J^{Na}(net)/I_{sc}	0.53 ± 0.08	0.89 ± 0.05
J^R/I_{sc}	0.20 ± 0.06	0.14 ± 0.03

[a] Control refers to symmetrical baths with K-H buffer and short-circuited tissue.

[b] 3-*O*-methylglucose (0.5 *M*) in K-H buffer placed in mucosal bath. Data represent means ± SEM (n = 12). I_{sc} is the short circuit current; J^{Na}(ms) and J^{Cl}(ms) are the sodium and chloride unidirectional fluxes from mucosa to submucosa. J^{Na}(sm) and J^{Cl}(sm) are similar unidirectional fluxes in the opposite direction. J^R refers to a net current not accounted by sodium and chloride net fluxes. The units of current and flow are μeq/cm² h. K-H buffer consists of (in mM): NaCl, 118; KCl, 5.6; $CaCl_2$, 1.9; $MgSO_4$, 1.2; NaH_2PO_4, 1.3; $NaHCO_3$, 25; glucose, 5.6. pH = 7.4 when bubbled with 95% O_2/5% CO_2.

From Mierson, S., Desimone, S. K., Heck, G. L., and DeSimone, J. A., *J. Gen. Physiol.*, 92, 87, 1988. With permission.

employed in recording responses from the taste nerves. The epithelial response begins with a rapid inward (first) current followed by an inflection point and then a slower rising (second) inward current that approaches an asymptotic value.[8-10] More than 90% of the second current is eliminated after treating the tissue with ouabain, evidence that it is mainly of transcellular origin. Lithium chloride gives a hyperosmotic response similar to that of NaCl and is blocked by amiloride to the same extent.[10] In the dog the ouabain-sensitive part of the current evoked by KCl is significantly smaller than that evoked by NaCl or LiCl, but like the other alkali metal chlorides, the hyperosmotic KCl current is also partially blocked by amiloride.

The basic structure of the NaCl-evoked current is seen in rat tongue epithelium *in vitro.*[18,19] It is also possible to make recordings of the stimulus-evoked current under voltage clamp from rat tongue *in vivo.* Under these conditions the two-component structure of the current is preserved and the temporal relation of the current to the neural response emerges clearly[20] (see Section IV.A).

For the canine lingual epithelium, *in vitro* decomposition of the SCC resulting from a mucosal 0.25 *M* NaCl concentration vs. a submucosa of KH buffer into its ionic components shows that enhanced Na influx accounts for most of the current and that Cl is now absorbed rather than secreted.[11] However, Na influx is nearly three times as great as that of Cl, a fact that emphasizes the predominance of cation-selective transport pathways in the lingual epithelium.

B. PASSIVE TRANSPORT PATHWAYS

In addition to ouabain-sensitive transcellular pathways, the lingual epithelium also contains cation-selective paracellular pathways through which ions can be driven by passive electrodiffusion.[9,10] Salt transport through these pathways is co-ion limited and this fact plays a central role in determining the influence of the anion on the neural response to salts[21,22] (see Section IV.B). The paracellular pathway is a shunt across the epithelium for nonelectrolytes as well as for some ions. In many epithelia the main shunt barriers exist across the apical tight junctional complex.[13,23] In the case of the canine tongue the paracellular pathway was shown to be cation selective with ionic flows governed by the electrodiffusion laws.[10] This conclusion generalizes to the rat.[19,22] The canine dorsal lingual epithelium is permeable to 3-*O*-methylglucose. The unidirectional fluxes of this sugar were unaffected by either submucosal addition of ouabain or mucosal addition of amiloride. Therefore, sugar transport was probably not by a transcellular route. In addition, the fluxes obeyed the Ussing flux ratio equation, further indication that transport occurred through a single passive barrier. In virtually all epithelia this barrier is in the tight junctions, the main diffusion barrier of the paracellular pathway. There is now good indication that ouabain is among the nonelectrolytes that can penetrate the canine lingual tight junctions. Simon and Verbrugge[24] have shown that the hyperosmotic response to 0.5 *M* NaCl is inhibited when ouabain is added to the mucosal solution. Because Simon et al.[17] have shown that the ouabain sensitive Na,K-ATPase is located exclusively on the basolateral membranes in lingual epithelial cells, it seems likely that ouabain placed on the mucosal side diffuses across the epithelium to effect its inhibition. This is probably by a paracellular route. It is interesting to note that rat lingual epithelial resistance is higher than that of dog, suggesting a paracellular pathway with somewhat lower permeability than that of the dog.

The permeability of the canine lingual frenulum to low molecular weight organic molecules has been determined.[25] Here, too, it appears that both transcellular and paracellular pathways are involved in the transport. Rat dorsal lingual epithelium also is permeable to a variety of nonelectrolytes.[26]

III. EFFECTS OF TASTE STIMULI ON THE LINGUAL EPITHELIUM *IN VITRO*

A. CANINE TONGUE

1. Salts

The most thoroughly studied salts are those most often used in taste electrophysiological experiments. The alkali metal chlorides stimulate an inwardly directed current that increases with concentration over the normal range of gustatory sensitivity.[8-11] For NaCl most of this current flows through a ouabain-sensitive (i.e., transcellular) pathway. At hyperosmotic mucosal concentrations there are both Na^+ and Cl^- influxes. The polarity of the ouabain-insensitive part of the flow indicates that the paracellular pathway is cation selective. The most anterior regions of the tongue are significantly more conductive to NaCl and LiCl than to KCl.[10,27] Responses to NaCl and LiCl are inhibited to the same extent by amiloride; KCl-evoked currents are also inhibited but to a lesser extent.[10] In this regard it is interesting to note (cf. Figure 1) that the canine chorda tympani response to KCl as well as that to NaCl can be inhibited by amiloride. The lack of specificity by amiloride in blocking alkali metal chloride-evoked transepithelial current, therefore, is also seen in the neural response. This would be anticipated if the current into taste cells reflected ion flow through nonspecific cation channels in the cell membranes. If such channels serve the function of primary receptor, both NaCl and KCl would then excite the same or nearly the same population of receptor cells. This suggests that the canine salt taste system is not tuned specifically to Na-Li, unlike the taste system of many rodents.[28,29]

Symmetrical addition of 0.15 *M* sodium acetate or sodium gluconate resulted in a 20 to 30% decrease in the short circuit and an increase in resistance relative to 0.15 *M* NaCl.[11] This is consistent with the chloride-dependent contribution to the current established through ion flux measurements (see Table 1). Replacement of sodium by tris or choline also reduced the current. When both sodium and chloride were replaced, currents were reduced to a greater extent, but never to zero. The possibility that relatively large ions such as tris might be current carriers at first seems remote. However, single channel recording and other methods have more recently demonstrated a class of ion channels (calcium-activated nonspecific cation channels) that are permeable to ions as large as tris and tetraethylammonium in a variety of cells.[30] It is conceivable that these or similar channels may exist in taste and/or epithelial cells and may account for some of the current in the absence of sodium or chloride. The decrease in current following symmetrical replacement of NaCl with another salt usually shows an undershoot. The slow partial recovery of current may signify the activation of cellular volume regulatory mechanisms,[31] an area that is largely unexplored. These are likely to be of considerable importance given the range of osmotic pressures over which the taste cells must remain functional.

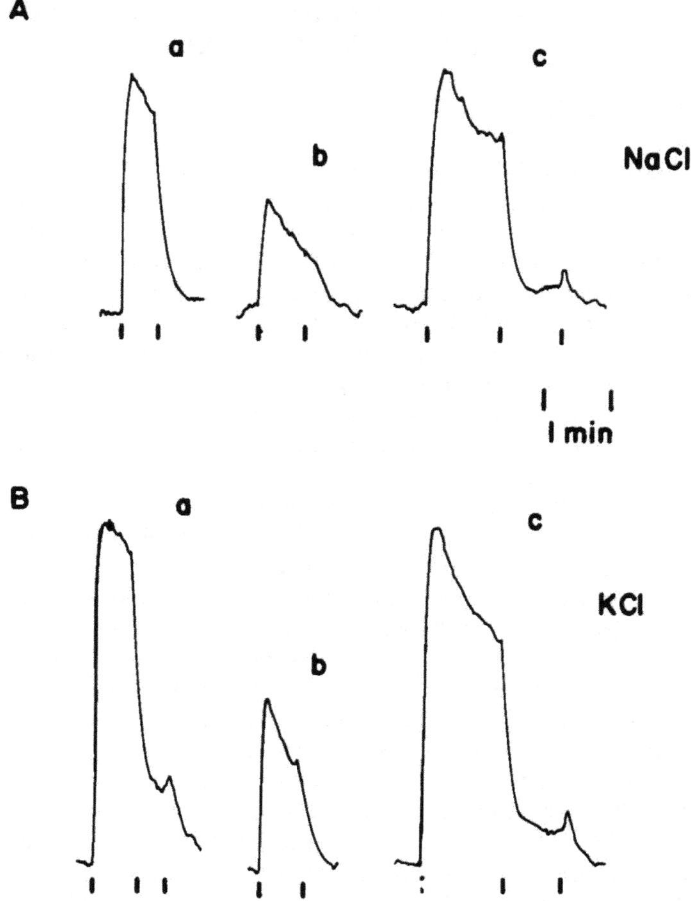

FIGURE 1. Integrated chorda tympani responses of the dog to 0.5 *M* NaCl (A) and 0.5 *M* KCl (B). (A). The tongue was adapted to 30 m*M* NaCl: (a) control response to 0.5 *M* NaCl, (b) response to 0.5 *M* NaCl following a 1 min rinse with 0.8 m*M* amiloride in 30 m*M* NaCl, (c) recovered response to NaCl. (B). Same as in (A) except the stimulus was KCl. Note amiloride inhibits KCl responses as effectively as NaCl responses. Recovery is complete for both salts.

2. Acids

The addition of HCl to the mucosal bathing solution of the *in vitro* canine preparation results in an inward current with threshold at pH 5.[9] The rise in current coincides with the range of sensitivity of the gustatory nerves. Both transcellular and paracellular pathways seem to play a role in the acid-evoked currents. The paracellular effect includes a titration of the fixed charges in the paracellular pathway effectively converting it from cation-selective to anion-selective.[9] This change in selectivity may be important in determining anion effects in acid stimulation.[9,32] The paracellular selectivity filter is also important in determining salt taste responses[22] and may account in part for salt-acid interactions (see also Chapter 6).[33]

3. Sugars

The canine dorsal lingual epithelium reacts to a variety of sugars with a dose-dependent increase in inward short circuit current.[10,34,35] Table 1 shows that the addition of 0.5 *M* 3-*O*-methylglucose to the mucosal K-H buffer solution results in a doubling of the inward sodium flux.[34] The sodium influx is inhibited both by ouabain and amiloride and the resistance changes are small, all indications that the ion influxes occur by a transcellular route. The sugar-evoked inward current is not strictly sodium selective. Simon et al.[35] determined that of the alkali metal chlorides tested at 50 m*M*, LiCl was most effective in evoking a saccharide-induced current. The rank ordering was Li > Na = K > Rb = Cs. However, at 0.15 *M*, NaCl was more effective than KCl in stimulating a current evoked by 0.5 *M* D-glucose. *N*-methyl-D-glucammonium chloride was ineffective.[34] At 0.5 *M* NaCl the presence of 0.5 *M* sucrose actually suppressed the baseline short circuit current. In this context it should be noted that a suppressive effect of 0.5 *M* NaCl on the chorda tympani response to 0.5 *M* sucrose had already been reported.[36]

The dependence of the canine chorda tympani response to sugars on electrolyte concentration has been thoroughly investigated by Kumazawa and Kurihara.[37] They have shown that the tonic response to 0.5 *M* sucrose varies non-monotonically with the NaCl concentration with the maximum response occurring at 100 m*M* NaCl. Simon et al.[35] showed a similar relation for glucose, also peaking at about 100 m*M* NaCl. Other ions such as K and tris also enhance the neural response to sucrose. There is then a notable parallel between the ion-dependent sugar-evoked current *in vitro* and the ion-dependent sugar-evoked neural response, suggesting that at least one sugar-evoked transduction mechanism involves a ligand-gated apical ion channel. On the other hand, Kumazawa and Kurihara[37] point out that the rank ordering of the currents evoked by various sugars *in vitro* is different from that observed in the chorda tympani. This might be explained, however, by the possible contribution of a population of nongustatory cells to the current or possibly by the presence of electrolyte-independent sugar transduction mechanisms. The fact that amiloride inhibits only 39.6% of the chorda tympani response of the dog to sucrose and 27.4% of that to fructose is consistent with this.[34] Figure 2 shows amiloride inhibition of these sugar responses at 1 *M* in 30 m*M* NaCl. In contrast to its effects on NaCl and KCl responses (cf. Figure 1), amiloride effects on sugar responses are only partially reversible. This may indicate that the sugar-sensitive cation pathway and the conductance pathways used by salts in the absence of sugars are not entirely the same. Nevertheless, the saccharide and the salt-stimulated ion pathways in canine lingual epithelia have a number of common properties such as inhibition by ouabain, amiloride,

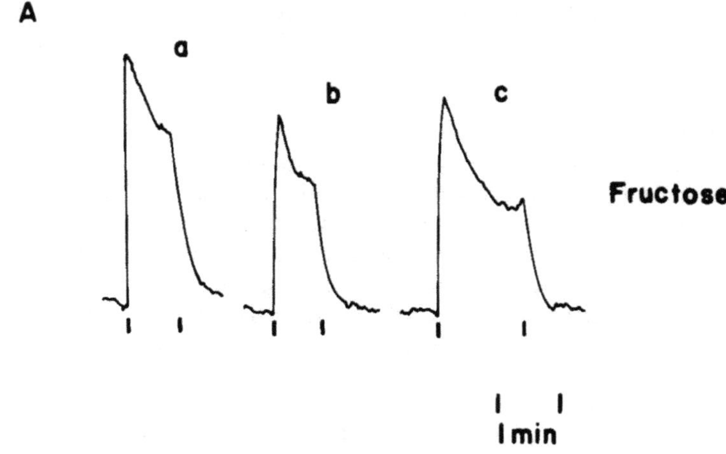

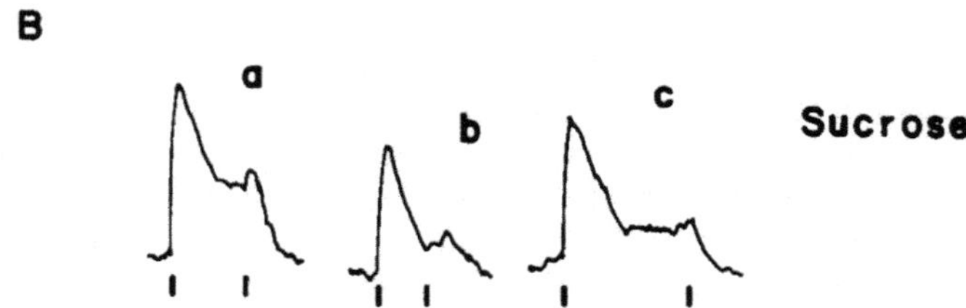

FIGURE 2. Integrated chorda tympani responses to 1 *M* fructose (A) and 1 *M* sucrose (B) in the same animal. The adapting solution was 30 m*M* NaCl and each sugar solution contained 30 m*M* NaCl. (A): (a) response to 1 *M* fructose, (b) response to 1 *M* fructose following a 1 min rinse with 0.8 m*M* amiloride in 30 m*M* NaCl, (c) recovered response. (B) Same as in (A) except the stimulus was 1 *M* sucrose. Amiloride partially inhibits sugar responses. Unlike salt responses recovery is incomplete.

$BaCl_2$, and $LaCl_3$.[24] Thus far, a clear demonstration of sugar-evoked lingual epithelial currents and corresponding ionic strength-dependent neural responses has been established only for the dog. This ligand-gated current mechanism does not appear to play a role in rats (unpublished data) or in frogs.[38]

4. Responses to Forskolin and Br-cAMP

Biochemical studies on rat lingual tissue indicate that sucrose stimulation results in cAMP production by a G protein-mediated process.[39,40] These results strongly suggest that sucrose-evoked transduction is cAMP mediated. Direct injection of cAMP into mouse taste cells results in depolarization with decreased conductance, a response also observed with sucrose.[41] There is evidence that a cAMP-dependent protein kinase mediates a decrease in taste cell

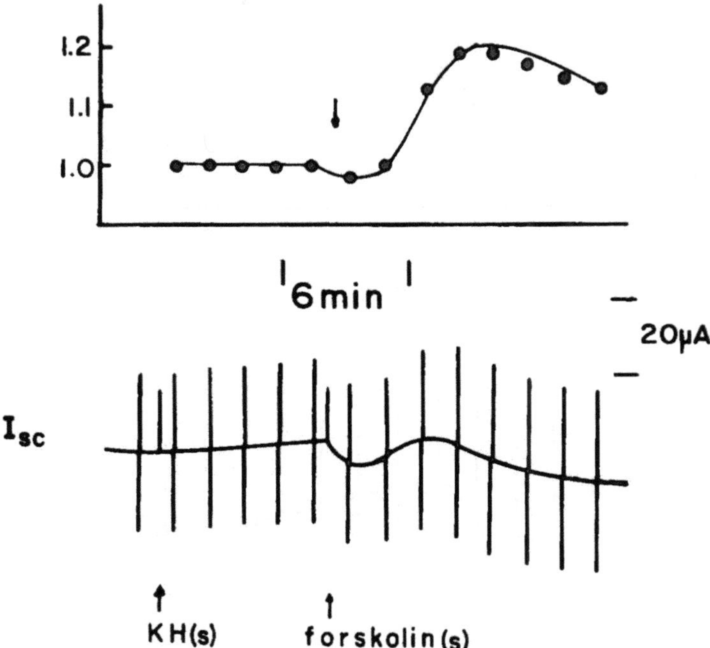

FIGURE 3. Effect of 10 μ*M* forskolin on the short circuit current (I_{sc}) across an isolated canine dorsal lingual epithelium. Lower trace shows the current following addition to the submucosal bath to produce the final concentration of 10 μ*M*. The complex current transient is accompanied by a transient increase in conductance (upper trace). For composition of KH see Table 1.

K ion conductance[42] and that this is the mechanism by which the cells are depolarized. It would appear, therefore, that saccharide-evoked transduction mechanisms in dogs and rodents are different in that in the former a cation influx is stimulated by saccharides while in the latter a K outflux is inhibited.

The situation is complicated, however, by experiments that show that the short circuit current and the conductance of the canine lingual epithelium *in vitro* are affected by forskolin and Br-cAMP. Figure 3 shows that 10 μ*M* forskolin added to the submucosal side of a canine lingual epithelium, symmetrically bathed in K-H buffer, produces a complex current transient. This consists of a current decrease followed by a rebound and then a much slower decrease. The transient current change is accompanied by a transient increase in conductance. The forskolin effect is qualitatively the same from the mucosal side (Figure 4); however, the magnitude is smaller and develops more slowly. This suggests that the effective site of action is more accessible from the submucosal side. One millimolar Br-cAMP added to the submucosal side (Figure 5) mimics the effect of forskolin on the current. Br-cAMP also causes an increased conductance, but unlike the effect of forskolin, the conductance increase persists.

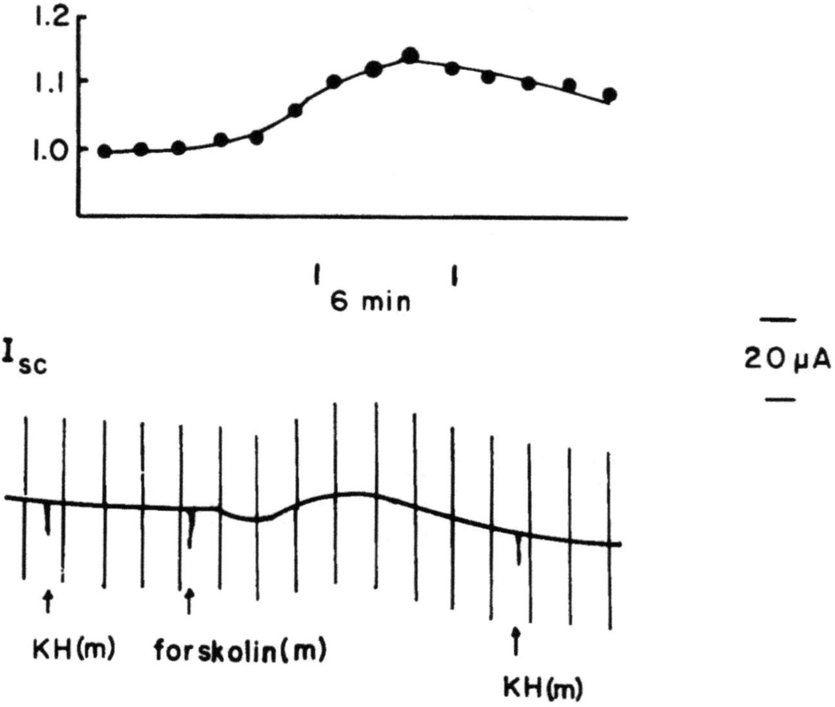

FIGURE 4. Similar, if somewhat attenuated, effect of adding forskolin to the mucosal side of the same preparation.

These results suggest that cells in the canine lingual epithelium contain adenylyl cyclase that can be activated pharmacologically by forskolin, resulting in a cAMP-evoked increase in cell ion conductance. The transient nature of the forskolin-evoked conductance increase may be due to the initial synthesis of cAMP and its subsequent breakdown by phosphodiesterases *in situ*. Br-cAMP, which is more phosphodiesterase resistant, does not result in a conductance decrease, consistent with this view. The identity of the cell types activated and the actual stimuli that promote cAMP synthesis are unknown. Saccharide binding to taste cell receptor sites remains a possible cause of cAMP production. Herness[43] reports that sucrose stimulation of isolated rat taste cells can cause an *increase* in conductance in potassium channels and that this increased conductance is mediated through cAMP production. Analogous coupling to conductance increases may also exist in the dog. However, other possibilities remain. For example, Simon and Baggett[44] report that acetylcholine may be a physiological stimulus leading to cAMP production by lingual epithelial cells outside the taste bud.

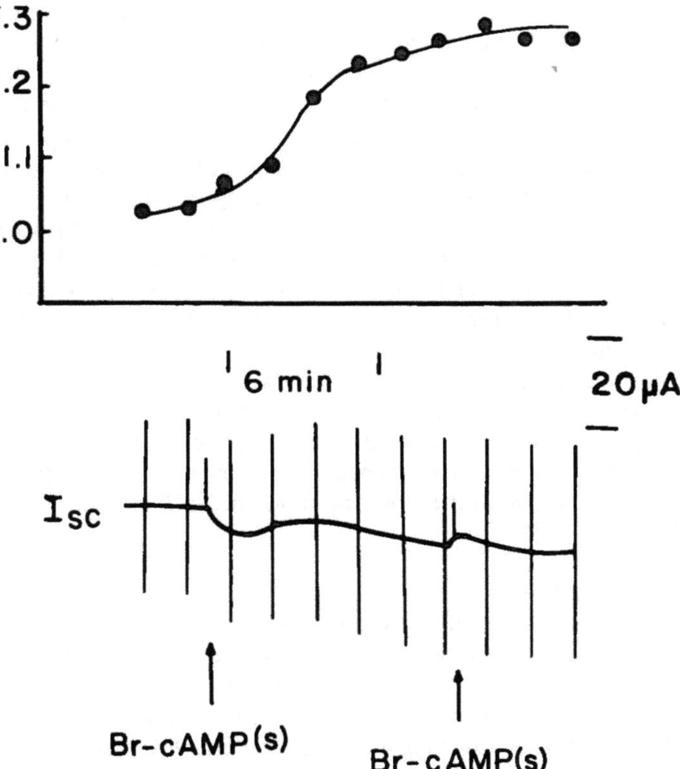

FIGURE 5. Effect of adding 1 m*M* Br-cAMP to the submucosal side of the same preparation. Note the conductance increases to a new plateau and shows no spontaneous recovery as in Figure 3 and 4.

B. RAT TONGUE

The ion transport properties of rat tongue *in vitro* have not been characterized as fully as those of the dog, mainly because the canine tongue is a more convenient size for ion flux studies. However, since much more taste neurophysiology is done on rats than on dogs, it is important to understand the rat ion transport systems in order to establish possible correlations with the neurophysiology.

Like canine tongue, rat tongue actively transports ions.[18,19,45] Amiloride inhibits the short circuit current evoked by NaCl but has no effect on that due to KCl.[19,20] A similar amiloride specificity, i.e., separation of Na from K-evoked responses, is also seen in rat neural recordings.[19,20,45,46] Two types of K channels exist in the rat lingual epithelium, distinguishable by differential blocking effects on the short circuit current by 4-aminopyridine (4-AP) and $BaCl_2$. The 4-AP-blockable sites exist on the apical membranes while the

$BaCl_2$-blockable sites exist on the submucosal side. Whereas part of the short circuit current across canine tongue is attributable to chloride secretion, transcellular chloride ion pathways in rat tongue have not been demonstrated. However, there is increasing evidence that the paracellular pathway in the lingual epithelium is chloride permeable,[19] and this may have important consequences for taste responses.[21,22,47] As in the case of the dog, most of the short circuit current evoked by NaCl across rat tongue is ouabain sensitive.[18,19] The amount of ouabain-sensitive NaCl current in rats may be strain dependent. Mierson et al.[18] found that control Wistar-Kyoto rats have a higher ouabain-sensitive NaCl current over the hyperosmotic range than spontaneously hypertensive rats. Strain differences such as these may help to uncover the influence of nutrition and metabolism as sources of regulation in taste reception. Reduction of the measurement of lingual currents under voltage clamp *in vitro* to the single papilla level has recently been achieved.[48] Under these conditions, NaCl evokes a slow inward current and monophasic transient currents (probably related to action potentials in the taste cells) superimposed on the slow current. Both the slow and transient NaCl currents are blocked by amiloride.

C. FROG TONGUE

In comparison with canine and rat tongue, the isolated frog lingual epithelium shows relatively little electrogenic active ion transport. Soeda and Sakudo[49] report mean open circuit potentials of 1.72 mV (referenced to mucosa) and short circuit currents of 3.26 $\mu A/cm^2$ in amphibian Ringer's solution. The potential or short circuit current nevertheless changes in characteristic ways, suggestive of a role for ion transport in taste reception, when various tastants are applied to the mucosa.[38,49] The transport hypothesis is strengthened by the establishment of amiloride-blockable nonselective cation channels in frog taste cells.[50] Herness'[51] observation that amiloride can inhibit both NaCl and KCl-evoked frog glossopharyngeal nerve responses is further support. Stimulation of frog KCl and NaCl neural responses by $CoCl_2$ suggests the presence of an apical membrane ion channel and parallels the effect of $CoCl_2$ on other amphibian ion-transporting tissue.[52] The hormone, aldosterone, selectively increases the frog neural response to NaCl following injection into the lingual circulation.[53] Therefore, Na-sensitive taste cells may also be targets for aldosterone. Perhaps this is a mechanism by which Na-taste acuity may be regulated at the cellular level.[54]

IV. RESPONSES OF THE LINGUAL EPITHELIUM *IN VIVO*

A. RESPONSES TO SALTS

There is an obvious advantage to measuring the transepithelial voltage or the current under voltage-clamp while simultaneously monitoring the neural response of the same lingual region. Rapuzzi and Ricagno[55] described a method for simultaneously recording the lingual potential and the neural response due to substances injected into the lingual circulation in the frog. They observed a slow potential and a phasic neural discharge in response to lingual injection of acetylcholine. Heck et al.[20] introduced a chamber fitted with voltage-sensing and current-passing electrodes that could be used in both voltage- and current-clamp mode. Responses in the rat made under voltage-clamp *in vivo* show that the first inward current coincides with the excitation phase of the neural response to NaCl. The slow current develops with the same exponential kinetics as slow adaption in the neural response.[56] Because studies *in vitro* show that the second current is ouabain-sensitive, it is probable that long term adaptation of the NaCl response involves taste cell repolarization by means of the sodium pump.

B. VOLTAGE SENSITIVITY AND THE SALT ANION EFFECT

The method of Heck et al.[20] permits the neural response to tastants to be studied under conditions of voltage-clamped field potential. This effectively adds a new, variable dimension to the study of peripheral taste responses. This is important because the field potential can directly influence the magnitude of the receptor potential. The direct agent of taste cell excitation is, of course, the depolarizing receptor potential expressed across the taste cell basolateral membranes. However, in the structured environment of the lingual epithelium where the taste cell apical and basolateral membranes make contact with different compartments, and where permselective tight junctions provide a paracellular pathway for ion flow, the receptor potential is constrained by the relation:

$$V_r = V_a - V_f \tag{1}$$

where V_r is the receptor potential referenced to the submucosal side, V_a is the potential across the taste cell apical membrane referenced to the mucosal side, and V_f is the transepithelial field potential also referenced to the mucosal side. From this relation it follows that two stimuli that result in identical potential across the apical cell membrane may nonetheless evoke different receptor potentials in the cell if at the same time the stimuli give rise to different field potentials. The ion selectivity of lingual paracellular ion exchange pathways has been established experimentally by DeSimone et al.[10] and Simon and Garvin.[9] The data have been mathematically modeled by DeSimone et al.[57] and Fidelman and Mierson.[58] The results indicate that the taste cell and its paracellular microenvironment form a functional taste sensory unit.

One consequence of the ion exchanger properties of the paracellular pathway is the difference in field potential set up across the lingual epithelium by different sodium salts. The shunt behaves like a leaky cation exchanger.

That is, cations have higher permeabilities than anions, but anions, if sufficiently mobile, can penetrate to a limited degree. The anions with limited mobility in the shunt should produce the largest field potentials, consequently, the smallest receptor potentials and smallest neural responses. This mechanism for the anion effect was proposed independently by both Harper[59] and Elliott and Simon.[21] Ye et al.[22] proved that anion-modulated field potentials exist *in situ* exerting an influence on receptor potentials. First, they measured simultaneously the field potentials and neural responses to different sodium salts in the rat. At any sodium concentration there was the predicted inverse relation between field potential magnitude and neural response. Then they voltage-clamped the field potential. This made the neural response independent of the field potential, proving that the anion effect is exerted mainly through the paracellular route.

Among anions, chloride shows some deviation from this model. This deviation becomes important above about 100 m*M* NaCl (Ye, Heck, and DeSimone, unpublished work). This chloride anomaly may be due to penetration of some NaCl through the tight junctions between taste cells. The sodium stimulus would then have access to basolateral membrane ion channels. This "extra" chloride-dependent stimulation would appear to be field potential independent. It would take place in a nearly isopotential region because most of the transepithelial potential is dropped across the taste cell apical membranes and the tight junctions.

More generally, if a given taste stimulus produces a neural response that is independent of field potential, one may conclude that either the conductance changes associated with transduction do not occur in the apical membrane or that transduction occurs without the generation of a receptor potential. In cases where a receptor potential is known to occur, a study of neural responses at various clamped transepithelial potentials should then locate the transduction conductance changes in either the apical or basolateral membrane of the taste cell.

V. STRUCTURAL CORRELATES OF LINGUAL EPITHELIAL FUNCTION

While the existence of a cation-selective paracellular shunt pathway in the lingual epithelium has been established on functional grounds, until recently little was known about the morphology of the shunts. Similarly, there has been a paucity of information on the extent to which the lingual epithelium behaves as a syncytium, i.e., whether the epithelial cells form communicating interfaces or gap junctions. The situation has been clarified considerably by Holland et al.[16] using both ultrastructural and electrophysiological methods.

A. TIGHT JUNCTIONS

With the canine lingual epithelium as a model, Holland et al.[16] used thin section electron microscopy to show that tight junctions are found only in the stratum corneum (uppermost cornified layer) and in the underlying stratum granulosum. Freeze-fracture methods along with lanthanum infiltration further establish these tight junctions as zonulae occludentes, the belt-like strands that separate apical and basolateral membranes. In the case of the cells of the fungiform papillae, Holland et al.[16] describe the freeze-fracture patterns of the tight junctions as "composed of ridges (and grooves) of variable widths." The variation appears as multiples of the unitary strand thickness of 6 to 8 nm. The occluding tight junctions are confined to the superficial granulosa region, but focal tight junctions can also be found in lower granulosa regions (see also Chapter 9).[16]

The addition of lanthanum chloride to either the mucosal or submucosal bath of isolated canine lingual epithelium results in about a 70% increase in resistance.[16] Because the transepithelial resistance is dominated by the shunt resistance, the localization of lanthanum complexes in the tight junctions is further confirming evidence that they are the permeability barrier of the paracellular shunt. Lanthanum, when added to the mucosal bath, penetrates the epithelium to the interface between the stratum corneum and the stratum granulosum. When added to the submucosal bath, lanthanum penetrates to the same level.[16] It appears that the most complete network of zonulae occludentes lies in the interface of the two superficial strata.

There is evidence that the electrical resistance of the tight junctions increases exponentially with the number of strands comprising the complexes.[60,61] Cereijido et al.[62] proposed that conductance pathways between strands are compartmentalized due to frequent crosslinking between strands. This arrangement would severely restrict the available conductance pathways through the zonula occludens. If the conductance at a given strand were to be reduced by a factor, α, of the conductance at the preceding strand, then the conductance, κ, for a junction of n strands would be:

$$\kappa = \kappa_o(1 - \alpha)^{n-1}, \qquad n = 1, 2, \ldots \tag{2}$$

where κ_0 is the inherent conductance of a strand. With the resistance, r, defined as $r = 1/\kappa$, the exponential growth of resistance with strand number follows.

Holland et al.[16] point out that on this basis the tight junction network in the nongustatory lingual epithelium will offer far greater resistance to ions (and presumably other tastants) than the taste pore. The taste pore would appear then to be a low resistance pathway in an otherwise high resistance surround, accounting for the highly efficient application of tastants as electric currents, i.e., electric taste. The tight junctions of the taste bud may be permeable enough to permit the penetration of sufficient NaCl to stimulate from the basolateral side of the taste cells as well the apical side. The presence of immunostaining for the amiloride-blockable sodium channel on the basolateral membranes of canine taste cells renders this a possibility.[63]

Tight junction formation appears to be triggered in neighboring cells by calcium binding to the extracellular domain of the transducer molecule, uvomorulin.[64] The process is mediated by two G proteins linked to phospholipase C and involves a phosphoinositide cascade leading to exocytotic fusion with the plasma membrane of a cytoplasmic vesicular compartment containing the junctional components.

B. GAP JUNCTIONS

Gap junctions are specialized regions of contact between cells that couple the cells electrically, i.e., they are low resistance pathways for the spread of current between cells. In many tissues the coupling is so strong as to render the tissue a functional syncytium. Gap junctions have been shown between taste bud cells in the mudpuppy[65,66] and the catfish.[67] Holland et al.[16] have observed gap junctions in the canine lingual epithelium between all extragemmal epithelial cells except the stratum corneum and therefore conclude that lingual epithelial cells form an electrical syncytium (see also Chapter 9). In this regard it is similar to many other transporting epithelia. Whether gap junctions between taste bud cells have important consequences for sharing of taste information among cells, and thus play a key role in taste sensory neural coding, is a topic of much speculation. Current ongoing research should begin to clarify this intriguing area.

VI. PENETRATION OF THE LINGUAL EPITHELIUM BY CHEMICALS

The fungiform papillae receive trigeminal innervation in addition to sensory innervation through the chorda tympani nerve. Trigeminal innervation accounts for about 75% of the total[68] and causes the sensations of pain, warm, and cold, all of which may be evoked by classes of chemicals. The chemical sensitivity of trigeminal nerve endings has been referred to as the "common chemical sense", but the current view is that trigeminal chemoreception arises from nociceptive fibers and not distinct classes of chemoreceptors. The term chemesthesis is now preferred.[69]

Because anatomically distinct trigeminal chemoreceptor cells probably do not exist, the mode of stimulation of the lingual trigeminal nerve endings is unclear (see also Chapter 9). The high thresholds for stimulation of the lingual nerve by salts[70] suggest that these stimuli must travel longer distances through a relatively low permeability barrier to provide an adequate stimulus.[16] Lanthanum, which was shown to localize in tight junctions,[16] inhibits lingual nerve responses to salts suggesting that paracellular diffusion plays a role in trigeminal nerve stimulation by electrolytes.[71] On the other hand, nonpolar trigeminal stimuli may penetrate the epithelial barrier through the cellular route.[72]

VII. NEUROTRANSMITTERS AND NEUROMODULATORS

Immunocytochemical methods have unleashed a veritable plethora of candidate neurotransmitters and neuromodulators both for gustation and the "common chemical sense". Roper[73] has recently summarized progress in this important area (see also Chapter 11). One of the more interesting observations is the presence of reciprocal synapses between basal cells and receptor cells of the mudpuppy taste bud.[72] This suggests that more than one receptor cell can be involved in the transduction and coding of a taste stimulus. Immunocytochemical methods show that the both γ-aminobutyric acid (GABA) and glutamate are localized in the nerve terminals that surround and synapse with taste buds.[74] Monoamine-containing cells[75] have been localized in taste bud cells, including basal cells.[73] In addition, the peptides vasoactive intestinal peptide,[76] calcitonin gene-related peptide (CGRP),[77] substance P,[78,79] cholecystokinin, and others[73] have been localized in taste buds or in their nerve supply. Substance P and CGPR immunoreactivity have been found in the trigeminal nerve endings where they may play a role in nociception.[69] Overall there appear to be a variety of releasible substances found in both taste cells and their nerve supply. The physiological conditions under which release occurs are presently unknown, but ongoing work in this area should provide clear insight into what are likely to be the first stages of taste quality coding.

ACKNOWLEDGMENTS

The authors thank Dr. Steven Price for helpful suggestions. Supported by NIH grant DC00122.

REFERENCES

1. **Schneyer, L. H. and Emmelin, N.,** Salivary secretion, in *International Review of Physiology: Gastrointestinal Physiology,* Vol. 1, Jacobson, E. D. and Shanbour, L. L., Eds., University Park Press, Baltimore, 1974, 184.
2. **Young, J. A., Cook, D. I., Van Lennep, E. W., and Roberts, M.,** Secretion by the major salivary glands, in *Physiology of the Gastrointestinal Tract,* Vol. 1, Johnson, L. R., Christensen, J., Jackson, M. J., Jacobson, E. D., and Walsh, J. H., Eds., Raven Press, New York, 1987, 773.
3. **Widdicombe, J. H. and Welsh, M. J.,** Ion transport by dog tracheal epithelium, *Fed. Proc.,* 39, 3062, 1980.
4. **Wiesmann, U. N., Boat, T. F., and di Sant'Agnese, P. A.,** Flow-rates and electrolytes in minor-salivary-gland saliva in normal subjects and patients with cystic fibrosis, *Lancet,* II, 510, 1972.
5. **Snyder, S. H. and Narahashi, T.,** Receptor-channel alterations in disease: many clues, few causes, *FASAB J.,* 4, 2707, 1990.
6. **Bradley, R. M.,** Salivary secretion, in *Smell and Taste in Health Disease,* Getchell, T. V., Doty, R. L., Bartoshuk, L. M., and Snow, J. B., Jr., Eds., Raven Press, New York, 1991, 127.

7. **Schmale, H., Holtgreve-Grez, H., and Christiansen, H.,** Possible role for salivary gland protein in taste reception indicated by homology to lipophilic-ligand carrier proteins, *Nature*, 343, 366, 1990.
8. **DeSimone, J. A., Heck, G. L., and DeSimone, S. K.,** Active ion transport in dog tongue: a possible role in taste, *Science*, 214, 1039, 1981.
9. **Simon, S. A. and Garvin, J. L.,** Salt and acid studies on canine lingual epithelium, *Am. J. Physiol.*, 249, C398, 1985.
10. **DeSimone, J. A., Heck, G. L., Mierson, S., and DeSimone, S. K.,** The active ion transport properties of canine lingual epithelia *in vitro:* implications for gustatory transduction, *J. Gen. Physiol.*, 83, 633, 1984.
11. **Mierson, S., Heck, G. L., DeSimone, S. K., Biber, T. U. L., and DeSimone, J. A.,** The identity of the current carriers in canine lingual epithelium *in vitro*, *Biochim. Biophys. Acta*, 816, 283, 1985.
12. **Ussing, H. H. and Zerahn, K.,** Active transport of sodium as the source of electric current in short-circuited isolated frog skin, *Acta Physiol. Scand.*, 23, 110, 1951.
13. **Moreno, J. H. and Diamond, J. M.,** Cation permeation mechanisms and cation selectivity in "tight junctions" of gallbladder epithelium, in *Membranes*, Vol. 3, Eisenman, G., Ed., Marcel Dekker, New York, 1975, 383.
14. **Frömter, E. and Diamond, J.,** Route of passive ion permeation in epithelia, *Nature New Biol.*, 235, 9, 1972.
15. **Palmer, L. G.,** The epithelial sodium channel, in *New Insights into Cell and Membrane Transport Processes*, Poste, G. and Crooke, S. T., Eds., Plenum Press, New York, 1986, 327.
16. **Holland, V. F., Zampighi, G. A., and Simon, S. A.,** Morphology of fungiform papillae in canine lingual epithelium: location of intercellular junctions in the epithelium, *J. Comp. Neurol.*, 279, 13, 1989.
17. **Simon, S. A., Holland, V. F., and Zampighi, G. A.,** Localization of Na,K-ATPase in lingual epithelia, *Chem. Senses*, 16, 283, 1991.
18. **Mierson, S., Welter, M. E., Gennings, C., and DeSimone, J. A.,** Lingual epithelium of spontaneously hypertensive rats has decreased short-circuit current in response to NaCl, *Hypertension*, 11, 519, 1988.
19. **Simon, S. A., Robb, R., and Schiffman, S. S.,** Transport pathways in rat lingual epithelium, *Pharmacol. Biochem. Behav.*, 29, 257, 1988.
20. **Heck, G. L., Persaud, K. C., and DeSimone, J. A.,** Direct measurement of translingual epithelial NaCl and KCl currents during the chorda tympani taste response, *Biophys. J.*, 55, 843, 1989.
21. **Elliott, E. J. and Simon, S. A.,** The anion in salt taste: a possible role for paracellular pathways, *Brain Res.*, 535, 9, 1990.
22. **Ye, Qing, Heck, G. L., and DeSimone, J. A.,** The anion paradox in sodium taste reception: resolution by voltage-clamp studies, *Science*, 253, 724, 1990.
23. **Lewis, S. A. and Diamond, J. M.,** Na^+ transport by rabbit urinary bladder, a tight epithelium, *J. Membr. Biol.*, 28, 1, 1976.
24. **Simon, S. A. and Verbrugge, J.,** Transport pathways in canine lingual epithelium involved in sweet taste, *Chem. Senses*, 15, 1, 1990.
25. **Siegel, I. A., Izutsu, K. T., and Watson, E.,** Mechanisms of nonelectrolyte penetration across dog and rabbit oral mucosa in vitro, *Arch. Oral Biol.*, 26, 357, 1981.
26. **Mistretta, C. M.,** Permeability of tongue epithelium and its relation to taste, *Am. J. Physiol.*, 220, 1162, 1971.
27. **Garvin, J. L., Robb, R., and Simon, S. A.,** Spatial map of salts and saccharides on dog tongue, *Am. J. Physiol.*, 255, R117, 1988.
28. **Boudreau, J. C., Hoang, N. K., Oravec, J., and Do, L. T.,** Rat neurophysiological taste responses to salt solutions, *Chem. Senses*, 8, 131, 1983.
29. **Frank, M. E., Contreras, R. J., and Hettinger, T. P.,** Nerve fibers sensitive to ionic taste stimuli in chorda tympani of the rat, *J. Neurophysiol.*, 50, 941, 1983.

30. **Partridge, L. D. and Swandulla, D.,** Calcium-activated non-specific cation channels, *TINS*, 11, 69, 1988.
31. **Lewis, S. A. and Donaldson, P.,** Ion channels and cell volume regulation: chaos in an organized system, *NIPS*, 5, 112, 1990.
32. **Beidler, L. M. and Gross, G. W.,** The nature of taste receptor sites, in *Contributions to Sensory Physiology*, Vol. 5, Neff, W. D., Ed., Academic Press, New York, 1971, 97.
33. **Ogawa, H.,** Effect of pH on the chorda tympani nerve of rats, *Jpn. J. Physiol.*, 19, 670, 1969.
34. **Mierson, S., DeSimone, S. K., Heck, G. L., and DeSimone, J. A.,** Sugar-activated ion transport in canine lingual epithelium, *J. Gen. Physiol.*, 92, 87, 1988.
35. **Simon, S. A., Labarca, P., and Robb, R.,** Activation by saccharides of a cation-selective pathway on canine lingual epithelium, *Am. J. Physiol.*, 256, R394, 1989.
36. **Andersen, H. T., Funakoshi, M., and Zotterman, Y.,** Electrophysiological responses to sugars and their depression by salt, in *Olfaction and Taste*, Vol. 1, Zotterman, Y., Ed., Pergamon Press, Oxford, 1963, 177.
37. **Kumazawa, T. and Kurihara, K.,** Large enhance of canine taste responses to sugar by salts, *J. Gen. Physiol.*, 95, 1007, 1990.
38. **Soeda, H. and Sakudo, F.,** Electrical responses to taste chemicals across the dorsal epithelium of bullfrog tongue, *Experientia (Basel)*, 41, 50, 1985.
39. **Striem, B. J., Pace, U., Zehavi, U., Naim, M., and Lancet, D.,** Sweet tastants stimulate adenylate cyclase coupled to GTP binding protein in rat tongue membranes, *Biochem. J.*, 260, 121, 1989.
40. **Striem, B. J., Naim, M., and Lindemann, B.,** Generation of cyclic AMP in taste buds of the rat circumvallate papilla in response to sucrose, *Cell. Physiol. Biochem.*, 1, 46, 1991.
41. **Tonosaki, K. and Funakoshi, M.,** Cyclic nucleotides may mediate taste transduction, *Nature*, 331, 354, 1988.
42. **Avenet, P. and Lindemann, B.,** Amiloride-blockable sodium currents in isolated taste receptor cells, *J. Membr. Biol.*, 105, 245, 1988.
43. **Herness, M. S.,** Sucrose stimulation of rat taste cells can increase membrane potassium conductance, *Neurosci. Abstr.*, 16, 25, 1990.
44. **Simon, S. A. and Baggett, H. C.,** Identification of muscarinic acetylcholine receptors in isolated canine lingual epithelia via voltage clamp measurements, *Arch. Oral Biol.*, 36, 805, 1991.
45. **Heck, G. L., Mierson, S., and DeSimone, J. A.,** Salt taste transduction occurs through an amiloride-sensitive sodium transport pathway, *Science*, 223, 403, 1984.
46. **Schiffman, S. S., Lockhead, E., and Maes, F. W.,** Amiloride reduces the taste intensity of Na^+ and Li^+ salts and sweeteners, *Proc. Natl. Acad. Sci. U.S.A.*, 80, 6136, 1983.
47. **Formaker, B. K. and Hill, D. L.,** An analysis of residual NaCl taste response after amiloride, *Am. J. Physiol.*, 255, R1002, 1988.
48. **Avenet, P. and Lindemann, B.,** Noninvasive recording of receptor cell action potentials and sustained currents from single taste buds maintained in the tongue: the response to mucosal NaCl and amiloride, *J. Membr. Biol.*, 124, 33, 1991.
49. **Soeda, H. and Sakudo, F.,** Epithelial responses to glycinamide and glycylglycinamide in the frog (Rana Catesbeiana) tongue, *Comp. Biochem. Physiol.*, 100A, 315, 1991.
50. **Avenet, P. and Lindemann, B.,** Amiloride-blockable sodium currents in isolated taste receptor cells, *J. Membr. Biol.*, 105, 245, 1988.
51. **Herness, M. S.,** Are apical membrane ion channels involved in frog taste transduction?, *Ann. N.Y. Acad. Sci.*, 510, 362, 1987.
52. **Herness, M. S.,** Specificity of mono- and divalent salt transduction mechanisms in frog gustation evidenced by cobalt chloride treatment, *J. Neurophysiol.*, 66, 580, 1991.
53. **Okada, Y., Miyamoto, T., and Sato, T.,** Aldosterone increases gustatory neural response to NaCl in frog, *Comp. Biochem. Physiol.*, 97A, 535, 1990.

54. **Garty, H. and Benos, D. J.,** Characteristics and regulatory mechanisms of the amiloride-blockable Na^+ channel, *Physiol. Rev.*, 68, 309, 1988.
55. **Rapuzzi, G. and Ricagno, G.,** Potenziali elettrici lenti della superficie linguale durante l'attivazione dei recettori gustativi di rana, *Boll. Soc. Ital. di Biol. Sper.*, 45, 59, 1969.
56. **Smith, D. V., Steadman, J. W., and Rhodine, C. N.,** An analysis of the time course of gustatory neural adaptation in the rat, *Am. J. Physiol.*, 229, 1134, 1975.
57. **DeSimone, J. A., Heck, G. L., Persaud, K. C., and Mierson, S.,** Stimulus-evoked transepithelial lingual currents and the gustatory neural response, in *Chemical Senses*, Vol. 1, Brand, J. G., Teeter, J. H., Cagan, R. H., and Kare, M. R., Eds. Marcel Dekker, New York, 1989, 13.
58. **Fidelman, M. L. and Mierson, S.,** Network thermodynamic model of rat lingual epithelium: effects of hyperosmotic NaCl, *Am. J. Physiol.*, 257, G475, 1989.
59. **Harper, H. W.,** A diffusion potential model of taste receptors, *Ann. N.Y. Acad. Sci.*, 510, 349, 1987.
60. **Claude, P. and Goodenough, D. A.,** Fracture features of zonulae occludentes from "tight" and "leaky" epithelia, *J. Cell Biol.*, 58, 390, 1973.
61. **Claude, P.,** Morphological factors influencing transepithelial permeability: a model for the resistance of zonula occludens, *J. Membr. Biol.*, 39, 219, 1978.
62. **Cereijido, M., Gonzalez-Mariscal, L., and Contreras, G.,** Tight junction: barrier between higher organisms and environment, *NIPS*, 4, 72, 1989.
63. **Simon, S. A., Holland, V. F., Benos, D. J., and Zampighi, G. A.,** Transcellular and paracellular pathways in lingual epithelia and their influence in taste transduction, *J. Elect. Micros. Tech.*, in press.
64. **Contreras, R. G., González-Mariscal, L., Balda, M. S., Garciá-Villegas, M. R., and Cereijido, M.,** The role of calcium in the making of a transporting epithelium, *NIPS*, 7, 105, 1992.
65. **West, C. H. K. and Bernard, R. A.,** Intracellular characteristics and responses of taste bud and lingual cells of the mudpuppy, *J. Gen. Physiol.*, 72, 305, 1978.
66. **Yang, J. and Roper, S. D.,** Dye-coupling in taste buds in the mudpuppy, *Necturus maculosus*, *J. Neurosci.*, 7, 3561, 1987.
67. **Teeter, J. H.,** Dye coupling in catfish taste buds, in Proc 19th Japanese Symp. Taste and Smell, Kimura, S., Miyoshi, A., and Shimada, I., Eds., Japanese Association for the Study of Taste and Smell (JASTS), Gifu, Japan, 1985, 29.
68. **Farbman, A. I. and Hellekant, G.,** Quantitative analysis of the fiber population in rat chorda tympani nerves and fungiform papillae, *Am. J. Anat.*, 153, 509, 1978.
69. **Silver, W. L. and Finger, T. E.,** The trigeminal system, in *Smell and Taste in Health and Disease*, Getchell, T. V., Bartoshuk, L. M., Doty, R. L., and Snow, J. B., Jr., Eds., Raven Press, New York, 1991, 97.
70. **Hellekant, G.,** The effect of ethyl alcohol on nongustatory receptors of the tongue of the cat, *Acta Physiol. Scand.*, 65, 243, 1965.
71. **Sostman, L. A. and Simon, S. A.,** Trigeminal nerve responses in the rat elicited by chemical stimulation of the tongue, *Arch. Oral Biol.*, 36, 95, 1991.
72. **Simon, S. A. and Sostman, A. L.,** Electrophysiological responses to nonelectrolytes in lingual nerve of rat and in lingual epithelia of dog, *Arch. Oral Biol.*, 36, 805, 1991.
73. **Roper, S. D.,** The microphysiology of peripheral taste organs, *J. Neurosci.*, 12, 1127, 1992.
74. **Jain, S. and Roper, S. D.,** Immunocytochemistry of gamma-aminobutyric acid, glutamate, serotonin, and histamine in *Necturus* taste buds, *J. Comp. Neurol.*, 307, 675, 1991.
75. **Hirata, K. and Nada, O.,** A fluorescence histochemical study of the monoamine-containing cell in the developing frog taste organ, *Histochemistry*, 67, 65, 1980.
76. **Herness, M. S.,** Vasoactive intestinal peptide-like immunoreactivity in rodent taste cells, *Neuroscience*, 33, 411, 1989.

77. **Silverman, J. D. and Kruger, L.,** Analysis of taste bud innervation based on glycoconjugate and peptide neuronal markers, *J. Comp. Neurol.*, 292, 575, 1990.
78. **Lundberg, J. M., Hökfelt, T., Änggård, A., Pernow, B., and Emson, P.,** Immunohistochemical evidence for substance P immunoreactive nerve fibers in the taste buds of the cat, *Acta Physiol. Scand.*, 107, 389, 1979.
79. **Yamasaki, H., Kubota, Y., Takagi, H., and Tohyama, M.,** Immunoelectron-microscopic study on the fine structure of substance-P-containing fibers of the taste buds of the rat, *J. Comp. Neurol.*, 227, 380, 1984.

Chapter 9

CHEMICAL RESPONSES OF LINGUAL NERVES AND LINGUAL EPITHELIA

S. A. Simon and Y. Wang

TABLE OF CONTENTS

0-8493-5341-6/93/$0.00+$.50

I. INTRODUCTION

The trigeminal, or fifth, cranial nerve carries most of the sensory information from the face scalp, mucous membranes of oral and nasal cavities, and from the surface of the eye. It consists of three major subdivisions; ophthalmic, maxillary, and mandibular, all of which branch extensively before terminating into very different types of epithelia (Figure 1). Trigeminal sensory receptors respond to mechanical, thermal, and chemical stimuli. Here, two aspects of chemical stimulation of trigeminal nerves will be discussed. The first pertains to its role in taste and the second to the influence of epithelia in modulating responses of trigeminal fibers to chemical stimuli.

Chemical stimuli introduced into the oral cavity may elicit responses from special sensory fibers associated with taste and from general sensory fibers associated with tactile, thermal, and painful stimuli. The responses elicited from these two types of nerve fibers depend on the phase of the stimuli (liquid, emulsion, suspension, solid), pH, concentration, temperature, and osmolality. Thus, the information elicited by chemical and/or physical stimuli involves an understanding of the responses of special sensory fibers associated with taste cells and general sensory trigeminal fibers associated with oral epithelia. These two types of nerves also provide information on the identity, concentration, and texture of the stimuli and also whether it should be chewed, swallowed, or expelled from the oral cavity. Many of these functions are modulated by the trigeminal nerve.

This chapter will focus on the response of the lingual branch of the trigeminal nerve to chemical stimuli. This nerve branch, for the most part, innervates the anterior two thirds of the tongue (Figure 1). There are three primary reasons to believe that lingual nerve fibers may contribute to the overall sensation of taste. First, they respond to a wide variety of chemical stimuli. Second, they completely surround and are even present in taste buds

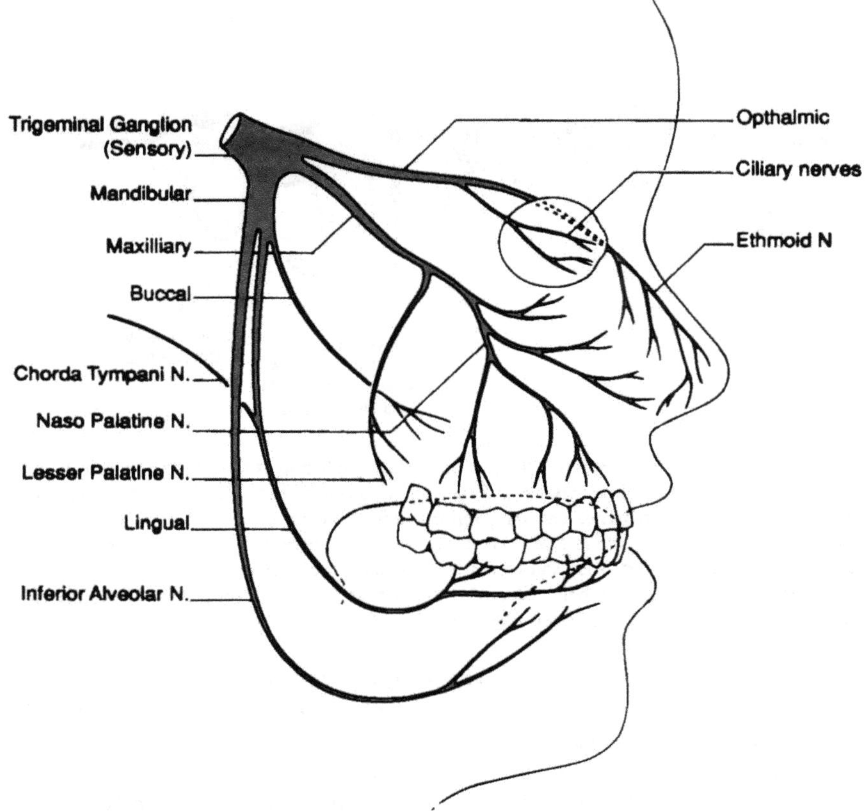

FIGURE 1. Illustration of the three sensory branches of the trigeminal nerve (V) and the chorda tympani nerve. The lingual nerve branch innervates the anterior two thirds of the tongue and the chorda tympani nerve innervates taste cells in this region of the tongue. (Adapted from Silver, W. L. and Finger, T. E., *The Trigeminal System: Smell in Health and Disease*, Getchell, T., Ed., Raven Press, New York, 1991. With permission.)

(Figure 2). Finally, some primary afferent lingual trigeminal and taste (chorda tympani) fibers terminate centrally in the same region of the nucleus of the solitary tract (see Chapter 13).

The role of the epithelia in modulating trigeminal nerve responses has not been well explored. Historically, most investigators treated the epithelia as a barrier that either delayed or even prevented stimuli from reaching sensory receptors. However, as will be shown, both lingual nerve fibers and lingual epithelia may respond to the same chemical stimuli, and, in some cases, at similar concentrations. Therefore, the influence of lingual epithelia in generating responses from lingual nerve fibers must be considered and the following questions addressed. How do hydrophilic (e.g., NaCl) and

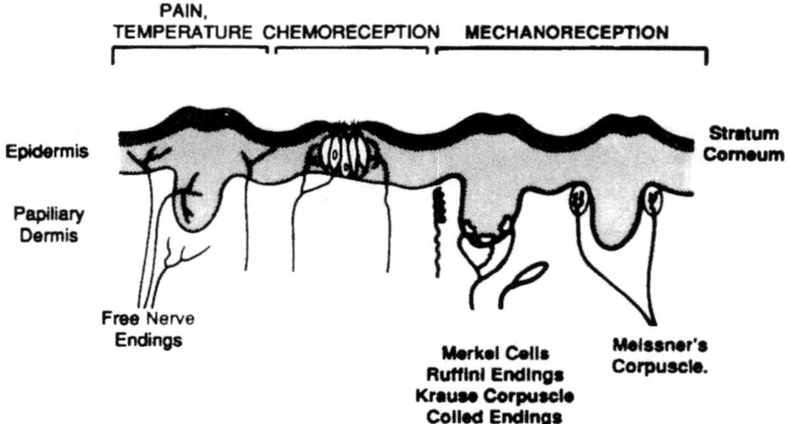

FIGURE 2. Sketch of a section of lingual epithelium containing a taste bud. The dark layer on the top represents the stratum corneum. Beneath it reside the other epithelia strata (not shown). The line between the epidermis and the papillary dermis represents the basement membrane. Lingual nerves may terminate as free nerve endings in the epidermis, in and around taste buds, and in the papillary dermis. Also shown are coiled endings and mechanosensitive fibers terminating in special capsules. It is believed that warm fibers are positioned deeper into the epithelial tissue than cold fibers. Intraepithelial fibers are of the C and Aδ class.

hydrophobic (e.g., menthol, phenyl ethanol) compounds elicit responses from different classes of trigeminal receptors embedded in *lingual* epithelia? Before proceeding to address this question, it is first necessary to review the morphology and electrophysiological responses of lingual nerve and lingual epithelia to chemical stimuli. When information on lingual nerve or epithelia is not available, data obtained from other trigeminal branches will be used.

II. PROJECTIONS OF TRIGEMINAL NERVES

A. PERIPHERY

The pseudo-unipolar cell bodies of the trigeminal (Gasserian or Semilunar) ganglion are located in Meckel's cave in the floor of the middle cranial fossa.[1] After the three sensory divisions exit the skull, they branch extensively (Figure 1). In humans, there is little spatial overlap in the distributions of the three sensory divisions and the trigeminal sensory root contains about 124,000 fibers with about half being unmyelinated.* The sensory components of the mandibular division (V3) include the buccal nerve (mucous membranes of mouth and gums), auriculotemporal nerve (head and scalp), inferior alveolar nerve (lower jaw, teeth, gums), meningeal nerve (anterior and middle cranial fosse), and the lingual nerve (anterior two thirds of the tongue and lower

* The cell bodies of sensory fibers that carry propriceptive information from the muscles of mastication reside in the mesencephalic nucleus of the brainstem.

gingivae). Also carried in V3 are efferent fibers to mastication muscles salivary glands[2] and specialized afferent taste fibers (chorda tympani, see Figure 1).

Before entering the tongue, the lingual nerve divides into several smaller branches[3] and terminates in filiform and fungiform papillae and in the epithelium between them. Fungiform papillae contain chorda tympani fibers associated with taste cells as well as lingual nerve fibers. In rat fungiform papillae about 75% of the total number of fibers (≈400) originate from the lingual nerve.[4] Some of these lingual fibers completely surround taste buds and a few even terminate in them[5,6,7] (Figure 2). However, lingual nerve fibers do not seem to form synapses with taste cells.[8]

B. PROJECTIONS

Centripetally from the trigeminal ganglion, most fibers bifurcate and send an ascending branch into the main sensory nucleus of V (SNV) and a descending branch into the spinal tract and nucleus of V. The ascending fibers are generally larger in diameter and respond to low intensity, steady and vibratory stimuli, whereas the descending fibers respond primarily to thermal and nociceptive stimuli.[9] While most sensory fibers from the lingual nerve terminate in the dorsal third of the spinal tract nucleus,[10] some lingual nerve fibers project into the lateral solitary nucleus where they intermingle with afferents from chorda tympani fibers, suggesting a role for trigeminal lingual fibers in taste (see Chapter 13 for additional details). Extracellular recordings from rat lateral solitary nucleus revealed that second order solitary neurons respond to both gustatory and tactile stimuli, thereby supporting the anatomical data regarding the convergence of sensory information from taste and sensory nerves.[11]

III. MORPHOLOGY OF ORAL EPITHELIA

All the epidermal epithelia in the oral cavity are classified as stratified squamous because they contain a stratum basale, stratum spinosum, stratum granulosum, and, in some cases, a stratum cornium (Figure 2). There is over a sixfold difference in thickness among the epithelia in the oral cavity.[12] Epithelia that do not contain corneocytes are classified as nonkeratinized (e.g., in humans these include buccal epithelium, lingual frenulum, floor of mouth, and soft palate). Epithelia containing a stratum corneum are classified as either para- or ortho-keratinized depending, respectively, on whether or not the corneocytes retain or lose their nuclei.* Keratinized epithelia include the dorsum of mammalian tongue (Figure 3), the hard palate, and gingiva.[12,13] These keratinized epithelia come in direct contact with a wide class of stimuli of various pHs, temperatures, and osmolalities, and a cornified layer is required to protect the underlying epithelial cells and fibers from damage.

* The particular classification of epithelia in the oral cavity may be species dependent. In mice, for example, the entire oral cavity is covered by ortho-keratinized epithelia.

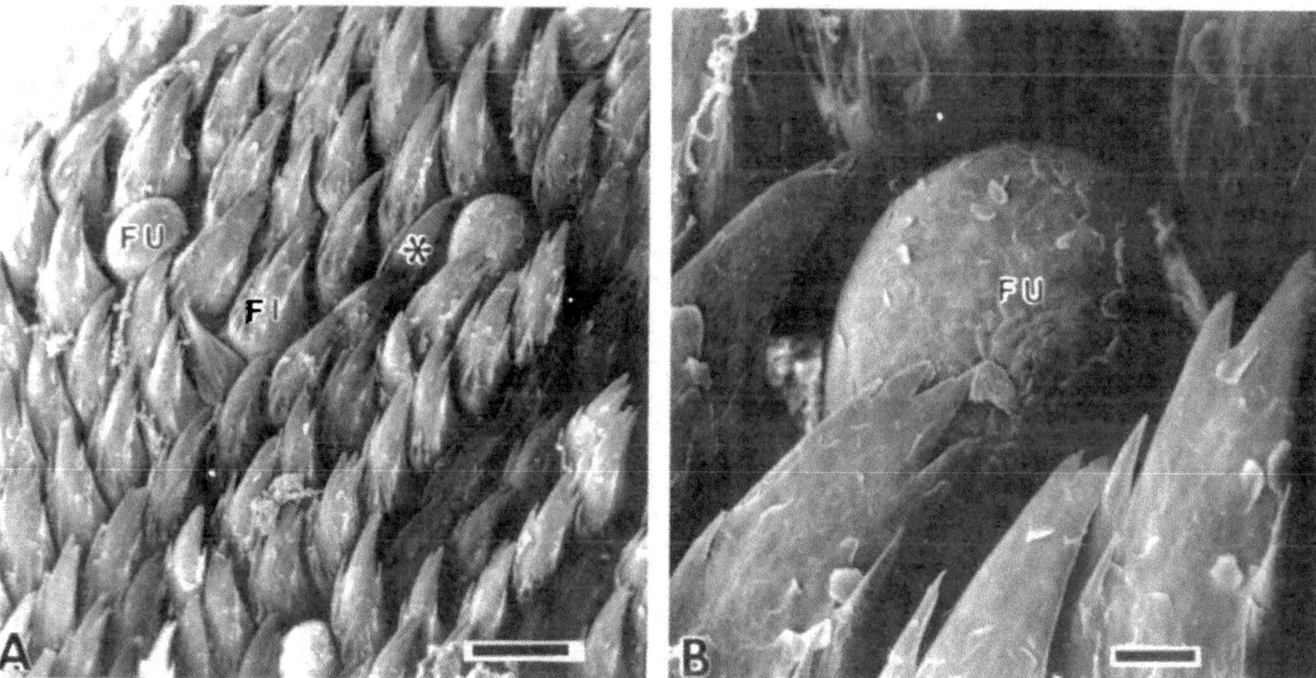

FIGURE 3. Scanning electron micrographs of the mucosal surface of canine lingual epithelium. (A) Low magnification view of a region of the epithelium anterior to the circumvallate papillae. The epithelium contains three rounded fungiform papillae (FU) and numerous pointed filiform papillae (FU). The region near the * is shown at higher magnification in B. Bar = 500 μm. (B) A view that shows sheets of cornified cells sloughing off from both filiform and fungiform papillae. Bar = 100 μm. (Adapted from Holland, V. F., et al., *J. Comp. Neurol.*, 279, 13, 1989. With permission.)

IV. NERVE TERMINATIONS IN ORAL EPITHELIA

Virtually all categories of nerve terminals (noncorpuscular, coiled, and corpuscular) have been identified in epithelia in the oral cavity (Figure 2 and see Chapter 3 for additional details).[14-16] The responses to chemical stimuli (other than anesthetics) have been investigated only in the noncorpuscular endings.

A. NONCORPUSCULAR

This category of nerve endings includes free nerve endings of Aδ and C fibers, and unencapsulated end bulbs. The majority of C and Aδ fibers terminate in papillary dermis (Figure 2) and the remainder penetrate into the epithelium but always terminate below the level of the tight junctions. In Figure 2, C and Aδ fibers are represented by those associated with temperature and pain as well as those surrounding the taste bud. Intraepithelial axons may lose their Schwann cells, but epithelial cells may, in some cases, form a "mesaxon" around them.[14] In fungiform papillae, intraepithelial fibers have their greatest density around taste buds.[7,17] Cutaneous C and Aδ fibers respond to temperature and/or nociceptive stimuli.[18] Intraepithelial fibers often contain substance P (SP) or CGRP, and in some fibers these two peptides are colocalized.[7,19] Similar fiber types are also present in the cornea and olfactory epithelium.[20,21] It has been reported that warm-sensitive fibers are located deeper in the connective tissue than cold-sensitive fibers.[22]

Unencapsulated end bulbs have been found in human hard palate and in rat incisive papillae.[13] These are small fibers and it is not clear whether they represent a separate entity or a variation of free nerve endings.[15]

B. COILED TERMINATIONS

In the oral cavity there are both loosely coiled terminations that are found throughout connective tissue papillae and tightly coiled terminations which are found coiled near the basal cell layer in the papillae.[23] Their function(s) have not been determined.

C. CORPUSCULAR TERMINATIONS

Merkel cells, or more precisely the nerve endings synapsed to them, are mechanoreceptors[24] which have been identified in fungiform papillae of monkeys[25] and in other regions of oral epithelia (see Chapters 2, 3, and 11 for additional information on Merkel cells). Miessner's corpusules, which are rapidly adapting mechanoreceptors, have been identified in the hard palate[13] and elsewhere in the oral cavity. Krause end organs are rapidly adapting, high threshold mechanoreceptors which are located in the mucosa of lips, hard palate, and tongue.[26] They are most frequently found at the tip of the tongue.

V. GAP AND TIGHT JUNCTIONS IN LINGUAL EPITHELIA

Tight junctions regulate the passage of small (and sometimes large) hydrophilic compounds into the extracellular space between tissues forming polarized epithelia.[27] In canine lingual epithelia, the tight junctions are cation selective but also permit the diffusion of Cl^-.[28] It has been recently shown that tight junctions play an important role in salt taste transduction (see also Chapter 8).[29,30] In mammalian lingual epithelia, tight junctions of the zonula type are found throughout the stratum corneum and also at the interface between the stratum corneum and stratum granulosum[31] (Figure 4). Tight junctions may be visualized via electron microscopy by using precipitates of lanthanum salts or horseradish peroxidase, since these electron dense compounds have a very small junctional permeability and accumulate at tight junctions (Figure 4).[31-34] When lanthanium salts are added to the dorsal surface of oral epithelia, very little diffuses beyond the first few layers of the stratum corneum (Figure 4). However, precipitates of lanthanium are observed at, but never beyond, the interface between the strata cornea and granulosa (Figure 4). When lanthanium salts are injected into oral epithelia, or are added to the solution bathing the serosal membranes, they diffuse throughout the extracellular space up to, but never beyond, the tight junctions at the interface between the strata granulosa and cornea.[31,33] In taste buds, tight junctions are located at their apical end[35,36] and are likely selective to cations but also permit the diffusion of small anions.[29] (See Chapter 8 for additional details). However, the tight junctions between taste cells are different from those of the surrounding epithelia since they are permeable to lanthanium.[38]

In lingual epithelia, the total measured transepithelial resistance (Rt) can be expressed as $Rt = Rc \times Rp/Rc + Rp$ where Rc = transcellular resistance and Rp = paracellular resistance.[37] In canine lingual epithelia, $Rc >> Rp$ and hence $Rt \approx Rp$.[38] Therefore, the increase in the Rt that results from the addition of lanthanium, reflects an increase of the paracellular resistance pathway.[38] *The primary reason for stressing the interaction of lanthanium with tight junctions in lingual epithelia is that lanthanium salts can block the trigeminal nerve responses elicited by salts.*[39] This point will be discussed in further detail below.

Lamellar or Orland bodies are also present in the extracellular space in the stratum corneum (if present) and upper layers of the stratum granulosum of buccal, palatial,[33] and lingual[31] epithelia. These extracellular organelles prevent water loss from epithelia and also provide a pathway for hydrophobic compounds introduced into the oral cavity to diffuse across the stratum corneum where they may partition into the plasma membranes of the "living" epithelial cells and subsequently into fibers which terminate in the epithelium or papillary layer.[40]

Gap junctions are extracellular organelles that provide low resistance pathways between epithelial cells through which molecules having mol wts

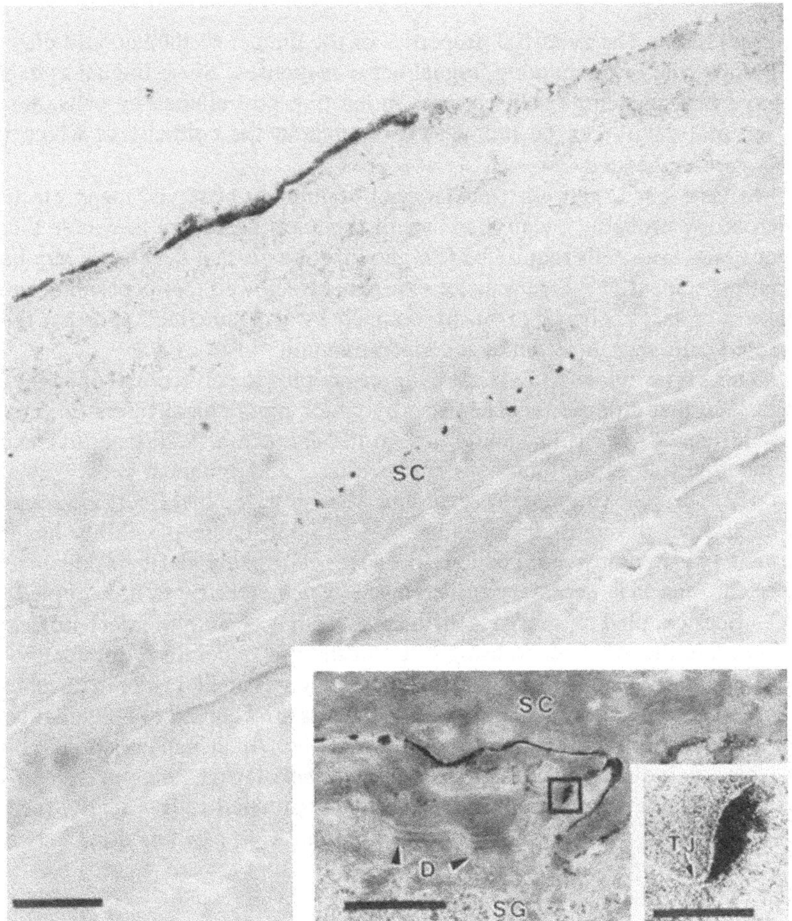

FIGURE 4. Transmission electron micrograph of an unstained longitudinal section from canine fungiform papillae that was infiltrated with lanthanum from the mucosal solution. The electron dense deposits are precipitates of lanthanum. In the upper figure, lanthanum penetrated across the stratum disjunctum and accumulated in the extracellular space between the first and second layers of the stratum corneum (SC). Bar = 1 μm. Inset: results of an experiment where lanthanum added to the mucosal solution penetrated to the interface between the stratum corneum and stratum granulosum (SG) where it was stopped by a tight junction (TJ) which prevented lanthanum from penetrating into the stratum granulosum (SG). The arrows point to modified desmosomes. Bar = 0.5 μm. The tight junction which prevented lanthanum from penetrating into the stratum granulosum is shown at higher magnification in the inset in the corner. Bar = 100 nm. (From Holland, V. F., et al., *J. Comp. Neurol.*, 279, 13, 1989. With permission.)

$<$ 800 can diffuse between cells.[41] Negative stain and freeze-fracture electron microscopy showed that gap junctions are present between all canine lingual epithelial cells except those in the stratum corneum.[31] *Therefore, the living epithelial cells of lingual epithelia are coupled and thus behave functionally as a syncitum*. The syncytial properties of the lingual epithelia could play an important role in modulating lingual nerve responses. Since lingual epithelia behave as a giant sinle cell, changes in ion transport elicited by cells closest to the oral cavity can be propagated throughout the epithelia in which the fibers are embedded.

In taste buds, gap junctions are not commonly observed using electron microscopy probably because so few of them are necessary to couple them. The reason taste cells require so few gap junctions is that they have very high input resistances.[42,43] Dye transfer experiments showed that occasional small clusters of taste cells (2 to 4) are coupled by gap junctions and that these coupled cells may function as a cooperative unit.[44,45]

There is no evidence that taste cells are coupled to perigemmal or epithelial cells,[45] so that proposed mechanisms by which perigemmal fibers (Figure 2) may influence taste cells should not, in the absence of additional evidence, include the role of gap junctions between taste and epithelial cells.

There is some controversy regarding whether trigeminal fibers are coupled to each other through gap junctions.[46,47] This intriguing possibility has important physiological implications because, if trigeminal fibers are indeed coupled, then it is necessary to determine which fiber types are coupled. A "C" fiber coupled to a large Aβ fiber should not be able to inject sufficient current into the Aβ fiber to bring it to threshold.* In contrast, stimulation of a large fiber coupled to a smaller fiber may drive the smaller fiber to threshold. It is also not clear whether intraepithelial fibers are coupled to epithelial cells via gap junctions.[47] If indeed they are coupled, then gap junctions could provide an indirect mechanism for intraepithelial trigeminal nerves to be activated by chemical stimuli interacting with epithelial cells. Clearly, further investigation is needed to clarify the importance of gap junctions between nerves and/or epithelial cells.

A. PERMEABILITY PATHWAYS IN ORAL EPITHELIA AND NOCICEPTION

Epithelial and nerve cells contain numerous proteins involved in ion and nonelectrolyte transport (channels, pumps, and exchangers) whose activity may be affected by changes in their local chemical environment, temperature, osmotic pressure, transmitters, and trophic factors. The central issue is to determine how chemicals, classified as irritants, can elicit responses from trigeminal fibers. In a comprehensive review of sensory irritants, Alarie[48] classified them in seven categories. Included are such chemically diverse compounds as acetic acid, benzene, capsaicin, ethanol, formaldehyde, sulfur dioxide, chloracetophenone, and acrolein. He did not include electrolytes like

* We are aware that this process may depend on the frequency and duration of stimulation.

NaCl, which, at high concentrations (≥0.5 *M*), are also irritants when introduced into the oral cavity[49] nor did he include compounds like menthol, which are not irritants, but stimulate trigeminal nerve cold fibers. To elicit responses from intraepithelial fibers or fibers in the connective tissue, these stimuli must initially diffuse into the epithelium and induce their sensation by *directly* associating with nociceptors or other fiber types and/or by *indirectly* influencing these fibers by eliciting responses from epithelial cells. One factor that makes this problem somewhat intractable is that we found that several nonpolar chemical stimuli elicit responses from trigeminal fibers and from intact epithelia at about the same concentrations.[50] Therefore, to obtain an understanding of how trigeminal nerve fibers are activated by chemical stimuli, it is also necessary to understand the responses of the intact epithelium to these same stimuli. Ideally, one would like to find a compound that did not elicit responses from epithelial cells but activates trigeminal fibers. Unfortunately, there are few such compounds.

B. PERMEABILITY OF NONELECTROLYTES INTO ORAL EPITHELIA

The two principal pathways for nonelectrolytes to diffuse into stratified epithelia are the transcellular and paracellular pathways. The transcellular pathway contains the plasma membranes of epithelial and nerve cells. The paracellular pathway contains the tight junctions and the extracellular space. There have been many studies on the permeability of nonelectrolytes across epithelia in the oral cavity[51-55] including lingual epithelia.[56] Since most nonelectrolytes enter epithelia by passive diffusion, the permeability coefficient (Pm) can be written: $Pm = K\ Dm/\tau$ where K is the oil/water partition coefficient, Dm is the diffusion constant, and τ is the thickness of the epithelium. For nonelectrolytes having molecular volumes greater than about 80 cm^3/mol (e.g., phenyl ethanol), *Pm is directly proportional to K.* Consistent with this solubility-diffusion paradigm, the concentration necessary to elicit responses from ethmoid fibers, lingual fibers, and lingual epithelia decreases with increasing stimulus oil/water partition coefficient.[57,58]

Nonelectrolytes having molecular volumes ≤ 80 cm^3/mol (e.g., urea) also diffuse across paracellular pathways.[51] Water flux across oral epithelia is determined by both active salt transport and osmotic gradients.[53]

C. NONELECTROLYTE DIFFUSION INTO EPITHELIA AND LATENCY

Latency is the difference in time between the stimulus onset and the elicited response. For chemically stimulated responses from lingual nerves, the latency can take up to tens of seconds (see Reference 59). Part of the latency can be attributed to the time required for nonelectrolytes to diffuse into the epithelia[60] and the remainder to attain a threshold concentration in epithelial or nerve cells. Therefore, the latency will depend on the stimulus concentration, stimuli's partition coefficient, flow rate, thickness of epithelium, thickness of unstirred layer, diameter of fiber, and distance of the fiber's terminals from the epithelial surface. Analytical solutions to various aspects of this problem have been obtained.[61-63] These analyses showed that the latency *decreases* the closer the fiber is to the surface, the smaller the fiber diameter, the smaller the unstirred layer, the larger the flow rate, the thinner the epithelium, the larger the stimuli's partition coefficient, and the larger the concentration gradient. Epithelia in the oral cavity differ markedly in their overall thickness.[12] All of these differences should be refected in the response latency. For example, filiform papillae have a very thick cornified layer[64] (Figure 3) and thus the latency should be much longer for fibers terminated in epithelia in these papillae than for fibers terminated in thinner epithelia such as the floor of the mouth, the lingual fenulum, the mucus-lined epithelia in the nasal cavity (respiratory and olfactory epithelia), and the cornea. Therefore, only considering the thickness of the epithelia, the latencies for trigeminal responses are smaller in olfactory epithelia and cornea then in lingual epithelia for a given stimulus.[57-59,65]

D. PERMEABILITY OF ELECTROLYTES

The study of electrolytes on trigeminal nerves is important since many common electrolytes (e.g., NaCl, KCl, NH_4Cl, $CaCl_2$) produce pain when applied directly to trigeminal nerves.[66] However, the pathway(s) by which electrolytes enter epithelia cannot be as easily generalized as for nonelectrolytes. The pathways depend on the type of epithelia (i.e., tight or leaky; keratinized or nonkeratinized), transport proteins present, type(s) and concentrations of salts present, temperature, permeability and selectivity of tight junctions, and concentration gradients. The various epithelia innervated by trigeminal fibers differ both structurally and in some of the membrane proteins regulating electrolyte and nonelectrolyte transport. Lingual, buccal, and corneal epithelia are Na^+ absorbing in the sense that ion transport is inhibited by the epithelial Na^+ channel inhibitor, amiloride,[28,67,68] In contrast, olfactory epithelia are Cl^--secreting epithelia.[69] Given the wide range of electrolytes (e.g., NaCl, KCl, $CaCl_2$, NH_4OH, HCl, HAc) that activate trigeminal fibers, and that relatively high concentrations are required to stimulate them, it is difficult to generalize about the transcellular and paracellular pathways taken by electrolytes as they partition into and around these various epithelia.

The commonality of olfactory,[69] corneal,[68] buccal,[67] and lingual[28,70] epithelia is that under symmetrical salt solutions, ion transport is markedly inhibited by the specific inhibitor of (Na^+, K^+)-ATPase, ouabain. In isolated mammalian lingual epithelia bathed in symmetrical Ringer's buffer, ouabain completely inhibits ion transport.[28,71] This enzyme, which is localized beyond the tight junctions in most stratified epithelia,[72,73] maintains the electrochemical gradients for Na^+ and K^+ by "pumping" Na^+ into and removing K^+ from the extracellular space. Clearly, inhibiting (Na^+, K^+)-ATPase by lowering the temperature or with a specific inhibitor ouabain will increase the K^+ concentration in the extracellular space thus depolarizing intraepithelial trigeminal fibers. The influence of (Na^+, K^+)-ATPase in modulating primary sensory afferent fibers from skin was noted when the addition of ouabain increased the rate of discharge of cold fibers.[74] In this case, as with most others, it was impossible to determine whether ouabain interacted directly with cold fibers[74] or indirectly with them by inhibiting transport across epithelial cells.

When hyperosmotic concentrations of salts are introduced into the oral cavity, processes will occur that may influence trigeminal nerves. One response to such stimuli will be a volume regulatory osmotic response that will alter the ion and water fluxes across the epithelia. The particular response is dependent on the specific salts and the specific transport pathways present on the apical membranes. In isolated dog tongue about 60 to 70% of the increase in the short circuit current (Isc) induced by the addition of hyperosmotic concentrations of NaCl is inhibited by either amiloride or ouabain. These data suggest that, with NaCl, a large percentage of transepithelial transport occurs via transcellular pathways,* since neither amiloride (at μM concentrations) nor ouabain directly affect paracellular pathways.[75] In addition, there will be an increase in the flux of salts (both anions and cations) across tight junctions (the ouabain-insensitive current) into the extracellular space that will change its composition. The diffusion of salts into the extracellular space will depend on the salt type and concentration gradient, solution pH, and ion selectivity of the tight junctions. Tight junctions select among ions on the basis of size and charge[76] and, therefore, large anions such as methanesulfonate and small highly charged cations such as La^{3+} (Figure 4) will not readily diffuse across tight junctions between epithelial cells.[77]

VI. SPATIAL RESPONSES OF LINGUAL FIBERS

Before entering the anterior of the tongue, the lingual nerve ramifies into several branches that terminate in the papillary layer and epithelium (Figures 1 and 2). In cat, there are three branches (anterior, medial, and posterior) whereas monkeys have at least five.[78] The tip of human tongues is more densely innervated then other parts[23,79] and is most sensitive to tactile,[15] cold,[80] and irritating stimuli.[81] Ishiko[78] simultaneously measured tactile, thermal, and chemical whole nerve responses from the lingual nerve branches in cats and monkeys. In cat, where there is very little overlap between sensory responses from the three lingual nerve branches, all three branches responded to tactile stimuli, but the anterior branch responded the most robustly. In contrast, only the medial branch responded to thermal stimuli. Thus, it appears that, in some

* The transcellular pathways include both the stratified lingual epithelia and taste cells. Analysis of the magnitude of short circuit current measured across *canine* lingual epithelia[28,75] and the distribution of (Na^+, K^+)-ATPase[70] shows that most of the ouabain- and amiloride-inhibitable current occurs across epithelial cells. In other species, such as rat, amiloride-inhibitable transport may traverse lingual epithelia primarily via taste cells.

species, lingual nerves indeed exhibit spatial differences. Additional studies are required to determine the spatial selectivity of trigeminal fibers to chemical and physical stimuli.

VII. RESPONSES OF LINGUAL NERVE TO CHEMICAL STIMULI

Many of the chemicals that elicit responses from taste and olfactory neurons also elicit responses from trigeminal neurons.* However, the concentrations necessary to elicit responses from trigeminal fibers to many, but by no means all, stimuli are higher than the concentrations necessary to elicit taste or olfactory responses.[83] These data suggest that the transduction mechanisms by which the same chemical stimuli elicit responses from trigeminal and special sensory fibers may be different, and, hence, one cannot generally seek to understand transduction mechanisms of trigeminal nerves through transduction mechanisms pertaining to taste or olfaction.

Therefore, to understand the mechanisms by which lingual fibers respond to chemical stimuli, three fundamental questions must be addressed: (1) How do hydrophilic and hydrophobic chemical stimuli enter lingual epithelia? (2) Do trigeminal nerves exhibit specificity to chemical stimuli? (3) What influence may the lingual epithelium have in modulating trigeminal nerve responses? In this section we show how vastly different classes of chemical stimuli elicit responses from lingual nerves and lingual epithelia.

A. HOW DO CHEMICAL STIMULI ENTER LINGUAL EPITHELIA?

A recent study of whole rat lingual nerve responses to high (molar) concentrations of NaCl, KCl, and $CaCl_2$ found that 2.5 m*M* $LaCl_3$ inhibited the responses of these salts (Figure 5).[39] When the tight junctions were occluded by La^{3+} (Figure 4), salts in the oral cavity were inhibited from diffusing into the extracellular space of lingual epithelia. This interpretation was supported by the observation that presentation of 1 *M* $LaCl_3$ to the dorsal surface of rat tongue did not elicit lingual nerve responses,[39] demonstrating that lingual nerve responses are not produced by salts that cannot enter the epithelium nor can they be produced by any solution having a large osmotic pressure. In contrast, whole lingual nerve recordings to nonpolar nonelectrolytes, including classical trigeminal stimulants like phenyl ethanol and amyl acetate, were not inhibited by $LaCl_3$,[57] suggesting that hydrophobic trigeminal stimulants enter lingual epithelia, not via tight junctions, but rather by initially partitioning into the lamellar bodies in the stratum corneum (if present) and then into the plasma membranes of epithelial cells. The ability of lanthanium to inhibit responses to salts of single lingual nerve C and Aδ fibers has also been demonstrated (Wang, Erickson, and Simon, unpublished observations).

* The trigeminal nerve contain all subclasses of Aβ, Aδ, and C fibers.[82]

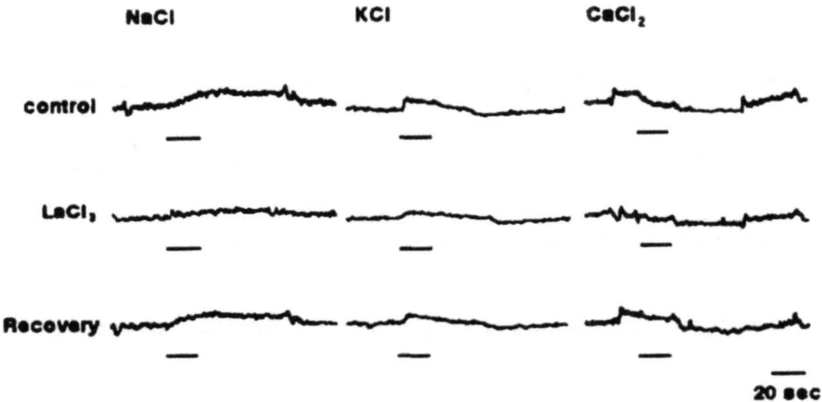

FIGURE 5. Whole nerve recordings from rat lingual nerve. Lanthanum reversibly inhibited responses to NaCl (2.5 *M*), KCl (2.5 *M*), and $CaCl_2$ (2 *M*). Traces recorded under control conditions (upper), immediately following 30 s incubation with 2.5 m*M* $LaCl_3$ (center), and 20 to 40 min after incubation with $LaCl_3$. Salt stimulation is indicated by solid line. (Adapted from Sostman, A. L. and Simon, S. A., *Arch. Oral Biol.*, 36, 95, 1991.)

B. DO TRIGEMINAL NERVES EXHIBIT SPECIFICITY TO CHEMICAL STIMULI?

Responses of the lingual nerve of rats elicited by salts exhibit different response patterns (Figure 5), suggesting that different fibers types may be activated by these compounds.[39] That is, salts other than $CaCl_2$ initially increased the lingual nerve activity whereas $CaCl_2$ initially decreased its activity. The activity of intradental nerves also increased upon addition of NaCl or KCl and were depressed by $CaCl_2$,[66,84,85] suggesting a similarity in the mechanisms by which different trigeminal nerve branches respond to salts. In neither study were the fiber types specified, although all the aforementioned salts caused pain when placed on dentine, suggesting that, in part, they activate nociceptive fibers.[66]

1. $CaCl_2$

Single fiber recordings from cat infraorbital nerve showed that $CaCl_2$ increased the activity of warm fibers and depressed the activity of cold fibers.[86] Although the whole nerve recordings reported above suggest that $CaCl_2$ inhibited cold fibers, calcium's action is not specific for cold fibers. Given calcium's effects on excitable membranes, this might be expected.[87]

2. KCl

In human skin, the application of KCl to a blister base causes a painful sensation,[88] suggesting that KCl activates nociceptive fibers. The addition of KCl to the pulp surface of teeth activates intradental C fibers[89] and the addition of KCl to rat tongue activated C fibers in lingual fibers (Figure 6). All single

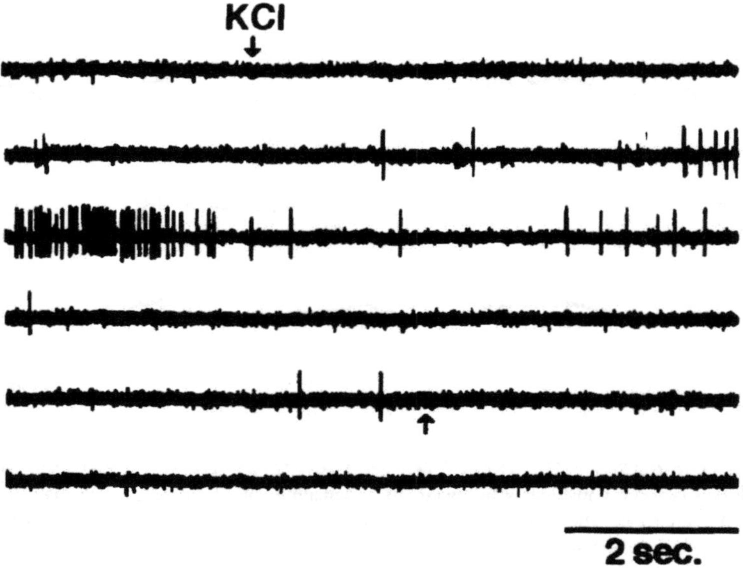

FIGURE 6. Extracellular recording from rat trigeminal ganglion in response to 2.5 *M* KCl applied at the arrow. After a several second latency there was a brief increase of activity that eventually diminished to sporadic activity. The upward pointing arrow represents the end of the KCl application. This figure is a digitized representation of the voltage responses.

neuron responses to KCl exhibit, after a latency, a brief burst of activity which is followed by a cessation of activity. It is not evident whether KCl will activate other fiber types, although it might be expected to, given that increasing the extracellular KCl concentration should depolarize all intraepithelial fibers.

3. Menthol

It is well known that menthol applied to lingual and other epithelia evokes a cooling sensation. Single fiber recordings of warm and cold fibers in cat lingual nerve showed that menthol (at low [μM] concentrations) increased the activity of cold fibers but did not affect warm fibers.[90] In contrast, more recent single fiber studies from rat lingual nerve showed a class of cold fibers whose activity was depressed after oral application of menthol.[91] It is difficult to reconcile these two data sets since in the earlier study menthol acted to sensitize cold fibers whereas in the latter it acted to desensitize them. It is not established whether menthol, at low concentrations, interacts with other fiber types although recent studies in our laboratory recording from trigeminal ganglia suggest that it does not. The mode of action of menthol on *cold fibers* is not understood. Structure-function studies using analogs of menthol suggest that menthol interacts with receptors,[92] perhaps associated with calcium channels.[93] At higher concentrations, menthol has an "anesthetic" effect in that it inhibits responses from pain fibers and mechanoreceptors.[90]

4. Ethanol and Other Alcohols

Ethanol produces a drying and/or burning sensation when applied to human lingual epithelia. Ethanol has been shown to activate lingual nerve fibers in cats and rats.[39,59] In a single fiber study of the effects of ethanol on rat lingual fibers, Hellekant identified a class of normally quiescent Aδ fibers (judged Aδ by their conduction velocity) which responded to ethanol, tactile, and cooling stimuli but did not respond to a solution of 0.2 *M* NaCl, 0.02 *M* acetic acid, and 5 m*M* quinine. Increasing the ethanol concentration (to 4 *M*) increased the fiber activity and decreased the latency[59] as would be expected from our previous discussion. At even higher ethanol concentrations the fiber activity decreased although the latency remained constant (at ≈2 s). These fibers did not respond to warming nor to temperatures that activate nociceptive fibers (53°C). Ethanol also did not stimulate high threshold mechanoreceptors. Hence, ethanol, at least at the lower concentrations, appears to activate a specialized class of cat lingual neurons. At the higher concentrations, ethanol has anesthetic properties. No mechanism was presented to suggest how ethanol may stimulate lingual nerve fibers.

Lingual and ethmoid nerves are activated by longer chain length alcohols at concentrations that depend on their partition coefficient.[57,58] That is, the higher the membrane/water partition coefficient, the lower the concentration necessary to activate these fibers. Analysis of the threshold concentrations for lingual nerve responses reveals that alcohols partition into a hydrophobic environment having the solubility characteristics of plasma membranes,[57] suggesting that alcohols traverse epithelia via by partitioning into plasma membranes of epithelial and or nerve cells. Whole nerve lingual or ethmoid responses to these stimuli exhibit an initial (phasic) increase in activity that reaches a maximum and then declines to a steady state (tonic) activity (Figure 7).[57,58] At higher concentrations, the phasic response remains, but the tonic response decreases to below its basal activity, suggesting that the anesthetics inhibited the spontaneously active cold fibers.[57] Moreover, responses to all mechanical stimuli are eliminated at high alcohol concentrations. The inhibition of responses to mechanical and thermal stimuli suggests that these compounds diffuse across the epithelium and into the papillary layer where many of the thermally sensitive-free nerve endings and encapsulated mechanoreceptors are found (Figure 2). Given that alcohols are anesthetics, inhibition of sensory stimuli is indeed expected. What is difficult to explain in terms of anesthetics directly interacting with lingual or ethmoid fibers is the initial increase in activity (see Figure 7). One possibility, which will be discussed below, is that the initial increase in activity may be a consequence of anesthetics interacting with epithelial cells.

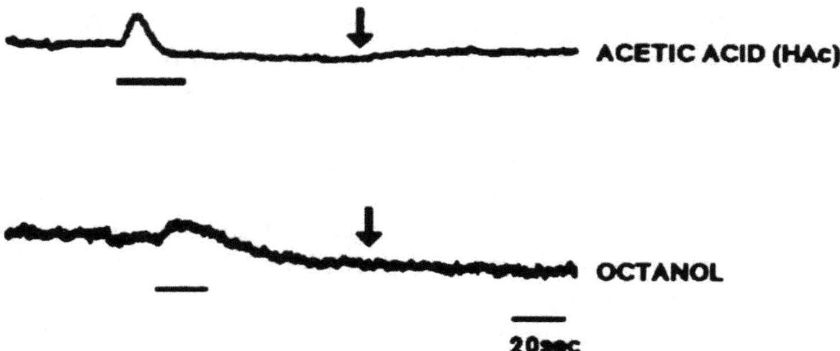

FIGURE 7. Integrated responses from rat lingual nerve to 1 *M* acetic acid (HAc) and 4 m*M* octanol. (Adapted from Sostman, A. L. and Simon, S. A., *Arch. Oral Biol.*, 36, 95 and 36(11), 805, 1991. With permission.)

5. Capsaicin

Capsaicin, the active ingredient of hot peppers, acts as an irritant in all animal species. When introduced into the oral cavity it produces a burning sensation. Recent psychophysical experiments found that application of capsaicin (albeit at relatively high concentrations; 20 ppm) to lingual epithelia did not eliminate the response to 30% ethanol and slightly reduced the responses to 5 *M* NaCl.[94] These data suggest that ethanol and NaCl may activate capsaicin-insensitive fibers.[94] One possibility is that ethanol and NaCl can be interacting with other fibers indirectly by modulating epithelial transport whereas capsaicin selectively eliminates a specific class of C and Aδ fibers.[95]

Capsaicin excites a class of polymodal nociceptive fibers of the C and Aδ type.[95] It also activates warm receptors but does not activate other C, Aδ, and Aβ fiber types.[95] On neurons, the receptor for capsaicin is a cation-selective channel that has a dissociation constant for capsaicin of about 300 n*M*.[95] At higher concentrations, capsaicin can also affect other channels[95] and therefore at high concentrations it cannot be used as a marker of polymodal nociceptors. Capsaicin-sensitive neurons include those containing SP, tachykinins, and CGRP and other peptides. Nevertheless, the presence of tachykinins, CGRP, or somatostatin in fibers cannot be used as markers for nociceptive afferents.[96] Finally, it is important to note that capsaicin, even at micromolar concentrations, affects some but not all nonneuronal plasma membranes.[97]

In a study of single rat lingual fibers, Okuni[98] found that capsaicin (at unknown concentrations) did not elicit responses from cold fibers or mechanoreceptors but did elicit a response from a single warm fiber and some nociceptors. Moreover, the fibers that were activated by capsaicin became unresponsive to other nociceptive stimuli. In a companion study, Okuni[99] also found that capsaicin also activated many chorda tympani fibers. However, no explanation was given for this effect. In a subsequent section we discuss possible mechanisms by which trigeminal lingual fibers may influence taste.

In a recent electrophysiological investigation of whole ethmoid nerve response to trigeminal stimuli, it was found that the nerve responses disappeared 40 days after injecting rat pups with capsaicin.[100] Under these same conditions, immunohistochemical investigations showed that intraepithelial CGRP- and SP-labeled fibers, normally present in control olfactory epithelia, were absent in the capsaicin injected rats. These data suggest that a large fraction of the trigeminal responses to irritants such as amyl acetate are generated by CGRP- and/or SP-containing neurons.

6. Acetylcholine (ACh)

Application of ACh to blister bases on human skin produces a painful sensation.[88] However, ACh failed to elicit a painful sensation after application of capsaicin to blister bases, suggesting that ACh interacts with nociceptive fibers. Single fiber recordings from cat lingual nerve revealed that a low concentration of ACh injected into cat tongue increased the activity of cold fibers whereas injection of higher ACh concentrations decreased cold fiber activity.[101] Thus, ACh appears to stimulate at least two distinct types of fibers.

Single, or few, unit recordings from frog glossopharyngeal nerve (which contains both special and general sensory fibers) revealed that application of 100 μg/ml ACh on the tongue for 3 min induced an initial short increase in activity.[102] Similar responses were obtained with cholinesterase inhibitors and ACh antagonists.[102] It was presumed that ACh activated nicotinic acetylcholine receptors (nAChr)[101,102] *on nerve terminals* (the epithelium was not considered as a possibility). Indirect evidence for the presence of nAChrs was obtained with psychophysical experiments which showed that the nAChr antagonist, mecamylamine (albeit at high concentrations), reduced the burning sensation produced by nicotine.[103] Patch-clamp recordings from isolated trigeminal ganglion cell bodies revealed the presence of both nicotinic acetylcholine receptors (Ma, Pugh, and Simon, in press). Thus ACh, in principle, can interact directly with trigeminal fibers to produce its response on cold or nociceptive fibers.

There is some evidence for ACh interacting with taste cells; Hwang et al.[103b] showed that carbachol increased phosphatidyl inositiol turnover in rat taste cells, However, intracellular recordings from taste cells from rat fungiform papilla showed that application of 1 mg/ml ACh to the tongue did not produce any changes in resting potential or in the taste cells response to NaCl.[104] The role of ACh in taste, however, remains puzzling since acetylcholine esterase (AChEase) is present throughout taste buds (see Chapter 11).[105]

C. What Influence May the Lingual Epithelia Have in Modulating Trigeminal Fiber Responses?

The influence of epithelia on mechanisms of trigeminal nerve stimulation is generally not considered to be more than influencing the latency. Most investigators tacitly assume that the chemical stimuli interact directly with the trigeminal nerves. This is a reasonable assumption since compounds like capsaicin induce painful stimuli when in contact with all types of epithelia and therefore its action (at least at low concentrations) appears to depend *only* on the presence of specific classes of nociceptive fibers. To this point, the effects of 1 μM capsaicin on isolated canine lingual epithelia are minimal, although higher concentrations inhibit ion transport (Simon, unpublished observation). However, for less specific trigeminal nerve stimuli (NaCl, HAc, phenol ethanol, ACh, and ethanol), the case for nerve specificity cannot be so clearly made. Indeed, lingual epithelia respond to the same stimuli at similar concentrations as lingual nerve fibers[57] and hence, for such stimuli, the role of the epithelia has to be considered. Isolated mammalian lingual epithelia respond to a wide variety of electrolytes (including NaCl, KCl, HCl) by increasing the net inward current flowing into the tongue (see Chapter 8 for additional details). Most of the current flows across the stratified epithelia (rather than across taste cells) and hence measurements of transport across isolated lingual epithelia (as given by the short circuit current [Isc]) primarily reflect transport across the stratified regions of lingual epithelia.[28,38,70]

There are two primary mechanisms by which epithelia can influence the activity of fibers associated with them: (1) *If* the fibers are coupled to epithelial cells, then stimulation or inhibition of epithelial transport could influence the fiber's resting potential (and therefore its activity) since current can flow to or from the epithelial cell to the fiber. However, as no experimental evidence for this mechanism exists, no further discussion of it is warranted. (2) Changing the ionic composition of the extracellular space in which the fibers are embedded can alter their activity. The parameter that would most obviously affect the activity of embedded neurons is the extracellular K^+ concentration since it determines, to a great extent, the resting potential. Increasing the extracellular KCl concentration would produce nerve depolarization and cause transmitters and peptides to be released from nerve terminals.[113] Release of transmitters (such as ACh) from nerve terminals can be produced by any mechanism that causes fibers to depolarize such as lowering the extracellular pH (Figure 7) or increasing the osmolality of the extracellular solutions.[114] Changing the extracellular calcium concentration can also influence the sensitivity of C[106] and other fiber types.[87] Changes in the ionic composition of the extracellular space can occur by electrolyte diffusion across tight junctions or by modifying the activity of (Na^+, K^+)-ATPase or other "pumps" or cotransporters present in epithelia. To this point, the Isc measured across isolated lingual epithelia is inhibited by trigeminal nerve stimulants such as phenol ethanol (PEA) (Figure 8A). The decrease in the (Na^+, K^+)ATPase-dependent Isc should increase the K^+ concentration in the extracellular space that should decrease the resting potential of lingual fibers bringing them closer to threshold.[85] *However, simply bringing the fibers to threshold cannot, by itself, explain the pattern of responses to the different salts.* For example, the response of the lingual C fiber to KCl shown in Figure 6 reveals a pattern

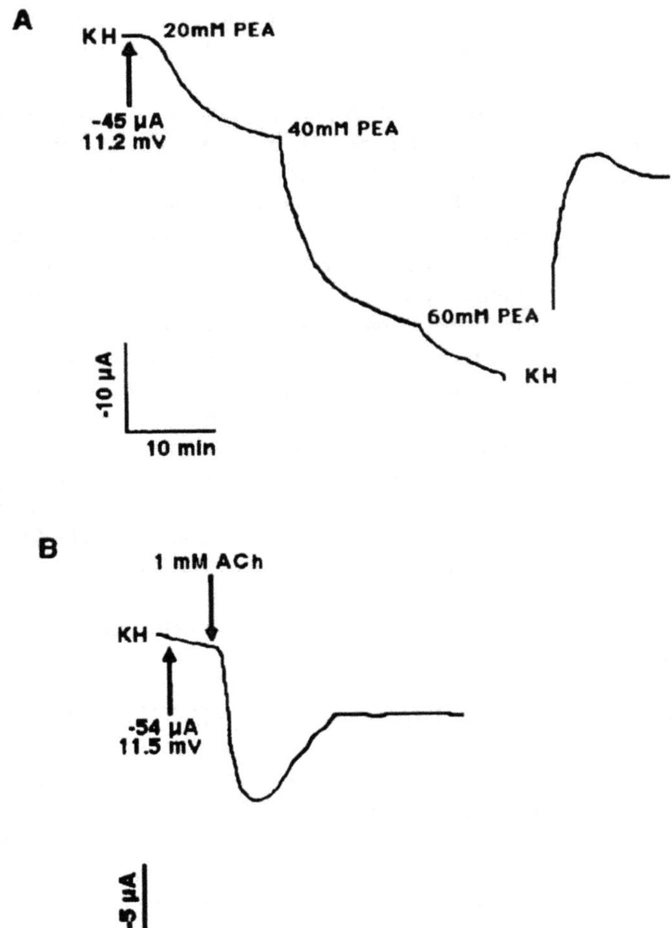

FIGURE 8. Short circuit current (Isc) measured across isolated canine lingual epithelia bathed in symmetrical solutions of Krebs-Henseleit buffer (KH) at 35°C. (A) Addition of phenol ethanol (PEA) reversibly inhibited Isc. (Adapted from Simon, S. A. and Sostman, A. L., *Arch. Oral Biol.*, 36(11), 805, 1991.) (B) Response of Isc to bolus of 1 m*M* acetylcholine (ACh) added to the serosal solution. Values of the open circuit potential and short circuit current in KH buffer are given in the left-hand corners of both panels. Chamber area = 3.1 cm^2.

where, after a latency, the activity initially increases, then decreases, and finally ceases. Without detailed knowledge of the number and types of channels, pumps, and other transporters in the nerve terminal and axon it is impossible to predict the response pattern. In addition, the response pattern will depend on the volume of the extracellular space and the rate that its ionic composition can return to its normal composition.[107] In the future it will be important to determine whether all fiber types will be activated by KCl or other monovalent salts.

Changes in ion transport across lingual epithelia can also be produced by neurotransmitters like AChRs (Figure 8B). In canine lingual epithelia, ACh induces its effect on transepithelial ion transport by binding to muscarinic acetylcholine receptors (AChrs).[108] Although the function of these (and other putative) receptors on epithelial cells is not well understood, it is evident from these data that ACh could induce an effect on nociceptive and thermally sensitive fibers by binding to receptors on epithelial cells which in turn could change the composition of the extracellular space in much the same manner as does phenyl ethanol (Figure 8A). Lingual epithelia are also responsive to menthol, octanol, and ethanol.[57] These compounds are anesthetics and hence would be expected to *inhibit* neuron activity if they interacted directly with them. Thus, the initial *increase* in activity found in whole lingual and ethmoid nerve recordings elicited by anesthetics such as octanol (Figure 8) or phenyl ethanol[57,58,109] may also result from an indirect effect of these compounds influencing transport across epithelial cells. Thus, it is extremely difficult to determine the underlying mechanisms by which *in situ* lingual fibers respond to chemical stimuli. What is clearly needed to sort out this issue is to investigate the responses of cultured trigeminal ganglion cells and epithelial cells to chemical stimuli.

VIII. ROLE OF LINGUAL FIBERS IN TASTE

Trigeminal nerve fibers convey tactile, thermal, and chemical information from stimuli (food) introduced into the oral cavity. Trigeminal fibers are found in taste buds and, in fact, 82% of the taste buds in the hard palate contain trigeminal fibers.[12] The number of trigeminal lingual fibers containing SP or CGRP present in taste buds is, however, comparatively few. In contrast, a relatively large number of fibers containing these peptides is found around taste buds.[5,7,17,110] These, as well as other, neuropeptides may be released into the extracellular space in response to nociceptive stimuli where they may interact with taste cells, taste fibers, epithelial cells, and other trigeminal fibers. There is little evidence for this paradigm in the taste system, but in the olfactory system SP has been shown to alter receptor cell responses to amyl acetate, alter the secretory activity of sustantacular cells, and increase the spontaneous activity of receptor cells.[111]

Virtually nothing is known about the role of trigeminal fibers in modulating mechanisms of taste transduction. We previously noted that perigemal fibers contain SP and other peptides. These peptides are known to affect proteins involved in transport in nervous and epithelial tissues. If perigemal fibers respond to nociceptive stimuli by releasing peptides such as SP, it follows directly that peptides may be released by stimuli that activate perigemal fibers. Such stimuli may include highly acidic and salty solutions and solutions containing nonspecific trigeminal stimuli (e.g., PEA). The release of peptides and/or transmitters from perigemal fibers could influence both taste and perigemmal cells.

Finally, the finding that capsaicin (as well as other trigeminal stimulants) activate chorda tympani fibers[99] brings up the issue of whether these compounds also interact with taste cells or taste fibers synapsed to taste cells and/or with intragemal or perigemal lingual fibers which do not form synaptic contacts. Clearly, additional work needs to be done to elucidate interactions between the special sensory system and general sensory systems regarding taste transduction mechanisms.

ACKNOWLEDGMENTS

This work was supported by NIH grant DC0165 and from the Smokeless Tobacco Research Council. We thank Mr. Larry Hawkey for assistance with the figures and Drs. W. Hall, W. Pugh, and S. Roper for criticisms.

REFERENCES

1. **Young, R. F.,** Fiber spectrum of trigeminal sensory root of frog, cat and man determined by electron microscopy, *J. Neurosurg.*, 21, 19, 1972.
2. **Hellekant, G. and Kashara, Y.,** Secretory fibers in the trigeminal part of the lingual nerve to the mandibular salivary glands of the rat, *Acta Physiol. Scand.*, 89, 198, 1973.
3. **Ishiko, N.,** Local taste specificity within lingual nerve fields of the cat, *Br. Res.*, 24, 34, 1970.
4. **Farbman, A. I. and Hellekant, G.,** Quantative analyses of the fiber population in rat chorda tympani nerves and fungiform papillae, *Am. J. Anat.*, 153, 509, 1978.
5. **Yamasaki, H., Kubota, Y., and Tohyama, M.,** Ontogeny of substance P-containing fibers in the taste buds and the surrounding epithelium. I. Light microscopic analysis, *Dev. Brain Res.*, 18, 301, 1985.
6. **Kinnman, E. and Aldskogius, H.,** Collateral reinnervation of taste buds after chronic sensory denervation: a morphological study, *J. Comp. Neurol.*, 270, 569, 1988.
7. **Silverman, J. D. and Kruger, L.,** Analysis of taste bud innervation based on glycoconjugate and peptide neuronal markers, *J. Comp. Neurol.*, 292, 575, 1990.
8. **Yamasaki, H., Kubota, Y., Takagi, H., and Tohyama, M.,** Immunoelectron-microscopic study on the fine structure of substance -P containing fibers in taste buds of the rat, *J. Comp. Neurol.*, 227, 380, 1984.
9. **Dubner, R. and Bennett, G. J.,** Spinal and trigeminal mechanisms of nociception, *Annu. Rev. Neurosci.*, 6, 381, 1983.
10. **Whitehead, M. C. and Frank, M. E.,** Anatomy of the gustatory system in the hamster: central projections of the chorda tympani and lingual nerve, *J. Comp. Neurol.*, 220, 378, 1983.
11. **Van Buskirk, R. L. and Erickson, R. P.,** Odorant responses in taste neurons of the rat, *Brain Res.*, 135, 287, 1977.

12. **Chen, S.-Y. and Squier, C. A.**, The ultrastructure of the oral epithelium, in *The Structure and Function of Oral Mucosa*, Meyer, J., Squier, C. A., and Gerson, S. J., Eds., Pergamon Press, Oxford, 1984.
13. **Chan, K. Y. and Byers, M. R.**, Anterograde axonal transport and intercellular transfer of WGA-HRP in trigeminal-innervated sensory receptors of rat incisive papilla, *J. Comp. Neurol.*, 234, 201, 1985.
14. **Farbman, A. I. and Allgood, J. P.**, Innervation of sensory receptors of the oral mucosa, in *Current Concepts of the Histology of Oral Mucosa*, Squire, C. A. and Meyer, J., Eds., Charles C Thomas, Springfield, IL, 1971.
15. **Holland, G. R.**, Innervation of oral mucosa and sensory perception, in *The Structure and Function of Oral Mucosa*, Meyer, J., Squier, C. A., and Gerson, S. J., Eds., Pergamon Press, Oxford, 1984.
16. **Halata, H. and Munger, B. L.**, The sensory innervation of primate facial skin. II. Vermillion boarder and mucosa of lip, *Brain. Res. Rev.*, 5, 81, 1983.
17. **Whitehead, M. C., Beeman, C. S., and Kinsella, B. A.**, Distribution of taste and general sensory nerve endings in fungiform papillae of the hamster, *Am. J. Anat.*, 173, 185, 1985.
18. **Perl, E. R.**, Unraveling the story of pain, in *Advances in Pain Research and Therapy*, Fields, H. L., Ed., Vol. 9, Raven Press, New York, 1985.
19. **Nagy, J. I., Goedert, M., Hunt, S. P., and Bond, A.**, The nature of the substance P-containing nerve fibers in taste papillae of the rat tongue, *Neuroscience*, 7, 3137, 1982.
20. **Kruger, L. and Mantyh, P. W.**, Gustatory and related chemosensory system, in *Handbook of Chemical Neuroanatomy: Integrated Systems of the CNS*, Part II, Bjorkland, A., Hokfelt, T., and Swanson, L. W., Eds., Elsevier, 1989, 105.
21. **Finger, T. E., St. Jeor, V. L., Kinnamon, J. C., and Silver, W. L.**, Ultrastructure of substance P- and CGRP-immunoreactive nerve fibers in the nasal epithelium of rodents, *J. Comp. Neurol.*, 294, 293, 1990.
22. **Hensel, H., Andres, K. H., and During, M. V.**, Structure and function of cold receptors, *Pfluegers Arch.*, 352, 1, 1974.
23. **Dixon, A. D.**, The position, incidence and origin of sensory nerve terminations in oral mucus membrane, *Arch. Oral Biol.*, 7, 39, 1962.
24. **Gottschaldt, K.-M. and Vahle-Hinz, C.**, Merkel cell receptors: structure and transducer function, *Science*, 214, 183, 1981.
25. **Toyoshima, K., Miyamoto, K., Itoh, A., and Shimamura, A.**, Merkel-neurite complexes in the fungiform papillae of two species of monkeys, *Cell Tissue Res.*, 250, 237, 1987.
26. **Spassova, I.**, Ultrastructural relationships between the receptor nerve fiber and surrounding lamellae in Krause end bulbs, *Acta Anat.*, 109, 360, 1981.
27. **Cereijido, M., Gonzales-Mariscal, M., Aliova, G., and Contreras, R. G.**, Tight Junctions, *CRC Crit. Rev. Anat. Sci.*, 1, 171, 1988.
28. **DeSimone, J. A., Heck, G. L., Mierson, S., and DeSimone, S. K.**, The active ion transport properties of canine lingual epithelia in vitro, *J. Gen. Physiol.*, 83, 633, 1984.
29. **Ye, Q., Heck, G. L., and DeSimone, J. A.**, The anion paradox in sodium taste reception: resolution by voltage-clamp studies, *Science*, 254, 724, 1991.
30. **Elliott, E. J. and Simon, S. A.**, The anion in salt taste: a possible role of tight junctions, *Brain Res.*, 535, 9, 1990.
31. **Holland, V. F., Zampighi, G. A., and Simon, S. A.**, Morphology of fungiform papillae in canine lingual epithelium: Location of intercellular junctions in the epithelium, *J. Comp. Neurol.*, 279, 13, 1989.
32. **Squier, C. A.**, The permeability of keratinized and nonkeratinized oral epithelium to horseradish peroxidase, *J. Ultrastruct. Res.*, 43, 160, 1973.
33. **Squier, C. A. and Rooney, L.**, The permeability of keratinized and nonkeratinized oral epithelium to lanthanum *in vivo*, *J. Ultrastruct. Res.*, 54, 286, 1976.

34. **Hashimoto, K.,** Intercellular spaces of the human epidermis as demonstrated with lanthanum, *J. Invest. Dermatol.*, 57, 17, 1971.
35. **Akisaka, T. and Oda, M.,** Taste buds in the vallate papillae of the rat studied with freeze-fracture preparation, *Arch. Hist. Jpn.*, 41, 87, 1978.
36. **Holland, V. F., Zampighi, G. A., and Simon, S. A.,** Tight junctions in taste buds: possible role in intravascular taste, *Chem. Senses*, 16, 69, 1991.
37. **Simon, S. A. and Verbrugge, J.,** Transport pathways in canine lingual epithelium involved in sweet taste, *Chem. Senses*, 15, 1, 1990.
38. **Simon, S. A., Holland, V. F., Benos, D. J., and Zampighi, G. A.,** Transcellular and paracellular pathways in lingual epithelia and their influence in taste transduction, *J. Electron Micros. Tech.*, in press.
39. **Sostman, A. L. and Simon, S. A.,** Trigeminal nerve responses in rat elicited by chemical stimulation of the tongue, *Arch. Oral Biol.*, 36, 95, 1991.
40. **Williams, M. L. and Elias, P. M.,** The extracellular matrix of the stratum corneum: role of lipids in normal and pathological function, in *CRC Critical Reviews in Therapeutic Drug Carrier Systems*, CRC Press, Boca Raton, FL, 1987, 95.
41. **Loewenstein, W. R.,** Junctional intercellular communication: the cell-to-cell membrane channel, *Physiol. Rev.*, 61, 829, 1981.
42. **Ramon, F., Zampighi, G., Simon, S. A., and Rivera, A.,** The number of gap junction channels related to the cell's input resistance, in *Cell Interaction in Gap Junctions*, Sperelakis, N. and Cole, W., Eds., CRC Press, Boca Raton, FL, 1989, 13.
43. **Roper, S. D.,** The cell biology of vertebrate taste receptors, *Annu. Rev. Neurosci.*, 12, 329, 1989.
44. **Teeter, J. H.,** Dye-coupling in catfish taste buds, in Proceedings of the 19th Japanese symposium on taste and smell, Kimura, S., Miyoshi, A., and Shimida, I., Eds., Vol., Ashai University, Gifu, Japan, 29, 1985.
45. **Yang, J. and Roper, S. D.,** Dye-coupling in taste buds in the mudpuppy, *Necturus maculosus*, *J. Neurosci.*, 7, 3561, 1987.
46. **Byers, M. R.,** Fine structure of trigeminal receptors in rat molars, in *Pain in the Trigeminal System*, Anders, K. and Rottery, L., Eds., Elsevier, Amsterdam, 1977, 13.
47. **Holland, G. R.,** Microtubule and microfilament populations of cell processes in the dental pulp, *Anat. Rec.*, 198, 421, 1980.
48. **Alarie, Y.,** Sensory irritation by airborne chemicals, *CRC Crit. Rev. Toxicol.*, 2, 299, 1973.
49. **Green, B. G. and Gelhard, B.,** Salt as an oral irritant, *Chem. Senses*, 14, 259, 1989.
50. **Simon, S. A., Hall, W. L., and Schiffman, S. S.,** Astringent-tasting compounds alter ion transport across isolated canine lingual epithelia, *Physiol. Behav.*, 43, 271, 1992.
51. **Siegel, I. A., Izutsu, K. T., and Watson, E.,** Mechanisms of nonelectrolyte penetration across dog and rabbit oral mucosa *in vitro*, *Arch. Oral Biol.*, 26, 357, 1981.
52. **Siegel, I. A.,** Permeability of the oral mucosa, in *The Structure and Function of Oral Mucosa*, Meyer, J., Squier, C. A., and Gerson, S. J., Eds., Pergamon Press, Oxford, 1984.
53. **Kaaber, S.,** The permeability and barrier functions of the oral mucosa with respect to water and electrolytes, *Acta Odont. Scand.*, 32(66), 10, 1974.
54. **Squier, C. A., Fejerskov, O., and Jepsen, A.,** The permeability of a keratinizing squamous epithelium in culture, *J. Anat.*, 126, 1030, 1978.
55. **Squier, C. A.,** Penetration of nicotine and nitrosonornicotine across porcine oral mucosa, *J. Appl. Toxicol.*, 6, 123, 1986.
56. **Mistretta, C. M.,** Permeability of tongue epithelium and its relation to taste, *Am. J. Physiol.*, 220, 1162, 1971.
57. **Simon, S. A. and Sostman, A. L.,** Electrophysiological responses to non-electrolytes in lingual nerve of rat and lingual epithelia of dog, *Arch. Oral Biol.*, 36(11), 805, 1991.

58. **Silver, W. L., Mason, J. R., Adams, M. A., and Smeraski, C. A.,** Nasal trigeminal chemoreception: responses to *n*-aliphatic alcohols, *Brain Res.*, 376, 221, 1986.
59. **Hellekant, G.,** The effect of ethyl alcohol on non-gustatory receptors of the tongue of the cat, *Acta Physiol. Scand.*, 65, 243, 1965.
60. **Burnette, R. R.,** A monte-carlo model for the passive diffusion of drugs across the stratum corneum, *Int. J. Pharmacol.*, 22, 89, 1984.
61. **Getchell, T. V. and Getchell, M. L.,** Early events in vertebrate olfaction, *Chem. Senses Flavor*, 2, 313, 1977.
62. **DeSimone, J. A. and Heck, G. L.,** An analysis of the effects of stimulus transport and membrane charge on the salt, acid and water-response of mammals, *Chem. Senses*, 5, 295, 1980.
63. **Cain, W. S.,** Perceptual characteristics of nasal irritation, in *Chemical Senses*, Green, B. G., Mason, J. R., and Kare, M. R., Eds., Marcel Dekker, New York, 1990.
64. **Hume, W. J. and Potten, C. S.,** The ordered columar structure of mouse filiform papilla, *J. Cell Sci.*, 22, 149, 1976.
65. **Dawson, W.,** Chemical stimulation of the peripheral trigeminal nerve, *Nature*, 196, 341, 1962.
66. **Orchardson, R.,** Ion sensitivity of intradental nerves in the cat, in *Pain in the Trigeminal Region*, Anderson, D. J. and Matthews, B., Eds., Elsevier/North-Holland, Amsterdam, 1977.
67. **Orlando, R. C., Tobey, N. A., Schreiner, V. J., and Readling, R. D.,** Active electrolyte transport in mammalian buccal mucosa, *Am. J. Physiol.*, 255, G286, 1988.
68. **Narula, P., Xu, M., Kuang, K., Akiyama, R., and Fishbarg, J.,** Fluid transport across bovine corneal epithelial cell monolayers, *Am. J. Physiol.*, 31, C98, 1992.
69. **Persaud, K. C., Heck, G. L., DeSimone, S. K., Getchell, T. V., and DeSimone, J. A.,** Ion transport across the frog olfactory mucosa: the action of cyclic nucleotides on the basal and odorant-stimulated states, *Biochim. Biophys. Acta*, 944, 49, 1988.
70. **Simon, S. A., Holland, V. F., and Zampighi, G. A.,** Localization of Na,K-ATPase in canine lingual epithelia, *Chem. Senses*, 16, 283, 1991.
71. **Simon, S. A., Robb, R., and Schiffman, S. S.,** Transport pathways in rat lingual epithelium, *Pharmacol. Biochem. Behav.*, 29, 257, 1988.
72. **Rodriguez-Boulan, E. and Nelson, W. J.,** Morphogenesis of the polarized epithelial cell phenotype, *Science*, 245, 718, 1989.
73. **Simon, S. A., Holland, V. F., and Zampighi, G. A.,** Localization of Na-K-ATPase in lingual epithelia, *Chem. Senses*, 16, 283, 1991.
74. **Pierau, F.-K., Torrey, P., and Carpenter, D.,** Effect of ouabain and potassium-free solution on mammalian thermosensitive afferents *in vitro*, *Pfluegers Arch.*, 359, 349, 1975.
75. **Simon, S. A. and Garvin, J. L.,** Salt and acid studies on canine lingual epithelia, *Am. J. Physiol.*, 249, C398, 1985.
76. **Diamond, J. M. and Wright, E. M.,** Molecular forces governing non-electrolyte permeation through cell membranes, *Proc. R. Soc.* B, 172, 273, 1969.
77. **Salas, P. J. I. and Moreno, J. H.,** Single-file diffusion multi-ion mechanism of permeation in paracellular epithelial channels, *J. Membr. Biol.*, 64, 103, 1982.
78. **Ishiko, N.,** Local gustatory functions associated with segmental organization of the anterior portion of cat's tongue, *Exp. Neurol.*, 45, 341, 1974.
79. **Dastur, D. K.,** The relationship between terminal lingual innervation and gustation: a clinical and histological study, *Brain*, 84, 499, 1961.
80. **Zotterman, Y.,** Special senses: thermal receptors, *Annu. Rev. Physiol.*, 15, 357, 1953.
81. **Lawless, H. T. and Stevens, D. A.,** Responses by humans to oral chemical irritants as a function of the locus of stimulation, *Percept. Psychophys.*, 43, 72, 1988.
82. **Cadden, S. W., Lisney, S. J. W., and Matthews, B.,** Thresholds to electrical stimulation in cat canine tooth-pulp with Aβ-, Aδ and C fiber conduction velocities, *Brain Res.*, 261, 31, 1983.

83. **Silver, W. L.,** The common chemical sense, in *Neurobiology of Taste and Smell*, Finger, T. E. and Silver, W. L., Eds., John Wiley & Sons, New York, 1987.
84. **Horiuchi, H. and Matthews, B.,** Responses of intradental nerves to chemical and osmotic stimulation of dentine in the cat, *Pain*, 2, 49, 1976.
85. **Markowitz, K., Bilotto, G., and Kim, S.,** Decreasing intradental nerve activity in the cat with potassium and divalent cations, *Arch. Oral Biol.*, 36, 1, 1991.
86. **Hensel, H. and Schäfer, K.,** Effects of calcium on warm and cold receptors, *Pfluegers Arch.*, 352, 87, 1974.
87. **Hille, B.,** *Ionic Channels of Excitable Membranes*, Sinauer Associates, Sunderland, MA, 426, 1984.
88. **Szolcsányi, J.,** A pharmacological approach to elucidation of the role of different nerve fibers and receptor endings in mediation of pain, *J. Physiol.*, Paris, 73, 251, 1977.
89. **Jyvasjarvi, E., Kniffki, K. D., and Mengel, M. K. C.,** Functional characteristics of afferent C fibers from tooth pulp and periodontal ligament, in *Progress in Brain Research*, Hamann, W. and Iggo, A., Eds., Elsevier, Holland, 1988.
90. **Hensel, H. and Zotterman, Y.,** The effect of menthol on the thermoreceptors, *Acta Physiol. Scand.*, 24, 27, 1951.
91. **Kosar, E. and Schwartz, G. W.,** Effects of menthol on peripheral nerve and cortical unit responses to thermal stimulation of the oral cavity in the rat, *Brain Res.*, 513, 202, 1990.
92. **Watson, H. R., Hems, R., Roswell, D. G., and Spring, D. J.,** New compounds, *J. Soc. Cosmet. Chem.*, 29, 185, 1978.
93. **Schaefer, K., Braun, H. A., and Isenberg, C.,** Effect of menthol on cold receptor activity. Analysis of receptor activity, *J. Gen. Physiol.*, 88, 757, 1986.
94. **Green, B. G.,** Capsaicin cross-desensitization on the tongue: physiophysical evidence that oral chemical irritation is mediated by more than one sensory pathway, *Chem. Senses*, 16, 675, 1991.
95. **Bevan, S. and Szolcsanyi, J.,** Sensory neuron-specific actions of capsaicin: mechanisms and applications, *TIPS*, 11, 330, 1990.
96. **Maggi, C. A. and Pierau, F.-K.,** Recent advances in research on sensory peptides and capsaician mechanisms, *Neurosci. Lett.*, 122, 199, 1991.
97. **Meddings, J. B., Hoagboam, C. M., Tran, K., Reynolds, J. D., and Wallace, J. L.,** Capsaicin effects on non-neural plasma membranes, *Biochim. Biophys. Acta*, 1070, 43, 1991.
98. **Okuni, Y.,** Response of lingual nerve fibers of the rat to pungent spices and irritants in pungent spices, *Shika Gakuho*, 78, 325, 1978.
99. **Okuni, Y.,** Response of chorda tympani fibers of the rat to pungent spices and irritants in pungent spices, *Shika Gakuho*, 77.9, 1323, 1977.
100. **Silver, W. L., Farley, L. G., and Finger, T. E.,** The effects of neonatal capsaician administration on trigeminal nerve chemoreceptors in the rat nasal cavity, *Brain Res.*, 561, 212, 1991.
101. **Dodt, E., Skouby, A. P., and Zotterman, Y.,** The effect of cholinergic substances on the discharges from thermal receptors, *Acta Physiol. Scand.*, 28, 101, 1953.
102. **Langren, S., Liljestrand, G., and Zotterman, Y.,** Chemical transmission in taste fibre endings, *Acta Physiol. Scand.*, 30, 105, 1954.
103. **Jarvik, M. E. and Assil, K. M.,** Mecamylamine blocks the burning sensation of nicotine on the tongue, *Chem. Senses*, 13, 213, 1988.

103b. **Hwang, P. M., Verma, A., Brodt, D. S., and Snyder, S. H.,** Localization of phosphotidylinositol signaling in rat taste cells; role in bitter taste transduction, *Proc. Nat'l. Acad. Sci. (U.S.A.)*, 87, 7395, 1990.

104. **Tateda, H. and Beidler, L. M.,** The receptor potential of the taste cell of the rat, *J. Gen. Physiol.*, 47, 479, 1964.
105. **Murray, R. G. and Fujimoto, S.,** Fine structure of gustatory cells in rabbit taste buds, *J. Ultrastruct. Res.*, 27, 444, 1969.

106. **Szolcanyi, J.,** Capsaicin, irritation and desensitization, *Chemical Senses*, Green, B. G., Mason, J. R., and Kare, M. R., Eds., Marcel Dekker, Basel, 1990.
107. **Scriven, D. R. L.,** Modeling repetitive firing and bursting in a small unmyelinated nerve fiber, *Biophys. J.*, 35, 715, 1981.
108. **Simon, S. A. and Baggett, H. C.,** Identification of muscarinic acetylcholine receptors in isolated canine lingual epithelia via voltage clamp measurements, *Arch. Oral Biol.*, 37, 685, 1992.
109. **Silver, W. L., Arzt, A. H., and Mason, J. R.,** A comparison of the discriminatory ability and sensitivity of the trigeminal and olfactory systems to chemical stimuli in the tiger salamander, *J. Comp. Physiol.* A, 164, 55, 1988.
110. **Finger, T. E.,** Peptide immunohistochemistry demonstrates multiple classes of perigemmal nerve fibers in the circumvallate papilla of the rat, *Chem. Senses*, 11, 135, 1986.
111. **Holly, A., Bouvet, J. F., and Delau, J. C.,** Evidence for interactions between trigeminal afferents and olfactory receptor cells in the amphibian olfactory mucosa, in *Chemical Senses, Vol. 2, Irritation*, Green, B. G., Mason, J. R., and Kare, M. R., Eds., Marcel Dekker, Bern, 1990, 61.
112. **Silver, W. L. and Finger, T. E.,** *The Trigeminal System Smell in Health and Disease*, Getchell, T., Ed., Raven Press, New York, 1991.
113. **Lapchak, P. A., Araujo, D. M., Quirion, R., and Collier, B.,** Presynaptic cholinergic mechanisms in the rat cerebellum: evidence for nicotinic but not muscarinic autoreceptors, *J. Neurochem.*, 53, 1843, 1989.
114. **Silva, N. L. and Boulant, J. A.,** Effects of osmotic pressure, glucose, and temperature on neurons in preoptic tissue slices, *Am. J. Physiol.*, 247, R335, 1984.

Chapter 10

GENETIC AND NEUROBEHAVIORAL APPROACHES TO THE TASTE RECEPTOR MECHANISM IN MAMMALS

Yuzo Ninomiya and Masaya Funakoshi

TABLE OF CONTENTS

0-8493-5341-6/93/$0.00 + $.50

I. INTRODUCTION

Genetic variation in mammalian gustatory sensitivity was first found in 1931[1] as human "taste blindness" for a bitter substance, PTC (phenylthiocarbamide). This finding provided a new insight into the study of taste reception and a good beginning of the genetic approach to the taste receptor mechanisms. Nevertheless, during the following four decades, no such approach had been successful. Studies[2-6] in rodents had provided considerable amounts of evidence for individual differences in responses to particular taste compounds. But they had determined neither a specific genetic model nor the physical basis for the differences. Starting in the 1970s, steady progress has been made in murine taste genetics. During the past two decades, genetic variants for bitter and sweet taste sensitivity have been isolated among inbred strains of mice,[7,8] and congenic strains, where specific taste genes from genetically different strains were transferred have been established.[9,10] These congenic strains are particularly valuable as models for studies of the taste receptor mechanisms.

In this chapter we review genetic studies during the past six decades on bitter, sweet, and salty taste sensitivity in mammals addressing the extent to which the basis of genetic variations can be associated with taste receptor events. We focus particularly on recent findings of single major genes which may control taste receptor sensitivity to D-phenylalanine (D-phe) and other sweet substances in mice.

II. GENETICS OF BITTER TASTE PERCEPTION

A. PHENYLTHIOCARBAMIDE (PTC) TASTE DIMORPHISM

The genetic study of taste sensitivity was first reported by Fox in 1931.[1] He found two groups of human subjects with markedly different taste thresholds for bitter substances containing an N–C=S group, such as PTC and several others. Subsequent genetic analyses[11-15] indicated that the difference depends on a single autosomal gene (*Ptc*) with one dominant (tasters or subjects with low PTC thresholds) and one recessive (nontasters or subjects with high PTC thresholds) allele. The percentage of nontasters varies considerably among ethnic groups sampled (e.g., 30 to 40% in the European and American Caucasians, 5 to 15% in the Japanese population, and less than 5% in the American Indians and the Ainus populations[16-19]). Behavioral measures for PTC taste sensitivity suggested the existence of similar polymorphism among individuals of other mammalian species, such as anthropoid apes,[20] Old World monkeys,[21] rats,[22] and mice.[5,23]

Most of the studies[13,15,24-29] have suggested that taste sensitivity to PTC or propylthiouracil (PROP), a structurally similar compound to PTC, is genetically different from the sensitivity to other bitter substances, such as urea, caffeine, sucrose octaacetate (SOA), quinine, diphenylguanidine, brucine, and denatonium benzoate, which are chemically unrelated to PTC. Some of the studies,[24-26] on the other hand, indicated correlations between PROP-sensitivity and perceived bitterness of caffeine, KCl, Na-benzoate, K-benzoate, and saccharin. The correlations, however, were derived from rather limited experimental conditions (only a few suprathreshold concentrations of test substances and selected subjects, etc.). It is unlikely that PTC or PROP sensitivity is strongly related to the sensitivity to other test compounds.

Thus, PTC taste deficit appears to have high selectivity in chemical structure. This suggests the possibility that the proposed *Ptc* gene affects the taste receptor for PTC. To clarify this possibility, utilizing other animal models,[30,31] neurophysiological examination of the receptor mechanisms must be undertaken. However, no such approach has been developed so far.

In human studies, some progress in identifying physical substrates for the PTC dimorphism has been made; the gene, *Ptc*, links to the Kell blood-group gene,[32-34] although the location for the *Kell* gene itself has not been established.

B. GENES AFFECTING AVOIDANCE OF SUCROSE OCTAACETATE (SOA) AND OTHER BITTER SUBSTANCES

The most successful genetic analysis of bitter taste has been made in SOA sensitivity of mice. Individual differences among mice to this bitter compound were initially reported by Warren.[4] Warren and Lewis[35] subsequently reported that in two-bottle preference tests one inbred strain (CFW/NIH) avoided SOA at concentrations of 1 μM to 1 mM to which several other inbred strains were indifferent. Nontaster-taster segregation ratios in F2 and backcross generations between the CFW/NIH (taster) and C57BL/6NIH or C57L/NIH (nontaster) strains were consistent with the expected Mendelian single locus model, with the taster allele dominant. Lush[36] surveyed further 31 inbred strains of mice and found that only one strain (SWR/Lac) avoided 100 μM SOA. His genetic analysis using the progeny of an SWR and LVC (nontaster) strain backcross also suggested the single-locus control of SOA avoidance. He designated two presumptive alleles as Soa^a (aversion-dominant) and Soa^b (blind-recessive). Recent studies by Whitney and Harder[37] have confirmed the earlier findings by using a set of recombinant inbred (RI) strains from SWR/J (taster) and C57L/J (nontaster) progenitor strains. The dichotomous strain distribution pattern (SDP) of SOA sensitivity among 6 of the SWXL/Ty RI strains was again consistent with single-locus expectation.

The *Soa* gene controlling the SOA sensitivity is reported to affect avoidance of other acetylated sugars, such as α- and β- galactose pentaacetate, β-lactose octaacetate,[38] and trehalose octaacetate,[10] and also a structurally different bitter substance, strychnine hydrochloride.[39] Whitney et al.[10,40] further suggested the possibility of nonspecific effects of *Soa* gene on aversive sensitivity to other bitter substances, raffinose undecaacetate (RUA), denatonium benzoate, and brucine, since their newly constructed congenic mice (B6.SW-Soa^a) showed avoidance to the above-mentioned bitter substances similar to the donor strain, SWR/J possessing *Soa^a* allele, but not to the inbred partner strain, C57BL/6J (B6), which possesses *Soa^b* allele. This B6.SW-*Soa^a* strain was produced through a successive series of backcrosses (11 times). This caused the genetic background of the congenic strain to be identical to that of the B6 strain except for a chromosome segment containing the *Soa* locus derived from the SWR strain. However, since the gene segment derived from the SWR strain at the 11th backcross generation is estimated to be about 18 cM (centimorgan) in length, differences between B6 and B6.SW *Soa^a* mice are possibly due not only to *Soa* locus but to other tightly linked loci as well. Furthermore, Lush[38] reported that the distribution of RUA sensitivity among 30 strains is different from that of SOA sensitivity, and his genetic analysis suggested that RUA sensitivity is determined by a different gene, *Rua*. If this is the case, there may be different receptors for RUA and SOA controlled by different genes, even if the genes would be tightly linked on a single chromosome.

Three additional loci affecting avoidance of quinine sulfate (*Qui*),[41] cycloheximide (*Cyx*),[42] and glycine (*Glb*)[42] have been proposed. The existence of these loci was determined by comparing SDPs (mentioned above) of sensitivities to these substances among sets of RI strains. Linkage studies by comparing between SDPs for bitter taste genes and *Prp Hind*III genomic fragment patterns among RI strains suggested that the following bitter taste genes, *Rua*,[10,43,44] *Qui*, *Cyx*, *Glb*, and *Soa*,[45] are closely linked to a cluster of *Prp* genes that specify the structure of salivary proline-rich proteins and located on distal chromosome 6.[46] Lush[8] estimated the linkage-distances (in centimorgans) between the former four genes to be as follows: *Cyx* - 2.78 - *Qui* - 0.88 - *Rua* - 0.91 - *Glb*. However, since the independence among these bitter taste genes is still controversial,[8,10] further investigations are needed to clarify this.

C. POSSIBLE SITE OF ACTION OF GENES AFFECTING AVOIDANCE OF SOA AND OTHER BITTER SUBSTANCES

Simple behavioral measurements have been very useful to discover taste variants from a large number of animals and have contributed to the first step of taste genetics as mentioned above. However, the measurement, especially the two-bottle preference test (usually consumption of the test solution for 24 to 48 h is compared with that of water), may not be suitable for the precise analysis of bitter taste sensitivity because responses obtained by this test inevitably involve postingestive influences. (Some of bitter substances, such as PTC, etc., are toxic[22].) To overcome this problem, Harder et al.[47] utilized a short-term measurement of number of licks (7.25 s test period) and successfully determined behavioral thresholds for SOA. In addition to such methodological improvement, the neural and molecular basis for the behavioral variation must be concomitantly examined in the peripheral taste system. However, unfortunately, only three such studies on bitter taste sensitivity have been reported so far. Shingai and Beidler[48] found strain differences, consistent with those observed in behavioral responses, in the magnitude of integrated whole nerve responses to SOA in both the chorda tympani and glossopharyngeal nerves. SOA thresholds of SWR/J mice (taster) of the taste nerves were 0.1 m*M* to 1.0 m*M*, whereas they were 100 m*M* in LP/J mice (nontaster). They found no significant difference among the strains in the neural response to quinine and PTC. Spielman et al.[49] biochemically examined SOA sensitivity using taste tissues of vallate and foliate papillae and found that in the presence of SOA the amount of accumulation of IP_3 (a candidate for the second messenger of bitter taste) of membrane preparations of the tissues was increased more sharply in B6.SW-*Soa*^{a} (taster) strain than in B6 (nontaster) strain. IP_3 was also increased when caffeine, strychnine, or denatonium benzoate was applied. In contrast, Gannon and Contreras[50] in their measurements of chorda tympani responses did not find high SOA sensitivity of B6.SW-*Soa*^{a} congenic mice. These results suggest the possibility that the *Soa* gene affects the receptor mechanism for SOA in the taste cell innervated by the glossopharyngeal nerve, but it does not affect quinine and PTC receptor sensitivity.

Hiji[51] proposed that the receptor molecules only for sweet substances are proteins, because the lingual treatment of a proteolytic enzyme, Pronase E, inhibits chorda tympani responses of rats only to sweet substances but not to bitter, sour, and salty compounds. If this is the case, the bitter taste genes may not be the structural genes for the receptor and they may regulate proteins affecting membrane lipid conformations[52] or other possible receptor elements. An alternative explanation is that the site of the gene action is not on the receptor membrane. In this respect, Azen et al.[44,45,53] suggested the possibility of the involvement of salivary protein composition because of close genetic linkage (possibly the same genes) between the bitter taste genes (*Qui*, *Rua*, *Cyx*) and the *Prp* genes (salivary proline-rich proteins). They[53] found PRP RNAs in von Ebner's gland of mice, which is reported to be an important gland that secretes substances possibly modifying bitter taste sensitivity (e.g., VEG protein[54]). Glendinning[55] reported that the increase of PRP concentration in saliva induced by β-agonist, isoproterenol, increased ingestion of a bitter and astringent substance, tannic acid in BALB mice, whereas this was not evident in SW, C3H, and B6 mice. Another possibility suggested is the difference in the number and distribution of taste buds. Miller and his colleagues[56,57] reported that the SWR/J mice (SOA taster) have more vallate and foliate and fewer fungiform taste buds than the C57BL/6J (SOA nontaster) strain. Even if so, its influence would not be limited to bitter sensitivity.

Thus, behavioral genetic studies of bitter taste have provided the first success in the genetics of taste. However, considerable difficulties may be expected in the elucidation of the molecular basis for the genetic variations.

III. GENETIC APPROACHES FOR SWEET TASTE RECEPTORS

A. PHYLOGENETIC DIFFERENCE IN SWEET TASTE RESPONSE

A variety of chemically distinct substances taste sweet to humans. They include common sugars (e.g., sucrose, fructose), amino acids (e.g., glycine, L-alanine, D-amino acids), peptides (e.g., aspartame: L-aspartyl-L-phenylalanyl methyl ester), glycoproteins (e.g., stevioside), proteins (e.g., monellin, thaumatin), and artificial sweeteners (e.g., saccharin, cyclamate). Monellin (mol wt 10,500) and thaumatin (mol wt 21,000) are each about 10^5 times sweeter than sucrose on a molar basis.[58,59] Psychophysical studies[60,61] showed that the taste of these proteins and other sweet substances is abolished by the lingual treatment of gymnemic acid, a selective inhibitor of sweet taste isolated from the leaves of *Gymnema sylvestre*.

Although many of the sweet substances are also preferred by most mammals, there are marked species differences in taste sensitivity and preference for some of these compounds. For example, rodents prefer sugars and saccharin, whereas cats do not prefer them.[62] A series of behavioral and electrophysiological studies by Hellekant and his colleagues[63-66] indicate phylogenetic difference in taste sensitivity to monellin, thaumatin, and aspartame. They showed that members of infraorder Catarrhina, such as chimpanzees and Old World monkeys, prefer solutions of these sweeteners, and their sweet-sensitive single fibers of the chorda tympani nerve respond to them. However, these sweeteners elicit neither behavioral nor neural responses in other classes of mammals, such as New World monkeys, prosimians, and nonprimates (rats, hamsters, guinea pigs, rabbits, dogs, and pigs). Hellekant et al.[67,68] further demonstrated that gymnemic acid suppressed chorda tympani responses to sweet substances in chimpanzees as well as in humans but not in other primates and nonprimates. In contrast, Imoto et al.[69] found that a newly isolated substance, gurmarin, reduced the chorda tympani responses to sweet substances in rats but it hardly affected sweet taste recognition in humans. Gurmarin is also isolated from the leaves of *Gymnema sylvestre*, but is chemically different from gymnemic acid (gurmarin is a peptide and gymnemic acid is a triterpene glycoside). These accumulated evidences clearly suggest a fundamental difference in the gustatory effects of the sweet receptors among mammals. This must have a genetic basis.

B. SWEET TASTE VARIANTS IN RODENTS

Genetic variation affecting the preference for sweet substances was first reported by Nachman.[2] He found that hooded rats preferred 0.25% sodium saccharin solution more than did Sprague-Dawley albino rats, with large individual differences in the latter group. His subsequent selective breeding experiments[3] suggested that saccharin preference in rats had a genetic basis, although the obtained set of data was not suitable for establishing a genetic model.

Mouse strain differences in saccharin preference have been reported by several investigators. Fuller[70] found C57BL/6J mice prefer 0.1% sodium saccharin whereas DBA/2J mice are indifferent to it. Genetic analysis using the F1 and F2 generations of these strains suggested that the saccharin preference in mice is determined by a single locus for which the designation *Sac* was proposed. The allele present in the C57BL/6J mice, Sac^b, is dominant over Sac^d, found in the DBA/2J mice and results in higher preference scores. Lush[71] examined preference of saccharin and acesulfame (another synthetic sweetener) among BXD RI strains and the backcross progeny from C57BL/6Ty and DBA/2Ty progenitor strains, and found segregation ratios in each progeny group that were consistent with a single locus model. He further demonstrated that distribution of preference scores among 26 inbred strains for saccharin is highly correlated not only with those for acesulfame but with dulcin and sucrose as well. This suggested that the *Sac* gene also affects preference for other sweeteners, acesulfame, dulcin, and sucrose. Correlations across strains for preferences within pairs of sweet substances, such as saccharin vs. sucrose,[72-74] sucrose vs. glucose,[74] and sucrose vs. maltose,[75] were also reported by other investigators.

Neither the chromosomal position nor the site of action of the *Sac* gene have been clarified yet. Lush[71] reported that *Sac* showed no linkage with coat color genes, brown (*b*, on chromosome 4) and dilute (*d*, on chromosome 9), and the quinine-tasting gene *Qui* (possibly located in a cluster of bitter taste genes on chromosome 6). Fuller[70] suggested that saccharin preference differences are attributed to a variation in central processing (hedonic response) rather than in the peripheral receptor sensitivity, since no strain difference was found in the detection threshold itself for saccharin. Yirmiya et al.[76] found that opioid receptor-deficient mice (CXBK) show lower saccharin preference than control (C57BL/6By) mice, and naltrexone, a specific opiate antagonist, reduces saccharin preference in both strains, almost completely abolishing preference in CXBK mice. These results suggest that genetic differences in brain opioid receptor density contribute to differences in the palatability of saccharin.

C. GENETIC APPROACHES TO THE SWEET RECEPTOR SENSITIVITY IN MICE

1. Strain Differences

Genetic analyses on taste sensitivity to sweet amino acids, D-phenylalanine (D-phe) and L-proline (L-pro), in mice have been developed by Ninomiya and his colleagues. Using an electrophysiological approach with mice, they found prominent strain differences in whole nerve and single fiber responses of the chorda tympani nerve to sweet tasting amino acids, D-phe and L-pro;[77] whole chorda tympani responses to D-phe were greater in C57BL/6-CrSlc (C57) mice than in BALB/cCrSlc (BALB) and C3H/HeSlc (C3H) mice. Response patterns across single fibers (across neuron pattern) for D-phe and L-pro were correlated with that for sucrose only in C57 mice. They did not find such strain differences in response to another sweet tasting amino acid, L-alanine. Subsequent behavioral studies[78] by using a conditioned taste aversion technique demonstrated consistent strain differences in behavioral responses to these amino acids. A taste aversion conditioned to D-phe generalized to sugars, saccharin, and D-tryptophan in C57 mice, whereas in BALB and C3H mice it generalized to the isomer L-phe (a bitter taste to humans) but not to the sweet substances. Similarly, conditioned aversion to L-pro generalized to sugars and L-alanine only in C57 mice. This conditioned taste aversion technique permits one to evaluate the similarity of sensory quality. Presumably, D-phe and L-pro taste similar (sweet) to sugars in C57 mice (D-phe, L-pro sweet taster) but they taste different (bitter or sour) in BALB and C3H mice (D-phe, L-pro nonsweet taster). Comparable strain differences in preference responses to 0.01 to 0.1 *M* D-phe and 0.1 to 0.5 *M* L-pro (two-bottle preference tests) were also reported. An inconsistency between the two behavioral measurements appeared, in that BALB mice are indifferent to 1.0 *M* L-pro in the two-bottle situation (measuring the intake of the test solution and water for 48 h). Nevertheless, they show generalization from 1.0 *M* L-pro to sugars and quinine in the conditioned aversion test (measuring the number of licks to test solutions each for 10 s).

2. Genetic Analysis

Ninomiya et al.[79-82] further investigated the genetic basis for strain differences in behavioral and neural responses to D-phe. On the basis of whether conditioned aversion of 0.1 *M* D-phe would generalize to 0.1 to 0.5 *M* sucrose, phenotypes in F1, F2 and backcross generations from C57 (sweet taster) and BALB (nonsweet taster) strains were clearly segregated into two groups with phenotypic ratios consistent with the expected Mendelian single locus model. They designated the gene, *dpa*, which has a major effect on sweet taste sensitivity, to D-phe. Concomitant measurements of the neural responses obtained from single chorda tympani fibers of each generation suggested clear correspondence with behavioral responses. That is, response patterns (across neuron patterns) of sweet taster mice were comparable between D-phe and sucrose, whereas nonsweet tasters show similar response patterns to D-phe and quinine or HCl. The lingual treatment with the proteolytic enzyme, Pronase E, suppressed D-phe responses of sweet-best fibers as well as sucrose responses in sweet tasters but did not affect D-phe responses of acid and bitter-best fibers in nonsweet tasters. Sweet-best fibers of nonsweet tasters do not respond to D-phe. This indicates that the site of action of the *dpa* gene is on the taste cell membrane.

Ninomiya et al.[82-84] also tested generalization from 0.5 *M* L-pro to sucrose in the backcross generation and found 2 out of 86 mice showing the L-pro phenotype are different from the D-phe phenotype. Although there exist only two recombinants, their neural responses to these amino acids, carefully recorded from the chorda tympani nerve, apparently corresponded to their differential behavioral responses; the threshold of sweet-best fibers (their responses are inhibited by Pronase E) of the two mice for D-phe is around 0.01 *M*, being comparable to that of C57 mice, whereas the threshold for L-pro is 1.0 *M*, similar to that of BALB mice. From these results, they proposed that L-pro sensitivity is determined by a different gene, *psr* (proline sweet response). Linkage tests in backcross generation indicate that *dpa* and *psr* genes are linked to a coat color gene, *b* (brown), and the *Mup-1* gene (major urinary proteins) on chromosome 4. The linkage-distances (in centimorgan) between the four genes estimated are as follows: *psr* - 2.3 ± 1.6 (S.E.) - *dpa* - 8.7 ± 2.8 - *Mup-1* - 5.8 ± 2.3 - *b*. Thus, these genetic and neurobehavioral approaches provided a strong evidence for the existence of multiple genes, each coded for different sweet receptor sites.

3. Independence of the *dpa* Gene

Further neurobehavioral comparisons between C57 and BALB mice revealed the relationship of D-phe sensitivity to the sensitivity to other sweet substances. Ninomiya et al.[82,85] examined behavioral and neural responses to 8 D-amino acids (D-alanine, D-serine, D-valine, D-leucine, D-methionine, D-histidine, D-tryptophan, and D-phe) and found that the generalization pattern across various taste substances for each of the D-amino acids except D-phe in C57 mice was significantly correlated with that in BALB mice. Comparably, Pronase E suppressed chorda tympani response of both strains to D-valine, D-leucine, D-methionine, D-histidine, and D-tryptophan and did not affect responses of either strains to D-alanine and D-serine, whereas this enzyme reduced D-phe response only in C57BL mice. These results suggest that the *dpa* gene selectively affect D-phe sensitivity.

Recently, another strain difference in the synergistic effect of saccharin on D-phe response was found.[86] In C57 mice, D-phe response was enhanced 2 to 5 times when D-phe was applied after saccharin (interrupted by water rinsing between stimulations) or it was mixed with 0.1 m*M* saccharin (this solution itself only slightly, if at all, produces the neural response). This synergistic effect of saccharin on D-phe response was not observed in BALB mice. Saccharin also enhanced 1.5 to 3 times D-tryptophan and D-histidine responses of both strains, but it did not affect responses of either strains to L-phe, L-pro, L-alanine, and other D-amino acids. Saccharin, D-phe, D-tryptophan, and D-histidine each possess a particular ring structure (a benzene ring for saccharin and D-phe, an indole ring for D-tryptophan, and an imidazole ring for D-histidine). This structural similarity probably relates to the occurrence of the synergism at the receptor level. The synergism between saccharin and D-phe observed only in C57 mice implies that receptor sites at least for D-phe and saccharin are different, but the saccharin site is able to interact with the D-phe site. The synergism between saccharin and D-tryptophan and

D-histidine in both two strains implies that the *dpa* gene does not affect receptor sites for saccharin, D-tryptophan, and D-histidine.

4. A *dpa* Congenic Line

To further evaluate genes controlling taste responses to D-phe and L-pro, a *dpa* congenic line was established using standard techniques.[86] Here the segment of the chromosome of the sweet taster strain, C57 (donor strain), carrying the genes responsible for tasting, as determined by behavioral tests, was transferred onto the BALB (nonsweet taster, congenic partner strain) background yielding the BALB/cCrSlc-*Dpa* congenic strain. Behavioral responses of this congenic strain to these amino acids as well as other sweeteners were not different from those of the donor strain.

Neurophysiological studies[86] were then undertaken to further characterize these three strains. In both congenic and C57 mice, chorda tympani responses to 0.1 *M* D-phe and 0.3 *M* L-pro was inhibited by Pronase E, and D-phe responses were enhanced with the addition of saccharin, whereas these response characteristics were not observed in BALB mice. Relative response magnitudes to 0.5 *M* sucrose and 0.02 *M* saccharin of these strains were about twice greater than those of BALB mice. The greater response to sucrose was evident at the concentration range between 0.03 and 1.0 *M*. Responses to other sweet tasting amino acids, L-alanine and D-tryptophan, and to HCl and quinine were not different among three strains. However, congenic mice are rather similar to BALB mice when tested with salt. Unlike the C57 mice these two strains showed no inhibition of NaCl responses by the sodium channel blocker, amiloride.

The segment of chromosome 4 derived from C57 mice, therefore, contains not only the *dpa* gene but also the *psr* gene and the genes controlling other sweetener receptor sites. It appears not to contain possible genes affecting amiloride sensitivity of the Na receptor system. Although the length is not clear, the chromosomal segment does not involve the *Mup-1* locus which is approximately 8.7 cM apart from the *dpa* gene, because the *Mup-1* phenotype was different between the congenic and C57 mice. These results suggest that multiple genes which cluster in a closely linked fashion on chromosome 4 may code for receptor proteins that recognize sweet tasting molecules.

D. SWEET TASTE SENSITIVITY AND A DIABETIC MUTANT GENE

It is known that a diabetic gene, *db*[87], producing diabetic mice with higher body weights and blood glucose levels, is also located on chromosome 4 (23.5 cM apart from *dpa*). Recently, Sako et al.[88] examined the chorda tympani responses of diabetic (db: C57BL/KsJ-*db*/*db*) and parent inbred strains (control: C57BL/KsJ), and found that diabetic mice showed about 1.3 to 2.0 times greater responses to 0.3 *M* sugars (sucrose, fructose, glucose, and maltose) than those of the control mice. Thresholds for sucrose and fructose in db mice were about 1.0 to 1.5 log unit lower than those in control mice. Responses to NaCl, HCl, and quinine were not different between the two strains. Comparably, taste preference scores for sugars at suprathreshold concentrations except 1.0 *M* were higher in db mice than in control mice. Neonatal mice of 7 to 9 d of age possessing a genotype, db/db, also showed lower sugar thresholds in the chorda tympani responses. In contrast, Streptozotocin-inducing adult diabetic mice possessing the genotype +/+ did not show an increase of sugar responses of the chorda tympani nerve. These findings indicate that high sugar sensitivity of the db mice is genetically determined. Diabetic mice induced by the *db* gene are characterized by higher insulin levels around 3 weeks of age (about 10 times higher than the normal level),[89] although it decreases to a normal level in adulthood. Taking this into consideration, they proposed that the *db* gene affects either the secretory system of pancreatic beta cells or sugar reception mechanisms of taste cells in mice.

Sako and Ninomiya[90] subsequently examined concentration-response relationships of the chorda tympani response to sugars in these strains and found that a double reciprocal plot for sucrose, glucose, and maltose in db mice shows a higher dissociation constant, but the maximum response magnitude, as compared with that in control mice, did not show any difference. From these results, they speculated that high sensitivities to sugars induced by *db* gene is not due to an increase of the number of sugar receptors. The *db* gene may affect the intracellular event which commonly leads to insulin secretion of pancreatic beta cells and neuro-transmitter secretion in taste cells. However, the data obtained are quite indirect and scarce. Further studies are needed to elucidate the precise function and the site of action of the *db* gene.

IV. GENETICS OF SALT TASTE SENSITIVITY

A. STRAIN DIFFERENCES IN SALT PREFERENCE AND ITS POSSIBLE NEURAL BASIS

Chemicals that taste salty to humans are thought to be limited to sodium salts. This is a remarkable difference from bitter and sweet tastes that are associated with a wide variety of complex organic molecules. Nevertheless, comparatively few studies have reported genetic variation in salt responses. In mice, Hoshishima et al.[5] initially reported slight difference in aversion threshold to NaCl among five inbred strains. Ninomiya et al.[91] found more segregative differences in NaCl preference/aversion between C57BL/6CrSlc and C3H/HeSlc vs. DBA/2CrSlc and BALB/cCrSlc. The former two strains avoid 0.1 and 0.15 *M* NaCl whereas the latter are indifferent to them. In these earlier studies, no strain was found to prefer NaCl over water at moderate concentrations. However, Lush[8] recently found that TO and 129/Sv mice prefer 0.03 to 0.15 *M* NaCl, while DBA/2 and A2G mice are indifferent, and C57BL/6Ty mice avoid. Similarly, Beauchamp[92] reported that 129/J mice prefer 0.075 *M* NaCl, but C57BL/6J mice avoid it (see Table 1). Genetic studies on these strain differences in salt responses are now in progress.[8]

TABLE 1
Mouse Strain Differences in Salt Taste Responses

Strain	Preference[a]	Amiloride sensitivity[b]
TO	+[1]	
129/Sv	+[1]	
129/J	+[2]	0[3]
DBA/2Cr	0[4]	0[4]
A2G	0[1]	
BALB/cCr	0[4]	0[4]
C3H/He	−[4]	+[4]
C57BL/6Ty	−[1]	
C57BL/6Cr	−[4]	+[4]
C57BL/6J	−[2]	

Note: [1]: Lush, 1991; [2]: Beauchamp, 1990; [3]: Ninomiya et al., 1991; [4]: Ninomiya et al., 1989.

[a] Preference for 0.075, 0.1, and/or 0.15 *M* NaCl solution measured by two-bottle preference test in each strain. +, preferent; 0, indifferent; −, avoidant.

[b] Amiloride inhibition on NaCl responses of the chorda tympani nerve of each strain. 0, no amiloride inhibition was observed; +, amiloride inhibition was observed.

The peripheral neural basis for NaCl preference differences was examined by Ninomiya et al.[91] (see Table 1). They found that the lingual treatment with amiloride, an inhibitor of salt taste responses in several mammalian species, suppressed responses of the chorda tympani nerve to NaCl at a concentration 0.1 *M* or more in C57 and C3H strains showing a lower NaCl aversion threshold (0.1 *M*), whereas the drug hardly affected Na responses in DBA and BALB strains with a higher aversion threshold (0.3 *M*). Amiloride did not affect KCl responses of all four strains. The NaCl/KCl response ratio (around 1.0) became very similar among the four strains after amiloride treatment. These results suggest that an additional amiloride-sensitive receptor component of the response to NaCl in C57 and C3H mice increases overall taste information about NaCl, and thereby decreases the aversion threshold. Recently, Ninomiya et al.[93] found that Na-preferring 129/J mice show no amiloride suppression in NaCl responses, similar to the Na indifferent strains. This indicates that amiloride sensitivity is not responsible for the behavioral segregation between preference and indifference. Therefore, amiloride sensitivity may relate only to the opposite side of behavioral response, aversion.

Strain differences in NaCl preference are reported among rats as well. Midkiff et al.[94,95] reported that Fischer-344 (F-344) inbred rats did not show NaCl preference at any concentration. This strain avoids NaCl solutions even around isotonic concentrations that are strongly preferred by other inbred strains, Munich-Wistar and Buffalo (BF), and by outbred Wistar rats. Despite preference difference, Midkiff and Bernstein[96] found no marked differences between F-344 and Wistar rats in sensitivity to low concentrations of NaCl and generalization pattern from NaCl to other substances in their conditioned taste aversion study. The genetic study of differences in NaCl preference was performed by Soller et al.[97] by using F1 and F2 generations between F-344 (avoidant) and BF (preferent) strains. They found that segregative ratios of two phenotypes of F2 generations were not significantly different from those predicted by the single-locus model with NaCl avoidance as a recessive trait. However, their phenotypic classification between avoidant and preferent seems to be not rigidly performed (30% of F-344 mice actually are indifferent to NaCl around isotonic concentrations) and, besides, the sample number was very small. Further studies are needed to determine the genetic basis for NaCl preference differences in rats.

Unfortunately, only one study has been made on the neural basis for NaCl preference differences in rats so far. Bernstein and Longley[98] reported that F-344 rats showed greater relative responses of the chorda tympani nerve to NaCl than the Wistar rats. After amiloride treatment, this strain difference in NaCl response disappeared. Again, variation in amiloride-sensitive receptor components of NaCl is suggested to play some role for differences in NaCl preference in rats.

B. FACTORS INFLUENCING ON GENETIC VARIATION IN SALT TASTE RESPONSE

It is proposed that sodium salt taste reception is initiated by sodium ion entry into the receptor cell via amiloride blockable ion channels.[99,100] It is, therefore, probable that amiloride sensitivity in each animal directly relates to salt taste sensitivity, and hence determines salt preference-aversion response as mentioned above. However, in rats, amiloride sensitivity itself is reported to be changeable under various environmental influences. Hill[101] reported that the chorda tympani response to NaCl was decreased in rats fed by a sodium-deficient diet during early pre- and postnatal periods. In these rats, amiloride did not suppress NaCl responses. The decreased NaCl response and amiloride sensitivity were, however, recovered after feeding them with the NaCl-containing normal diet for at least 15 d. Ninomiya et al.[102] found that removal of sublingual glands in neonated animals increased the magnitude of amiloride inhibition of NaCl responses of the chorda tympani and abolished behavioral preference for 0.1 *M* NaCl in Wistar rats. Herness[103] reported that aldosterone-injected rats showed greater amiloride inhibition of NaCl responses than intact rats did. Hormonal and dietary influences on behavioral NaCl preference were also suggested in rats (see also Chapter 6). This salt preference increased by adrenalectomy,[104] administration of deoxycorticosterone acetate (DOCA),[105] and depletion of dietary sodium.[106] Furthermore, Rowland and Fregley[107] demonstrated that the spontaneous aversion of F-344 rats to isotonic NaCl solution is either partially or completely reversed by any of these treatments mentioned above. Therefore, in rats, it is suggested that salt preference (not aversion) is strongly regulated by the renin-angiotensin-aldosterone system (especially brain angiotensin II[108]), no matter how the hormonal system affects peripheral amiloride-sensitive NaCl receptor components.

Unlike rats, most mice have no spontaneous preference for NaCl over water, and intake of NaCl was not influenced by any of the treatments, including adrenalectomy, administration of DOCA, and depletion of dietary sodium.[5,109,110] This suggests a remarkable species difference between rodents in the hormonal effects or influences on behaviors related to hydromineral homeostasis. Mice probably possess high susceptibility in NaCl responses to environmental conditions, and thereby exhibit relatively stable amiloride sensitivity of NaCl receptors. Therefore, future genetic approaches to the salt receptor mechanisms may include identification of distinct variants of mice in amiloride sensitivity of the chorda tympani nerve (NaCl responses of the glossopharyngeal nerve are not inhibited by amiloride both in mice[111] and rats[112]) and behavioral aversion threshold for NaCl. These approaches may lead to the identification of the structural gene for the amiloride-sensitive sodium channel on the taste cell membrane.

V. FUTURE DIRECTIONS

Murine genetic studies have isolated taste variants among inbred strains and have determined single-locus control for taste sensitivities to five bitter (SOA, quinine, raffinose acetate, cycloheximide, and glycine) and three sweet substances (saccharin, D-phe, and L-pro). However, because of possible influences of the genetic background other than taste genes, the specificity of effects of most of the proposed genes on the taste sensitivity has not been thoroughly established. To overcome this problem, congenic strains (mentioned above) for each taste gene must be produced. Two different congenic strains are now available: one for a bitter taste gene on chromosome 6, *Soa* (B6.SW-*SOA*^{a}), and the other for a sweet taste gene on chromosome 4, *dpa* (BALB/CrSlc-*Dpa*). These congenic strains offer valuable models for studies of taste receptor mechanisms. Unfortunately, however, neural responses of the *Soa* congenic strain have not been thoroughly studied so far. This must be accomplished to elucidate the function of the *Soa* gene on the bitter receptor mechanism. Accumulated data of neural responses in the *dpa* congenic strain suggest the possibility that the transferred chromosomal segment of this strain contains genes affecting receptor sensitivities to D-phe, L-pro, saccharin, and other sweet substances. The next step is, therefore, to examine individuality of each sweet taste gene and elucidate the molecular basis for the difference in sweet receptor sensitivity between the congenic and parent strains. The molecular genetic studies for constructing and screening of cDNA library from taste tissues of these strains must proceed. For the screening strategy, technologies utilizing the difference in a short chromosomal segment between the congenic and parent stocks are available. A newly isolated specific inhibitor for rodents' sweet taste responses, gurmarin,[69] will provide another powerful tool for screening and identifying the sweet receptor molecules in the mouse molecular genetic studies.

Genetic approaches to the salt receptor mechanisms are just beginning and must follow earlier works on sweet taste with similar strategy to find out the structural gene code for amiloride-sensitive sodium channel on the taste cell membrane. No genetic studies have been undertaken on sour and the unique glutamate (umami) taste. For these two tastes, recent electrophysiological studies proposed that voltage-sensitive apical potassium channels are involved in sour taste transduction[113] and that proteinaceous receptors producing a synergistic action are responsible for umami taste perception.[114] Therefore, these postulated specific receptor elements for sour and umami tastes may also be the next targets of genetic study which starts with identification of their distinct variants.

ACKNOWLEDGMENTS

We thank Drs. Y. Fukami and B. Bryant for their valuable comments on the manuscript.

REFERENCES

1. **Fox, A. L.**, Taste blindness, *Science*, 73, 14, 1931.
2. **Nachman, M.**, The influence of diet and age on saccharin preference in rats, *Am. Psychol.*, 12, 461, 1957.
3. **Nachman, M.**, The influence of saccharin preference, *J. Comp. Physiol. Psychol.*, 52, 451, 1959.
4. **Warren, R. P.**, Preference aversion in mice to bitter substance, *Science*, 140, 808, 1963.
5. **Hoshishima, K., Yokoyama, S., and Seto, K.**, Taste sensitivity in various strains of mice, *Am. J. Physiol.*, 202, 1200, 1962.
6. **Fuller, J. L. and Cooper, C. W.**, Saccharin reverses the effect of food deprivation upon fluid intake in mice, *Anim. Behav.*, 15, 403, 1967.
7. **Funakoshi, M., Tanimura, T., and Ninomiya, Y.**, Genetic approaches to the taste receptor mechanisms, *Chem. Senses*, 12, 285, 1987.
8. **Lush, I. E.**, The genetics of bitterness, sweetness, and saltiness in strains of mice, in *Chemical Senses, Vol. 3, Genetics of Perception and Communication*, Wysocki, C. J. and Kare, M. R., Eds., Marcel Dekker, New York, 1991, 227.
9. **Whitney, G., Harder, D. B., and Gannon, K. S.**, The B6.SW bilineal congenic SOA-Taster mice, *Behav. Genet.*, 19, 409, 1989.
10. **Whitney, G., Harder, D. B., Gannon, K. S., and Maggio, J. C.**, Congenic lines differing in ability to taste sucrose octaacetate, in *Chemical Senses, Vol. 3, Genetics and Perception and Communication*, Wysocki, C. J. and Kare, M. R., Eds., Marcel Dekker, New York, 1991, 243.

11. **Snyder, L. H.,** Inherited taste deficiency, *Science*, 74, 151, 1931.
12. **Fox, A. L.,** The relationship between chemical constitution and taste, *Proc. Nat. Acad. Sci. U.S.A.*, 18, 115, 1932.
13. **Blakeslee, A. F.,** Genetics of sensory thresholds:taste for phenyl-thio-carbamide, *Proc. Nat. Acad. Sci. U.S.A.*, 18, 120, 1932.
14. **Blakeslee, A. F. and Salmon, M. R.,** Odor and taste blindness, *Eugen. News*, 16, 105, 1931.
15. **Blakeslee, A. F. and Salmon, M. R.,** Genetics of sensory thresholds: individual taste reactions for different substances, *Proc. Nat. Acad. Sci. U.S.A.*, 21, 84, 1935.
16. **Levine, P. and Anderson, A. S.,** Observations on taste blindness, *Science*, 75, 497, 1932.
17. **Parr, L. W.,** Taste blindness and race, *J. Hered.*, 25, 187, 1934.
18. **Rikumaru, J.,** Taste deficiency of Japanese and other races in Formosa, *Am. J. Psychol.*, 48, 649, 1936.
19. **Sato, T. and Sata, O.,** Taste thresholds of Japanese dental students to phenylthiocarbamide, *Chem. Senses*, 14, 847, 1989.
20. **Fischer, R. A., Ford, E. B., and Huxley, J.,** Taste testing the anthropoid apes, *Nature*, 144, 750, 1939.
21. **Chiarelli, B.,** Sensitivity to P.T.C. (phenyl-thio-carbamide) in primates, *Folia Primatol.*, 1, 88, 1963.
22. **Richter, C. P. and Clisby, K. H.,** Phenylthiocarbamide taste thresholds of rats and human beings, *Am. J. Physiol.*, 134, 157, 1941.
23. **Klein, T. W. and DeFries, J. C.,** Similar polymorphism of taste sensitivity to PTC in mice and men, *Nature*, 225, 555, 1970.
24. **Hall, M. J., Bartoshuk, L. M., Cain, W. S., and Stevens, J. C.,** PTC taste blindness and the taste of caffeine, *Nature*, 253, 442, 1975.
25. **Bartoshuk, L. M.,** Bitter taste of saccharin related to the genetic ability to taste the bitter substance 6-*n*-propylthiouracil, *Science*, 205, 934, 1979.
26. **Bartoshuk, L. M., Rifkin, B., Mark, L. E., and Hooper, J. E.,** Bitterness of KCl and benzoate:related to genetic status for sensitivity to PTC/PROP, *Chem. Senses*, 13, 517, 1988.
27. **Frank, R. A. and Korchmar, D. L.,** Gustatory processing differences in PTC tasters and non-tasters: a reaction time analysis, *Physiol. Behav.*, 35, 239, 1985.
28. **Leach, E. J. and Noble, A. C.,** Comparison of bitterness of caffeine and quinine by a time-intensity procedure, *Chem. Senses*, 11, 339, 1986.
29. **Mela, D. J.,** Bitter taste intensity: the effect of tastant and thiourea taster status, *Chem. Senses*, 14, 131, 1989.
30. **Tobach, E., Bellin, J. S., and Das, D. K.,** Differences in bitter taste perception in three strains of rats, *Behav. Genet.*, 4, 405, 1974.
31. **Whitney, G. and Harder, D. B.,** PTC preference among laboratory mice: understanding of a previously "unreplicated" report, *Behav. Genet.*, 16, 605, 1986.
32. **Chautard-Freire-Maia, E. A.,** Linkage relationships between 22 autosomal markers, *Ann. Hum. Genet.*, 38, 191, 1974.
33. **Conneally, P. M., Dumont-Driscoll, M., Huntzinger, R. S., Nance, W. E., and Jackson, C. E.,** Linkage relations of the loci for Kell and phenylthiocarbamide taste sensitivity, *Hum. Hered.*, 26, 267, 1976.
34. **Spence, M. A., Falk, C. T., Neiswanger, K., Field, L. L., Marazita, M. L., Allen, F. H., Siervogel, R. M., Roche, A. F., Crandall, B. F., and Sparkes, R. S.,** Estimating the recombination frequency for the PTC-Kell linkage, *Hum. Genet.*, 67, 183, 1984.
35. **Warren, R. P. and Lewis, R. C.,** Taste polymorphism in mice involving a bitter sugar derivative, *Nature*, 227, 77, 1970.
36. **Lush, I. E.,** The genetics of tasting in mice. I. Sucrose octaacetate, *Genet. Res.*, 38, 93, 1981.

37. **Whitney, G. and Harder, D. B.**, Single-locus control of sucrose octaacetate tasting among mice, *Behav. Genet.*, 16, 559, 1986.
38. **Lush, I. E.**, The genetics of tasting in mice. IV. The acetates of raffinose, galactose and -lactose, *Genet. Res.*, 47, 117, 1986.
39. **Lush, I. E.**, The genetics of tasting in mice. II. Strychnine, *Chem. Senses*, 7, 93, 1982.
40. **Whitney, G., Maggio, J. C., and Harder, D. B.**, Manifestations of the major gene influencing SOA tasting among mice: classic taste qualities, *Chem. Senses*, 15, 243, 1990.
41. **Lush, I. E.**, The genetics of tasting in mice. III. Quinine, *Genet. Res.*, 44, 151, 1984.
42. **Lush, I. E. and Holland, G.**, The genetics of tasting in mice. V. Glycine and cycloheximide, *Genet. Res.*, 52, 207, 1988.
43. **Capeless, C. G., Whitney, G., Gannon, K. S., Harder, D. B., Azen, E. A., Beamer, W. G., and Taylor, B. A.**, The sucrose octaacetate taste gene *(Soa)* is on distal mouse chromosome 6 and is closely linked (or identical) to salivary proline-rich protein genes *(Prp)*, *Chem. Senses*, 15, 559, 1990.
44. **Azen, E. A.**, Linkage studies of genes for salivary proline-rich proteins and bitter taste in mouse and human, in *Chemical Senses, Vol. 3, Genetics of Perception and Communication*, Wysocki, C. J. and Kare, M. R., Eds., Marcel Dekker, New York, 1991, 279.
45. **Azen, E. A., Lush, I. E., and Taylor, B. T.**, Close linkage of mouse genes for salivary proline-rich proteins (PRPs) and taste, *Trends Genet.*, 2, 199, 1986.
46. **Azen, E. A., Davisson, M. T., Cherry, M., and Taylor, B. A.**, *prp* (proline-rich protein) genes linked to markers *Es-12* (esterase 12), *Ea-10* (erythrocyte alloantigen 10) and loci on distal mouse chromosome 6, *Genomics*, 5, 415, 1989.
47. **Harder, D. B., Whitney, G., Frye, P., Smith, J. C., and Rashotte, M. E.**, Strain differences among mice in taste psychophysics of sucrose octaacetate, *Chem. Senses*, 9, 311, 1984.
48. **Shingai, T. and Beidler, L. M.**, Interstrain differences in bitter taste responses in mice, *Chem. Senses*, 10, 51, 1985.
49. **Spielman, A. I., Huque, T., Brand, J. G., and Whitney, G.**, The mechanism of sucrose octaacetate (bitter taste) signal transduction, *Chem. Senses*, 16, 585, 1991.
50. **Gannon, K. S. and Contreras, R. J.**, Electrophysiological responses of the chorda tympani nerve to bitter and salt solutions in inbred strains of mice, *Chem. Senses*, 16, 525, 1991.
51. **Hiji, Y.**, Selective elimination of taste responses to sugars by proteolytic enzymes, *Nature*, 256, 427, 1975.
52. **Kurihara, K., Yoshii, K., and Kashiwayanagi, M.**, Transduction mechanisms in chemoreception, *Comp. Biochem. Physiol.*, 85, 1, 1986.
53. **Azen, E. A., Hellekant, G., Sabatini, L. M., and Warner, T. F.**, mRNAs for PRPs, statherin and histatins in von Ebner's gland tissues, *J. Dent. Res.*, 69, 1724, 1990.
54. **Schemale, H., Holtgreve-Grez, H., and Christansen, H.**, Possible role for salivary gland protein in taste reception indicated by homology to lipophilic-ligand carrier proteins, *Nature*, 343, 366, 1990.
55. **Glendinning, J. I.**, Effects of isoproterenol treatment on ingestive responses to tannic acid in mice, *Chem. Senses*, 16, 527, 1991.
56. **Miller, I. J., Jr. and Whitney, G.**, Sucrose octaacetate-taster mice have more vallate taste buds than non-tasters, *Neurosci. Lett.*, 360, 271, 1989.
57. **Krimm, R. F. and Miller, I. J., Jr.**, Taste buds of the foliate and fungiform pappila compared for two strains of mice, *Chem. Senses*, 14, 739, 1989.
58. **Morris, J. A., Martenson, R., Deibler, G., and Cagan, R. H.**, Characterization of monellin, a protein that tastes sweet, *J. Biol. Chem.*, 248, 534, 1973.
59. **Van del Wel, H. and Loeve, K.**, Isolation and characterization of thaumatin I and II, the sweet-tasting proteins from *Thaumatococcus danielli* Benth, *Eur. J. Biochem.*, 31, 221, 1972.

60. **Bartoshuk, L. M., Dateo, G. P., Vandenbelt, D. J., Buttrick, R. C., and Long, L., Jr.**, Effects of *Gymnema sylvestre* and *Synsepalum dulcificum* on taste in man, in *Olfaction and Taste*, Vol. 3, Pfaffmann, C., Ed., Rockefeller University Press, New York, 1969, 436.
61. **Van del Wel, H. and Arvidson, K.**, Qualitative psychophysical studies on gustatory effects of the sweet-tasting proteins thaumatin and monellin, *Chem. Senses*, 3, 291, 1978.
62. **Carpenter, G. A.**, Species differences in taste preferences, *J. Comp. Physiol. Psychol.*, 49, 139, 1956.
63. **Brouwer, J. N., Hellekant, G., Kasahara, Y., Van del Wel, H., and Zotterman, Y.**, Electrophysiological study of the gustatory effects of the sweet proteins monellin and thaumatin in monkey, guinea pig, and rat, *Acta Physiol. Scand.*, 89, 550, 1973.
64. **Glaser, D., Hellekant, G., Brouwer, J. N., and Van del Wel, H.**, The taste responses in primates to the proteins thaumatin and monellin and their phylogenetic implications, *Folia Primatol.*, 29, 56, 1978.
65. **Hellekant, G.**, On the gustatory effects of monellin and thaumatin in dog, hamster, pig and rabbit, *Chem. Senses Flav.*, 2, 97, 1976.
66. **Hellekant, G., Glaser, D., Brouwer, J., and Van del Wel, H.**, Gustatory responses in three prosimian and two simian primate species to six sweeteners and miraculin and their phylogenetic implications, *Chem. Senses*, 6, 165, 1981.
67. **Hellekant, G., Hard Af Segerstad, C., Robert, T., Van del Wel, H., Brouwer, J. N., Glaser, D., Haynes, R., and Eichberg, J. W.**, Effects of gymnemic acid of the chorda tympani proper nerve responses to sweet, sour, salty and bitter taste stimuli in the chimpanzee, *Acta Physiol. Scand.*, 124, 399, 1985.
68. **Hellekant, G. and Ninomiya, Y.**, On the taste of umami in chimpanzee, *Physiol. Behav.*, 49, 927, 1991.
69. **Imoto, T., Miyasaka, A., Ishima, R., and Akasaka, K.**, A novel peptide isolated from the leaves of *Gymnema Sylvestre*. I. Characterization and its suppressive effect on the neural responses to sweet taste stimuli in the rat, *Comp. Biochem. Physiol.*, 100, 309, 1991.
70. **Fuller, J. L.**, Single-locus control of saccharin preference in mice, *J. Hered.*, 65, 33, 1974.
71. **Lush, I. E.**, The genetics of tasting in mice. VI. Saccharin, acesulfame, dulcin and sucrose, *Genet. Res.*, 53, 95, 1989.
72. **Peltz, W. E., Whitney, G., and Smith, J. C.**, Genetic influences on saccharin preference of mice, *Physiol. Behav.*, 10, 263, 1973.
73. **Ramirez, I. and Fuller, J. L.**, Genetic influence on water and sweetened water consumption in mice, *Physiol. Behav.*, 16, 163, 1976.
74. **Stockton, M. D. and Whitney, G.**, Effects of genotype, sugar, and concentration on sugar preference of laboratory mice *(Mus musculus)*, *J. Comp. Physiol. Psychol.*, 86, 62, 1974.
75. **Harder, D. B., Maggio, J. C., and Whitney, G.**, Assessing gustatory detection capabilities using preference procedures, *Chem. Senses*, 14, 547, 1989.
76. **Yirmiya, R., Lieblich, I., and Liebeskind, J. C.**, Reduced saccharin preference in CXBK (opioid receptor-deficient) mice, *Brain Res.*, 438, 339, 1988.
77. **Ninomiya, Y., Higashi, T., Katsukawa, H., Mizukoshi, T., and Funakoshi, M.**, Qualitative discrimination of gustatory stimuli in three different strains of mice, *Brain Res.*, 322, 83, 1984.
78. **Ninomiya, Y., Mizukoshi, T., Higashi, T., Katsukawa, H., and Funakoshi, M.**, Gustatory neural response in three different strains of mice, *Brain Res.*, 302, 305, 1984.
79. **Ninomiya, Y. and Funakoshi, M.**, Inheritance of taste sensitivity to D-phenylalanine in mice, *Proc. Jpn. Symp. Taste and Smell*, 19, 138, 1985.
80. **Ninomiya, Y. and Funakoshi, M.**, Taste receptor mechanisms for a sweet substance, D-phenylalanine in mice, *Proc. Jpn. Symp. Taste and Smell*, 20, 53, 1986.

81. **Ninomiya, Y., Higashi, T., Mizukoshi, T., and Funakoshi, M.,** Genetics of the ability to perceive sweetness of D-phenylalanine in mice, *Ann. N.Y. Acad. Sci.*, 510, 527, 1987.
82. **Ninomiya, Y., Sako, N., Katsukawa, H., and Funakoshi, M.,** Taste receptor mechanisms influenced by a gene on chromosome 4 in mice, in *Chemical Senses, Vol. 3, Genetics of Perception and Communication*, Wysocki, C. J. and Kare, M. R., Eds., Marcel Dekker, New York, 1991, 267.
83. **Ninomiya, Y., Sako, N., and Funakoshi, M.,** Genetic control for taste receptor mechanisms in mice, in *Proc. Int. Uni. Physiol. Sci. XVII*, Hirvonen, L., Timisjarvi, J., Niiranen, S., and Leppaluoto, J., Eds., Oy Liitto, Oulu, Finland, 1989, 141.
84. **Ninomiya, Y.,** Genetic analysis on multiple sweetener receptor sites in mice, *Proc. Jpn. Symp. Taste Smell*, 23, 23, 1989.
85. **Ninomiya, Y., Sako, N., and Funakoshi, M.,** Selective effects of the *dpa* gene on the ability to taste D-phenylalanine in mice, *Proc. Jpn. Symp. Taste and Smell*, 21, 153, 1987.
86. **Ninomiya, Y.,** Genes controlling receptor mechanisms for sweet taste, *Jpn. J. Physiol*, 41, 218, 1991.
87. **Hummel, K. P., Dickie, M. M., and Coleman, D. L.,** Diabetes, a new mutation in the mouse, *Science*, 153, 1127, 1966.
88. **Sako, N., Ninomiya, Y., and Funakoshi, M.,** The effect of a diabetes mutant gene, *db*, on sugar taste sensitivity in mice, *Proc. Jpn. Symp. Taste and Smell*, 24, 215, 1990.
89. **Turman, R. W. and Doisy, R. J.,** The influence of age on the development of hypertriglyceridaemia and hypercholesterolaemia in genetically diabetic mice, *Diabetologia*, 13, 7, 1977.
90. **Sako, N. and Ninomiya, Y.,** Concentration-response relationships of the chorda tympani nerve for sugars in diabetic mutant mice (C57BL/KsJ-db/db), *Proc. Jpn. Symp. Taste and Smell*, 25, 309, 1991.
91. **Ninomiya, Y., Sako, N., and Funakoshi, M.,** Strain differences in amiloride inhibition of NaCl responses in mice, *Mus musculus, J. Comp. Physiol.*, 166, 1, 1989.
92. **Beauchamp, G. K.,** Genetic control over salt preference in inbred strains of mice, *Chem. Senses*, 15, 551, 1990.
93. **Ninomiya, Y., Beauchamp, G. K., and Yamazaki, K.,** unpublished data, 1991.
94. **Midkiff, E. E., Fitts, D. A., Simpson, J. B., and Berstein, I. L.,** Absence of sodium chloride preference in Fischer-344 rats, *Am. J. Physiol.*, 249, R438, 1985.
95. **Midkiff, E. E., Fitts, D. A., Simpson, J. B., and Berstein, I. L.,** Attenuated sodium appetite in response to sodium deficiency in Fischer-344 rats, *Am. J. Physiol.*, 252, R562, 1987.
96. **Midkiff, E. E. and Bernstein, I. L.,** Generalization of conditioned taste aversion to sodium chloride in Fischer-344 and Wistar rats, *Ann. N.Y. Acad. Sci.*, 510, 498, 1987.
97. **Sollars, S. I., Midkiff, E. E., and Berstein, I. L.,** Genetic transmission of NaCl aversion in the Fischer-344 rat, *Chem. Senses*, 15, 521, 1990.
98. **Bernstein, I. L. and Longley, A.,** Amiloride-sensitivity of the chorda tympani response to NaCl in Fischer-344 and Wistar rats, *Chem. Senses*, 15, 553, 1990.
99. **DeSimone, J. A., Heck, G. L., Mierson, S., and DeSimone, S. K.,** The active ion transport properties of canine lingual epithelia *in vitro*. Implications for gustatory transduction, *J. Gen. Physiol.*, 83, 633, 1984.
100. **Heck, G. L., Mierson, S., and DeSimone, J. A.,** Salt transduction occurs through an amiloride-sensitive sodium transport pathway, *Science*, 223, 403, 1984.
101. **Hill, D. V.,** Susceptibility of the developing rat gustatory system to the physiological effects of dietary sodium deprivation, *J. Physiol.*, 393, 413, 1987.
102. **Ninomiya, Y., Katsukawa, H., and Funakoshi, M.,** Alteration of salt taste sensitivity by the neonatal removal of sublingual glands in the rat, *Comp. Biochem. Physiol.*, 94, 89, 1989.
103. **Herness, M. S.,** Aldosterone increases amiloride sensitive sodium channels in rat taste cells:implications for sodium appetite, *Chem. Senses*, 16, 534, 1991.

104. **Richter, C. P.,** Increased salt appetite in adrenalectomized rats, *Am. J. Physiol.*, 115, 155, 1936.
105. **Rice, K. K. and Richter, C. P.,** Increased sodium chloride and water intake of normal rats treated with deoxycorticosterone acetate, *Endocrinology*, 33, 106, 1943.
106. **Stricker, E. M. and Wilson, N. E.,** Salt-seeking behavior in rats following acute sodium deficiency, *J. Comp. Physiol. Psychol.*, 72, 416, 1970.
107. **Rowland, N. E. and Fregley, M. J.,** Induction of an appetite for sodium in rats that show no spontaneous preference for sodium chloride solution — The Fischer 344 strain, *Behav. Neurosci.*, 102, 961, 1988.
108. **Fregly, M. J. and Rowland, N. E.,** Role of renin-angiotensin-aldosteron system in NaCl appetite of rats, *Am. J. Physiol.*, 248, R1, 1985.
109. **Kutscher, C. L. and Steilen, H.,** Increased drinking of a nonpreferred NaCl solution during food deprivation in the C3H/HeJ mouse, *Physiol. Behav.*, 10, 29, 1973.
110. **Rowland, N. E. and Fregly, M. J.,** Characteristics of thirst and sodium appetite in mice *(Mus musculus)*, *Behav. Neurosci.*, 102, 969, 1988.
111. **Ninomiya, Y., Tanimukai, T., Yoshida, S., and Funakoshi, M.,** Gustatory neural responses in preweanling mice, *Physiol. Behav.*, 49, 913, 1991.
112. **Formaker, B. K. and Hill, D. V.,** Lack of amiloride sensitivity in SHR and WKY glossopharyngeal taste responses to NaCl, *Physiol. Behav.*, 50, 765, 1991.
113. **Kinnamon, S. C. and Roper, S. D.,** Evidence for a role of voltage-sensitive apical K^+ channels in sour and salt taste transduction, *Chem. Senses*, 13, 115, 1988.
114. **Ninomiya, Y., Kurenuma, S., Nomura, T., Uebayashi, H., and Kawamura, H.,** Taste synergism between monosodium glutamate and 5′-ribonucleotide in mice, *Comp. Biochem. Physiol.*, 101, 97, 1992.

Coding

Chapter 11

SYNAPTIC INTERACTIONS IN TASTE BUDS

Stephen D. Roper

TABLE OF CONTENTS

0-8493-5341-6/93/$0.00 + $.50

Many peripheral sensory organs carry out some degree of information processing before transmitting their signals to higher centers in the brain and spinal cord. Perhaps the retina is the most familiar example where complex signal modulation and processing occur. Signal processing in the retina involves electrical and chemical synaptic interactions between several strata of cells before ganglion cells are activated and transmit their messages to the lateral geniculate.[1] Several lines of evidence, including results from morphological, physiological, pharmacological, and immunocytochemical studies, indicate that signal processing also occurs in taste buds. The intent of this chapter is to review current concepts on how signal processing is thought to occur in the peripheral organs of taste.

I. EARLY EVENTS IN TASTE RECEPTION

Three important events take place when a taste stimulus interacts with the apical, chemosensitive tips of taste receptor cells (Figure 1). First, the chemical potential inherent in a solution of molecules or in an electrochemical gradient of ions is converted (transduced) into a change in the membrane voltage at the apical tip of the cell. That is, a receptor current flows across the apical membrane and a receptor potential is generated.* This initial event is termed *transduction*. The underlying membrane mechanisms for transduction are discussed in Chapters 6 and 8.

Second, the receptor potential is conducted from the tip to the base of the receptor cell where synapses are located. It is unclear at present what proportion of the apical receptor potential is conducted *electrotonically* to synaptic sites; whether *action potentials* are important in the propagation of signals throughout the receptor cell; or whether some combination of active and passive mechanisms conducts the receptor potential to the base of the taste receptor cell. Certainly, taste cells are capable of generating action potentials and do so in response to chemical stimulation under many conditions.[4-7] Taste cells also have a high input resistance (0.15 to 1.6 GΩ).[8-10] Consequently, the space constant is long, and a significant portion of an apically-generated electrical signal is electrotonically propagated to the base of the cell.[11] The relative importance of passive (electrotonic) *versus* active propagation of the receptor potential from its site of initiation to synaptic sites is not established to date. Clearly, however, the conductance of the basolateral membrane will readily influence the propagation of receptor currents throughout the taste cell and the ability of the cell to generate action potentials. Specifically, the conductance along the basolateral membrane represents a drain on the receptor currents, shunting the electrotonic spread of the signal

* This represents the general case. There are indications in rat taste buds that for some stimuli, ligand binding activates a G protein which results in phophatidylinositol turnover and the formation of intracellular inositol 1,4,5-triphosphate (IP_3).[2,3] IP_3 in turn releases Ca^{2+} from intracellular stores.[2] In principle, this could bypass transduction steps that involve apical transmembrane current flow.

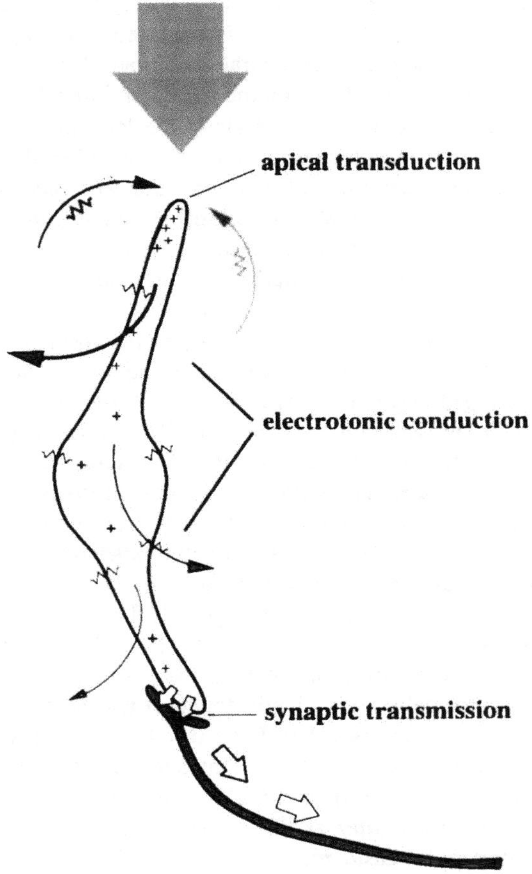

FIGURE 1. Schematic drawing of a taste receptor cell, showing the three steps in taste reception: (1) apical transduction; (2) electrotonic conduction; (3) synaptic transmission. For clarity, active propagation of receptor currents through the receptor cell by impulses is not shown (see text). A taste stimulus (large stippled arrow at top) is shown producing a depolarizing receptor potential (+'s) in the tip of the receptor cell. The excitatory receptor currents (arrows) underlying this receptor potential are propagated through the length of the receptor cell and across the basolateral membrane resistance. These currents complete a paracellular loop that includes the junctional complex between adjacent taste cells near the taste pore (lightly stippled region). At the basal end of the cell, the excitatory currents activate voltage-dependent Ca^{2+} channels. This allows an influx of Ca^{2+} which, in turn, elicits neurotransmitter release (open arrows at synapse). The end result is a series of impulses in the sensory afferent nerve fiber (open arrows along axon).

throughout the cell (see Figure 1). It is noteworthy that the depolarization-activated K^+ conductance of taste cells is confined to the apical membrane and does not occur along the basolateral membrane.[12,13] The absence of a voltage-dependent K^+ conductance on the basolateral membrane reduces passive losses as receptor currents spread along the long narrow taste receptor cells. This assures that receptor currents are conducted to the basal portions of the cell. These considerations also suggest that any factors that alter the properties of the basolateral membrane will influence the amplitude of receptor potentials that reach synapses at the base of the cell. That this may be an important modulatory mechanism during taste reception is discussed below regarding the actions of serotonin.

The presence of receptor current loops passing through the basolateral membrane (shown in Figure 1) also bears upon the recent findings that paracellular resistance pathways may play an important role in taste reception (see Chapters 8 and 9 and References 14 to 16). The currents generated at the apical tip return via pathways that include the basolateral membrane resistance in series with the resistance of paracellular spaces. In many excitable tissues, extracellular resistances are ignored since they are low and constant. This situation may not apply to taste buds since the paracellular current pathway includes junctional complexes between adjacent cells near their apical tips,[17-19] represented by the stippled region in Figure 1. This paracellular resistance may vary depending on the ionic environment (see Chapters 8 and 9).

Third, the conducted receptor potential evokes neurotransmitter release. Synapses on receptor cells are typically found at and below the level of the nucleus, i.e., towards the middle to bottom half of the taste bud.[20-22] It is here that depolarizing receptor potentials presumably trigger voltage-dependent Ca^{2+} channels and an influx in Ca^{2+}. Calcium ion influx leads to the release of the neurotransmitter at the synapse and completes the chain of events during peripheral taste reception — transduction, conduction, and transmission.

It has already been pointed out that electrotonic spread of current down the long narrow taste cells is a potential site for modulation of taste reception. Other potential sites for modulation are the apical chemosensitive membrane and taste cell synapses. Modulation of apical chemosensitivity is known to occur, such as that 5′ ribonucleotides enhance the action of monosodium glutamate.[23,24] Another example is that protons (low pH) influence Na^+ transduction.[25] However, the remainder of this chapter will focus on the modulation of peripheral taste reception and signal processing at the second and third steps of the aforementioned sequence, electrotonic conduction and synaptic transmission.

II. SYNAPSES IN TASTE BUDS

There are two broad categories of cells in taste buds — elongate receptor cells and oblate basal cells. The latter, basal cells, do not extend processes to the taste pore and therefore cannot participate directly in taste reception. Within these two broad groups there are cell types, such as light and dark receptor cells (see Chapter 2). Recent studies on the synaptic and cellular organization of *Necturus* taste buds indicate that there are two types of basal cells. One type of basal cell appears to be a stem cell and is responsible for maintaining the population of cells within the taste bud (e.g., References 26 to 28). Besides stem basal cells, another type of basal cell is found in taste buds in fish, amphibia, and reptiles. Investigations of taste buds in *Necturus* indicate that both types of basal cells coexist in the taste bud.[28,29]* This second type of basal cell makes synaptic contacts with adjacent receptor cells as well as with sensory fibers. In fact, the majority of ultrastructurally identified synapses in taste buds from *Necturus* are found on basal cells of this class.[22] This second class of basal taste cell has many features that are characteristic of cutaneous Merkel cells, and a number of investigators have pointed out these similarities (however, cf. discussion by Munger in Chapter 3). Among these similarities are the presence of dense core vesicles and short cytoplasmic spines, and immunoreactivity for serotonin and neuron-specific enolase.[22,29,31-38] For clarity and to distinguish this type of basal cell from the stem cell, I will refer to it as a Merkel-like basal cell (see also Chapter 2).**

The presence of synapses on Merkel-like basal cells suggests that synaptic interactions occur among cells within the taste bud. This thesis was advanced as early as 1966 by Hirata[43] and in 1971 by Reutter[44] and has been discussed in recent reviews.[45,46] There is now electrophysiological evidence that receptor cells are synaptically coupled to basal cells, confirming earlier electron microscopic findings. In *Necturus*, Ewald and Roper[47] recorded postsynaptic potentials have been recorded in identified Merkel-like basal cells in response to chemical excitation of receptor cells (Figure 2).*** Postsynaptic responses (excitatory postsynaptic potentials, EPSPs) in Merkel-like basal cells were inhibited by bathing the tissue in Cd^{2+} which blocks Ca^{2+} currents and hence blocks transmitter release. However, there are many unusual features of these

* Korte[30] also noted that there are two types of basal cells in taste buds in the turtle. Recent findings from *Necturus* taste buds now provide specific roles for each of these types of basal cells.[22,28,29]

** In mammalian taste buds, one type of elongate receptor cell also possesses dense core vesicles and makes synaptic contacts with axons. This type of cell is particularly evident in taste buds from rabbits and has been classified as a type III receptor cell.[39] Some authors have drawn a comparison between this type of receptor cell and Merkel cells based upon ultrastructural features, especially the presence of dense-core vesicles (e.g., References 27, 40), and based upon immunoreactivity for serotonin.[41,42] The similarity between the "Merkel-like" receptor cell in mammalian taste buds and Merkel-like basal cells in taste buds from lower vertebrates is intriguing. It remains to be established whether there is any homology or analogy between the nonmammalian and mammalian "Merkel-like" cells.

***Cell identification in these experiments was accomplished by including Lucifer Yellow, an intensely fluorescent dye, in the recording pipette and marking the cell for subsequent light microscopic visualization.[47] Although cell identification was not at the electron microscopic level, the cells were presumed to be the Merkel-like basal cells since Merkel-like basal cells, and not basal stem cells, possess synaptic contacts.

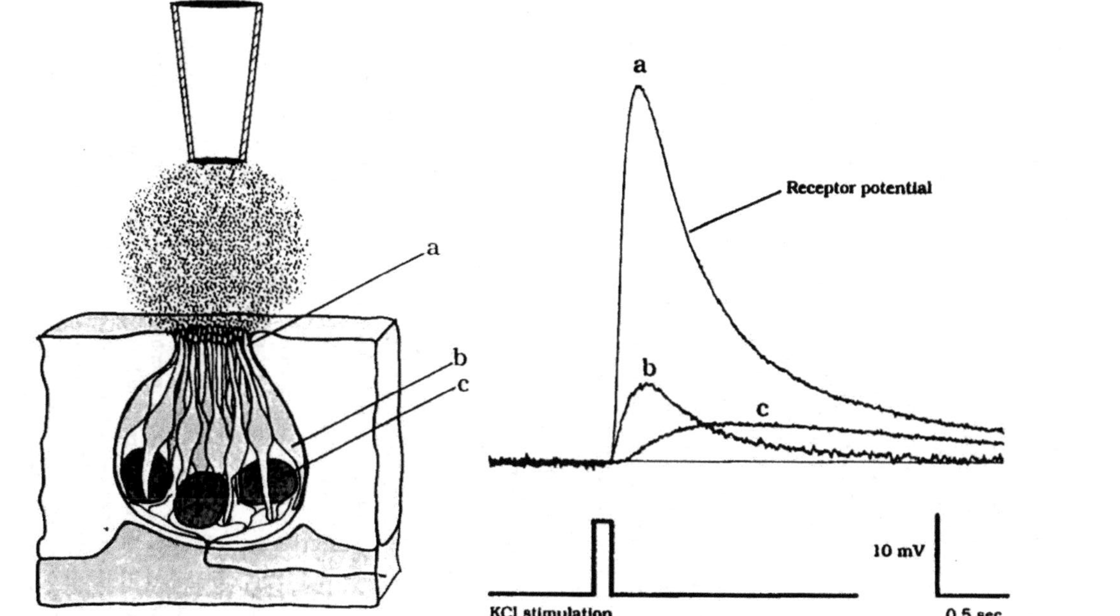

FIGURE 2. Taste receptor cells are synaptically coupled to basal cells in *Necturus*. *Left*, schematic diagram of the lingual slice preparation with a taste bud embedded in the thin tissue. A stimulating pipette placed above the taste pore delivers a brief pulse of KCl (140 m*M*) to the apical tips of receptor cells. a,b,c refer to the sites of the intracellular microelectrode recordings (right). *Right*, Superimposed responses elicited by KCl stimulation of 3 cells (a,b,c) in a single taste bud. Largest response (a) is the receptor potential recorded from an impalement in the apical tip of a receptor cell (−67 mV resting potential). Remaining traces are responses from 2 cells impaled in the basal region of the taste bud. The onset of the response in cell b (−40 mV resting potential) had no latency relative to the receptor potential (a) and was subsequently shown to be an impalement of a basal process of a taste receptor cell. The response in c (−46 mV resting potential) had a 100 msec latency relative to the receptor potential. Responses such as shown in c were subsequently shown to be from basal cells. Bottom trace, KCl stimulus. (Adapted from Ewald, D. E. and Roper, S. D., *J. Neurophysiol.*, 67, 1316, 1992.)

EPSPs. For example, synaptic delays for EPSPs in Merkel-like basal cells are 10s to 100s of msec (Figure 2) and the postsynaptic responses are quite labile. Repetitive stimulation soon causes the postsynaptic potentials to run down. These features suggest that second messengers are involved in generating the EPSPs, but these details remain to be investigated. In principle, there is no requirement for fast synaptic transmission in taste buds. The nature of the stimulus presentation in taste (dissolved foodstuffs) is not an inherently rapid event, unlike in vision and audition.

III. NEUROTRANSMITTERS AT TASTE CELL SYNAPSES

A. AMINO ACIDS

The identity of neurotransmitters at synapses between receptor cells and Merkel-like basal cells, and at other synapses in taste buds, is not yet known. Glutamate and GABA occur in high concentration in axons that innervate taste buds in *Necturus*, raising the possibility that amino acid neurotransmission occurs in taste buds.[38,48] Curiously, these transmitter candidates are found in the sensory axons, that is, in the presumed *post*synaptic elements. On the other hand, ultrastructural descriptions of synapses in *Necturus* show that synaptic vesicles are often clustered near membrane densities in pre- and postsynaptic elements of a synapse,[22] as if transmission occurs in both directions.* There is a precedence for bidirectional synapses in other tissues,[50,51] but whether this occurs in taste buds is mere speculation. Investigators have reported that GABA, when applied to the taste bud, influences responses of the glossopharyngeal nerve in the frog,[52] and glutamate has been shown to depolarize taste cells when it is focally applied.[53] Thus, there are data from immunocytochemical and electrophysiological studies supporting the involvement of amino acids in synaptic transmission in taste buds. The evidence for glutamate and/or GABA being a neurotransmitter, however, is far from compelling yet.

Over the past several years, researchers have identified the presence of a number of other potential neurotransmitters or key enzymes related to transmitters in taste buds, including acetylcholine (ACh), peptides, and biogenic amines (for example, serotonin, noradrenaline, adrenaline). These data are reviewed in the following pages.

B. ACETYLCHOLINE

The strongest evidence that has been cited for cholinergic mechanisms is that acetylcholine esterase is localized to taste buds in a wide variety of species (see review Reference 46 and Welton et al.[82]). ACh esterases can be associated with noncholinergic neurons, however, so this argument for cholinergic trans-

* Yoshie et al.[49] also comment on the possibility of bidirectional or reciprocal synapses involving type III cells and nerve terminals in guinea pig taste buds.

mission in taste buds is not particularly strong.[54] In taste buds from the rat tongue, phosphatidylinositol (PI) turnover was enhanced by the cholinergic mimetic, carbachol.[3] These findings are consistent with cholinergic mechanisms in taste buds, and indeed Hwang et al.[3] interpreted their data as suggesting efferent cholinergic innervation of taste buds. Early reports on responses of the chorda tympani nerve to chemical stimulation also implicated cholinergic transmission; injecting cholinergic agonists and antagonists into the tongue appeared to modify responses recorded in the chorda tympani nerve.[55] However, those findings have been challenged on the basis that the high concentrations of agents used had nonspecific actions unrelated to cholinergic transmission.[56] We have attempted to localize choline acetyl transferase (ChAT) with immunocytochemical methods in mudpuppy (*Necturus maculosus*) taste buds, but were unable to detect this key synthetic enzyme for ACh there.[82] A caveat is that the absence of ChAT may merely represent the failure of immunocytochemical techniques on this enzyme rather than represent the absence of cholinergic mechanisms. Alternatively, the synthetic enzyme may not be present in sufficiently high concentrations for immunocytochemical detection if it were found in efferent cholinergic nerve fibers, as suggested by Hwang et al.[3] This interpretation presupposes, though, that there is efferent innervation of taste buds, itself a controversial topic (cf. References 45, 83). It is troubling that the data for cholinergic mechanisms in taste buds is very contradictory and very incomplete, and that no consistent picture has yet emerged (see also Reutter, Chapter 2).

C. NEUROPEPTIDES

A number of peptides have been found in taste buds, especially in fibers innervating taste buds or coming near taste buds. Typically, peptidergic fibers are few compared with the dense glutamatergic and GABAergic innervation of taste buds when the two have been compared in the same tissue.[82] Axons containing substance P, VIP, CGRP, gastrin-releasing peptide, NPY, peptide histidine isoleucine, neurokinin A, bombesin, and somatostatin innervate taste buds in many species (see Table 2 in Reference 82 for references). Electron microscopic studies in the rat indicate that fibers that contain substance P do not actually form synapses in taste buds, but merely pass nearby taste cells.[57,58] This might indicate that substance P, and possibly other peptides, act as neuromodulators of taste cell function or are involved in neurotrophic mechanisms, rather than mediate transmission at discrete synaptic sites (see also Oakley, Chapter 4).

The effects of peptides on taste bud function have not been examined extensively at the cellular level. Depleting substance P from axons that innervate taste buds had few notable effects on taste in rats.[59] Intraperitoneal injection of CCK in the rat enhanced chorda tympani responses elicited by NaCl and sucrose.[60] Injecting vasopressin into the lingual artery in frogs increased taste responses recorded in the glossopharyngeal nerve, especially NaCl and HCl responses.[61] Applying vasopressin to isolated hamster taste cells enhanced currents evoked by salty and acid stimuli.[84]

D. CATECHOLAMINES AND MONOAMINES

Biogenic amines offer the most likely candidates for neurotransmitters and neuromodulators. Many of the data from different approaches are now beginning to converge and form a coherent story, at least regarding the monoamine, serotonin. DeHan and Graziadei[62] used the Falk-Hillarp histofluorescence technique to visualize amines in frog taste disks. Initial reports suggested that some taste cells contained catecholamines such as norepinephrine or epinephrine. However, subsequent investigations using more refined histofluorescence analyses found that serotonin, not epinephrine or norepinephrine, was more likely to be present in frog taste cells.[63]* Since then, the preponderance of data, including immunohistochemical and electron microscopic observations, point to serotonin being a major constituent of taste buds in several species. Specifically, Merkel-like basal cells in fish and amphibian taste buds are serotonergic. Further, one of the types of elongate receptor cells in mammalian taste buds (the type III cell) is reported to contain serotonin.[41]

E. SEROTONIN

Serotonin exerts powerful effects on taste receptor cells. Early investigations on the pharmacological actions of serotonin in taste buds led to contradictory and confusing results, possibly because tests were conducted in the intact animal and agents were applied topically or injected into the lingual arterial supply. Since monoamines and catecholamines affect blood vessels and may have other sites of action apart from taste buds, results may have been contaminated by pharmacological actions unrelated to taste reception, per se. More recently, though, serotonin has been tested at the cellular level in simpler tissues. The findings clearly show that serotonin can modulate the conduction of signals and synaptic transmission in taste cells.[53,68,81]

In thin slices of *Necturus* lingual epithelium that contain taste buds, it is possible to bath apply low concentrations of serotonin (1 to 100 μM) and test its effects on receptor cells. When this is done, one of the most lasting effects of serotonin is that *Necturus* taste receptor cells are hyperpolarized and their input resistance increases.[68] As a consequence of the increased input resistance, one would predict that the electrotonic propagation of signals from the apical to basal end of the taste cell would be enhanced (cf. Figure 1). That this indeed occurs has recently been shown. Serotonin (100 μM) had no direct

* Catecholamines may yet be discovered to be important neuromodulators in taste reception. Although norepinephrine and epinephrine do not appear to be found in taste cells, a number of researchers have observed adrenergic *nerve fibers* in the connective tissue immediately below taste buds, or in some cases actually penetrating into the taste bud (e.g., Reference 64; see Takeda[64] and Welton et al.[82] for additional references). Adrenergic innervation may represent sympathetic modulation of taste reception either by direct action on taste cells or indirectly, for example, by influencing the local blood supply to taste buds. There are reports in the older literature that stimulating the sympathetic nerve supply to the tongue in frogs and rats alters taste responses.[65-67] These reports remain to be confirmed and extended to the cellular level.

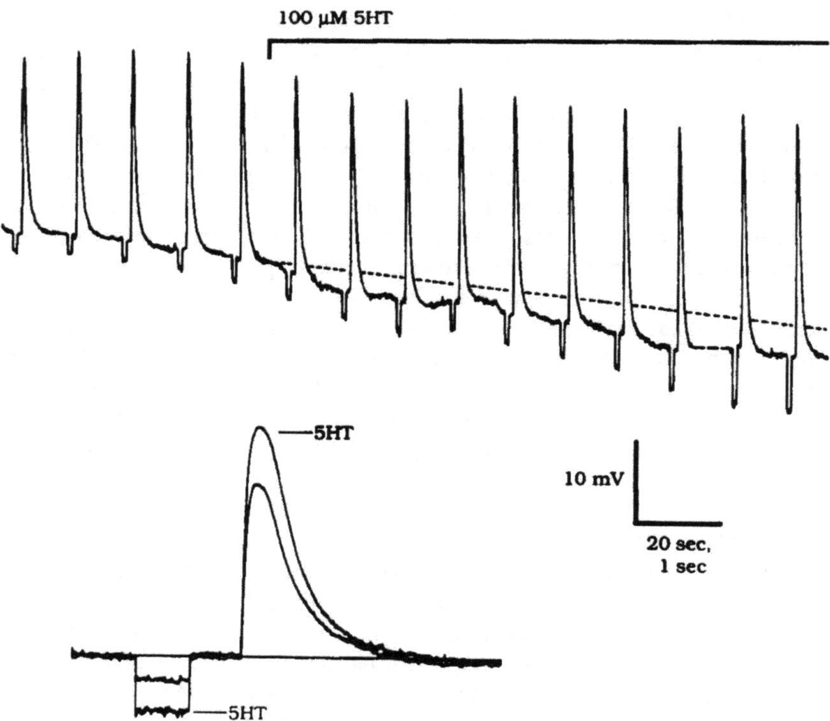

FIGURE 3. Receptor potentials recorded in the basal process of *Necturus* receptor cells are increased by serotonin. Brief pulses of 140 m*M* KCl were applied to the apical tip of a taste cell in the lingual slice preparation[47] to elicit receptor potentials. The recording microelectrode was situated in the basal process of a receptor cell. Each receptor potential was preceded by a 1 s hyperpolarizing pulse through the recording microelectrode (balanced bridge circuit) to monitor the input resistance of the receptor cell. *Upper trace*, continuous intracellular recording taken 1 min before and 2 min during bath application of 100 µ*M* serotonin (5HT). The dotted line represents the (extrapolated) resting potential in the absence of 5HT (resting potential = −60 mV when 5HT was added to the bath). *Lower traces*, receptor potentials and input resistance measurements shown at a faster time base. Two traces are superimposed. The smaller responses were recorded 1 min prior to adding 5HT to the bath. The larger responses (labeled 5HT) were recorded 1.5 min after 5HT was introduced in the bath. Both the amplitude of the receptor potential and the input resistance were increased ~30% by 5HT. (From Roper, S. D., and Ewald, D. A., *Sensory Transduction*, Corey, D. P. and Roper, S. D., Eds., Rockefeller University Press, New York, 1992, chap. 18. With permission.)

effect on the apical receptor potential, i.e., on apical transduction. However, serotonin increases the amplitude of receptor potentials recorded in the basal ends of *Necturus* taste cells (Figure 3).[68,81]

Postsynaptic responses in Merkel-like basal cells in *Necturus* are also increased by serotonin, presumably due to the enhanced amplitude of the presynaptic (taste receptor cell) signal. One would expect that transmission at synapses between taste receptor cells and afferent nerve fibers would also be enhanced, for the same reasons (increased presynaptic signal).[81]

Serotonin also may affect the release of neurotransmitter directly. This is suggested by patch recordings from isolated *Necturus* taste receptor cells.[53,53a] In many taste receptor cells, focal application of serotonin (1 to 100 μM) increases Ca^{2+} currents evoked by depolarizing voltage pulses.[53,53a] This effect can be quite striking, with the peak Ca^{2+} current sometimes doubling in amplitude. Transmitter release is elicited by Ca^{2+} influx, presumably through voltage dependent Ca^{2+} channels situated near synapses. Thus, this effect of serotonin should lead to an increased transmitter output. The enhancement of Ca^{2+} currents would act synergistically with the enhancement of electrotonic propagation of receptor currents. At present, it has not been possible to test the effect of serotonin on the output of taste buds (for example, responses in chorda tympani nerve fibers) in an isolated preparation. Consequently, these proposed effects of serotonin on transmitter release represent a working hypothesis that remains to be tested.

Serotonin also produces inhibitory effects on Ca^{2+} currents recorded from taste receptor cells in *Necturus*.[53,53a] Typically, higher concentrations of serotonin (100 μM) are necessary to produce this effect. It is possible that inhibition by serotonin is a property of a different population of receptor cells than those in which serotonin enhances Ca^{2+} currents. Alternatively, two different classes of serotonin receptors, with differing affinities for the agonist, may exist on one cell. Inhibition of Ca^{2+} currents would be expected to reduce transmitter output at synapses.

Figure 4 presents a summary of the neurotransmitter candidates that have been shown to occur in taste buds and that are discussed above.

IV. SYNAPTIC TRANSMISSION IN THE TASTE BUD: A WORKING MODEL

The above studies on neurotransmitters in taste buds, taken together, suggest a model for information processing and synaptic interactions in the peripheral taste organs. Merkel-like basal cells contain serotonin, as discussed above. These cells have also been shown to take up and release serotonin, and the release is Ca-dependent.[69] Serotonin exerts important effects on adjacent taste receptor cells. These data are all consistent with the proposal that serotonin is a neurotransmitter or neuromodulator of taste receptor activity. One model that seems to fit these data is depicted in Figure 5. The initial event is chemostimulation of the taste bud, shown diagrammatically by the stippled arrow at the top of the figure. This produces a depolarizing receptor potential in the apical tip of taste receptor cells that is then propagated electrotonically (and perhaps via action potentials) to the base of the taste cell

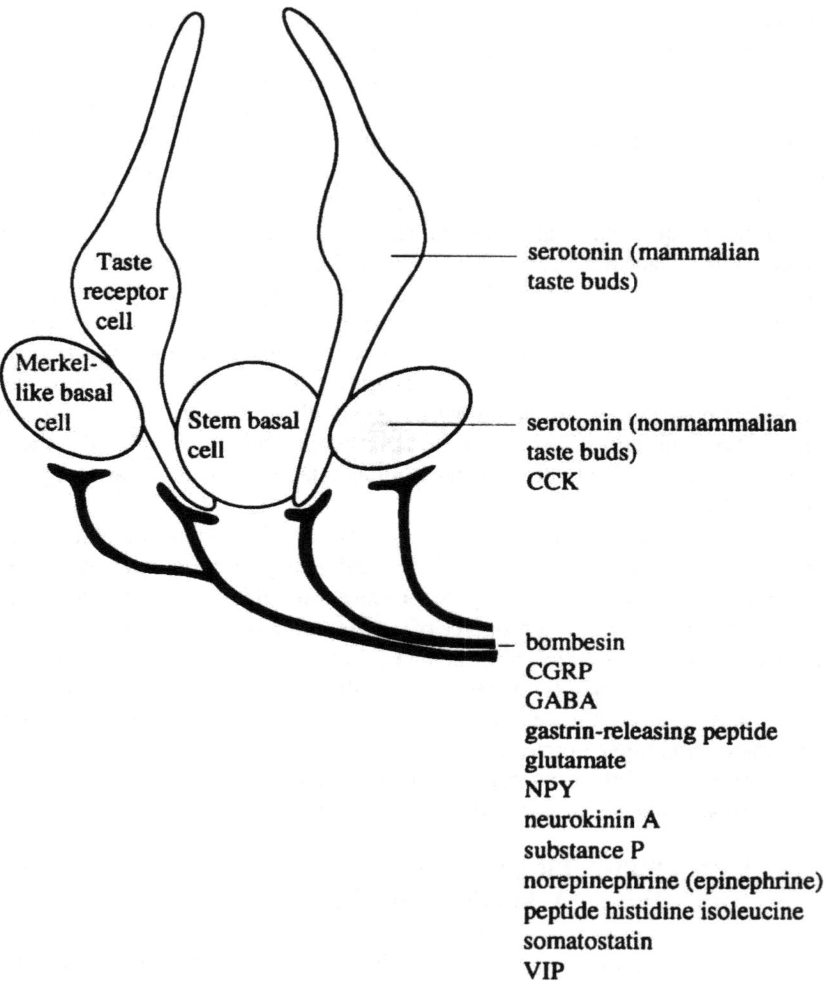

FIGURE 4. Schematic drawing of a taste bud summarizing the neurotransmitter and neuromodulator candidates that have been reported to date. Although the axons that innervate taste buds are shown contacting basal cells (stem basal cells and Merkel-like basal cells) in parallel with taste receptor cells, the precise wiring diagram is not actually known. In mammalian taste buds, serotonin is localized to taste receptor cells, possibly type III cells (see text). However, in taste buds from lower vertebrates, serotonin is found in Merkel-like basal cells. A number of neuroactive peptides are found in axons that innervate taste buds, as shown in this figure (Table 2 in Reference 82).

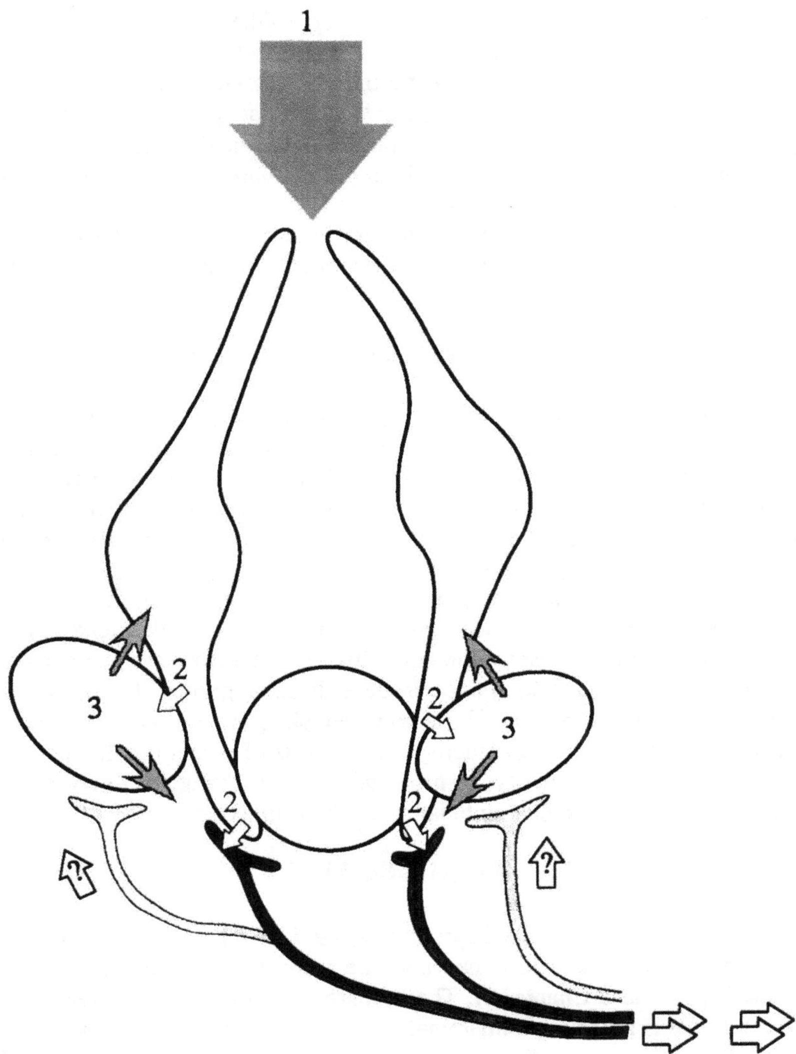

FIGURE 5. Schematic illustration of proposed model for synaptic interactions in the taste bud and serotonergic neuromodulation of taste receptor cells. (1) Chemostimulation (stippled arrow at top) stimulates taste receptor cells; (2) receptor cells synaptically activate Merkel-like basal cells (open arrows, stippled cells) and sensory afferent axons in parallel; (3) Merkel-like basal cells release serotonin (stippled arrows). Serotonin acts at receptor cell synapses as well as more generally on the basolateral membranes of receptor cells, as shown here. Alternately, Merkel-like basal cells might be stimulated via efferent fibers or axon collaterals, shown here by open arrows containing question marks (see text).

where synapses are located. Secondarily, Merkel-like basal cells are synaptically activated (◊), triggering them to release serotonin (➞).* Serotonin acts on taste receptor cells, increasing the electrotonic spread of the apical depolarization as well as modulating Ca^{2+} currents at synapses. These effects appear to represent positive feedback, but they are self-limiting by the nature of the changes. That is, increasing the input resistance of the taste receptor cell has a natural limitation, i.e., when all the serotonin-sensitive conductances are blocked. Furthermore, as mentioned above, at higher concentrations (100 μM), serotonin begins to exhibit inhibitory effects on Ca^{2+} currents in some cells.

V. ELECTRICAL SYNAPSES

This discussion has left out the consideration of lateral synaptic interactions in the taste bud that are mediated by electrical synapses. A number of researchers have shown that taste cells in fish and amphibia are electrically coupled. (Electrical coupling has not yet been investigated in mammalian taste buds.) The principal observation is that two or more taste cells are dye-coupled when Lucifer yellow is injected into a single receptor cell.[70-73] There have also been electrophysiological observations in *Necturus* that indicate taste cells are coupled.[70,74,75] Cell-to-cell coupling is not widespread in taste buds. It has only been observed to occur between receptor cells and does not appear to involve basal cells. Furthermore, only small groups of receptor cells (typically, 2 to 3) are coupled.[72] Electrical coupling in taste buds could serve to spread receptor potentials among small subsets of receptor cells, to coordinate their electrical activity, to promote repetitive firing, or to coordinate their metabolism, or some combination of these functions.

VI. SUMMARY

This chapter has described synaptic interactions that are now believed to occur in taste buds and that may underlie some form of peripheral signal processing (see also Chapter 2). These interactions focus on the Merkel-like basal cell that is found in nonmammalian taste buds, although analogous cells may also exist in mammalian taste buds. The model describes a specific role for the Merkel-like basal cell that has been suggested before in general terms, namely that the Merkel-like basal cell is a paracrine cell or paraneuron.[76-78]

* The question as to whether there can be efferent nerve activation of Merkel-like basal cells is unanswered, and the arrows with question marks in Figure 5 are intended to represent this uncertainty. That is, centrifugal activation from higher centers in the brain might also trigger serotonin release independent of, or even prior to, chemostimulation at the taste pore. Also, axon reflexes from afferent sensory fibers might lead back to the same or adjacent taste buds and stimulate Merkel-like basal cells. It is not possible to rule out these routes for exciting Merkel-like basal cells, but neither is there strong evidence favoring such efferent or antidromic activation.

Paracrine cells or paraneurons are cells that "release their product into the extracellular space for simple diffusion to target cells in the immediate vicinity" (p. 83 in Reference 79). One of the neuroactive substances released by the Merkel-like basal cell is serotonin. Serotonin acts as a neuromodulator of taste receptor cell activity. Since the Merkel-like basal cell also contains other neuroactive substances such as CCK,[82] it would not be surprising if other neuromodulators are core-leased by this cell. An analogy between Merkel-like basal cells and cells that mediate lateral interactions in the retina has made before.[80] Perhaps it is now appropriate to consider that signal processing occurs in the peripheral sensory organs in taste as it does in vision.

ACKNOWLEDGMENTS

I thank Albertino Bigiani, Rona Delay, Doug Ewald, Sue Kinnamon, and Sid Simon for their helpful discussions and comments. This work was supported, in part, by NIH grants 2 P01 DC00244, 2 R01 DC00374, and 5 R01 DC01238.

REFERENCES

1. **Dowling, J. E.,** *The Retina: An Approachable Part of the Brain,* Harvard University Press, Cambridge, 1987.
2. **Akabas, M. H., Dodd, J., and Al-Awqati, A. Q.,** A bitter substance induces a rise in intracellular calcium in a subpopulation of rat taste cells, *Science,* 242, 1047, 1988.
3. **Hwang, P. M., Verma, A., Bredt, D. S., and Snyder, S. H.,** Localization of phosphatidylinositol signaling components in rat taste cells: role in bitter taste transduction, *Proc. Natl. Acad. Sci. U.S.A.,* 87, 7395, 1990.
4. **Roper, S.,** Regenerative impulses in taste cells, *Science,* 220, 1311, 1983.
5. **Kashiwayanagi, M., Miyake, M., and Kurihara, K.,** Voltage-dependent Ca^{2+} channel and Na^{+} channel in frog taste cells, *Am. J. Physiol.,* 244, C82, 1983.
6. **Avenet, P. and Lindemann, B.,** Action potentials in epithelial taste receptor cells induced by mucosal calcium, *J. Membr. Biol.,* 95, 265, 1987.
7. **Avenet, P. and Lindemann, B.,** Non-invasive recording a receptor cell action potentials and sustained currents from single taste buds maintained in tongue: the response to mucosal NaCl and amiloride, *J. Membr. Biol.,* in press.
8. **Kinnamon, S. C. and Roper, S. D.,** Passive and active membrane properties of mudpuppy taste receptor cells, *J. Physiol. (London),* 383, 601, 1987.
9. **Avenet, P. and Lindemann, B.,** Patch-clamp study of isolated taste receptor cells of the frog, *J. Membr. Biol.,* 97, 223, 1987.
10. **Spielman, A. I., Mody, I., Brand, J. G., Whitney, G., MacDonald, J. F., and Salter, M. W.,** A method for isolating and patch-clamping single mammalian taste receptor cells, *Brain Res.,* 503, 326, 1989.
11. **Roper, S. D.,** *Ion Channels and Taste Transduction,* Marcel Dekker, New York, 1989, 137.
12. **Kinnamon, S. C., Dionne, V. E., and Beam, K. G.,** Apical localization of K^{+} channels in taste cells provides the basis for sour taste transduction, *Proc. Natl. Acad. Sci. U.S.A.,* 85, 7023, 1988.

13. **Roper, S. D. and McBride, D. W. Jr.,** Distribution of ion channels on taste cells and its relationship to chemosensory transduction, *J. Membr. Biol.*, 109, 29, 1989.
14. **Elliott, E. J. and Simon, S. A.,** The anion in salt taste: a possible role for paracellular pathways, *Brain Res.*, 535, 9, 1990.
15. **Ye, Q., Heck, G. L., and DeSimone, J. A.,** The anion paradox in sodium taste reception: resolution by voltage-clamp studies [see comments], *Science*, 254, 724, 1991.
16. **Simon, S. A., Holland, V. F., Benos, D. J., and Zampighi, G. A.,** Transcellular and paracellular pathways in lingual epithelia and their influence in taste transduction, *J. Electron Microsc. Tech.*, in press.
17. **Holland, V. F., Zampighi, G. A., and Simon, S. A.,** Morphology of fungiform papillae in canine lingual epithelium: location of intercellular junctions in the epithelium, *J. Comp. Neurol.*, 279, 13, 1989.
18. **Akisaka, T. and Oda, M.,** Taste buds in the vallate papilla of the rat studied with freeze-fracture preparation., *Arch. Histol. Jpn.*, 41, 87, 1978.
19. **Jahnke, K. and Baur, P.,** Freeze-fracture study of taste bud pores in the foliate papillae of the rabbit, *Cell Tissue*, 200, 245, 1979.
20. **Kinnamon, J. C., Taylor, B. J., Delay, R. J., and Roper, S. D.,** Ultrastructure of mouse vallate taste buds. I. Taste cells and their associated synapses, *J. Comp. Neurol.*, 235, 48, 1985.
21. **Kinnamon, J. C., Sherman, T. A., and Roper, S. D.,** Ultrastructure of mouse vallate taste buds. III. Patterns of synaptic connectivity, *J. Comp. Neurol.*, 270, 1, 1988.
22. **Delay, R. J. and Roper, S. D.,** Ultrastructure of taste cells and synapses in the mudpuppy *Necturus maculosus*, *J. Comp. Neurol.*, 277, 268, 1988.
23. **Cagan, R. H.,** Allosteric regulation of glutamate taste receptor function, in *Umami: A Basic Taste*, Kawamura, Y. and Kare, M. R., Eds., Marcel Dekker, New York, 1987, 155.
24. **Yamaguchi, S.,** Basic properties of umami and effects on humans, *Physiol. Behav.*, 49, 833, 1991.
25. **Gilbertson, T. A., Avenet, P., Kinnamon, S. C., and Roper, S. D.,** Proton currents through amiloride-sensitive Na channels in hamster taste cells: role in acid taste transduction, *J. Gen. Physiol.*, 100, 803, 1992.
26. **Fujimoto, S. and Murray, R. G.,** Fine structure of degeneration and regeneration in denervated rabbit vallate taste buds, *Anat. Rec.*, 168, 393, 1970.
27. **Ide, C. and Munger, B. L.,** The cytologic composition of primate laryngeal chemosensory corpuscles, *Am. J. Anat.*, 158, 193, 1980.
28. **Delay, R. J. and Roper, S. D.,** Cell lineage in the mudpuppy, *Necturus maculosus*, *Chem. Senses*, 14, 694, 1989.
29. **Taylor, R., Delay, R., and Roper, S.,** Are basal cells in taste buds identical to cutaneous Merkel cells?, *Assoc. Chemorecept. Sci. (Abstracts)*, XIV, 48, 1992.
30. **Korte, G. E.,** Ultrastructure of the tastebuds of the red-eared turtle, *Chrysemys scripta elegans*, *J. Morphol.*, 163, 231, 1980.
31. **Farbman, A. I. and Yonkers, J. D.,** Fine structure of the taste bud in the mudpuppy, *Necturus maculosus*, *Am. J. Anat.*, 131, 353, 1971.
32. **During, M. V. and Andres, K. H.,** The ultrastructure of taste and touch receptors of the frog's taste organ, *Cell Tissue Res.*, 165, 185, 1976.
33. **Jakubowski, M.,** New details of the ultrastructure (TEM, SEM) of taste buds in fishes, *Z. Mikrosk. Anat. Forsch.*, 97, 849, 1983.
34. **Toyoshima, K. and Shimamura, A.,** Monoamine-containing basal cells in the taste buds of the newt *Triturus pyrrhogaster*, *Arch. Oral Biol.*, 32, 619, 1987.
35. **Toyoshima, K. and Shimamura, A.,** An immunohistochemical demonstration of neuron-specific enolase in the Merkel cells of the frog taste organ, *Arch. Histol. Cytol.*, 51, 295, 1988.
36. **Kuramoto, H.,** An immunohistochemical study of cellular and nervous elements in the taste organ of the bullfrog, *Rana catesbeiana*, *Arch. Histol. Cytol.*, 51, 205, 1988.

37. **Toyoshima, K.,** Ultrastructure and immunohistochemistry of the basal cells in the taste buds of the loach, *Misgurnus anguillicaudatus, J. Submicrosc. Cytol. Pathol.*, 21, 771, 1989.
38. **Jain, S. and Roper, S. D.,** Immunocytochemistry of gamma-aminobutyric acid, glutamate, serotonin and histamine in *Necturus* taste buds, *J. Comp. Neurol.*, 307, 675, 1991.
39. **Murray, R. G.,** The mammalian taste bud type III cell: a critical analysis, *J. Ultrastruct. Mol. Struct. Res.*, 95, 175, 1986.
40. **Ciges, M., Diaz-Flores, L., Gonzalez, M., and Rama, J.,** Ultrastructural study of taste buds at rest and after stimulation, and comparative study between type III cell and Merkel cells, *Acta Otolaryngol.*, 81, 209, 1976.
41. **Uchida, T.,** Serotonin-like immunoreactivity in the taste bud of the mouse circumvallate papilla, *Jpn. J. Oral Biol.*, 27, 132, 1985.
42. **Fujimoto, S., Ueda, H., and Kagawa, H.,** Immunocytochemistry on the localization of 5-hydroxytryptamine in monkey and rabbit taste buds, *Acta Anat. (Basel)*, 128, 80, 1987.
43. **Hirata, Y.,** Fine structure of the terminal buds on the barbels of some fishes, *Arch. Histol. Jpn.*, 26, 507, 1966.
44. **Reutter, K.,** Die Geschmacknospen des Zwergwelses *Amiurus nebulosus*. Morphologische und histochemische Untersuchungen, *Z. Zellforsch. Mikrosk. Anat.*, 120, 280, 1971.
45. **Roper, S. D.,** The cell biology of vertebrate taste receptors, *Annu. Rev. Neurosci.*, 12, 329, 1989.
46. **Roper, S. D.,** The microphysiology of peripheral taste organs, *J. Neurosci.*, 12, 1127, 1992.
47. **Ewald, D. E. and Roper, S. D.,** Intercellular signalling in *Necturus* taste buds: chemical excitation of receptor cells elicits responses in basal cells, *J. Neurophysiol.*, 67, 1316, 1992.
48. **Lu, K.-S. and Roper, S. D.,** Electron microscopic immunocytochemistry of glutamate-containing nerve fibers in the taste bud of the mudpuppy *(Necturus maculosus), J. Electron Microsc. Tech.*, in press.
49. **Yoshie, S., Wakasugi, C., Teraki, Y., and Fujita, T.,** Fine structure of the taste bud in guinea pigs. I. Cell characterization and innervation patterns, *Arch. Histol. Cytol.*, 53, 103, 1990.
50. **Anderson, P. A. V.,** Physiology of a bidirectional, excitatory, chemical synapse, *J. Neurophysiol.*, 53, 821, 1985.
51. **Anderson, P. A. V. and Grunert, U.,** Three-dimensional structure of bidirectional, excitatory chemical synapses in the jellyfish *Cyanea capillata, Synapse*, 2, 606, 1988.
52. **Tateda, H. and Beidler, L. M.,** The receptor potential of the taste cell of the rat, *J. Gen. Physiol.*, 47, 479, 1964.
53. **Delay, R. J., Kinnamon, S. C., and Roper, S. D.,** Membrane properties and transmitter sensitivity of Merkel-like basal cells in *Necturus* taste buds, *Assoc. Chemoreception Sci. Abstracts*, XIV, 210, 1992.
53a. **Delay, R. J., Kinnamon, S. C., and Roper, S. D.,** Serotonin modulates voltage-actuated calcium currents in *Necturus* taste receptor cells, *Biophys. J.*, 64, A390.
54. **Rama Sastry, B. V. and Sadavongvivad, C.,** Cholinergic systems in non-nervous tissues. *Pharmacol. Rev.*, 30, 65, 1978.
55. **Landgren, S., Liljestrand, G., and Zotterman, Y.,** Chemical transmission in taste fiber endings, *Acta Physiol. Scand.*, 30, 105, 1954.
56. **Duncan, C. J.,** Synaptic transmission at taste buds, *Nature*, 203, 875, 1964.
57. **Yamasaki, H. and Tohyama, M.,** Ontogeny of substance P-like immunoreactive fibers in the taste buds and their surrounding epithelium of the circumvallate papillae of the rat. II. Electron microscopic analysis, *J. Comp. Neurol.*, 241, 493, 1985.

58. **Hirata, K., Miyahara, H., and Kanaseki, T.,** Substance-P-containing fibers in the incisive papillae of the rat hard palate. Light- and electron-microscopic immunohistochemical study, *Acta Anat. (Basel)*, 132, 197, 1988.
59. **Silver, W. L., Mason, J. R., Marshall, D. A., and Maruniak, J. A.,** Rat trigeminal, olfactory and taste responses after capsaicin desensitization, *Brain Res.*, 333, 45, 1985.
60. **Serova, O. N. and Esakov, A. I.,** [Activating effect of cholecystokinin-pancreozymin on the taste-receptor system of the rat], *Fiziol. Zh. SSSr.*, 71, 1271, 1985.
61. **Okada, Y., Miyamoto, T., and Sato, T.,** Vasopressin increases frog gustatory neural responses elicited by NaCl and HCl, *Comp. Biochem. Physiol.*, 100A, 693, 1991.
62. **DeHan, R. S. and Graziadei, P.,** The innervation of frog's taste organ: a histochemical study, *Life Sci.*, 13, 1435, 1973.
63. **Hirata, K. and Nada, O.,** A monoamine in the gustatory cell of the frog's taste organ, *Cell Tissue Res.*, 159, 101, 1975.
64. **Takeda, M.,** An electron microscopic study on the innervation in the taste buds of the mouse circumvallate papillae, *Arch. Histol. Jpn.*, 39, 257, 1976.
65. **Kimura, K.,** Factors affecting the response of taste receptors of rat, *Kuramoto Med. J.*, 14, 95, 1961.
66. **Chernetski, K. E.,** Sympathetic enhancement of peripheral sensory input in the frog, *J. Neurophysiol.*, 27, 493, 1964.
67. **Rapuzzi, G. and Ricagno, G.,** L'efferenza sinaptica dei recettori linguali di rana, *Arch. Fisiol.*, 70, 88, 1973.
68. **Ewald, D. A. and Roper, S. D.,** Basal cells modulate receptor cell function in *Necturus* taste buds by a serotonergic mechanism, *Soc. Neurosci. Abstr.*, 18, 596, 1992.
69. **Welton, J. and Roper, S.,** *In vitro* uptake of ^{3}H serotonin in taste buds of the mudpuppy, *Necturus maculosus, Assoc. Chemorecept. Sci. (Abstracts)*, XIV, 194, 1992.
70. **West, C. H. K. and Bernard, R. A.,** Intracellular characteristics and responses of taste bud and lingual cells of the mudpuppy, *J. Gen. Physiol.*, 72, 305, 1978.
71. **Teeter, J.,** Dye-coupling in catfish taste buds, in *Proc. 19th Japanese Symp. on Taste and Smell*, Kimura, S., Miyoshi, A., and Shimada, I., Eds., Asahi Univ., Hozumi-cho, Gifu, 1985, 29.
72. **Yang, J. and Roper, S. D.,** Dye-coupling in taste buds in the mudpuppy, *Necturus maculosus, J. Neurosci.*, 7, 3561, 1987.
73. **Sata, O. and Sato, T.,** Dye-coupling among cells in taste disk in frog, *Chem. Senses*, 14, 316, 1989.
74. **Bigiani, A. and Roper, S.,** Patch clamp recordings from cells in intact taste buds in thin lingual slices, *Chem. Senses*, 16, 501, 1991.
75. **Bigiani, A., Avenet, P., and Roper, S. D.,** Electrical coupling between cells in the taste buds of *Necturus maculosus, Pflügers Arch.*, 420, R157 1982.
76. **Fujita, T., Kanno, T., and Kobayashi, S.,** *The Paraneuron*, Springer-Verlag, Tokyo, 1988, 367.
77. **Toyoshima, K.,** Chemoreceptive and mechanoreceptive paraneurons in the tongue, *Arch. Histol. Cytol.*, 52(Suppl.), 383, 1989.
78. **Toyoshima, K., et al.,** Ultrastructural and histochemical changes in the frog taste organ following denervation, *Arch. Histol. Jpn.*, 47, 31, 1984.
79. **Fawcett, D. W.,** *A Textbook of Histology*, 11th ed., W. B. Saunders, Philadelphia, 1986.
80. **Uga, S. and Hama, K.,** Electron microscopic studies on the synaptic region of the taste organ of carps and frogs, *J. Electron Microsc.*, 16, 269, 1967.
81. **Roper, S. D. and Ewald, D. A.,** Peripheral events in taste transduction, in *Sensory Transduction*, Corey, D. P. and Roper, S. D., Eds., Rockefeller University Press, New York, 1992, chap. 18.
82. **Welton, J., Taylor, R., Porter, A. J., and Roper, S. D.,** Immunocytochemical survey of putative neurotransmitters in taste buds from *Necturus maculosus, J. Comp. Neurol.*, 323, 1, 1992.

83. **Farbman, A. I. and Hellekant, G.,** Quantitative analyses of the fiber population in rat chorda tympani nerves and fungiform papillae, *Am. J. Anat.*, 153, 1978.
84. **Gilbertson, T. A., Roper, S. D., and Kinnamon, S. C.,** Proton currents through amiloride-sensitive Na channels in isolated hamster taste cells: enhancement by vasopressin and cAMP, *Neuron*, in press, 1993.

Chapter 12

SENSORY CODING BY PERIPHERAL TASTE FIBERS

David V. Smith and Marion E. Frank

TABLE OF CONTENTS

0-8493-5341-6/93/$0.00 + $.50

The perception of saltiness, sweetness, sourness, or bitterness emerges from neural activity within the central nervous system. These psychological concepts are used by humans to describe the sensations arising from stimulation of gustatory receptors by a variety of chemical stimuli. Information necessary for these perceptions is carried to the brain by the activity in peripheral taste nerves. The response profiles of peripheral gustatory nerve fibers reflect the way in which the sensitivities of taste receptor cells are distributed among these first-order neurons. Transduction of specific chemical stimuli (e.g., sodium ions, protons, sugars, or alkaloids) by taste receptors may give rise to activity in several types of afferent nerve fibers. Understanding the neural coding of taste information must begin with knowledge about how chemical sensitivities, represented by specific transduction mechanisms, are distributed and organized among peripheral and central gustatory neurons.

Taste receptor mechanisms are distributed across several different subpopulations of taste buds, which are located on different regions of the tongue and oral epithelium and are innervated by one of several cranial nerves. The taste-responsive fibers in these nerves can be classified on the basis of their response spectra, and fibers in the VIIth, IXth, and Xth cranial nerves have very different profiles of sensitivity, which reflect differential input from several taste transduction mechanisms in the periphery. This chapter describes the organization of gustatory sensitivities in fibers of the chorda tympani (CT), glossopharyngeal (IXth), and superior laryngeal nerve (SLN), and the integrated responses of the greater superficial petrosal (GSP) nerve of the hamster. The differences among these gustatory nerves are discussed in relation to the representation of taste quality and to taste-mediated behavioral responses. Finally, labeled-line and pattern theories of taste quality coding are discussed and the role neuron types can play in the definition of the across-neuron patterns is demonstrated.

I. TASTE BUD DISTRIBUTION AND PERIPHERAL INNERVATION

A. TASTE BUD POPULATIONS

Taste receptors are distributed on the tongue and throughout the oropharyngeal and laryngeal epithelium of mammals within several subpopulations of taste buds innervated by one of three cranial nerves.[1] In the hamster, for example, about 18% of the taste buds are located in the fungiform papillae on the anterior portion of the tongue, 32% are within the foliate papillae on the posterior sides of the tongue, 23% in the single midline vallate papilla on the posterior tongue, about 14% on the palate, distributed between the nasoincisive papillae (2%) and the soft palate (12%), and about 10% on the laryngeal surface of the epiglottis and the aryepiglottal folds.[2] There are also a small number of taste buds within the sublingual organ, the buccal walls, the nasopharynx, and the upper reaches of the esophagus. Similar distributions are seen in the rat[3-5] and other mammalian species that have been examined.[1,6-10]

B. PERIPHERAL INNERVATION OF TASTE BUDS

1. Facial Nerve

Taste buds in the fungiform papillae on the anterior portion of the tongue and in the more rostral of the foliate papillae on the sides of the tongue are innervated by the CT branch of the facial (VIIth) nerve.[11-14] The GSP branch of the VIIth cranial nerve innervates taste buds on the soft palate and in the nasoincisor ducts.[4,5,15] The neurons giving rise to the gustatory fibers of the CT and GSP are located within the geniculate ganglion. These two branches of the VIIth nerve carry gustatory information to the rostral pole of the nucleus of the solitary tract (NST), where their afferent terminations are coextensive.[16,17] Thus, it may be reasonable to think about the gustatory input from the facial nerve as a functional entity, as it appears to be in aquatic vertebrates.[18-20]

2. Glossopharyngeal Nerve

The vallate and foliate papillae on the posterior tongue contain taste buds innervated by the lingual-tonsillar branch of the glossopharyngeal (IXth) nerve.[12,13,21,22] There is also some evidence that taste buds within the nasopharynx are innervated by the IXth cranial nerve.[23] Afferent fibers of the IXth nerve, the cell bodies of which lie in the petrosal ganglion, project into the medulla and terminate within the NST somewhat caudal to, but overlapping with, the termination of the VIIth nerve.[16,17,24] In rats and hamsters, over half of all the taste buds are distributed within the vallate and foliate papillae.[2,3,25]

3. Superior Laryngeal Nerve

Taste buds distributed on the laryngeal surface of the epiglottis, on the aryepiglottal folds, and in the upper reaches of the esophagus are innervated by the SLN, which is a branch of the vagus (Xth) nerve.[9,26] Fibers of the SLN, whose cell bodies lie within the nodose ganglion, project into the NST caudal to those of the VIIth and IXth cranial nerves, but with some overlap rostrocaudally, especially with the IXth nerve termination.[16,17,24,27] Epiglottal taste buds are found in all mammalian species that have been studied and are as numerous as those on the soft palate and almost as abundant as those in the fungiform papillae on the anterior tongue.[2,28]

II. ELECTROPHYSIOLOGY OF PERIPHERAL TASTE FIBERS

Although most gustatory electrophysiology has been conducted on fibers of the CT nerve or on cells in the central nervous system to which these fibers project,[29-36] several recent papers address the sensitivities of fibers innervating other subpopulations of taste buds, including those on the posterior tongue,[37-39] palate,[40-43] and epiglottis.[44-46] The narrative to follow describes the response properties of fibers in the peripheral gustatory nerves of the hamster, a species we have used extensively for the neurophysiological investigation of both peripheral and central gustatory processing; data from other species are integrated into this discussion where appropriate.

A. CHORDA TYMPANI FIBERS

The response properties of gustatory fibers in the CT nerve, which innervates taste buds on the anterior part of the tongue, have been characterized in a number of mammalian species.[29,32,47-49] Fibers of the hamster CT are differentially responsive to a wide array of gustatory stimuli[29,30] and input to brainstem gustatory nuclei from the anterior tongue of the hamster is sufficient to discriminate among several classes of stimuli[34] that are behaviorally discriminable by hamsters.[50-52]

1. Best-Stimulus Classification

Since the earliest electrophysiological studies of single taste fibers,[49,53] it has been recognized that individual fibers of the CT nerve are responsive to stimuli representing more than one of the classical four taste qualities (salty, sour, sweet, and bitter). It was the multiple sensitivity of CT fibers that first led Pfaffmann[49,53,54] to propose that taste quality is coded by the relative amount of activity across gustatory afferent fibers (i.e., by an across-fiber pattern). Nevertheless, fibers within the CT nerve of the hamster,[29] and of other species as well,[32,55] can be grouped into classes on the basis of their relative sensitivities to four midrange stimuli. When the stimuli 0.1 *M* sucrose, 0.03 *M* NaCl, 3 m*M* HCl, and 1 m*M* quinine-HCl (QHCl) are ordered hedonically along the abscissa from most to least preferred, the response spectra of hamster CT fibers show a single peak as sucrose- , NaCl- , or HCl-best fibers.[29] Few fibers (<1%) most responsive to this half-maximal concentration of QHCl are found in the CT nerve. Mean response profiles for each of these best-stimulus classes are shown in Figure 1, where it can be seen that CT fibers often respond to more than one of these four stimuli. Responses to stimuli other than the best stimulus are relatively larger in fibers

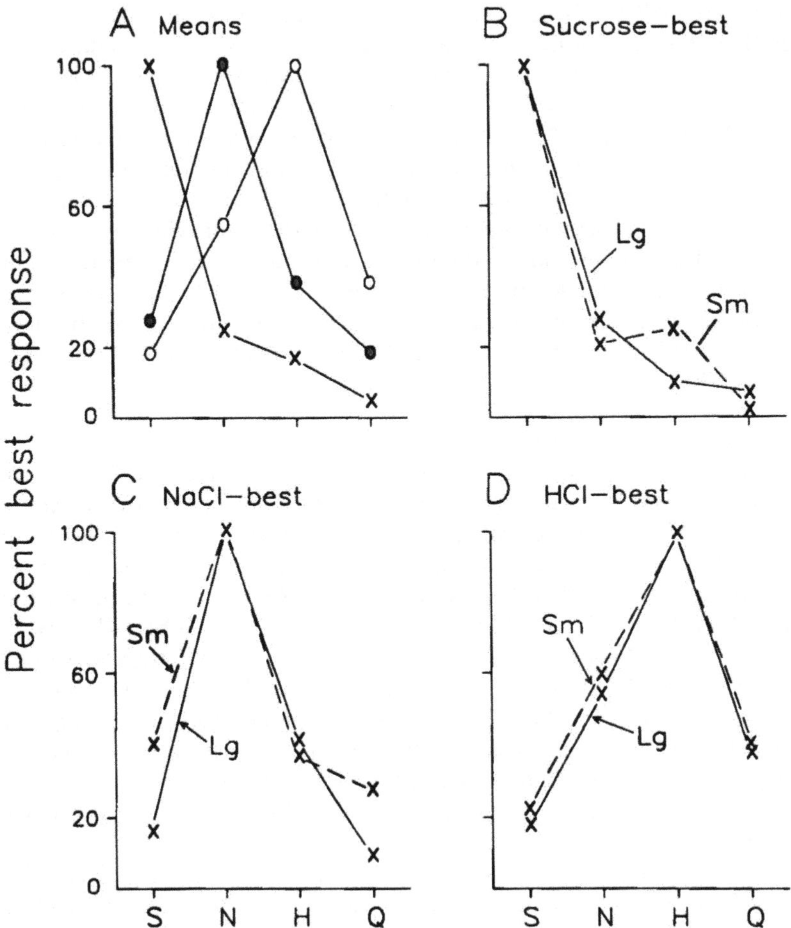

FIGURE 1. Mean relative response profiles of hamster CT fibers across the four basic tastes: 0.1 *M* sucrose (S), 0.03 *M* NaCl (N), 3 m*M* HCl (H), and 1 m*M* QHCl (Q). The fibers were divided into three groups: those which responded best to sucrose (n = 20), NaCl (n = 42), or HCl (n = 17). The responses of each fiber to the other three stimuli were expressed as percentages of the best response. The mean percentages for fibers which responded best to sucrose (x), NaCl (●), or HCl (o) are plotted in A. Each of the three groups of fibers, sucrose-best (B), NaCl-best (C), and HCl-best (D), were divided into more and less sensitive halves, that is into subgroups with the larger (Lg) and smaller (Sm) best responses in the group. The order of the stimuli along the abscissa is from most (S) to least (Q) preferred. (Adapted from Frank, M., *J. Gen. Physiol.*, 61, 588, 1973. With permission.)

showing lower response rates (Sm) than in fibers with higher response rates (Lg), i.e., the faster fibers are somewhat more narrowly tuned.

These data show that CT fibers that respond best to 0.1 *M* sucrose will likely respond second best to 0.03 *M* NaCl, their responses to 3 m*M* HCl and 1 m*M* QHCl will be even smaller, and they will be relatively more specific in their responsiveness than NaCl- or HCl-best fibers. Further, 75% of CT fibers that respond best to NaCl will respond second best to HCl and 25% will respond second best to sucrose, but in either case their responses to QHCl will be small. Finally, CT fibers that respond best to HCl are likely to have a large second-best response to NaCl, and a larger response to QHCl and a smaller response to sucrose than other fibers. Thus, fibers of the hamster CT nerve have an organization to their sensitivities, conveniently described by three classes based on sensitivities to four basic stimuli applied to the anterior portion of the tongue: sucrose- , NaCl- , and HCl-best fibers. This kind of best-stimulus classification has been used extensively in subsequent studies of the hamster gustatory system[30,33-36,46,52,56] and that of other species as well.[30,32,38]

2. Organization of Sensitivities

The appropriateness of this best-stimulus classification of CT fibers was examined further by analyzing the responses of 40 CT fibers to an array of 13 stimuli.[30] The response profiles of these 40 fibers are shown in Figure 2, which allows a comparison among them. Within this figure, the fibers are arranged according to their best stimulus, with *fibers 1 to 10* being sucrose-best, *fibers 11 to 32* NaCl-best, and *fibers 33 to 40* HCl-best. The cross-hatched bars in Figure 2 represent responses to four prototypical stimuli: sucrose (sweet), NaCl (salty), HCl (sour), and QHCl (bitter). The solid bars represent responses to other stimuli, shown along the abscissa. Although the response profiles within a best-stimulus group are not identical, inspection of Figure 2 gives the impression that there are essential similarities among profiles of fibers within a group and striking differences between profiles in different groups.

The response profiles shown in Figure 2 were subjected to a hierarchical cluster analysis, which addressed the question of whether it is reasonable to assume that these profiles were sampled from distinct subpopulations rather than from a single population.[57] In this analysis, the most similar pairs of profiles are clustered together first, followed by the clustering of pairs of profiles that are more and more dissimilar. Distance in this analysis was quantified by calculating the square root of chi-square, comparing impulse frequencies to the 13 stimuli for each pair of profiles. This measure of distance is insensitive to the obvious differences in absolute impulse frequency among the various fibers, but reflects the differences in the relative effects of the stimuli.[57] The resulting dendrogram depicting the hierarchical arrangement of this clustering is shown in Figure 3. At the right of the figure, the fiber numbers are indicated as they are specified in Figure 2. The analysis segregated the CT fiber profiles into three major clusters, members of which are connected by solid lines in Figure 3. This conclusion is based on a regular, stepwise increase in the intercluster distance as the linking proceeds, until, in moving from three clusters to two, a dramatic increase in the intercluster

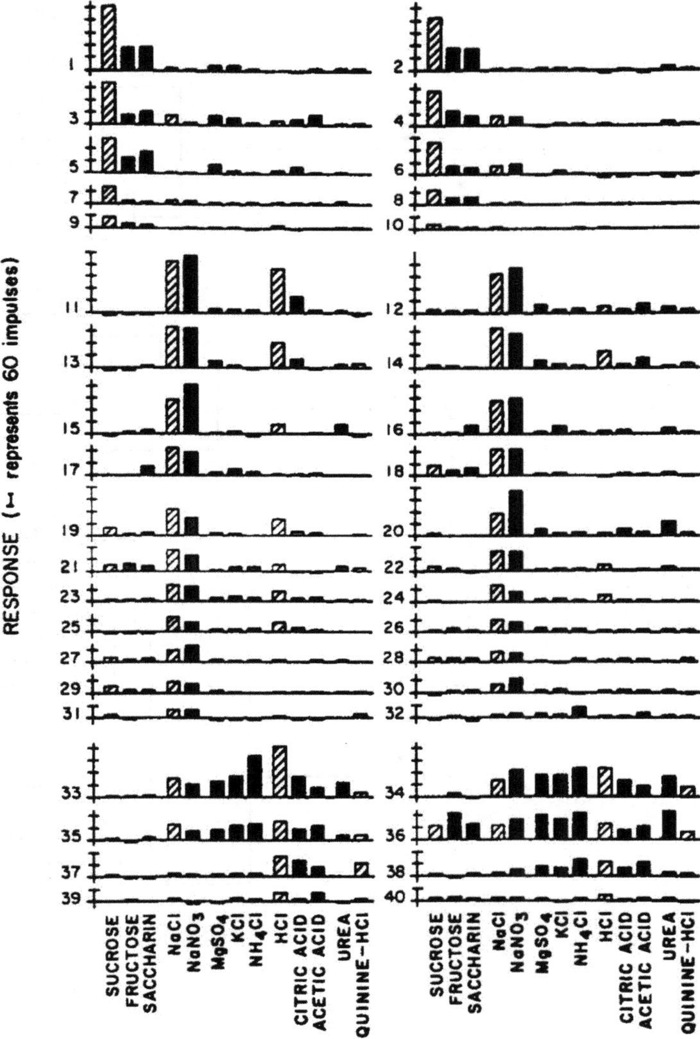

FIGURE 2. Response profiles of hamster CT nerve fibers. The numbers to the left of each profile identify the fibers. The left-hand column shows profiles for odd-numbered fibers and the right-hand column shows profiles for even-numbered fibers. *Fibers 1 to 10* are sucrose-best, *fibers 11 to 32* are NaCl-best, and *fibers 33 to 40* are HCl-best. Test stimuli are listed along the abscissa, beneath bars whose heights represent response rates. The response (nerve impulses) rates for 5 s are indicated. The long horizontal lines from which the bars project represent a rate of zero. Each short horizontal line sequentially crossing the profile's ordinate represents an additional rate increase of 60 impulses above the spontaneous rate. Response rates that are lower than the spontaneous rate are seen as bars extending below the horizontal zero line. The cross-hatched bars represent responses to prototypical stimuli (sucrose, NaCl, HCl, and QHCl) and the filled bars represent responses to other stimuli. (From Frank, M. E., et al., *J. Gen. Physiol.*, 91, 861, 1988. With permission.)

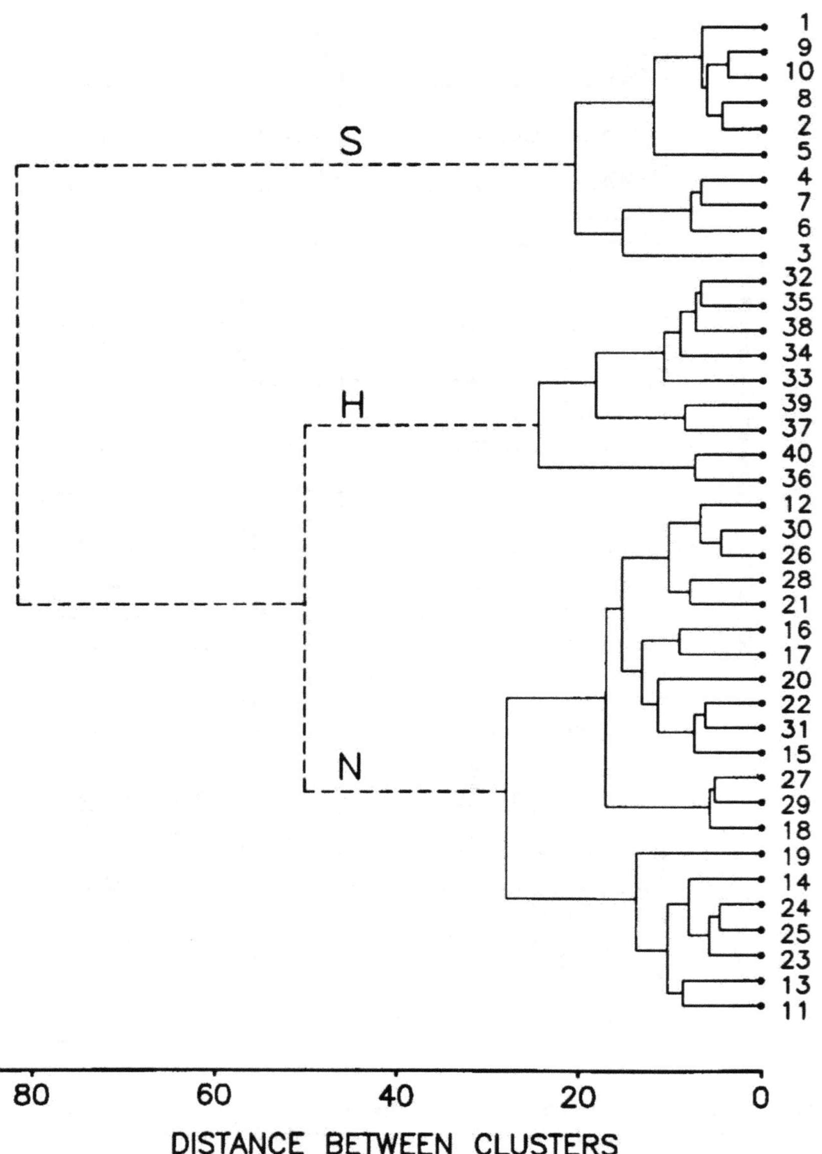

FIGURE 3. Cluster analysis of hamster CT fiber response profiles. The parallel horizontal lines of the dendrogram represent profiles, or groups of profiles, of fibers indicated at the right and numbered as in Figure 2. The major profile clusters (S, H, N) are identified to the left of the defining vertical lines. The distances between profiles (or groups of profiles) are obtained by projecting the vertical lines to the distance scale along the abscissa. (Modified from Frank, M. E., Bieber, S. L., and Smith, D. V., *J. Gen. Physiol.*, 91, 861, 1988. With permission.)

distance occurs. The fiber classes defined by the cluster analysis are labeled "S", "H", and "N" in Figure 3, which correspond, with one exception, to the sucrose- , HCl- , and NaCl-best fibers, respectively. Thus, within the hamster CT nerve, classification of fibers by using 4 prototypical stimuli produces the same classification as a hierarchical cluster analysis based on response profiles across 13 stimuli. Similar classes emerged from hierarchical clustering of second- and third-order brainstem neurons in the hamster when as few as three[31] or as many as 18 stimuli were applied to the anterior portion of the tongue.[33] The stimulus array must, however, include at least one example of three stimulus classes: (1) sweeteners, (2) sodium salts, and (3) nonsodium salts and acids (see below).

3. Across-Fiber Patterns

When an array of 13 stimuli was applied to the hamster's fungiform papillae, the responses of the CT fibers reflected the perceptual similarities and differences among the stimuli, as reflected in studies of behavioral generalization among taste stimuli by hamsters.[50-52] As previously demonstrated for rat CT fibers,[58,59] stimuli with similar taste quality produce patterns of activity across hamster CT fibers that are highly correlated. The across-fiber patterns for 13 stimuli are shown in Figure 4, where it can be seen that stimuli with similar taste produce similar patterns of activity. For example, the correlation between the patterns produced by sucrose and Na-saccharin was +0.89, between NaCl and $NaNO_3$ was +0.93, and between KCl and NH_4Cl was +0.91. Within this stimulus array, three distinctly different patterns are seen. One is elicited by sucrose, fructose, and Na-saccharin. A second pattern is elicited by NaCl and $NaNO_3$. A third, more variable pattern is evoked by the remaining stimuli, including the acids, nonsodium salts, urea, and QHCl. Within these patterns, the most responsive neurons for a particular group of stimuli tend to fall within one of the best-stimulus classes of fibers. For example, for the patterns evoked by sucrose, fructose, and Na-saccharin, the sucrose-best fibers *(fibers 1 to 10)* are the most responsive. For the sodium salts, the NaCl-best fibers *(fibers 11 to 32)* are most responsive; however, the HCl-best fibers are often quite responsive as well. For the nonsodium salts and acids the HCl-best fibers *(fibers 33 to 40)* are most responsive, although HCl also activates NaCl-best fibers. Thus, within the activity elicited in the CT nerve, the responses of particular sets of fibers (S, N, or H fibers) typically dominate the patterns evoked by particular sets of stimuli (sweet-tasting, sodium salts, or nonsodium salts and acids).

4. Amiloride Sensitivity

The organization of sensitivities within the CT nerve suggests that at least three kinds of receptor mechanisms provide input to fibers of this nerve. These transduction mechanisms might be expected to define the spectrum of chemicals to which sucrose- , NaCl- , and HCl-best nerve fibers respond.

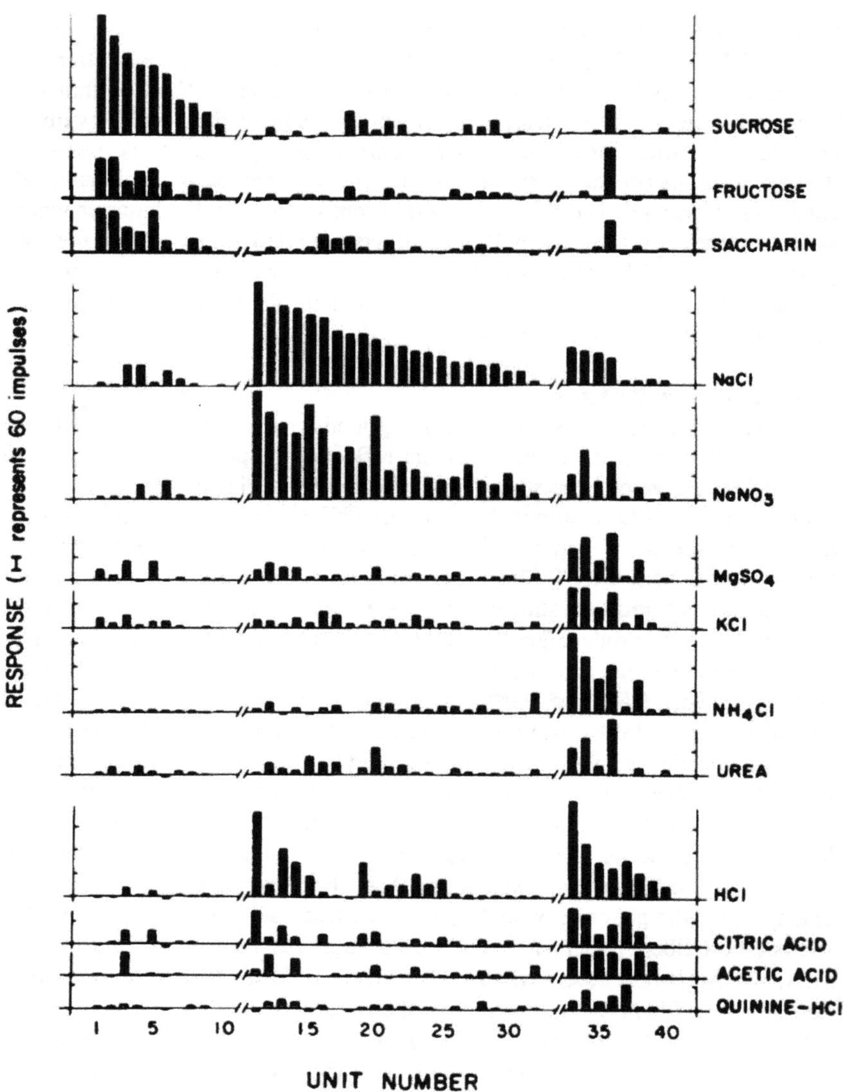

FIGURE 4. Across-fiber patterns for hamster CT fibers. The response rates elicited for 5 s in CT *fibers 1 to 40* (Figure 2) are represented by consecutive filled bars from left to right. Each row depicts the pattern elicited by the stimulus indicated to the right of the horizontal line representing the zero-response rate. These horizontal lines are broken twice with two short parallel diagonals at the transitions between sucrose-best and NaCl-best (between *fibers 10* and *11*) and NaCl-best and HCl-best (between *fibers 32* and *33*) fibers. Fiber numbers are indicated along the abscissa. The short horizontal lines parallel to the zero-rate line interrupting the ordinates indicate successive increases in the response rate of 60 nerve impulses. The bars extending below the zero line indicate response rates lower than the spontaneous rate. (From Frank, M. E., et al., *J. Gen. Physiol.*, 91, 861, 1988. With permission.)

There is a growing body of literature that suggests that one receptor mechanism for sodium salts is an amiloride-blockable sodium channel on the apical membrane of taste receptor cells.[60-65] Application of amiloride to the tongue reduces the response to NaCl in single CT nerve fibers of hamsters and rats[66-68] and also responses of the whole CT nerve in several mammalian species.[63,66,69-72] These data imply that taste stimulation by sodium salts may involve entry of the stimulus into taste receptor cells through an amiloride-sensitive epithelial ion channel.[61]

Recent studies have shown that afferent input from the transduction of sodium salts by this amiloride-sensitive channel is distributed to one class of taste fibers in both the rat[68] and hamster.[66] Although not seen in a study by Hill,[67] possibly due to the nonspecific effects of higher amiloride concentrations, amiloride primarily affected NaCl responses in NaCl-best fibers. Responses to NaCl in HCl-best fibers, which are often substantial[55] (see Figures 1, 2, and 4 above), were unaffected by amiloride treatment. In the hamster, the response to NaCl after amiloride treatment was at a mean of 5.1% of control rates in 16 NaCl-best fibers, but the response of 10 HCl-best fibers to NaCl was unaffected by amiloride treatment.[66] Figure 5 shows the mean response of these samples of NaCl- and HCl-best hamster CT fibers to NaCl before and after amiloride treatment. Even though the mean response to 0.1 *M* NaCl in HCl-best fibers was about half that evoked in NaCl-best fibers, amiloride suppressed the response only in the NaCl-best units. Thus, the receptors that utilize the amiloride-blockable transduction mechanism appear to activate NaCl-best fibers nearly exclusively. Recent studies of cells in the rat NST show that this segregation of amiloride-sensitive input to NaCl-best cells is maintained in the second-order neurons.[73,74] This segregation of a particular receptor input to one class of afferent fibers and to the cells in the brainstem to which these fibers project provides additional support for gustatory neuron types at both peripheral and brainstem levels.

B. GLOSSOPHARYNGEAL FIBERS

Studies of the responsiveness of the whole glossopharyngeal (IXth) nerve in a number of mammalian species have suggested that its sensitivities to gustatory stimuli are different from those of the CT nerve.[75-80] For example, the IXth nerve of the rat contains many fibers that are tuned to respond best to QHCl or to sucrose, which are poor stimuli for the rat CT nerve.[38] Responses of cells in the rat petrosal ganglion, which contains the cell bodies of the IXth nerve afferent fibers, also suggest a greater responsiveness to QHCl and sucrose than is seen in the CT nerve.[37] Similarly, the sensitivities of hamster IXth nerve fibers are strikingly different from those of the hamster CT nerve.[39]

1. Organization of Sensitivities

Fibers of the hamster's IXth cranial nerve are predominantly responsive to HCl and QHCl and show much less responsiveness to sucrose and NaCl

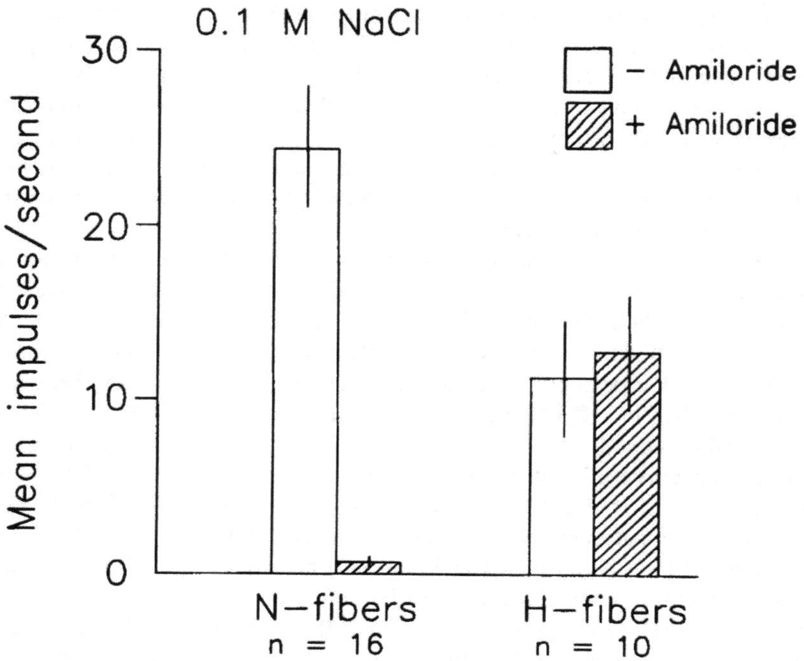

FIGURE 5. Mean impulses per second in 16 NaCl-best and 10 HCl-best fibers of the hamster's CT nerve evoked by 0.1 *M* NaCl after treatment of the tongue with distilled water (– amiloride, open bars) or 10 μ*M* amiloride (+ amiloride, hatched bars). Error bars represent ± 1 SEM. (Data from Hettinger, T. P. and Frank, M. E., *Brain Res.*, 513, 24, 1990. With permission.)

over a wide range of concentrations.[39] The mean responses of 56 hamster IXth nerve fibers innervating the foliate papillae and of 27 fibers innervating the vallate papilla are shown in Figure 6. The responses of each of these 83 fibers were recorded to stimulation of the appropriate receptive field with an array of 20 stimuli: 5 concentrations each of sucrose, NaCl, HCl, and QHCl. A single IXth nerve fiber innervated either the vallate or the foliate papillae, but never both, since flowing solutions into the trenches of these separate papillae never stimulated the same fiber; this is also true of rat IXth nerve fibers.[38] Sensitivity to QHCl is greater in the IXth nerve than in the CT, but the threshold for QHCl in hamster IXth nerve fibers is relatively high (between 1 and 3 m*M*). This threshold, however, corresponds to the point where a clear aversion for QHCl emerges in two-bottle preference experiments on hamsters.[81] The stimulus concentrations that were half maximal for the CT nerve (0.1 *M* sucrose, 0.03 *M* NaCl, 3 m*M* HCl, and 1 m*M* QHCl) were not very effective stimuli for the IXth nerve, which in itself suggests that the taste receptors differ between the anterior and posterior tongue. The response measure employed to examine the organization of IXth nerve sensitivities was the sum of the responses to all five concentrations (total response), which is

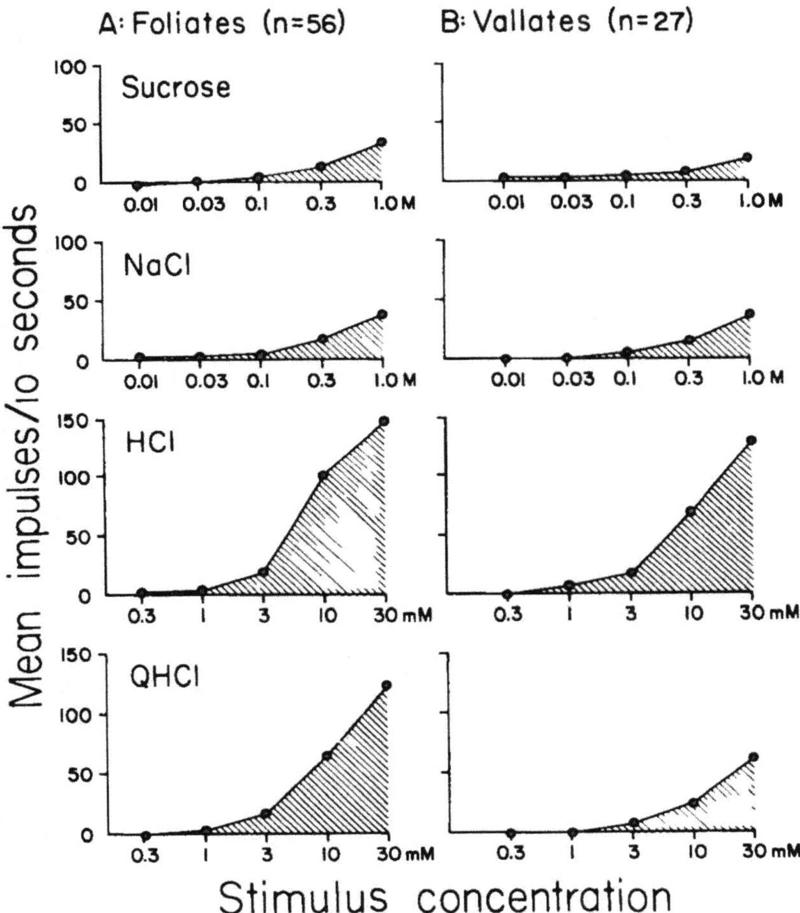

FIGURE 6. Mean concentration-response functions (nerve impulses per 10 s) for 56 hamster IXth nerve fibers innervating the foliate papillae and 27 IXth nerve fibers innervating the vallate papilla. Responses are shown across five concentrations. The shaded area is proportional to the sum of the responses across these concentrations (total response), which was used as a response measure for each fiber. (Reprinted from Hanamori, T., et al., *J. Neurophysiol.*, 60, 478, 1988. With permission.)

proportional to the area under the concentration-response function (shaded in Figure 6). This measure ("best" total response) likely gives a more accurate characterization of the sensitivities of a neuron than the response to a single concentration, since any best-stimulus classification depends upon the choice of concentrations.[35,36,55,82] Classification of IXth nerve fibers using this total response measure resulted in 8 sucrose- , 4 NaCl- , 52 HCl- , and 19 QHCl-best fibers. Most units (>90%) were similarly classified using responses to

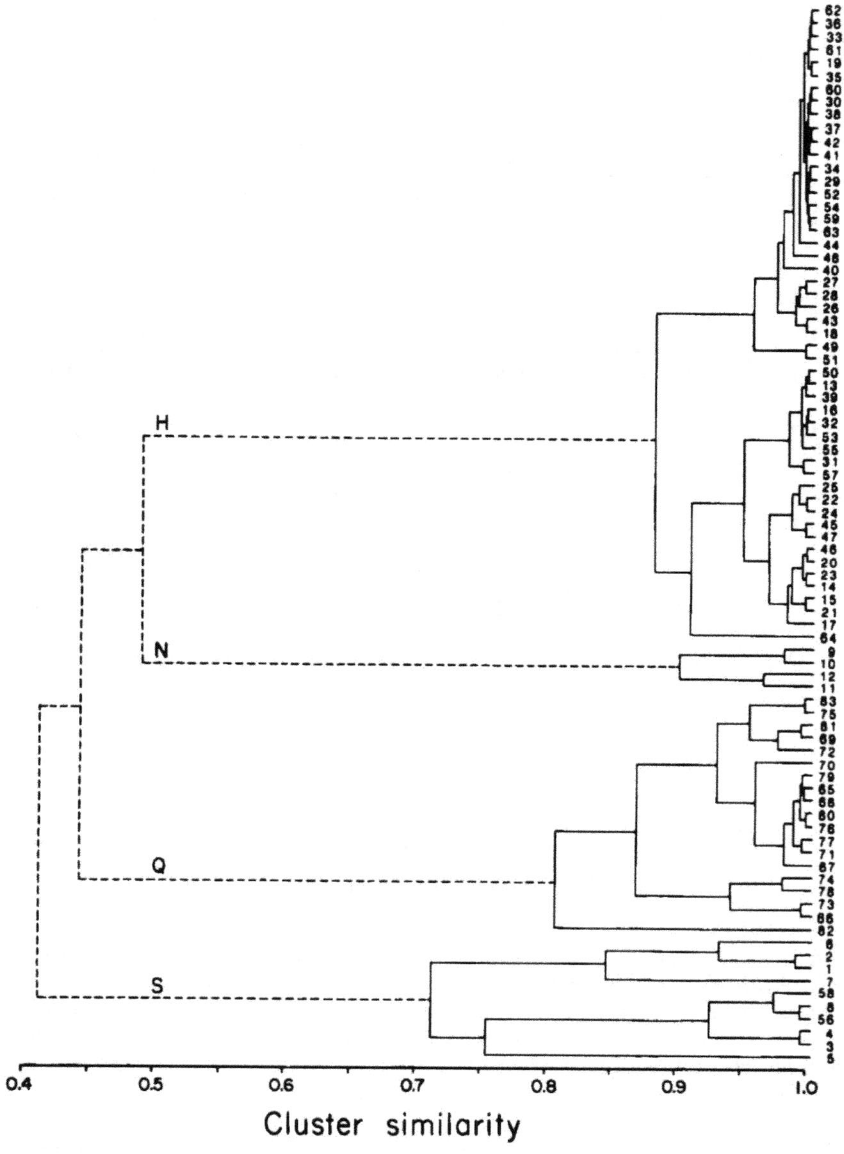

FIGURE 7. Cluster analysis of hamster IXth nerve fiber response profiles. The parallel horizontal lines of the dendrogram represent profiles, or groups of profiles, of fibers indicated at the right by fiber numbers. *Fibers 1 to 8* are sucrose-best, *fibers 9 to 12* are NaCl-best, *fibers 13 to 64* are HCl-best, and *fibers 65 to 83* are QHCl-best. The major profile clusters (H, N, Q, S) are identified on the left side of the dendrogram. The amalgamation of the major clusters is indicated with dashed lines. The similarities between profiles (or groups of profiles) are obtained by projecting the vertical lines to the similarity scale along the abscissa. (Modified from Hanamori, T., Miller, I. J., Jr., and Smith, D. V., *J. Neurophysiol.*, 60, 478, 1988. With permission.)

0.3 *M* sucrose, 0.3 *M* NaCl, 10 m*M* HCl, and 10 m*M* QHCl, which are midrange for the hamster's IXth nerve (see Figure 6).

The results of a hierarchical cluster analysis of the fibers' response profiles (total responses) are shown in Figure 7. This dendrogram depicts the order of clustering of the 83 IXth nerve fibers, proceeding from those with the most similar profiles to those that are most different. Fiber numbers indicate the best stimulus for each fiber, with *fibers 1 to 8* being sucrose-best, *9 to 12* NaCl-best, *13 to 64* HCl-best, and *65 to 83* QHCl-best. The cluster similarities (correlations) shown along the abscissa decrease gradually during the initial stages of the clustering process and then become markedly smaller as more dissimilar groups of fibers are clustered together. On the basis of these similarities, four groups of fibers (labeled H, N, Q, and S) are suggested. These clusters consist of one group composed entirely of HCl-best fibers (H, n = 50), one entirely of NaCl-best fibers (N, n = 4), one entirely of QHCl-best fibers (Q, n = 19), and another containing all of the sucrose-best fibers and two HCl-best fibers (S, n = 10). Thus, as with the CT data, cluster analysis of response profiles and best-stimulus designation yield nearly identical results. The four clusters are joined in the clustering process only at correlations <0.5 and these amalgamations are depicted in Figure 7 with dashed lines. The mean response profiles for the best-stimulus classes based on the best total response for these 83 IXth nerve fibers are shown below in Figure 8A.

2. Comparison of IXth Nerve Fibers to CT Fibers

Based on stimulation of the anterior tongue with a single midrange concentration of each of four stimuli, Frank[29] classified 79 hamster CT fibers into 1 of 4 classes: sucrose-best (n = 20), NaCl-best (n = 42), HCl-best (n = 17), or QHCl-best (n = 1). Thus, the majority of hamster CT fibers respond best to sucrose or NaCl. On the other hand, most hamster IXth nerve fibers respond best to HCl or QHCl. A comparison of the mean response profiles of CT and IXth nerve fibers is shown in Figure 8, which demonstrates differences in responsiveness in terms of both the numbers of fibers in the various best-stimulus classes and in their relative responsiveness to the four stimuli. Fibers in the IXth nerve are mostly responsive to HCl or QHCl (Figure 8A), whereas those in the CT nerve respond predominantly to sucrose or NaCl (Figure 8B). The response profiles shown in Figure 8 are not based on stimulation by the same stimulus concentrations. The total response of IXth nerve fibers across all five concentrations is compared in Figure 8 with the response to a single concentration of each stimulus for CT fibers. Note that maximal mean rates for the total response of IXth nerve fibers are nearly four times the rates seen to single concentrations in the CT. Nevertheless, the relative differences between these two nerves in their responsiveness to sucrose, NaCl, HCl, and QHCl are quite clear.

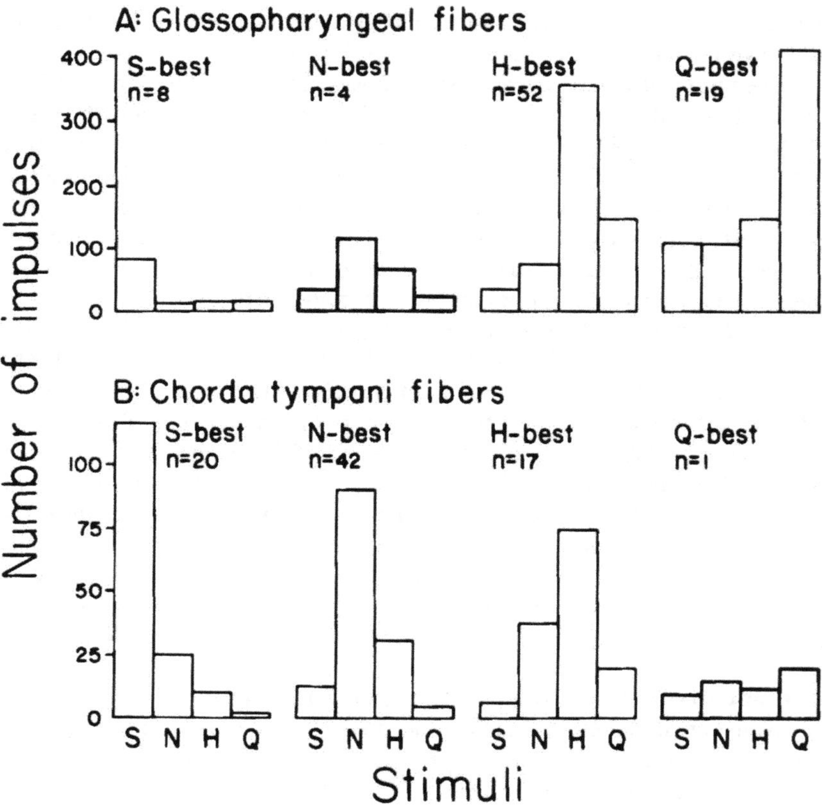

FIGURE 8. Mean response profiles for 83 fibers in the hamster IXth nerve (A) and 79 fibers in the hamster CT nerve (B). Number of impulses is the total number of impulses in 10 s across 5 concentrations in A and number of impulses in 5 s to a single midrange concentration in B. (From Hanamori, T., et al., *J. Neurophysiol.*, 60, 478, 1988. With permission.)

In several mammalian species there is relatively poor sensitivity to QHCl in the CT nerve.[29,30,32,48] On the other hand, the IXth nerve is predominantly responsive to QHCl in several species, including rats,[38] gerbils,[77] mice,[78] and sheep.[75] This predominant QHCl sensitivity in the IXth nerve is also evident in fish and amphibians. The IXth and Xth nerves in catfish supply taste buds in the oral and pharyngeal epithelium and are similar to VIIth nerve fibers of the catfish in their response to amino acids.[83] However, fibers in the IXth and Xth nerves are much more sensitive than VIIth nerve fibers to QHCl.[83] Similarly, IXth nerve fibers of several amphibian species show very low thresholds for quinine.[84-86] Thus there is a general trend across vertebrate species for fibers in the IXth nerve to be relatively more responsive to quinine than those in the VIIth nerve.

3. Amiloride Insensitivity

The hamster IXth nerve contains very few neurons that are specifically sensitive to sodium salts; only 4 of 83 fibers were classified as NaCl-best (Figures 6 to 8). Similarly, there were no fibers seen in the rat IXth nerve specifically tuned to NaCl,[38] although NaCl was effective in driving units most responsive to acids and nonsodium salts. The ability of rats and hamsters to detect[87] and recognize[88] sodium salts is disrupted if the CT nerves are cut, but these behaviors toward NaCl are not disturbed following bilateral transection of only the IXth nerve.[89] Treatment of the tongue with amiloride during conditioning to avoid NaCl interferes with the ability of rats to make subsequent discriminations between sodium and nonsodium salts.[90] Mixing NaCl with amiloride also increases the intake of normally nonpreferred NaCl solutions in hamsters.[66] Amiloride has a specific effect on NaCl-best units in the CT nerve of these species.[66,68] Consistent with the failure of IXth nerve transection to affect the behavior toward sodium salts and also with the paucity of NaCl-best fibers in the IXth nerve is the fact that amiloride treatment has no effect on the rat's IXth nerve response to NaCl.[91] Taken together, these data imply that the sensory input necessary for the discrimination of sodium salts from other stimuli may be carried primarily in the CT nerve. In fact, the specific effects of amiloride on NaCl-best CT fibers suggests that the information carried by NaCl-best fibers is necessary for the discrimination of sodium salts from other stimuli. Furthermore, the specific effects of amiloride on NaCl-best NST neurons have been taken as evidence that these cells constitute a labeled line for the coding of the salty taste.[73,74,92] At the very least, amiloride-sensitive afferent information about sodium stimuli is segregated to a subset of NST cells, even though convergence occurs between peripheral fibers and NST neurons.[42,93] The implications of this segregation for the coding of taste quality will be discussed below.

C. SUPERIOR LARYNGEAL FIBERS

The responsiveness of afferent fibers in the SLN to chemical stimulation of the larynx has been studied in a number of species. Electrophysiological studies that have directly compared the responsiveness of the SLN to that of other gustatory nerves have shown some striking differences in their sensitivities.[44-46,78,94,95] These differences are described below for fibers of the hamster's SLN.[45,46]

1. Organization of Sensitivities

Responses of 65 single fibers in the SLN of the hamster were recorded following stimulation of the laryngeal epithelium with the same array of stimuli employed in the study of the IXth nerve (see Figure 5 above) plus distilled water.[46] Because of the significant response to distilled water in these fibers, each stimulus except the NaCl concentration series was dissolved in 0.154 *M* NaCl, which produces a minimal discharge in SLN fibers. The mean

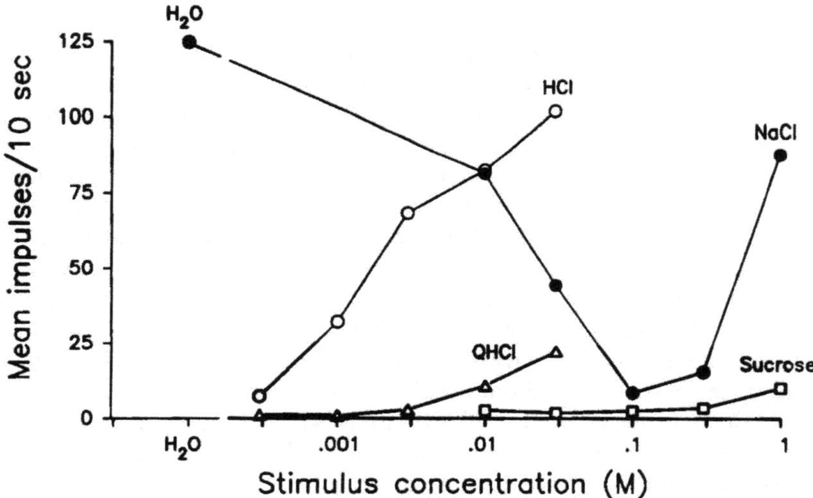

FIGURE 9. Mean concentration-response functions (impulses per 10 s) for 65 fibers in the hamster SLN. All 65 fibers were stimulated with every stimulus concentration. Response to distilled H_2O is shown as the lower extreme of the NaCl series, which was dissolved in distilled H_2O. All other stimulus series were dissolved in 0.154 *M* NaCl, which was used as a rinse before and after each stimulus presentation. (From Smith, D. V. and Hanamori, T., *J. Neurophysiol.*, 65, 1098, 1991. With permission.)

responses to these 21 stimuli across these 65 fibers are shown in Figure 9, where it is evident that distilled water, HCl, and high concentrations of NaCl are much more effective stimuli than QHCl or sucrose. These fibers responded most to distilled water and as the concentration of NaCl was increased (i.e., as water was "diluted") there was a decreasing response of SLN fibers, reaching a minimum around the adapting concentration of 0.154 *M* NaCl. Further increases in NaCl concentration produced increasing levels of response in these fibers. Similar concentration-dependent relationships have been shown for SLN fibers in other species.[96-99] Since increasing the concentration of NaCl but not of non-chloride sodium salts decreases the response to water, it has been proposed that the response to water is mediated through the outward movement of Cl^- ions through the receptor membrane.[96,97] However, there are as yet no direct electrophysiological studies on transduction by laryngeal chemoreceptors to support this hypothesis.

Responses of fibers in the hamster's SLN primarily reflect the input from three receptor mechanisms: water, acid, and sodium. Classifying these 65 SLN fibers according to which of five stimuli (distilled water, 1.0 *M* NaCl, 0.03 *M* HCl, 0.03 *M* QHCl, or 1.0 *M* sucrose) was the most effective stimulus resulted in 26 water- , 17 NaCl- , 20 HCl- , and 2 QHCl-best fibers. These concentrations were chosen for the best-stimulus classification because of the complex relationship in the water-NaCl function (Figure 9) and the general lack of responsiveness to QHCl and sucrose at all but the strongest concentrations. At these intensities, most fibers responded best to water, HCl, or NaCl rather than to QHCl or sucrose. Within these best-stimulus classes, fibers were not specifically tuned to their best stimulus. Water-best fibers (which also responded to hypotonic NaCl solutions) were also responsive to HCl and hypertonic NaCl. Similarly, NaCl-best fibers also responded moderately to water (and hypotonic NaCl) and to strong HCl. Fibers responding best to HCl also responded moderately to water and NaCl and somewhat to strong QHCl. These best-stimulus classes of cells were not as distinct from one another as those in the hamster CT or IXth nerve.[46]

2. Continual Distribution of Fiber Types

Given the range of sensitivities shown by most chemosensory fibers, classification by their best stimulus will always provide some grouping of the fibers (see also Reference 100). For fibers in the CT and IXth nerves, the best-stimulus classification is corroborated by hierarchical cluster analysis, which shows that similarities and differences in the shapes of the response profiles are consistent with such a classification (see Figures 3 and 7 above). Although it has been argued that hierarchical clustering will always produce clusters,[100] fibers of the SLN provide a case in which no distinct clusters of fibers are evident. The results of a hierarchical cluster analysis of the response profiles of these 65 SLN fibers are shown in the dendrogram of Figure 10. In contrast to the hierarchical clustering of CT and IXth nerve fibers, the intercluster distances in this amalgamation gradually increase, but there is never a sharp increase in the magnitude of this distance, even at the last few steps of the clustering process. This gradual progression of the cluster distances indicates that these clusters are no more distinct than are individual fibers within the clusters. However, for purposes of discussion, some of the clusters within this hierarchy have been identified (1, 2, and 3). The best-stimulus designation of each fiber is indicated to the right of each fiber number. Within the hierarchical arrangement of the fibers, they tend to group together on the basis of the patterns of their sensitivities, as expected. For example, cluster 1 includes 2 subclusters, one of which includes only water-best fibers. The other subcluster contains mostly HCl- and water-best fibers. Cluster 2 contains only HCl-best fibers, and cluster 3 contains 2 subclusters, one of which contains only NaCl-best fibers and the other of which contains both NaCl- and HCl-best fibers. The two QHCl-best fibers *(fibers 64 and 65)* are separate from these three clusters and are grouped together very late in the hierarchy, indicating that they are not very similar to one another. Because of the gradual stepwise increase in the cluster distances shown in this dendrogram, one would conclude that these clusters of fibers are not markedly different. Rather, they appear to represent a continuum of response profiles, each somewhat different from the others. A hierarchical cluster analysis of 52 hamster SLN fibers to a wider array of 20 stimuli resulted in a similar lack of distinct fiber types.[45]

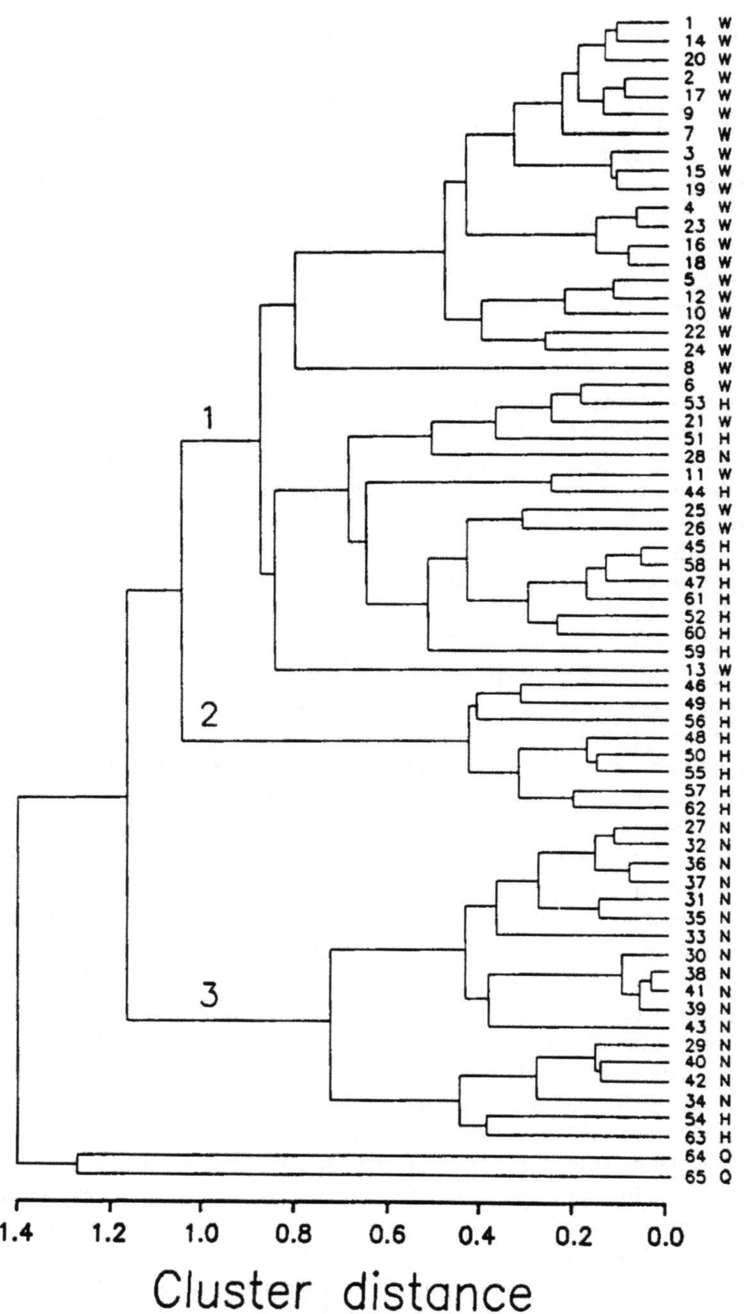
1
2
3
1 W
14 W
20 W
2 W
17 W
9 W
7 W
3 W
15 W
19 W
4 W
23 W
16 W
18 W
5 W
12 W
10 W
22 W
24 W
8 W
6 W
53 H
21 W
51 H
28 N
11 W
44 H
25 W
26 W
45 H
58 H
47 H
61 H
52 H
60 H
59 H
13 W
46 H
49 H
56 H
48 H
50 H
55 H
57 H
62 H
27 N
32 N
36 N
37 N
31 N
35 N
33 N
30 N
38 N
41 N
39 N
43 N
29 N
40 N
42 N
34 N
54 H
63 H
64 O
65 O
1.4
1.2
1.0
0.8
0.6
0.4
0.2
0.0
Cluster distance

FIGURE 10.

The organization of sensitivities reflected in the response profiles of these 65 SLN fibers was examined further using principal axis factor analysis, from which three factors emerged that accounted for 96.5% of the data variance.[46] These factors represented idealized profiles showing specific responses to water, HCl, or NaCl. However, examination of the correlations of each individual profile with these three factors demonstrated that many of them were highly correlated with more than one factor. This is in marked contrast to a similar analysis of the profiles of CT fibers, where most fibers correlated predominantly with only a single factor.[30] This factor analysis and the hierarchical cluster analysis (Figure 10) both support the notion that chemosensory SLN fibers cannot be easily categorized into fiber types, as can chemosensory fibers in the CT and IXth nerves of the hamster.

3. Inability to Discriminate Among Taste Qualities

Studies of generalizations of learned aversions have begun to define taste quality for a number of mammalian species, including rats and hamsters.[50-52,58,90,101-105] These rodents easily discriminate among sucrose, NaCl, HCl, and QHCl; they group the tastes of other sugars and Na-saccharin with sucrose, other acids, and some nonsodium salts like NH_4Cl with HCl, other sodium salts with NaCl, and some bitter-tasting salts like $MgSO_4$ with QHCl. Input from the fungiform papillae of the hamster is sufficient to allow neural and behavioral discrimination among sugars, sodium salts and acids.[30,50] In the rat, discrimination of sugars from other stimuli may depend primarily on input from the GSP nerve.[41,87,106] However, neural activity in the CT nerve does not discriminate well between QHCl and the nonsodium salts and acids; this discrimination is much more dramatic in fibers of the IXth nerve.[38] Thus the behavioral distinctions among stimuli with different taste qualities may rely somewhat on input from different cranial nerves. Fibers of the SLN innervating laryngeal chemoreceptors, however, do not appear to discriminate among stimuli with different taste qualities.[45] A multidimensional scaling analysis of the similarities among the across-fiber patterns generated by 20 stimuli, each at a single concentration, did not produce a separation between stimuli of different quality, as seen in the responses of CT[30] or IXth nerve[38] fibers. Rather, the predominant dimension separated stimuli that were excitatory for SLN fibers, such as KCl, acids, and urea, from those that were inhibitory, such as $CaCl_2$ and the sugars.[45] Thus chemosensitive fibers of the SLN appear to be suited to a role in airway protection, by signaling deviations from the normal pH and ionic milieu of the larynx,[46,96,107,108] rather than in the discrimination among gustatory qualities.

FIGURE 10. Cluster tree (dendrogram) depicting the order of clustering for 65 hamster SLN fibers resulting from a hierarchical cluster analysis. Fiber numbers and best-stimulus designations (W, H, N, Q) are shown on the right. Distances between fiber profiles or clusters of profiles joined at each step are indicated along the abscissa. For purposes of discussion, 3 major clusters (1, 2, 3) are indicated; a scree analysis[57] of the cluster distances did not, however, suggest that these clusters are distinct. Modified from a figure in Smith and Hanamori (1991), with permission.

D. GREATER SUPERFICIAL PETROSAL NERVE

Taste buds in the nasoincisor ducts, along the "Geschmacksstreifen" (in the rat), and on the soft palate are innervated by fibers of the GSP nerve.[4,5] Although it has not yet been possible to analyze the responses of single gustatory fibers of the GSP, recent studies have examined the responsiveness of the whole GSP nerve in the rat[41] and in the hamster.[40] In addition, the responses of single neurons in the rat's NST to stimulation of the nasoincisor ducts and soft palate have been examined.[42,43] In general, the palate of both rats and hamsters appears to be more responsive to sweet-tasting stimuli than is the anterior tongue.

1. Integrated GSP Responses

Since the GSP nerve lies predominantly within bone, it is difficult to dissect for electrophysiological studies. To date, only integrated responses from the whole nerve have been examined.[40,41] The integrated responses of the hamster GSP and CT nerves to 0.1 *M* NaCl and to a series of 0.5 *M* sugars are shown in Figure 11. The tonic response to 0.1 *M* NaCl was a much larger proportion of the initial phasic discharge in the CT than in the GSP. By comparison, the responses to the sugars were 2 to 3 times greater in the GSP relative to the phasic response to NaCl (which is approximately equal for the CT and GSP in this figure); relative to the tonic response to NaCl, the GSP responses to the sugars are, of course, even greater. The integrated tonic responses of the hamster GSP were larger than those in the CT to these sugars, but were somewhat smaller for NaCl and for QHCl when averaged over their concentration ranges.[40]

2. Greater Responsiveness to Sugars

The relative difference in the response of the GSP to NaCl and sucrose, shown in Figure 11 for the hamster, is even more striking in the rat.[41] The CT nerve of the rat is relatively insensitive to sucrose and to other compounds that are described as sweet by humans.[49,55] This has traditionally been puzzling to taste physiologists, since rats show an avid preference for sweet-tasting compounds.[109-111] However, both the palate[41-43] and the posterior tongue[37,38] of the rat are more sensitive than the anterior tongue to sucrose and other sweet stimuli. Even in the hamster, whose CT nerve is relatively responsive to sucrose, the GSP is more responsive to sweet-tasting stimuli.[40]

E. DIFFERENTIAL GUSTATORY INPUTS TO THE BRAINSTEM

The data presented above for the four gustatory nerves of the hamster show that the various populations of taste buds differ in their sensitivities and contribute different kinds of afferent information to the brainstem. Although there are species differences in the distributions of some of these sensitivities, the general conclusion is that the various taste bud populations provide different kinds and amounts of gustatory information.[38-40,42,46] It is likely that these variable inputs are important for different kinds of taste-mediated behavior.

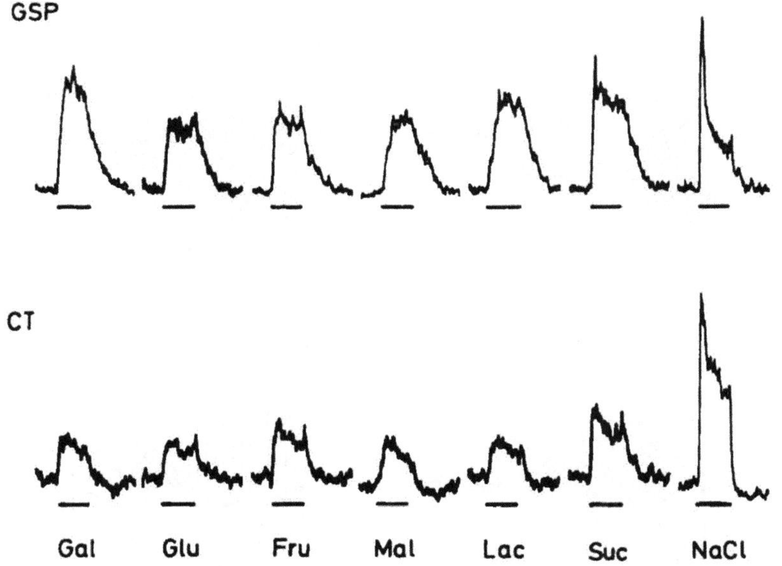

FIGURE 11. Integrated responses of the hamster GSP and CT nerves to six 0.5 *M* sugars and to 0.1 *M* NaCl. Stimuli were applied for a 10-s duration, shown by the horizontal mark below each integrated response. A distilled water rinse began at the end of the stimulus mark. Integrator time constant (RC) = 470 msec. (From Harada, S. and Smith, D. V., *Chem. Senses*, 17, 37, 1992. With permission.)

1. Comparison Among Gustatory Nerves

The relative responsiveness of the hamster's four gustatory nerves (CT, IXth, GSP, and SLN) to 0.3 *M* sucrose, 0.3 *M* NaCl, 10 m*M* HCl, 10 m*M* QHCl, and distilled water is shown in Figure 12. These particular concentrations were chosen for comparison because they were common to the several studies described above.[39,40,46] The figure depicts the response of each stimulus as a proportion of the total response to all five; thus the comparison among the nerves is relative. For the CT and GSP, the responses were the tonic integrated responses of the whole nerve.[40] The responses for the IXth nerve and the SLN were mean impulses per 10 s across 83 IXth nerve fibers[39] and 65 SLN fibers.[46] The differences in sucrose sensitivity are highlighted by the exploded portion of the pie charts. Receptors on the anterior tongue (CT) and palate (GSP) provide relatively more information about NaCl and sucrose than receptors on the posterior tongue (IXth) or in the larynx (SLN). Sensitivity to HCl is relatively similar in every gustatory nerve; QHCl clearly has its greatest relative effect in the IXth nerve. Only the SLN of the hamster responds to distilled water; this response occurred after adaptation to 0.154 *M* NaCl, but neither the CT, the IXth, nor the GSP nerve responds to water after NaCl adaptation (e.g., see Figure 11).

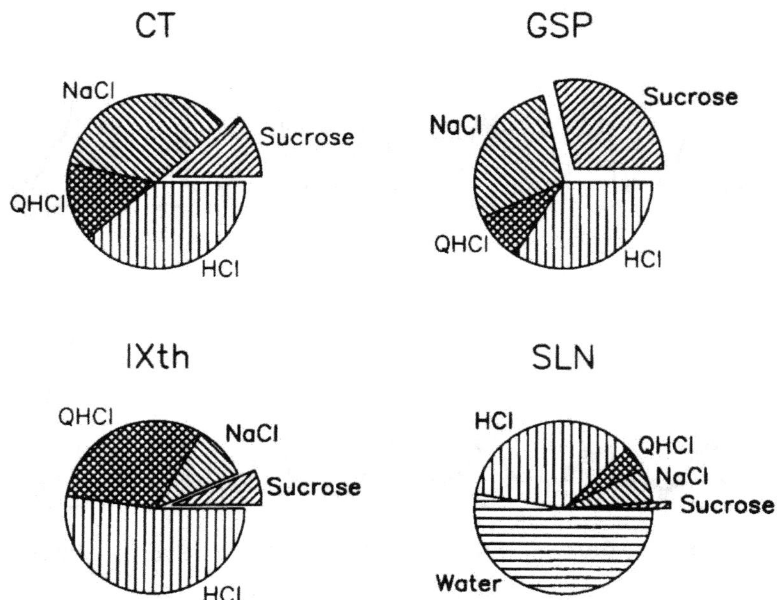

FIGURE 12. Proportional responses to 0.3 *M* sucrose, 0.3 *M* NaCl, 10 m*M* QHCl, 10 m*M* HCl, and distilled water in four gustatory nerves of the hamster. Data for the CT and GSP were integrated tonic responses;[40] those for the IXth and SLN are mean firing rates over a 10-s response period in 83 IXth nerve fibers[39] and 65 SLN fibers.[46] Each pie chart represents the response to each stimulus as a proportion of the sum of the responses of that nerve to all five stimuli. (From Harada, S. and Smith, D. V., *Chem. Senses*, 17, 37, 1992. With permission.)

2. Differential Inputs to the NST and Taste-Mediated Behaviors

Information arising from these various gustatory nerves projects into the nucleus of the solitary tract (NST) of the medulla (Travers, Chapter 13). Here second-order neurons give rise to ascending projections to the parabrachial nuclei (PbN) of the pons, which in turn sends projections to the thalamus and insular cortex and to widespread areas of the limbic forebrain.[112-114] There are also numerous connections of NST neurons to the oral motor nuclei via interneurons in the reticular formation.[115,116] These anatomical relationships, the differential sensitivities of the VIIth, IXth, and Xth nerves, and the contribution of gustatory afferent input to taste-mediated behaviors are summarized in the schematic diagram of Figure 13. Whereas taste physiologists have focused largely on the role of gustatory afferent fibers and central neurons in taste quality perception, there are a number of taste-mediated behaviors ranging from tongue movements to salivation to preabsorptive insulin release that have their neuronal substrate within the brainstem.[117-120] Input from the various gustatory nerves contributes differentially to these diverse taste-mediated behaviors.

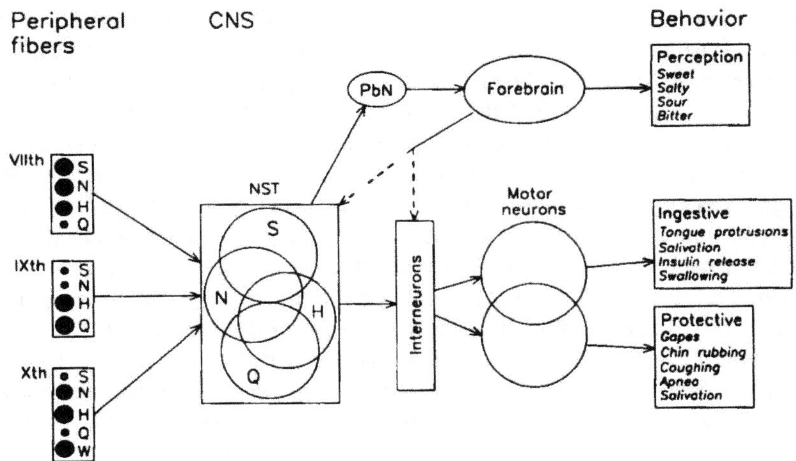

FIGURE 13. Schematic diagram of the chemosensory inputs of three cranial nerves to the taste-responsive portion of the nucleus of the solitary tract (NST) and their putative role in taste-mediated behaviors. The size of the filled circles for each of the peripheral nerves (VIIth, IXth, and Xth) depicts the relative responsiveness of these nerves to sucrose (S), NaCl (N), HCl (H), QHCl (Q), and water (W). Sensitivities of NST cells are largely overlapping, with each cell type somewhat responsive to two or three of the basic stimuli.[35] Sucrose and QHCl stimulate few of the same NST cells, however. Output from the NST ascends in the classic taste pathway to give rise to perceptions of sweetness, saltiness, sourness, and bitterness and to hedonic tone (not depicted). Local reflex circuits within the brainstem control ingestive and protective responses evoked by taste stimulation. Behavioral data suggest that both ingestive and protective responses can be triggered in parallel, depending upon the quality of the stimulus.[123,125]

One way of viewing taste is as the rostral component of a visceral afferent system, which includes gustatory, respiratory, cardiovascular, and gastrointestinal functions.[121] Taste buds innervated by the VIIth, IXth, and Xth nerves contribute differentially to this visceral continuum, with VIIth nerve fibers responsive primarily to preferred stimuli like sucrose and NaCl, IXth nerve fibers most sensitive to aversive stimuli like HCl and QHCl, and Xth nerve fibers responsive to stimuli that deviate from the normal pH and ionic milieu of the larynx (Figures 12 and 13). Sucrose, for example, stimulates predominantly fibers of the VIIth nerve. These fibers project into the NST, where cells that are sucrose-best are more broadly tuned than CT fibers in the hamster; many of these cells also respond to NaCl and to HCl, but are often inhibited by QHCl.[35,122] Ultimately, the output of these second-order neurons ascends to the forebrain to give rise to the perception of sweetness (Figure 13). At the same time, these cells provide input to motor systems that drive the ingestive components of feeding behavior, including rhythmic mouth movements, tongue protrusions, lateral tongue protrusions, salivary secretion, insulin release, and swallowing.[117,120,123] Conversely, QHCl stimulates predominantly fibers of the IXth nerve. These fibers project into the NST, where they drive cells that are also responsive to HCl and NaCl but not to sucrose.[73,93] Quinine-sensitive cells of the NST send ascending projections to the forebrain to give rise to sensations of bitterness (Figure 13), but they also provide input to motor systems that drive protective behaviors like gaping, chin rubbing, forelimb flailing, locomotion, and fluid rejection.[117,123] In the rat, the number of gapes elicited by quinine stimulation is reduced by almost one half after bilateral transection of the IXth nerve.[124] Sucrose and quinine produce very different patterns of ingestive and protective taste reactivity,[117] and a combination of these behaviors can be triggered by mixtures of sucrose and quinine.[125] Similarly, in the hamster the patterns of taste reactivity to sucrose and quinine are quite different, whereas sodium salts and acids produce patterns consisting of combinations of both ingestive and protective behaviors.[123] The taste quality of the stimulus is very directly related to the specific pattern of taste reactivity that is elicited in the hamster.[123]

The superior laryngeal branch of the Xth nerve is involved in swallowing, airway protection, and a number of other visceral reflexes. Respiratory apnea is produced by laryngeal stimulation with water,[96,108] and chemosensory fibers of the rat SLN have been shown to mediate diuresis in response to stimulation of the laryngeal mucosa with water.[126] Besides its obvious role in controlling ingestive behavior, taste also triggers a number of metabolic responses, including salivary, gastric, and pancreatic secretions,[120] although the specific contributions of particular cranial nerves to these responses are not well understood. Thus, in addition to their mediation of gustatory sensation, taste buds may have a number of roles related to gustatory-visceral regulation, depending upon their peripheral distribution and innervation.

The differential role of these cranial nerves in controlling ingestive behavior is seen most clearly in studies of fish.[18,127] Catfish that were blinded and had their olfactory epithelium destroyed could localize a source of food, suggesting that taste can serve in fish as a "distance" receptor, much as olfaction does in terrestrial vertebrates.[127] This distance sense is mediated by the VIIth nerve, as demonstrated in selective ablation experiments.[18] Removal of the entire sensory area of the facial lobes produces a fish that is alert to the presence of food in its environment, but cannot localize the food or pick it up. An ablated fish will sometimes hover over food and make swallowing movements, but it cannot pick up the food into its mouth. If food is placed in the mouth, swallowing is normal. In fish with the vagal lobes ablated, food-seeking and pickup behavior are completely normal. However, the fish is unable to swallow food once it is in the mouth. Thus, the facial taste system apparently controls the localization and pickup phases of feeding, whereas the vagal system controls the swallowing phase. These differential roles are mediated in part by the brainstem connections of the facial and vagal lobes.[19,20] The situation in mammals is anatomically more complex, involving considerable overlap among the brainstem projections of the VIIth, IXth, and Xth nerves and location of all the taste buds within the oropharyngeal cavity, unlike catfish where the facially innervated taste buds are all external. Nevertheless, it is likely that these separate cranial nerves provide differential control over various aspects of ingestive behavior, even in mammals.[128]

III. CODING OF TASTE QUALITY

A. TASTE CODING THEORIES

The nature of the neural representation of taste quality has been debated for many years.[49,53,54,59] Recently, there has been considerable discussion about the way in which the nervous system extracts and codes information about taste quality[30,32-34,52,56,74,92,100,129-132] and about the nature of the taste qualities themselves.[51,92,133-136] The coding issue rests on whether the activity in a given sensory neuron is an unambiguous representation of the quality of the stimulus applied to its receptors or whether this activity is meaningful only in the context of activity in other afferent fibers. The labeled-line hypothesis of taste quality coding[32,74,92,131] suggests that activity in a particular fiber type represents a specific taste quality, whereas the across-fiber pattern theory[49,53,54,59,100,129,130] holds that a particular pattern of activity across the entire ensemble of afferent fibers represents a taste quality. A variant of the across-fiber pattern theory suggests that quality is coded by a comparison of activity across more than one fiber type.[34,56]

1. Multiple Sensitivities of Gustatory Fibers

Both peripheral and central gustatory neurons in a variety of mammalian species typically respond to more than one of the stimuli representing the sweet, salty, sour, or bitter taste qualities.[30,33-36,38,39,48,49,53,59,74] For example, in Figure 4 (above), which shows the patterns of activity generated by 13 stimuli across 40 hamster CT fibers, it can be seen that sucrose-best cells *(fibers 1 to 10)* are sometimes driven by salts or acids, and NaCl-best cells *(fibers 11 to 32)* are sometimes driven by sugars or acids; this multiple sensitivity occurs to an even greater degree in HCl-best cells *(fibers 33 to 40)*. Hamster taste neurons are somewhat more broadly tuned in the NST and PbN than in the CT nerve,[35,36,122] and, even in the periphery, taste fibers become more broadly tuned as the stimulus concentrations are increased.[39] Since the responses of taste fibers can be modulated by both quality and intensity, the response of any one neuron alone is ambiguous with respect to either parameter.[35,36,49,54] Gustatory neurons are also sometimes responsive to temperature[48,137] and tactile stimuli.[138,139] Thus, it is important to remember that impulse traffic in a single neuron may be related to a number of stimulus parameters, making the unambiguous interpretation of that signal impossible without comparing it to activity in other cells.[129,130,140,141] Therefore, in thinking about the way in which sensory information is coded in the gustatory system, it is important to keep in mind that cells at all levels of the pathway are broadly responsive to stimuli that vary in perceptual quality, are more broadly responsive at high than at low intensities, and are often sensitive to other modalities, like touch and temperature.

2. Across-Fiber Patterns

It was the multiple sensitivity of fibers in the CT nerve that first led Pfaffmann[49,53,54] to propose that taste quality is coded by the pattern of activity across taste fibers. With this coding hypothesis, taste quality can be invariant with increased intensity even though any single neuron would show an increased breadth of responsiveness. The pattern of activity generated across the entire array of taste neurons at a higher concentration would be similar in shape, varying primarily in amplitude.[59,100,129,130,142] The across-fiber patterns to an array of 13 stimuli are shown above in Figure 4 for 40 hamster CT fibers. It is obvious in this figure that stimuli with similar tastes, such as the sweeteners or the sodium salts, generate highly similar patterns of activity across these fibers. These similarities are typically measured by calculating the across-fiber correlation between pairs of stimuli,[59] although other indices have been proposed.[143,144] Several behavioral investigations have demonstrated that stimuli that evoke highly correlated neural patterns are judged by experimental animals to have similar tastes.[51,52,58,101,103] This across-fiber pattern or ensemble view of quality coding makes the multiple sensitivity of gustatory neurons an essential part of the neural code for taste quality. This theoretical view stresses that the code for quality is given in the response of the entire population of cells,[100,129,130,141,145] placing little or no emphasis on the role of an individual neuron and does not require the existence of neuron types. In fact, in its purest form the ensemble theory insists that there be no fiber types.[100,130] Erickson[100,129,130,141,145] has argued that such a coding mechanism could operate for a number of sensory systems, particularly for nontopographic modalities employing neurons that are broadly tuned across their stimulus array.

When the across-neuron correlations are calculated for an array of gustatory stimuli across either peripheral or central neurons, stimuli with similar tastes correlate highly and those with different tastes correlate less. Almost every neurophysiological study that has taken this approach to analyzing the responses of gustatory cells has shown that the across-neuron patterns reflect the qualitative similarities among taste stimuli.[30,32,34-36,38,52,56,59,73,74,100,129,130,141,145] Often the across-neuron correlations are used as input to a multivariate statistical procedure in order to generate a "taste space" which represents the neural similarities and differences among the stimuli. A three-dimensional representation of the across-neuron correlations among 18 stimuli applied to the anterior tongue of the hamster is shown in Figure 14. This particular taste space was derived from the responses of 31 neurons in the hamster PbN and was generated using multidimensional scaling (KYST). Within this space, there is clear separation between the sweet-tasting stimuli (open circles), the sodium salts (Xs), the nonsodium salts and acids (solid circles), and the two

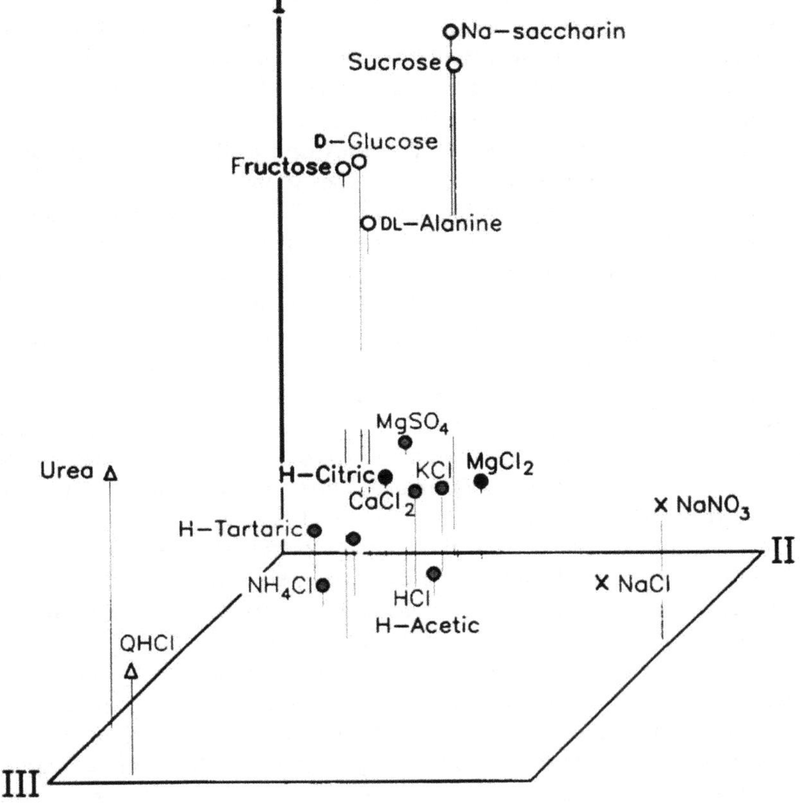

FIGURE 14. Three-dimensional "taste space" showing the similarities and differences among 18 stimuli delivered to the anterior tongue of the hamster. This space was derived from multi-dimensional scaling (KYST, Bell Laboratories) of the across-neuron correlations among these stimuli recorded from neurons in the parabrachial nuclei (PbN) of the hamster. Four groups of stimuli are indicated by different symbols: sweeteners, sodium salts, nonsodium salts and acids, and bitter-tasting stimuli. (Modified from Smith D. V., Van Buskirk, R. L., Travers, J. B., and Bieber, S. L., *J. Neurophysiol.*, 50, 541, 183. With permission.)

bitter-tasting stimuli (open triangles). This arrangement of stimuli based on similarities among their across-neuron patterns suggests that there is sufficient information within these patterns to discriminate among these four groups of stimuli, even though any one cell in the hamster PbN is likely to respond to stimuli of more than one group.[33,34,36]

3. Labeled Lines

Even though mammalian taste neurons are broadly tuned, many investigators have attempted to group them into functionally meaningful categories.[29,30,33,35-37,39,49,53,146,147] It is obvious from the discussion above about the organization of sensitivities in the hamster's CT and IXth nerves that it is possible to group these taste fibers into classes on the basis of their best stimulus if four stimuli representing different taste qualities are used. Indeed, a hierarchical cluster analysis of the similarities and differences in the shapes of their response profiles across a broader array of stimuli results in distinct groups of CT fibers[30] (Figure 3) that correspond to these "best-stimulus" groups. The implication of distinct fiber types in the coding of taste quality began with Frank's[29] categorization of hamster CT fibers into "best-stimulus" groups (see above). This categorization became the focus of an ensuing controversy over the neural representation of taste quality when it was suggested that these fiber types coded taste quality in a labeled-line fashion.[32,131] That is, that "sweetness" is coded by activity in sucrose-best neurons, "saltiness" by activity in NaCl-best neurons, etc. Thus, in contrast to a "population" approach to the understanding of gustatory neurobiology (across-neuron patterns), this labeled-line position advocates a "feature extraction" approach, in which particular neurons (or groups of neurons) play specific roles in the representation of taste quality. Pfaffmann[131] first proposed a labeled-line code for sweetness because activity in sucrose-best fibers in the squirrel monkey CT nerve correlated better with the animal's preference behavior toward a number of sugars than did activity in the whole nerve. Recently, on the basis of the specific effects of amiloride on the responses to sodium and lithium salts in rat and hamster NaCl-best CT fibers[66,68] and in rat NaCl-best NST neurons,[73,74] and on what appear to be the specific effects of sodium deprivation on NaCl-best fibers in the rat CT,[148] Scott and his colleagues[73,74,92] have argued that saltiness is coded by activity in NaCl-best neurons. This argument is based on the fact that following amiloride treatment of the tongue, rats cannot discriminate sodium from nonsodium salts[90] and that the across-neuron patterns within the NST are not distinct between sodium and nonsodium salts and acids.[73,74] These investigators argue, therefore, that activity in NaCl-best cells signals "saltiness" in a labeled-line fashion.

B. GUSTATORY NEURON TYPES: IS TASTE CODED BY LABELED LINES?

The labeled-line approach to taste quality coding requires the existence of neuron types, whereas the across-neuron pattern approach as originally proposed[129,130,141,145] does not. The number of labeled lines would equal the number of discrete taste qualities, which would each be signaled by activity in these separate afferent channels. Consequently, the question of the existence of gustatory neuron types has received a lot of recent attention. However, it is a misconception that the existence of fiber types (defined by their best stimulus, similarities in their profiles, or by other criteria such as their amiloride sensitivity) necessarily implies that these classes of cells comprise labeled lines. A classic example where receptor types are evident, but where there is general agreement about the existence of a pattern code, is in vertebrate color vision.[56,130,140] Color vision is based on three photopigments segregated into three types of cone receptors, which are broadly tuned across the visible spectrum.[149-151] A particular color is coded by the pattern of activation of the three types of cones. Although the original formulation of the across-neuron pattern theory for taste quality excluded neuron types, a pattern code is clearly not incompatible with their existence.

1. Controversy Over Fiber Types

Following the introduction of the best-stimulus classification of CT fibers,[29] Woolston and Erickson[132] correctly noted that the subdivision of neurons into groups on the basis of their response to one or a few stimuli could be an arbitrary division of a continuous population of cells. Further, the experimenter's choice of stimuli and their concentrations could greatly influence the resulting classification. These considerations are particularly applicable to a sensory system such as taste, in which no stimulus continuum has been identified. As a consequence of such considerations, these investigators[132] argued for an approach to the classification of gustatory neurons based on traditional taxonomic procedures.[152,153] Using this approach, Erickson and his colleagues examined the classification of neurons in the rat CT nerve[154] and NST[132] using hierarchical cluster analysis. The solutions obtained in each case gave no evidence for distinct neuron types, suggesting that the subdivision of these cells into best-stimulus categories might be an arbitrary exercise. However, the rat CT primarily possesses only two sets of sensitivities: those to sodium salts and those to acids and other electrolytes, which overlap considerably in NaCl- and HCl-best fibers.[55] These two overlapping systems, however, are readily separated pharmacologically with amiloride at both the CT[66,68] and NST[73,74] levels.

As seen above in Figures 3 and 7, fiber types are readily discernible within the CT and IXth nerves of the hamster, based on the relative similarities and differences among their response profiles.[30,39] Further, application of multivariate statistical techniques to neurons in the hamster NST and PbN strongly suggests similar neuron types.[31,33,56] Some investigators have argued that the recognition of neuron types depends upon the strictness of one's criteria when examining a cluster dendrogram.[92] We believe that the dendrograms for the hamster CT and IXth nerves (Figures 3 and 7) are strongly suggestive of fiber types. These hierarchical arrangements are strikingly different from that for SLN fibers (Figure 10), which suggests a continuous distribution of fiber profiles. Contrary to what others have argued,[92,100] it does not appear that just any data set will provide evidence for clusters. Classifying the fibers in the CT and IXth nerves as members of distinct fiber types is a decision based on the relative similarities of the profiles within a cluster and the striking differences among the profiles across clusters. We believe that at all levels of the hamster gustatory system that have been examined, except in the SLN, there is strong evidence that taste sensitivities are organized into relatively distinct sets of fibers and neurons. The issue, of course, with respect to sensory coding is what role these fiber types play in the neural code for taste quality. The mere existence of a fiber-type organization of sensitivities is not enough to conclude that they function as labeled lines in the coding of quality.

2. Amiloride Sensitivity — NaCl-Best Fibers

As discussed above, it appears that one receptor mechanism for sodium salts is an amiloride-sensitive ion channel on the apical membrane of taste receptor cells.[60,61,63-65] Recent studies have shown that afferent input from this amiloride-sensitive channel appears to be distributed to NaCl-best fibers in the rat[68] and hamster[66] CT nerve (see Figure 5). Responses to NaCl in HCl-best fibers, which are often substantial,[29,30,55] were unaffected by amiloride treatment. Thus the receptors that utilize amiloride-blockable channels appear to activate NaCl-best fibers almost exclusively. Recent studies of cells in the rat NST show that this segregation of amiloride-sensitive input to NaCl-best cells is maintained in the second-order neurons.[73,74] These results are strong evidence that at least the sodium-sensitive neuron types described in the CT and NST are "natural" types,[155] rather than merely arbitrary classes.

The selective distribution of amiloride-blockable sensitivities into NaCl-best cells has been interpreted as evidence for the labeled-line coding of saltiness.[73,74,92] These investigators demonstrated that if sodium channels were blocked by amiloride the response to sodium salts was specifically decreased in NaCl-best cells in the rat NST and that a multidimensional taste space generated from the across-neuron correlations following amiloride treatment indicated that NaCl and LiCl were no longer discriminable from nonsodium salts and acids. These neurophysiological findings correspond to behavioral studies, which show that treatment of the tongue with amiloride during conditioning to avoid NaCl interferes with the ability of rats to make subsequent discriminations between sodium and nonsodium salts.[90] The conclusion drawn from these experiments was that the distinct taste of NaCl is coded by activity in NaCl-best neurons, i.e., by a labeled line or an "independent coding channel."[92] We agree that the selective distribution of amiloride-sensitive input into NaCl-best cells is supportive evidence for the concept of neuron types in taste and that it suggests a critical role for these cells in coding the taste of sodium salts, but we do not think that these data comprise definitive evidence that these neuron types function as labeled lines (see below).

3. The Role of Neuron Types in Defining the Across-Neuron Patterns

The conclusion of Scott and his colleagues[73,74,92] that NaCl-best cells in the rat NST comprise a labeled line for saltiness is based on the fact that blocking the response of NaCl-best cells with amiloride prevents the neural discrimination between sodium and nonsodium salts, i.e., their across-neuron patterns are no longer distinct. Several years ago, we examined the roles played by neuron types in the hamster brainstem in the definitions of the across-neuron patterns.[33,34,56] Our conclusions were that the discrimination among stimuli with different tastes (such as sodium and nonsodium salts) depended upon comparisons of the activity in different neuron types (such as NaCl- and HCl-best cells) and that one neuron type alone was insufficient to discriminate between stimuli with different taste qualities.

This approach to the understanding of the role of neuron groups in defining the similarities and differences among stimuli might best be appreciated by first considering a system that is better understood. In the coding of stimulus wavelength by the vertebrate visual system, three types of broadly sensitive photoreceptor pigments are involved.[151] The color of the wavelength of light falling on the retina can be accurately encoded by considering the relative activity in these three broadly sensitive photoreceptors, i.e., by a pattern.[145,149,150] Deficiencies in one or more of the photoreceptor pigments results in various forms of visual chromatic deficiency or "color blindness".[149,150] Using multidimensional scaling of the responses of color receptors, we can demonstrate how color vision would degenerate in the absence of one of the three primary photopigments. As data for this analysis, the three color matching functions ($\bar{x}$, $\bar{y}$, and $\bar{z}$) in the CIE system[149,150] were used as response profiles for "photoreceptors". Random noise was then added to each of these functions to generate a multivariate matrix of the "responses" of 33 "photoreceptors" (11 each of the three, $\bar{x}$, $\bar{y}$, and $\bar{z}$) at 12 wavelengths (from 420 to 680 nm). Using this matrix of data, the similarities and differences among the 12 stimulus wavelengths were then examined using multidimensional scaling (KYST). The results of this analysis are shown in the two-dimensional "color space" in Figure 15 (top), where the various wavelengths form the familiar color circle.[149,150,156] Thus, visual receptors that functioned like these three "photoreceptors" could code the similarities and differences among these wavelengths as judged psychologically (see Reference 156 for a multidimensional analysis of judgements of similarities among colors).

Given such a representation of the similarities and differences among these various wavelengths, one can then examine the sort of representation that would be possible in the absence of one set of receptors ($\bar{x}$, $\bar{y}$, or $\bar{z}$). The color space resulting from responses in the $\bar{y}$ and $\bar{z}$ "photoreceptors" in the absence of $\bar{x}$ is shown at the bottom of Figure 15. The various wavelengths, rather than being arranged in a circle, are segregated into two groups, dividing between 480 and 500 nm, suggesting that individuals lacking the $\bar{x}$ receptor type would have difficulty discriminating among wavelengths between 440 and 480 nm and among those between 500 and 660 nm, although they could easily discriminate between these two groups of stimuli. This is the kind of color deficiency seen in protanopia, where individuals lacking the "red" photopigment perceive long wavelength stimuli as one color (yellow) and short wavelength stimuli as another (blue), with a neutral (gray) point at 494 nm.[149,150,157] Thus, after creating a color space through the multidimensional

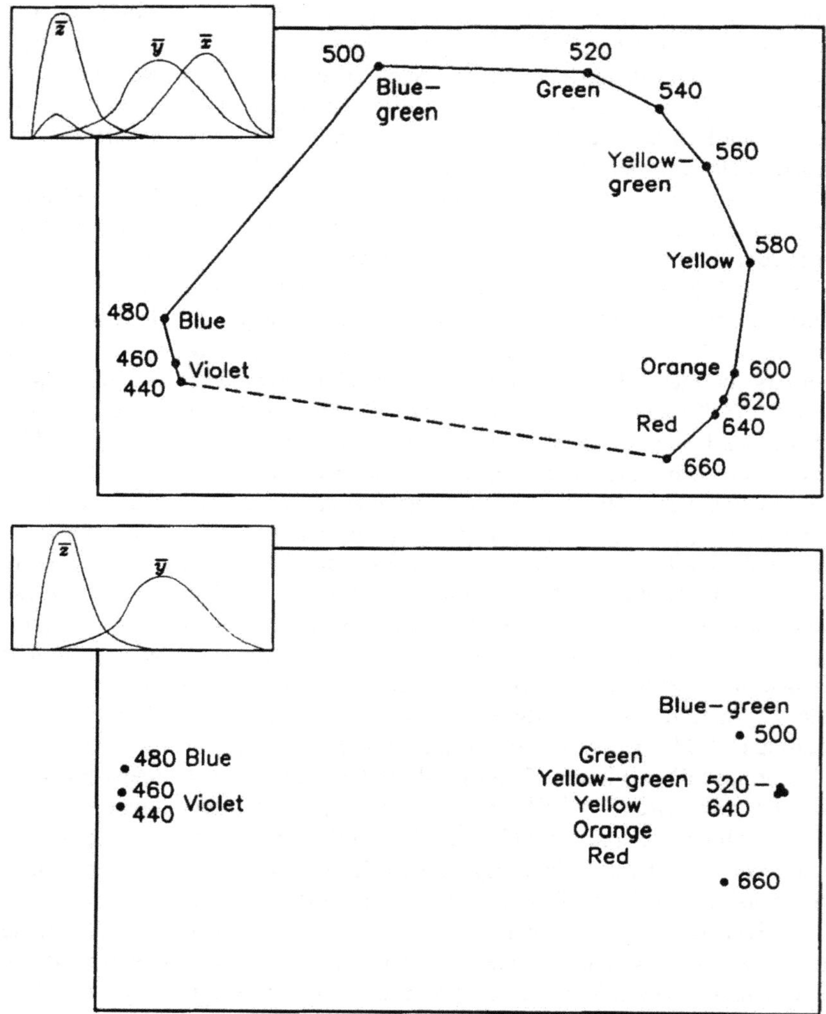

FIGURE 15. Locations of each of 12 stimulus wavelengths in a two-dimensional "color space" obtained through multidimensional scaling (KYST, Bell Laboratories) of the "responses" of 33 "photoreceptors" (11 each of the three types shown in the inset). (A) Color space generated from the across-"photoreceptor" correlations among the 12 wavelengths based on all three receptor types. (B) Color space generated from the across-"photoreceptor" correlations among the 12 wavelengths based on only two receptor types (without the $\bar{x}$ type, but with the two shown in the inset). (Modified from Smith, D. V., *Progress in Clinical and Biological Research, Contemporary Sensory Neurobiology*, Vol. 176, Alan R. Liss, New York, 1985, 75. With permission.)

scaling of "photoreceptor" functions, one can then examine the neural discriminability among stimuli in the absence of one of these receptor types. The results of such an analysis are predictive of the type of color discriminability seen in persons with chromatic deficiency,[149,150,157] even with regard to the location of the neutral point. Additional analyses (not shown) of these theoretical "data" in the absence of the $\bar{y}$ and $\bar{z}$ receptor types, respectively, resulted in a similar grouping of wavelengths, with neutral points corresponding to those seen in deuteranopia (499 nm) and tritanopia (570 nm), respectively. These data, and those on color blind individuals, show that the absence of one of three photoreceptor types results in the inability to discriminate among certain wavelengths. Thus the coding of color by the visual system requires the comparison among inputs from all three photopigments.[145,149-151]

There are no known analogies to color blindness in taste, but the experiments of Scott and his colleagues[73,74,92] on the rat provide an experimental demonstration of an analogous phenomenon. Blocking the "sodium receptor" (i.e., the NaCl-best neurons) with amiloride resulted in the inability of the remaining cells to discriminate between sodium and nonsodium salts, i.e., their across-neuron patterns were not distinct without the input from the NaCl-best cells. In our previous work on cells in the hamster PbN, we demonstrated that the across-neuron patterns for sodium salts were distinct from those for nonsodium salts and acids only if the activity of both NaCl-best and HCl-best cells was considered.[33,34,56] The taste space generated by multidimensional scaling of the across-neuron correlations among 18 stimuli in the hamster PbN was shown above in Figure 14, where it is evident that sodium salts are discriminable from nonsodium salts and acids, from sweet-tasting, and from bitter-tasting stimuli. Eliminating NaCl-best neurons (or the HCl-best neurons) from this analysis resulted in a lack of discrimination between the sodium and the nonsodium salts and acids[34,56] exactly like that reported after blocking NaCl-best NST cells with amiloride in the rat.[73,74,92] That is, the across-fiber patterns for sodium and nonsodium salts were not different from one another; without considering the differential response of two neuron types (NaCl- and HCl-best) to these two classes of stimuli they were coexistent within the taste space.[34,56] A two-dimensional representation of the taste space of Figure 14 is shown in Figure 16A. All four groups of stimuli are segregated within this space when the responses of three groups of neurons (S, N, H; see inset) are included in the multidimensional scaling analysis. Elimination of the N cells (NaCl-best) from the analysis resulted in a lack of discrimination between the sodium salts and the nonsodium salts and acids (Figure 16B). Similar results were obtained by the elimination of HCl-best cells, which were also necessary to define the differences in the across-neuron patterns between the sodium salts and the nonsodium salts and acids. All three neuron types (S, N, H) were necessary for the PbN to sort out three groups of stimuli (sweeteners, sodium salts, and nonsodium salts and acids) on the basis of their across-neuron patterns. We interpreted these results to imply that taste quality discrimination depends upon a comparison of activity across broadly tuned neuron types, comparable to the coding of color vision by broadly tuned photoreceptors.[34,56] We believe that the results of Scott and colleagues following amiloride treatment in the rat are compatible with this interpretation.

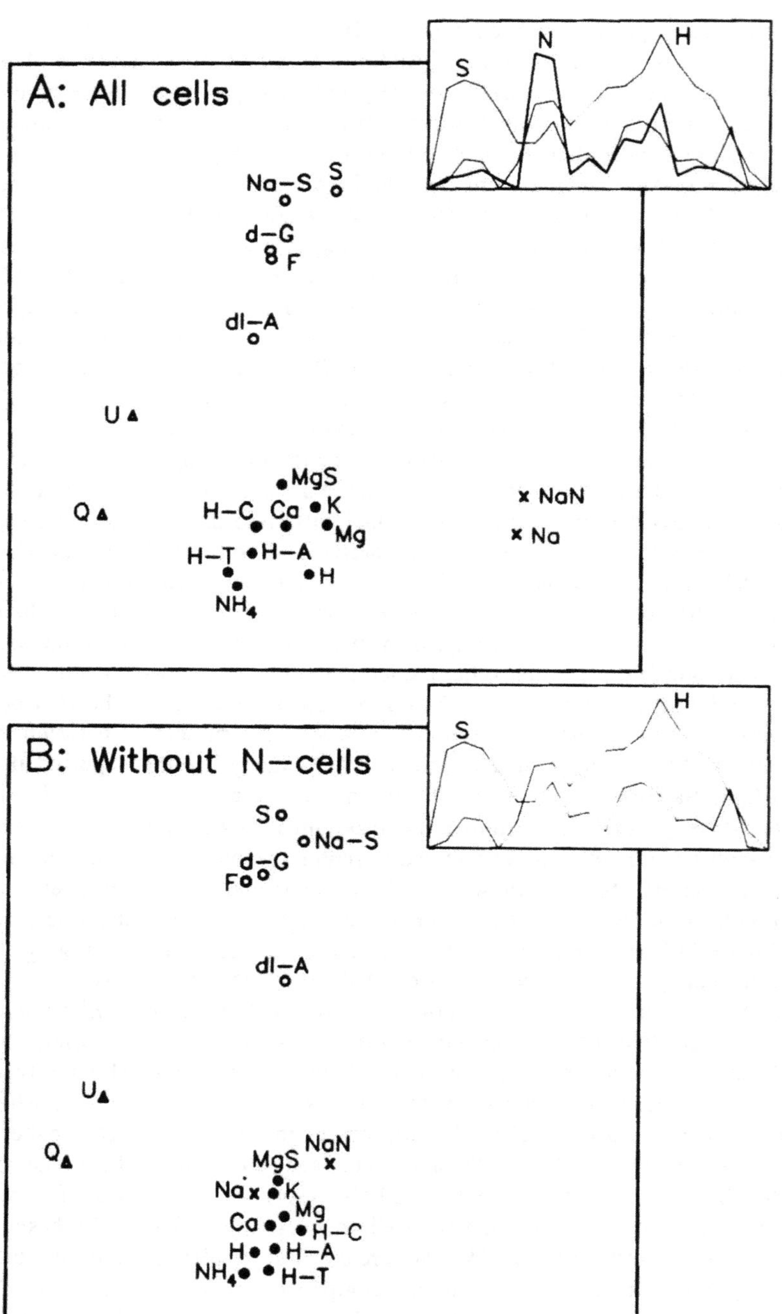
A: All cells
S
N
H
Na-S
S
d-G
F
dl-A
U
Q
MgS
H-C
Ca
K
Mg
H-T
H-A
H
NH4
NaN
Na
B: Without N-cells
S
H
S
Na-S
d-G
F
dl-A
U
Q
NaN
MgS
Na
K
Mg
Ca
H-C
H
H-A
NH4
H-T

FIGURE 16.

The conclusion from their work that NaCl-best cells are coding "saltiness" in a labeled-line fashion is analogous to concluding that the long wavelength photoreceptor ($\bar{x}$ in Figure 15) is coding "redness" in a labeled-line manner. In both taste and color vision, elimination of any one cell type results in a lack of separation between stimuli of different quality within multidimensional space[34,56,74] and a loss of behavioral discrimination among the same stimuli.[90,150,157] Thus we suggest that the coding of taste quality is analogous to color coding, i.e., that the code lies in the relative pattern of activity across several neuron types.

The segregation of amiloride-sensitive receptor input into NaCl-best fibers[66,68] and the hierarchical cluster analyses shown above for hamster CT and IXth nerve fibers suggests strongly that there are functional neuron classes in the peripheral gustatory system. Our earlier work on hamster brainstem cells[33,34,36,56] and the effect of blocking NaCl-best neuron responses in the rat NST with amiloride[73,74] demonstrate that the neural patterns of activity generated by taste stimuli are dominated by the responses of particular classes of neurons. No one neuron type alone is capable of providing information that can distinguish the across-neuron patterns evoked by dissimilar-tasting compounds. More than one neuron type must contribute to the pattern in order for the patterns evoked by unlike stimuli to be distinct. Thus, in this sense, the various neuron types (sucrose- , NaCl- , HCl- , or QHCl-best) are critically important for the discrimination of taste quality. NaCl-best cells or HCl-best cells can define the similarities among sodium salts, but activity in both neuron types are required to distinguish sodium salts from nonsodium salts and acids. Behavioral[90] and neural[73,74] data in rats support this requirement, but there is no evidence to date that NaCl-best cells are labeled lines signaling "saltiness". On the contrary, we think that NaCl-best cells provide a critical part of a pattern across neuron types that codes saltiness; these cells by themselves are not capable of distinguishing sodium salts from other stimuli.[34] Taste quality discrimination requires the comparison of activity

FIGURE 16. Two-dimensional representations of taste stimulus relationships obtained from multidimensional scaling of across-neuron correlations among 18 stimuli. (A) This is a two-dimensional representation of the space shown above in Figure 14, generated from the across-neuron correlations across 31 PbN neurons of three types (sucrose- , NaCl- , and HCl-best cells). The average response profiles for each of these best-stimulus neuron types (S, N, H) are shown in the inset; stimuli are arranged across the abscissa of the inset in the following order, which also gives the abbreviations used in the figure: S, sucrose; F, fructose; d-G, d-glucose; Na-S, Na-saccharin; dl-A, dl-alanine; Na, NaCl; NaN, $NaNO_3$; MgS, $MgSO_4$; K, KCl; NH_4, NH_4Cl; Mg, $MgCl_2$; Ca, $CaCl_2$; H, HCl; H-T, tartaric acid; H-C, citric acid; H-A, acetic acid; U, urea; Q, QHCl. (B) Two-dimensional space showing the similarities and differences among these 18 stimuli when the NaCl-best neurons are not included in the analysis. The across-neuron correlations were calculated across the sucrose- and HCl-best cells only (inset shows average response profiles for the S and H neuron types). Without the contribution of the NaCl-best cells, the hamster gustatory system cannot discriminate sodium salts from nonsodium salts and acids. (Modified from Smith, D. V., Van Buskirk, R. L., Travers, J. B., and Bieber, S. L., *J. Neurophysiol.*, 50, 541, 1983. With permission.)

across several neuron types, some of which (e.g., QHCl-best fibers) are segregated predominantly into different cranial nerves.

REFERENCES

1. **Miller, I. J., Jr. and Bartoshuk, L. M.,** Taste perception, taste bud distribution, and spatial relationships, in *Smell and Taste in Health and Disease*, Getchell, T. V., Doty, R. L., Bartoshuk, L. M., and Snow, J. B., Jr., Eds., Raven Press, New York, 1991, 205.
2. **Miller, I. J., Jr. and Smith, D. V.,** Quantitative taste bud distribution in the hamster, *Physiol. Behav.*, 32, 275, 1984.
3. **Hosley, M. A., Hughes, S. E., and Oakley, B.,** Neural induction of taste buds, *J. Comp. Neurol.*, 260, 224, 1987.
4. **Miller, I. J., Jr.,** Gustatory receptors of the palate, in *Food Intake and Chemical Senses*, Katsuki, Y., Sato, M., Takagi, S., and Oomura, Y., Eds., University of Tokyo Press, Tokyo, 1977, 173.
5. **Miller, I. J., Jr. and Spangler, K. M.,** Taste bud distribution and innervation on the palate of the rat, *Chem. Senses*, 7, 99, 1982.
6. **Bradley, R. M.,** Development of the taste bud and gustatory papillae in human fetuses, in *Third Symposium on Oral Sensation and Perception: The Mouth of the Infant*, Bosma, J. F., Ed., Charles C Thomas, Springfield, IL, 1972, 137.
7. **Bradley, R., Cheal, M., and Kim, Y.,** Quantitative analysis of developing epiglottal taste buds in sheep, *J. Anat.*, 130, 25, 1980.
8. **Elliott, R.,** Total distribution of taste buds on the tongue of the kitten at birth, *J. Comp. Neurol.*, 66, 361, 1937.
9. **Khaisman, E. B.,** Particular features of the innervation of taste buds of the epiglottis in monkeys, *Acta Anat.*, 95, 101, 1976.
10. **Nilsson, B.,** The occurrence of taste buds in the palate of human adults as evidenced by light microscopy, *Acta Odontol. Scand.*, 37, 253, 1979.
11. **Fish, H., Malone, P., and Richter, C.,** The anatomy of the tongue of the domestic Norway rat. I. The skin of the tongue; the various papillae; their number and distribution, *Anat. Rec.*, 89, 429, 1944.
12. **Oakley, B.,** Taste bud development in rat vallate and foliate papillae, in *Mechanoreceptors*, Hnik, P., Soukup, T., Vejsada, R., and Zelina, J., Eds., Plenum Press, New York, 1988, 17.
13. **Whiteside, B.,** Nerve overlap in the gustatory apparatus of the rat, *J. Comp. Neurol.*, 44, 363, 1927.
14. **Zalewski, A. A.,** Role of nerve and epithelium in the regulation of alkaline phosphatase activity in gustatory papillae, *Exp. Neurol.*, 23, 18, 1969.
15. **Cleaton-Jones, P.,** A denervation study of taste buds in the soft palate of the albino rat, *Arch. Oral Biol.*, 21, 79, 1971.
16. **Contreras, R. J., Beckstead, R. M., and Norgren, R.,** The central projections of the trigeminal, facial, glossopharyngeal and vagus nerves: an autoradiographic study in the rat, *J. Auton. Nerv. Syst.*, 6, 303, 1982.
17. **Hamilton, R. B. and Norgren, R.,** Central projections of gustatory nerves in the rat, *J. Comp. Neurol.*, 222, 560, 1984.
18. **Atema, J.,** Structures and functions of the sense of taste in the catfish *(Ictalurus natalis)*, *Brain Behav. Evol.*, 4, 273, 1971.
19. **Finger, T. E. and Morita, Y.,** Two gustatory systems: facial and vagal gustatory nuclei have different brainstem connections, *Science*, 227, 776, 1985.

20. **Morita, Y. and Finger, T. E.,** Reflex connections of the facial and vagal gustatory systems in the brainstem of the bullhead catfish, *Ictalurus nebulosus*, *J. Comp. Neurol.*, 231, 547, 1985.
21. **Guth, L.,** Effects of glossopharyngeal nerve transection on the circumvallate papilla of the rat, *Anat. Rec.*, 128, 715, 1957.
22. **Oakley, B.,** Reformation of taste buds by crossed sensory nerves in the rat's tongue. *Acta Physiol. Scand.*, 79, 88, 1970.
23. **Travers, S. P. and Nicklas, K.,** Taste bud distribution in the rat pharynx and larynx, *Anat. Rec.*, 227, 373, 1990.
24. **Hanamori, T. and Smith, D. V.,** Gustatory innervation in the rabbit: central distribution of sensory and motor components of the chorda tympani, glossopharyngeal, and superior laryngeal nerves, *J. Comp. Neurol.*, 282, 1, 1989.
25. **Miller, I. J., Jr. and Smith, D. V.,** Proliferation of taste buds in the foliate and vallate papillae of postnatal hamsters, *Growth Dev. Aging*, 52, 123, 1989.
26. **Feindel, W.,** The neural pattern of the epiglottis, *J. Comp. Neurol.*, 105, 269, 1956.
27. **Hanamori, T. and Smith, D. V.,** Central projections of the hamster superior laryngeal nerve, *Brain Res. Bull.*, 16, 271, 1986.
28. **Belecky, T. L. and Smith, D. V.,** Postnatal development of palatal and laryngeal taste buds in the hamster, *J. Comp. Neurol.*, 293, 646, 1990.
29. **Frank, M.,** An analysis of hamster afferent taste nerve response functions, *J. Gen. Physiol.*, 61, 588, 1973.
30. **Frank, M. E., Bieber, S. L., and Smith, D. V.,** The organization of taste sensibilities in hamster chorda tympani nerve fibers, *J. Gen. Physiol.*, 91, 861, 1988.
31. **McPheeters, M., Hettinger, T. P., Nuding, S. C., Savoy, L. D., Whitehead, M. C., and Frank, M. E.,** Taste-responsiveness neurons and their locations in the solitary nucleus of the hamster, *Neuroscience*, 34, 745, 1990.
32. **Pfaffmann, C., Frank, M., Bartoshuk, L. M., and Snell, T. C.,** Coding gustatory information in the squirrel monkey chorda tympani, in *Progress in Psychobiology and Physiological Psychology*, Vol. 6, Sprague, J. M. and Epstein, A. N., Eds., Academic Press, New York, 1976, 1.
33. **Smith, D. V., Van Buskirk, R. L., Travers, J. B., and Bieber, S. L.,** Gustatory neuron types in hamster brain stem, *J. Neurophysiol.*, 50, 522, 1983.
34. **Smith, D. V., Van Buskirk, R. L., Travers, J. B., and Bieber, S. L.,** Coding of taste stimuli by hamster brain stem neurons, *J. Neurophysiol.*, 50, 541, 1983.
35. **Travers, J. B. and Smith, D. V.,** Gustatory sensitivities in neurons of the hamster nucleus tractus solitarius, *Sensory Processes*, 3, 1, 1979.
36. **Van Buskirk, R. L. and Smith, D. V.,** Taste sensitivity of hamster parabrachial pontine neurons, *J. Neurophysiol.*, 45, 144, 1981.
37. **Boudreau, J. C., Do, L. T., Sivakumar, L., Oravec, J., and Rodriguez, C. A.,** Taste systems of the petrosal ganglion of the rat glossopharyngeal nerve, *Chem. Senses*, 12, 437, 1987.
38. **Frank, M. E.,** Taste-responsive neurons of the glossopharyngeal nerve of the rat, *J. Neurophysiol.*, 65, 1452, 1991.
39. **Hanamori, T., Miller, I. J., Jr., and Smith, D. V.,** Gustatory responsiveness of fibers in the hamster glossopharyngeal nerve, *J. Neurophysiol.*, 60, 478, 1988.
40. **Harada, S. and Smith, D. V.,** Gustatory sensitivities of the hamster's soft palate, *Chem. Senses*, 17, 37, 1992.
41. **Nejad, M. S.,** The neural activities of the greater superficial petrosal nerve of the rat in response to chemical stimulation of the palate, *Chem. Senses*, 11, 283, 1986.
42. **Travers, S. P. and Norgren, R.,** Coding the sweet taste in the nucleus of the solitary tract: differential roles for anterior tongue and nasoincisor duct gustatory receptors in the rat, *J. Neurophysiol.*, 65, 1372, 1991.
43. **Travers, S. P., Pfaffmann, C., and Norgren, R.,** Convergence of lingual and palatal gustatory neural activity in the nucleus of the solitary tract, *Brain Res.*, 365, 305, 1986.

44. **Bradley, R. M., Stedman, H. M., and Mistretta, C. M.,** Superior laryngeal nerve response patterns to chemical stimulation of sheep epiglottis, *Brain Res.*, 276, 81, 1983.
45. **Dickman, J. D. and Smith, D. V.,** Response properties of fibers in the hamster superior laryngeal nerve, *Brain Res.*, 450, 25, 1988.
46. **Smith, D. V. and Hanamori, T.,** Organization of gustatory sensitivities in hamster superior laryngeal nerve fibers, *J. Neurophysiol.*, 65, 1098, 1991.
47. **Fishman, I. Y.,** Single fiber gustatory impulses in rat and hamster, *J. Cell. Comp. Physiol.*, 49, 319, 1957.
48. **Ogawa, H., Sato, M., and Yamashita, S.,** Multiple sensitivity of chorda tympani fibers of the rat and hamster to gustatory and thermal stimuli, *J. Physiol.*, 199, 223, 1968.
49. **Pfaffmann, C.,** Gustatory nerve impulses in rat, cat and rabbit, *J. Neurophysiol.*, 18, 429, 1955.
50. **Frank, M. E. and Nowlis, G. H.,** Learned aversions and taste qualities in hamsters, *Chem. Senses*, 14, 379, 1989.
51. **Nowlis, G. H. and Frank, M. E.,** Quality coding in gustatory systems of rats and hamsters, in *Perception of Behavioral Chemicals*, Norris, D. M., Ed., Elsevier/North-Holland, Amsterdam, 1981, 59.
52. **Smith, D. V., Travers, J. B., and Van Buskirk, R. L.,** Brainstem correlates of gustatory similarity in the hamster, *Brain Res. Bull.*, 4, 359, 1979.
53. **Pfaffmann, C.,** Gustatory afferent impulses, *J. Cell. Comp. Physiol.*, 17, 243, 1941.
54. **Pfaffmann, C.,** The afferent code for sensory quality, *Am. Psychol.*, 14, 226, 1955.
55. **Frank, M. E., Contreras, R. J., and Hettinger, T. P.,** Nerve fibers sensitive to ionic taste stimuli in chorda tympani of the rat, *J. Neurophysiol.*, 50, 941, 1983.
56. **Smith, D. V.,** The neural representation of gustatory quality, in *Progress in Clinical and Biological Research, Contemporary Sensory Neurobiology*, Vol. 176, Correia, M. J. and Perrachio, A., Eds., Alan R. Liss, New York, 1985, 75.
57. **Bieber, S. L. and Smith, D. V.,** Multivariate analysis of sensory data: a comparison of methods, *Chem. Senses*, 11, 19, 1986.
58. **Erickson, R. P.,** Sensory neural patterns and gustation, in *Olfaction and Taste*, Zotterman, Y., Ed., Pergamon Press, Oxford, 1963, 205.
59. **Erickson, R. P., Doetsch, G. S., and Marshall, D. A.,** The gustatory neural response function, *J. Gen. Physiol.*, 49, 247, 1965.
60. **Avenet, P. and Lindemann, B.,** Amiloride-blockable sodium currents in isolated taste receptor cells, *J. Membr. Biol.*, 105, 245, 1988.
61. **DeSimone, J. A., Heck, G. L., Mierson, S., and DeSimone, S. K.,** The active ion transport properties of canine lingual epithelia *in vitro*. Implications for gustatory transduction, *J. Gen. Physiol.*, 83, 633, 1984.
62. **Formaker, B. K. and Hill, D. L.,** An analysis of residual NaCl taste response after amiloride, *Am. J. Physiol.*, 255, R1002, 1988.
63. **Heck, G. L., Mierson, S., and DeSimone, J. A.,** Salt taste transduction occurs through an amiloride-sensitive sodium transport pathway, *Science*, 223, 403, 1984.
64. **Kinnamon, S. C.,** Taste transduction: a diversity of mechanisms, *Trends Neurosci.*, 11, 491, 1988.
65. **Simon, S. A., Robb, R., and Schiffman, S. S.,** Transport pathways in rat lingual epithelium, *Pharmacol. Biochem. Behav.*, 29, 257, 1988.
66. **Hettinger, T. P. and Frank, M. E.,** Specificity of amiloride inhibition of hamster taste responses, *Brain Res.*, 513, 24, 1990.
67. **Hill, D. L.,** Development of amiloride sensitivity in the rat peripheral gustatory system: a single fiber analysis, in *Olfaction and Taste IX*, Roper, S. D. and Atema, J., Eds., New York Academy of Sciences, New York, 1987, 366.
68. **Ninomiya, Y. and Funakoshi, M.,** Amiloride inhibition of responses of rat single chorda tympani fibers to chemical and electrical tongue stimulations, *Brain Res.*, 451, 319, 1988.

69. **Brand, J. G., Teeter, J. H., and Silver, W. L.,** Inhibition by amiloride of chorda tympani responses evoked by monovalent salts, *Brain Res.*, 334, 207, 1985.
70. **DeSimone, J. A. and Ferrell, F.,** Analysis of amiloride inhibition of chorda tympani taste response of rat to NaCl, *Am. J. Physiol.*, 249, R52, 1985.
71. **Hellekant, G., DuBois, G. E., Roberts, T. W., and van der Wel, H.,** On the gustatory effects of amiloride in the monkey *(Macaca mulatta)*, *Chem. Senses*, 13, 89, 1988.
72. **Herness, M. S.,** Effects of amiloride on bulk flow and iontophoretic taste stimuli in the hamster, *J. Comp. Physiol. A*, 160, 281, 1987.
73. **Giza, B. K. and Scott, T. R.,** The effect of amiloride on taste-evoked activity in the nucleus tractus solitarius of the rat, *Brain Res.*, 550, 247, 1991.
74. **Scott, T. R. and Giza, B. K.,** Coding channels in the taste system of the rat, *Science*, 249, 1585, 1990.
75. **Mistretta, C. M. and Bradley, R. M.,** Developmental changes in taste responses from glossopharyngeal nerve in sheep and comparisons with chorda tympani responses, *Dev. Brain Res.*, 11, 107, 1983.
76. **Oakley, B.,** Altered taste responses from cross-regenerated taste nerves in the rat, in *Olfaction and Taste II*, Hayashi, T., Ed., Pergamon Press, London, 1967, 535.
77. **Oakley, B., Jones, L. B., and Kaliszewski, J. M.,** Taste responses of the gerbil IXth nerve, *Chem. Senses Flav.*, 4, 79, 1979.
78. **Shingai, T. and Beidler, L. M.,** Response characteristics of three taste nerves in mice, *Brain Res.*, 335, 245, 1985.
79. **Yamada, K.,** Gustatory and thermal responses in the glossopharyngeal nerve of the rat, *Jpn. J. Physiol.*, 16, 599, 1966.
80. **Yamada, K.,** Gustatory and thermal responses in the glossopharyngeal nerve of the rabbit and cat, *Jpn. J. Physiol.*, 17, 94, 1967.
81. **Carpenter, J. A.,** Species differences in taste preference, *J. Comp. Physiol. Psychol.*, 49, 139, 1956.
82. **Maes, F. W.,** Improved best-stimulus classification of taste neurons, *Chem. Senses*, 10, 35, 1985.
83. **Kanwal, J. S. and Caprio, J.,** An electrophysiological investigation of the oropharyngeal (IX-X) taste system in the channel catfish, *Ictalurus punctatus*, *J. Comp. Physiol. A*, 150, 345, 1983.
84. **Gordon, K. D. and Caprio, J.,** Taste responses to amino acids in the southern leopard frog, *Rana sphenocephala*, *Comp. Biochem. Physiol.*, 81A, 525, 1985.
85. **Yoshii, K., Kobatake, Y., and Kurihara, K.,** Selective enhancement and suppression of frog gustatory responses to amino acids, *J. Gen. Physiol.*, 77, 373, 1981.
86. **Yoshii, K., Yoshii, C., Kobatake, Y., and Kurihara, K.,** High sensitivity of *Xenopus* gustatory receptors to amino acids and bitter substances, *Am. J. Physiol.*, 243, R42, 1982.
87. **Spector, A. C., Schwartz, G., and Grill, H. J.,** Chemospecific deficits in taste detection following selective gustatory deafferentation in rats, *Am. J. Physiol.*, 258, R820, 1990.
88. **Nitabach, M., Schwartz, G., Spector, A., and Grill, H.,** The anterior tongue receptive field specifies sodium taste in the rat, *Chem. Senses*, 14, 734, 1989.
89. **Spector, A. C., Schwartz, G., and Grill, H. J.,** Chemospecific deficits in taste detection following selective gustatory deafferentation, *Soc. Neurosci. Abstr.*, 14, 1063, 1988.
90. **Hill, D. L., Formaker, B. K., and White, K. S.,** Perceptual characteristics of the amiloride-suppressed sodium chloride taste response in the rat, *Behav. Neurosci.*, 104, 734, 1990.
91. **Formaker, B. K. and Hill, D. L.,** Lack of amiloride sensitivity in SHR and WKY glossopharyngeal taste responses to NaCl, *Physiol. Behav.*, 50, 765, 1991.
92. **Scott, T. R. and Plata-Salaman, R.,** Coding of taste quality, in *Smell and Taste in Health and Disease*, Getchell, T. V., Doty, R. L., Bartoshuk, L. M., and Snow, J. B., Jr., Eds., Raven Press, New York, 1991, 345.

93. **Sweazey, R. D. and Smith, D. V.,** Convergence onto hamster medullary taste neurons, *Brain Res.*, 408, 173, 1987.
94. **Bradley, R. M.,** The role of epiglottal and lingual chemoreceptors: a comparison, in *Determination of Behavior by Chemical Stimuli*, Steiner, J. E. and Ganchrow, J. R., Eds., Information Retrieval, London, 1982, 37.
95. **Stedman, H., Bradley, R., Mistretta, C., and Bradley, B.,** Chemosensitive responses from the cat epiglottis, *Chem. Senses*, 5, 233, 1980.
96. **Boggs, D. F. and Bartlett, D., Jr.,** Chemical specificity of a laryngeal apneic reflex in puppies, *J. Appl. Physiol.*, 53, 455, 1982.
97. **Shingai, T.,** Ionic mechanisms of water receptors in the laryngeal mucosa of the rabbit, *Jpn. J. Physiol.*, 27, 27, 1977.
98. **Shingai, T.,** Water fibers in the superior laryngeal nerve of the rat, *Jpn. J. Physiol.*, 30, 305, 1980.
99. **Storey, A. T.,** A functional analysis of sensory units innervating epiglottis and larynx, *Exp. Neurol.*, 20, 366, 1968.
100. **Erickson, R. P.,** Grouping in the chemical senses, *Chem. Senses*, 10, 333, 1985.
101. **Morrison, G. R.,** Behavioral response patterns to salt stimuli in the rat, *Can. J. Psychol.*, 21, 141, 1967.
102. **Nachman, M.,** Learned aversion to the taste of lithium chloride and generalization to other salts, *J. Comp. Physiol. Psychol.*, 56, 343, 1963.
103. **Nowlis, G. H., Frank, M. E., and Pfaffmann, C.,** Specificity of acquired aversions to taste qualities in hamsters and rats, *J. Comp. Physiol. Psychol.*, 94, 932, 1980.
104. **Smith, D. V. and Theodore, R. M.,** Conditioned taste aversions: generalizations to taste mixtures, *Physiol. Behav.*, 32, 983, 1984.
105. **Spector, A. C. and Grill, H. J.,** Differences in the taste quality of maltose and sucrose in rats: issues involving the generalizations of conditioned taste aversions, *Chem. Senses*, 13, 95, 1988.
106. **Krimm, R. F., Nejad, M. S., Smith, J. C., Miller, I. J., Jr., and Beidler, L. M.,** The effect of bilateral sectioning of the chorda tympani and the greater superficial petrosal nerves on the sweet taste in the rat, *Physiol. Behav.*, 41, 495, 1987.
107. **Shingai, T. and Shimada, K.,** Reflex swallowing by water and chemical substances applied in the oral cavity, pharynx, and larynx of the rabbit, *Jpn. J. Physiol.*, 26, 455, 1976.
108. **Storey, A. T. and Johnson, P.,** Laryngeal water receptors initiating apnea in the lamb, *Exp. Neurol.*, 47, 42, 1975.
109. **Beebe-Center, J. G., Black, P., Hoffman, A. C., and Wade, M.,** Relative per diem consumption as a measure of preference in the rat, *J. Comp. Physiol. Psychol.*, 41, 239, 1948.
110. **Davis, J. D.,** The effectiveness of some sugars in stimulating licking behavior in the rat, *Physiol. Behav.*, 11, 39, 1973.
111. **Richter, C. P. and Campbell, K.,** Taste thresholds and taste preferences of rats for five common sugars, *J. Nutr.*, 20, 31, 1940.
112. **Norgren, R.,** Gustatory afferents to ventral forebrain, *Brain Res.*, 81, 285, 1974.
113. **Norgren, R.,** Taste pathways to hypothalamus and amygdala, *J. Comp. Neurol.*, 166, 17, 1976.
114. **Norgren, R. and Leonard, C. M.,** Ascending central gustatory pathways, *J. Comp. Neurol.*, 150, 217, 1973.
115. **Travers, J. B.,** Organization and projections of the orofacial motor nuclei, in *The Rat Nervous System*, Vol. 2, *Hindbrain and Spinal Cord*, Paxinos, G., Ed., Academic Press, Sydney, 1985, 111.
116. **Travers, J. B. and Norgren, R.,** Afferent projections to the oral motor nuclei in the rat, *J. Comp. Neurol.*, 220, 280, 1983.

117. **Grill, H. J. and Norgren, R.,** The taste reactivity test. I. Mimetic responses to gustatory stimuli in neurologically normal rats, *Brain Res.*, 143, 263, 1978.
118. **Grill, H. J. and Norgren, R.,** The taste reactivity test. II. Mimetic responses to gustatory stimuli in chronic thalamic and chronic decerebrate rats, *Brain Res.*, 143, 281, 1978.
119. **Kawamura, Y. and Yamamoto, T.,** Studies on neural mechanisms of the gustatory-salivary reflex in rabbits, *J. Physiol.*, 285, 35, 1978.
120. **Mattes, R. D.,** Sensory influences on food intake and utilization in humans, *Hum. Nutr. Appl. Nutr.*, 41A, 77, 1987.
121. **Norgren, R.,** Taste and the autonomic nervous system, *Chem. Senses*, 10, 143, 1985.
122. **Smith, D. V. and Travers, J. B.,** A metric for the breadth of tuning of gustatory neurons, *Chem. Senses Flav.*, 4, 215, 1979.
123. **Brining, S. K., Belecky, T. L., and Smith, D. V.,** Taste reactivity in the hamster, *Physiol. Behav.*, 49, 1265, 1991.
124. **Travers, J. B., Grill, H. J., and Norgren, R.,** The effects of glossopharyngeal and chorda tympani nerve cuts on the ingestion and rejection of sapid stimuli: an electromyographic analysis in the rat, *Behav. Brain Res.*, 25, 233, 1987.
125. **Berridge, K. C. and Grill, H. J.,** Isohedonic tastes support a two-dimensional hypothesis of palatability, *Appetite*, 4, 221, 1984.
126. **Shingai, T., Miyaoka, Y., and Shimada, K.,** Diuresis mediated by the superior laryngeal nerve in rats, *Physiol. Behav.*, 44, 431, 1988.
127. **Bardach, J. E., Todd, J. H., and Crickmer, R.,** Orientation by taste in fish of the genus *Ictalurus*, *Science*, 155, 1276, 1967.
128. **Nowlis, G. H.,** From reflex to representation: taste-elicited tongue movements in the human newborn, in *Taste and Development: The Genesis of Sweet Preference*, Weiffenbach, J. M., Ed., DHEW Publication No. (NIH) 77–1068, Bethesda, MD, 1977, 190.
129. **Erickson, R. P.,** The "across-fiber pattern" theory: an organizing principle for molar neural function, in *Contributions to Sensory Physiology*, Vol. 6, Neff, W. D., Ed., Academic Press, New York, 1982, 70.
130. **Erickson, R. P.,** On the neural basis of behavior, *Am. Sci.*, 72, 233, 1984.
131. **Pfaffmann, C.,** Specificity of the sweet receptors of the squirrel monkey, *Chem. Senses Flav.*, 1, 61, 1974.
132. **Woolston, D. C. and Erickson, R. P.,** Concept of neuron types in gustation in the rat, *J. Neurophysiol.*, 42, 1390, 1979.
133. **Erickson, R. P.,** The role of "primaries" in taste research, in *Olfaction and Taste VI*, LeMagnen, J. and MacLeod, P., Eds., Information Retrieval, London, 1977, 369.
134. **McBurney, D. H.,** Are there primary tastes for man?, *Chem. Senses Flav.*, 1, 17, 1974.
135. **McBurney, D. H. and Gent, G. F.,** On the nature of taste qualities, *Psychol. Bull.*, 86, 151, 1979.
136. **Schiffman, S. S. and Erickson, R.,** The issue of primary tastes versus a taste continuum, *Neurosci. Biobehav. Rev.*, 4, 109, 1980.
137. **Travers, S. P. and Smith, D. V.,** Responsiveness of neurons in the hamster parabrachial nuclei to taste mixtures, *J. Gen. Physiol.*, 84, 221, 1984.
138. **Hanamori, T., Ishiko, N., and Smith, D. V.,** Multimodal responses of taste neurons in the frog nucleus tractus solitarius, *Brain Res. Bull.*, 18, 87, 1987.
139. **Pfaffmann, C., Erickson, R. P., Frommer, G. P., and Halpern, B. P.,** Gustatory discharges in the rat medulla and thalamus, in *Sensory Communication*, Rosenblith, W. A., Ed., John Wiley & Sons, New York, 1961, 455.
140. **Crick, F. H. C.,** Thinking about the brain, *Sci. Am.*, 241, 219, 1979.
141. **Erickson, R. P.,** Parallel "population" neural coding in feature extraction, in *The Neurosciences: Third Study Program*, Schmitt, F. O. and Worden, F. G., Eds., MIT Press, Cambridge, MA, 1974, 155.
142. **Ganchrow, J. R. and Erickson, R. P.,** Neural correlates of gustatory intensity and quality, *J. Neurophysiol.*, 33, 768, 1970.

143. **Di Lorenzo, P. M.,** Across unit patterns in the neural response to taste: vector space analysis, *J. Neurophysiol.*, 62, 823, 1989.
144. **Gill, J. M., II and Erickson, R. P.,** Neural mass differences in gustation, *Chem. Senses*, 10, 531, 1985.
145. **Erickson, R. P.,** Stimulus coding in topographic and nontopographic afferent modalities: on the significance of the activity of individual sensory neurons, *Psychol. Rev.*, 75, 447, 1968.
146. **Boudreau, J. C. and Alev, N.,** Classification of chemoresponsive tongue units of the cat geniculate ganglion, *Brain Res.*, 54, 157, 1973.
147. **Boudreau, J. C., Sivakumar, L., Do, L. T., White, T. D., Oravec, J., and Hoang, N. K.,** Neurophysiology of geniculate ganglion (facial nerve) taste systems: species comparisons, *Chem. Senses*, 10, 89, 1985.
148. **Contreras, R. and Frank, M.,** Sodium deprivation alters neural responses to gustatory stimuli, *J. Gen. Physiol.*, 73, 569, 1979.
149. **Boynton, R. M.,** Vision, in *Experimental Methods and Instrumentation in Psychology*, Sidowski, J., Ed., McGraw Hill, New York, 1966, 273.
150. **Boynton, R. M.,** Color vision, in *Woodworth and Schlosberg's Experimental Psychology*, Kling, J. W. and Riggs, L. A., Eds., Holt, Reinhart and Winston, New York, 1971, 315.
151. **Marks, W. B., Dobelle, W. H., and MacNichol, E. F., Jr.,** Visual pigments of single primate cones, *Science*, 143, 1181, 1964.
152. **Rowe, M. H. and Stone, J.,** Naming of neurons: classification and naming of cat retinal ganglion cells, *Brain Behav. Evol.*, 14, 185, 1977.
153. **Tyner, C. F.,** The naming of neurons: applications of taxonomic theory to the study of cellular populations, *Brain Behav. Evol.*, 12, 75, 1975.
154. **Erickson, R. P., Covey, E., and Doetsch, G. S.,** Neuron and stimulus typologies in the rat gustatory system, *Brain Res.*, 196, 513, 1980.
155. **Rodieck, R. W. and Brenning, R. K.,** Retinal ganglion cells: properties, types, genera, pathways and trans-species comparisons, *Brain Evol. Behav.*, 23, 121, 1983.
156. **Shepard, R. N.,** Multidimensional scaling, tree-fitting, and clustering, *Science*, 210, 390, 1980.
157. **Graham, C. H., Sperling, H. G., Hsia, Y., and Coulson, A. H.,** The determination of some visual functions of a unilaterally color-blind subject: methods and results, *J. Psychol.*, 51, 3, 1961.

Chapter 13

OROSENSORY PROCESSING IN NEURAL SYSTEMS OF THE NUCLEUS OF THE SOLITARY TRACT

Susan P. Travers

TABLE OF CONTENTS

0-8493-5341-6/93/$0.00 + $.50

I. INTRODUCTION

The mouth is an exquisitely sensitive region of the body, containing sensory receptors for sapid chemicals as well as for innocuous and noxious mechanical and thermal stimuli impinging on both hard and soft tissues. The nucleus of the solitary tract (NST) is a unique recipient of gustatory information, but also receives significant innocuous mechanical and thermal input arising from the oral mucous membranes, a function the NST shares with the trigeminal system. In turn, the NST is the source of input to higher neural structures involved in processing gustatory and oral somatosensory information. This information is used for a variety of purposes, including the control of somatic and autonomic reflexes, the modulation of ingestive behavior, and the discrimination of sapid chemicals along intensive, hedonic, and qualitative "dimensions". Initial studies of the NST and related systems revealed only the most general outline of its neuroanatomical organization and of the chemosensitive profiles of individual gustatory cells. More modern neuroanatomical studies by Norgren and his colleagues[1-4] began to reveal the connectional complexities of this system, identifying a previously unsuspected brainstem relay and projections into the limbic, as well as the thalamo-cortical system. Over the last decade, increasingly precise and quantitative analyses have continued to disclose the complicated and heterogeneous nature of this system, but also have begun to delineate regularities in its organization.

This chapter focuses on recent studies and attempts to correlate neurophysiological, neuroanatomical, and neurochemical data to identify functional principles of the NST and its related neural systems. The most detailed review will concentrate on the NST itself, because that is the level at which the system has been most thoroughly studied. In addition, the organizational properties of the parabrachial nucleus (PBN), and the "lemniscal" forebrain systems receiving input from PBN will also be discussed. Finally, I will attempt to elucidate what common characteristics and systematic changes occur in the chemosensitive response profiles of individual gustatory neurons at different levels of the neuraxis.

II. NUCLEUS OF THE SOLITARY TRACT

A. CYTOARCHITECTURE AND AFFERENT INPUT

1. Cytoarchitecture

The rostral portion of the NST receives primary afferent terminations from oral, pharyngeal, and laryngeal gustatory and somatosensory receptors. The caudal portion receives primary afferent terminations from sensory receptors in more distal regions of the alimentary tract and from other viscera. Although the cytoarchitecture of the caudal region has received more attention and been studied in more species, recently, two laboratories have provided descriptions of NST neuronal morphology[5-7] and subnuclear organization[7] in the hamster that pay particular attention to the rostral region. Whitehead defines the rostral *division* in the hamster as that portion of the nucleus rostral to the level at which the NST first merges with the IVth ventricle. This division of NST includes almost all of the area which receives primary afferent input from the oral cavity *(vide infra)*. Primary afferent input from the pharynx and larynx extends into the immediately adjacent caudal division. Earlier investigators usually divided the rostral NST into only two subdivisions, medial and lateral.[8,9] Based on its cytoarchitecture, the more recent scheme suggests considerable additional heterogeneity. I will summarize the recent parcellation scheme of Whitehead[7] for the rostral division, in order to use it as a reference point for other anatomical and physiological observations when possible. It should be kept in mind, however, that the applicability of this parcellation to other species awaits systematic morphologic investigation.

The rostral division of the hamster NST can be further subdivided into lateral, central, medial, dorsal, and ventral subdivisions (Figure 1). The rostral

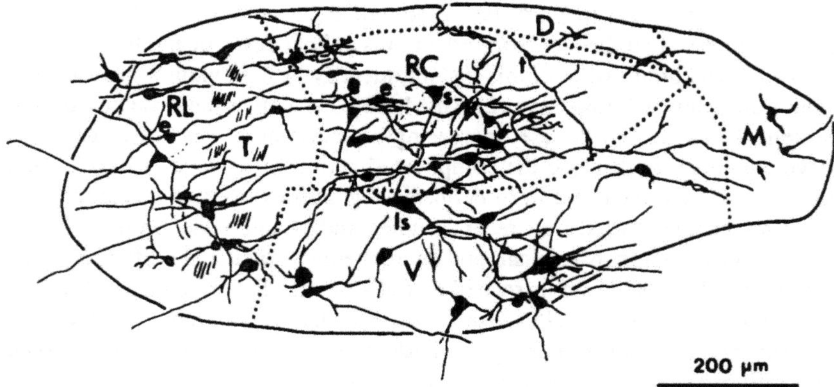

FIGURE 1. Rostral NST in the transverse plane reconstructed from two consecutive 80 μM sections near the rostral pole. The tissue was stained with the Golgi-Cox method. Upper case letters denote the various subdivisions: rostral lateral (RL), rostral central (RC), medial (M), dorsal (D), ventral (V), and the solitary tract (T). Lower case letters denote the different cell types: elongate (e), tufted (t), and stellate (s). A large stellate cell (ls) is present at the dorsal border of the ventral subdivision and its dendrites extend into the RC and M subdivisions. (From Whitehead, J., *Comp. Neurol.*, 276, 547, 1988. With permission.)

lateral division is adjacent to the solitary tract, bordering it medially at the rostral pole of the nucleus, but medially and laterally just caudal to the rostral pole (see Figure 1). The rostral central subdivision is immediately medial to the rostral lateral subdivision and is bordered medially by the medial subdivision. The dorsal subdivision overlies the rostral central and rostral lateral subdivisions, whereas the ventral subdivision lies below them. Three morphologically discrete cell types, elongate, stellate, and tufted, have been identified in the nucleus. Elongate cells possess only two primary dendrites, which are primarily oriented in the medio-lateral axis. Stellate cells possess more dendrites, which can radiate in all directions, though the medio-lateral axis is sometimes prominent. Tufted cells also may have several dendrites, but the length or branching of the dendrites defines a preferred orientation, usually in the dorso-ventral axis. Differences in the proportion, size, and packing density of their cellular content facilitate definition of the subdivisions mentioned above. Neurons in the rostral central subdivision are comprised of an assortment of all three cell types, predominantly small- or medium-sized and densely packed. Less densely packed elongate cells dominate the rostral lateral subdivision. Stellate cells dominate both the dorsal and ventral subdivisions, but the cells of the dorsal division are much smaller and have fewer dendrites. The medial division contains loosely packed stellate and tufted cells. Anatomical and physiological data discussed below suggest that at least some of these anatomical distinctions are accompanied by differences in function.

Although the rostral division contains an array of cell sizes and types, with dendrites oriented in all directions, the dendrites of the elongate cells, which are numerous in the rostral lateral and rostral central subdivisions, ramify prominently in the medio-lateral axis.[7] Another morphological description of the cell types in NST[6] classifies them somewhat differently, but, again, the medio-lateral axis is described as a prominent dendritic orientation for NST neurons. This could be functionally important since the terminal branches of primary afferent terminals are oriented in the same direction, which optimizes the opportunity for synaptic interaction.[6,7] When sections are cut in the horizontal plane,[5,6] it also becomes obvious that the dendrites of many NST neurons extend for long distances in the rostro-caudal axis. It is thus likely that dendritic processing in NST takes place over longer distances rostro-caudally and medio-laterally than dorso-ventrally.[5,6] Intranuclear dendritic processing, however, undoubtedly also occurs over short distances in the latter axis, e.g., from the ventral to the rostral central subdivision[7] (see Figure 1). To some extent, the preferred orientations of NST dendritic arborizations may simply reflect the dimensions of the nucleus, which is narrowest dorso-ventrally.

2. Primary Afferent Input

In mammalian species, branches of the Vth, VIIth, IXth, and Xth cranial nerves transmit gustatory and somatosensory information from the oral cavity, pharynx, and larynx to recipient neurons in NST. The NST appears to be the major, probably exclusive, central recipient of gustatory information, but somatosensory input from these regions also reaches the brainstem trigeminal nuclei. The projections of the primary afferent nerves carrying gustatory information (VII, IX, X) have a common pattern in the many mammals in which they have been studied, synapsing in the nucleus in a rostral to caudal sequence, which mirrors their rostral to caudal entry into the brainstem (reviewed in Reference 10). Terminations from the chorda tympani (CT) branch of VII, which innervates taste buds on the fungiform papillae of the anterior tongue, are heaviest in the rostral pole of NST. In the rat, terminations from the greater superficial petrosal (GSP) branch of VII, which innervates the taste buds associated with the nasoincisor ducts of the hard palate, as well as the majority on the soft palate,[11] overlap substantially with those occupied by the CT, but are probably centered more caudally (see Figure 6 in Hamilton and Norgren[12]). The heaviest terminations from the lingual-tonsillar branch of IX, which innervates taste buds of the foliate and circumvallate papillae on the posterior tongue and in the nasopharynx,[13] are caudal to the terminations of VII. In the hamster, the terminals of the VIIth and IXth nerve are both heaviest in the rostral central and the rostral lateral NST subdivisions,[7] albeit at different rostro-caudal levels. The most caudal gustatory projection is from the superior laryngeal (SLN) branch of the Xth nerve, which innervates laryngeal receptors. Terminations from this nerve appear to be centered approximately at the level where the NST merges with the IVth ventricle, so that many terminals extend into the caudal division, in contrast to the distributions from VII or IX, which terminate most heavily in the rostral division. The peripheral and central connections of these gustatory cranial nerve branches suggest a somatotopic arrangement of taste information in the NST in the rostral-caudal axis. The existence of substantial overlap in the projections from the three afferent nerves in this axis, however, also provides an anatomical substrate for integration of gustatory afferent information arising from different taste bud subpopulations, as does the extension of dendrites in this axis. Recent, somewhat more detailed reports of the primary afferent projections, however, indicate that there may be somewhat less overlap than appreciated previously. Thus, the CT and IXth nerves do send some projections to the same level of NST, but the dominant IXth nerve projection is more dorsal than the projection from the CT.[14] Similarly, in the lamb, both IX and SLN send projections to the region of the interstitial subnucleus of NST, but the IXth nerve projection is centered more dorsally and laterally than that from SLN.[15]

The existence of trigeminal nerve projections to NST is well documented.[8,9,12,16-18] Anterograde tracing studies of the projections of individual nerve branches suggest that only intraoral regions have primary connections in the nucleus and that certain intraoral structures are represented more heavily than others. Thus, the lingual nerve, which innervates somatosensory receptors on the anterior portion of the tongue, has heavy projections to NST,[12,19] while the inferior alveolar nerve, which innervates the mandibular teeth and periodontium, has been reported to have only very light[12,20] or no[21] solitary connections. Branches of the trigeminal nerve which innervate exterior surfaces, such as the corneal, supraorbital, infraorbital, and mental, appear not to project to NST (but see Reference 22). There are conflicting opinions as to whether trigeminal afferents innervating the nasal mucosa reach the nucleus.[23,24] The projections from trigeminal nerves overlap with projections from other afferent cranial nerves but the largest degree of overlap appears to be with the more lateral projections arising from IX.[7,12]

The synaptic relationships that primary afferent endings enter into in NST are not as well understood as are the analogous relationships for primary afferent endings in other sensory systems. One thorough study, however, provides some insights into these relationships for facial primary afferent endings.[25] Perhaps the most salient observation is that facial primary afferent fibers do not make synaptic contact with second order cell bodies in NST. Instead, these primary afferent endings synapse almost exclusively onto dendrites or dendritic spines. The most frequent relationship appears to be simple: an asymmetric contact from a primary afferent ending onto the postsynaptic process, apparently with no other endings involved. More complex relationships, however, are also frequent. For example, some dendritic spines receive input both from a primary afferent fiber and from an additional (unidentified) source. Because these electron microscopic observations were not serial reconstructions, it is likely that even more complex relationships would be apparent if all of the input onto a single neuron were examined. Thus, these initial observations of the synaptology of primary afferent gustatory endings in the NST offer evidence for an anatomical substrate for the complex processing of gustatory information at the first order afferent relay.[25]

3. Central Afferent Input

The NST also receives afferent input from spinal, brainstem, and forebrain sources. A recent retrograde tracing study documents strong projections from the dorsal horn of the spinal cord and the spinal trigeminal nucleus to NST.[26] In this study injections of a retrograde tracer were placed in NST, but they were centered in the caudal, visceral division, i.e., mainly caudal to sites receiving heavy primary afferent input from the oral cavity. It is unclear whether this projection extends to the rostral division of NST, although there were earlier suggestions that the projection is limited to caudal NST.[8] Similar uncertainty exists with regard to the reported raphe[27] projections to NST since evidence for these projections is based on retrograde tracer injections centered in the caudal division. A recent report suggests that a region of the central gray projects to NST, but these projections do not appear to extend rostral to the level of NST that is coincident with the hypoglossal nucleus.[28] Other central afferents to NST, however, do extend along the entire rostro-caudal axis, although the magnitude of the projection varies, and a topographic organization is evident for some projections. Van der Kooy et al.[29] carried out an extensive analysis of forebrain projections to NST, utilizing both anterograde and retrograde tracing techniques. This study demonstrated that the prefrontal cortex, insular cortex, bed nucleus of the stria terminalis, central nucleus of the amygdala, lateral hypothalamus, and paraventricular nucleus of the hypothalamus all project to the entire rostrocaudal extent of NST. Other anterograde tracing studies suggest that the projection from the insular cortex to NST is topographic, with the rostral insular cortex projecting to the rostral NST and the more caudal insular cortex projecting to the caudal NST.[30,31] Neurophysiological studies have suggested that the rostral insular cortex is the gustatory region, whereas the caudal insular cortex receives visceral afferent input.[32] Thus, the topographic organization of descending cortical projections suggests that they maintain functional segregation. Finally, the subjacent reticular formation may also project to NST, since reticular neurons are often filled following injection of horseradish peroxidase into NST.[33,34] Such labeling, however, may be due to anterograde transport via reticular dendrites that extend into NST.[7,33]

B. NEUROCHEMISTRY

In comparison with other sensory systems, our knowledge of the neurochemistry of the gustatory system is meager. Krueger and Mantyh[35] recently reviewed this topic for the *Handbook of Gustatory Neuroanatomy*. By necessity, however, the majority of the chapter actually deals with what is known about the physiological and anatomical organization of the gustatory system, and about the neurochemistry of CNS regions *adjacent* to gustatory areas. The adjacent regions often contain high concentrations of a variety of neuropeptides, whereas gustatory areas contain much lower amounts and fewer of these substances.[35] The neurochemical organization of the NST exemplifies this pattern. For example, a recent study characterized the distribution of cell bodies containing three neuropeptides: enkephalin, neurotensin, and neuropeptide Y, within NST.[36] Labeled cell bodies containing these peptides were numerous in caudal NST, but much less frequent in rostral NST. Indeed, judging from the location of the plotted cells, only enkephalin-containing neurons were in a location likely to overlap locations where our laboratory has observed orally responsive neurons, i.e., a few enkephalinergic cells were located in the most caudal and lateral portion of rostral NST, near to the most caudal limit of neurons responsive to oral mechanical stimulation. A small number of neurons containing the other two neuropeptides were also found at this level of NST, but were restricted to the most medial pole of the nucleus, medial to orosensory responses. Similarly, the distribution of cholecystokinin-, galanin-, and corticotropin-releasing factor containing neurons,[37] as well as several other peptides, also appears to be largely restricted to the caudal NST, as are certain peptide-containing fibers and receptors.[35] Both substance P-containing fibers and receptors, however, do appear to extend to rostral regions.[38] Indeed, using a slice preparation, substance P has recently been demonstrated to have depolarizing effects on 20/43 neurons tested.[39]

A few recent studies have begun to concentrate on the distribution of neurotransmitters and histochemical staining properties of neurons in the rostral, orosensory NST. Consistent with their rather ubiquitous distribution in the brain, amino acids appear to function as neurotransmitters in orosensory NST. Neurons containing gamma amino butryric acid (GABA), normally thought to be an inhibitory transmitter,[40] are located in gustatory regions of NST.[41] Interestingly, it appears likely that these neurons are local-circuit cells, since they project neither to PBN, nor to more caudal regions of the NST or reticular formation.[41] Recent *in vitro* recording studies have observed the effects of GABA on rostral NST neurons. Consistent with the results of the immunohistochemistry, in slice preparations, a large proportion of the neurons tested are inhibited by application of GABA.[42,43] The function of these cells in processing gustatory information *in vivo* is not known. Although response decrements evoked by gustatory stimuli have been reported in NST, their incidence is rather low (e.g., Reference 44). Thus, it seems likely that the numerous GABA-ergic neurons normally affect processing in a more subtle way. For example, they could contribute to the nonlinear integration in NST neurons, as has been observed for that summation of responses arising from spatially distinct gustatory receptors.[45] Although the distribution of glutaminergic cell bodies and terminals has not been specifically studied in the rostral NST, this amino acid, thought to function as an excitatory neurotransmitter,[40] is found in more caudal NST regions.[46] In addition, a recent study documented the distribution and function of this neurotransmitter in the vagal (intraoral gustatory) lobe of the fish.[47] It would appear that glutamate functions as an excitatory neurotransmitter in the vagal lobe, but it does not appear to be a primary afferent transmitter. The distribution of dopamine has also been studied in "gustatory" NST.[48] This transmitter is present in at least 10 to 15% of neurons in the gustatory NST in hamster, but in a smaller proportion in rat. The dopaminergic neurons were much more numerous in the "ventral" level of NST[48] defined in sections cut horizontal to the brainstem surface.[49] This level appears to coincide with the junction of the rostral and caudal NST divisions as defined by Whitehead,[7] i.e., the most caudal portion of orosensory NST, according to our electrophysiological studies *(vide infra)*.

In addition to studies of neurotransmitter distribution in NST, recent studies have also begun to elucidate the histochemical staining properties of NST neurons, as characterized by cell surface macromolecules and enzyme content. The plant lectin *Griffonia simplicifolia* I-B_4, which specifically binds to D-galactose residues present on neuronal cell membranes, stains presumed gustatory cell bodies in the geniculate and petrosal ganglia, as well as fibers in the rostral NST.[50] This lectin also stains certain classes of somatosensory fibers, so it is not a specific marker for gustatory cells. Staining for NADH-dehydrogenase (a mitochondrial enzyme) activity is also quite intense in gustatory NST.[51] Interestingly, this staining appears somewhat selective; it is most robust at the rostral pole of NST, in the regions receiving the heaviest input from the CT nerve. This correlates well with electrophysiological studies which indicate that gustatory neurons with receptive fields in the anterior oral cavity have higher spontaneous rates than such neurons with receptive fields in the posterior oral cavity.[52]

C. PHYSIOLOGY

1. Topographic Organization

a. Gustatory Orotopy

The topographic projections of primary afferent nerves into NST provide an anatomical basis for predicting that gustatory responses in the nucleus will be arranged according to the location of their receptive field, i.e., *orotopically*. On the other hand, overlap in primary afferent input,[9,12,19] intranuclear projections,[4,53] and dendritic arborization[5-7] preclude making a foregone conclusion. Both single-unit[45,52,54,55] and multi-unit studies[56,57] of the topographic organization of NST in rodents, however, do support this prediction. The organization of gustatory responses is obvious when the several spatially distinct subpopulations of oral taste buds* are stimulated individually, or when the relative magnitude of the responses arising from the anterior and posterior halves of the oral cavity stimulation are compared.[56,57] The observed physiological organization is consistent with the relative densities of primary afferent terminations and is quite simple. As depicted by the filled symbols in Figure 2, gustatory responses arising from the anterior oral cavity (AO) tend to be represented anterior and/or lateral to responses arising from the posterior oral cavity (PO).[52,54,55] It is interesting that spatially separate subpopulations that are innervated by the same afferent nerve branch are no exception to this organization. Thus, the taste buds of the nasoincisor ducts, which are located in the AO, and those on the soft palate, located in the PO, are both innervated by the greater superficial petrosal nerve,[11] but responses to sapid stimulation of the nasoincisor ducts are usually found anterior to responses to sapid stimulation of the soft palate in NST. A similar relationship exists for the representation of responses arising from stimulation of the anterior tongue and more posterior retromolar mucosa,[55] whose taste buds are both innervated by the chorda tympani nerve.[58] Studies which have attempted to stimulate all the oral gustatory receptors suggest that oral taste responses are found over most of the rostrocaudal extent of the rostral division of NST but do not appear to extend into the caudal division.[55,56,59,60] A recent study in the hamster which reconstructed the recording sites of single taste-responsive NST neurons, based upon extracellular horseradish peroxidase deposits and cytoarchitecture, further demonstrates that gustatory neurons are confined to the rostral central and the medial portion of the rostral lateral subnuclei of the rostral division.[61] Studies of lamb NST neurons that are responsive to chemical stimulation of the epiglottis demonstrate that they are centered at or just rostral to the junction of the rostral and caudal subdivisions of NST,[62,63] suggesting an overall rostral to caudal organization of gustatory responses in the NST that mirrors their rostral to caudal peripheral location.

b. Segregation by Modality

Somatosensory information from the oral cavity reaches the solitary nucleus via various branches of the trigeminal nerve and by way of the lingual branch of the glossopharyngeal nerve, which contains not only gustatory-responsive fibers, but also fibers responsive to somatosensory stimulation of

* Throughout the text, the term "taste bud subpopulation" refers to a group of taste buds that is spatially distinct from other such groups. For example in the rat oral cavity, taste buds form at least seven distinct groups: (1) on the fungiform papillae of the anterior tongue, (2) within the nasoincisor ducts that open into the incisal papilla of the hard palate, (3) on the sublingual organ near the openings of the submandibular salivary gland, (4) on the soft palate, (5) in the trench of the midline circumvallate papilla, (6) in the folds of the foliate papillae, and (7) on the retromolar mucosa adjacent to the foliate papillae (reviewed in Reference 228). A similar organization occurs for most mammalian species, with some variations. We consider the first three subpopulations listed to lie in the anterior oral cavity, and the remaining subpopulations to lie in the posterior oral cavity.

the posterior tongue[64] and palate.[65] Although many gustatory-responsive neurons in NST also respond to somatosensory stimulation *(vide infra)*, there is also a substantial population of NST neurons that appears to respond exclusively to somatosensory (usually mechanical) stimulation of the oral cavity.[54,55,59,60,66,67] The number of gustatory- and mechanically responsive NST neurons appear to be roughly equivalent,[55,60] suggesting that the solitary nucleus is equally important in processing information pertaining to these two modalities. Recent work in our laboratory suggests that, in the rat, neurons responsive to oral gustatory stimulation, and those exclusively responsive to oral mechanical stimulation, are not intermingled, but form segregated, albeit overlapping distributions in the rostral subdivision of NST (compare locations of open [somatosensory] vs. closed [gustatory] symbols in Figure 2). Unlike gustatory neurons, mechanically responsive neurons were never found at the rostral pole of NST, but were first encountered approximately 0.5 mm caudal to the border. They then dominated the lateral 1/3 of the nucleus for the remainder of the rostral subdivision, extending for 0.5 mm into the caudal subdivision, where orally responsive taste neurons were never found. As the density of orally responsive taste neurons began to diminish caudally, mechanically responsive neurons began to dominate the medial, as well as the lateral, portions of the rostral subdivision. Thus, the rostral and medial portions of the rostral subdivision appear specialized for processing oral gustatory signals whereas the lateral and caudal regions are specialized for processing comparable somatosensory signals. It is tempting to speculate that this differential distribution corresponds to cytoarchitectonic boundaries observed in the hamster, with oral mechanical responses dominating the rostral lateral, and oral taste responses dominating the rostral central subdivisions. This hypothesis, however, requires establishing that the rostral division of the rat NST has the same cytoarchitectonic subdivisions as the hamster, and relating physiologically defined responses to the boundaries.

Oral somatosensory responses exhibit orotopy that is organized in a fashion similar to that exhibited by taste responses. The orotopy is not precise and is obscured by the rather large size of the receptive fields, which often span the AO and PO.[55] When the receptive fields of mechanically responsive neurons are classified according to whether AO or PO stimulation evokes the most vigorous response, however, an orderly organization of those responses is very apparent, with the AO represented anterior and lateral to the PO (see Figure 2). Anatomical data from Altschuler et al.[68,69] suggests that the orotopic pattern of gustatory and somatosensory responses in NST is simply a specific instance of a general pattern of alimentary tract representation in the nucleus. These investigators made injections of wheat germ agglutinin conjugated to horseradish peroxidase or cholera toxin-HRP into several peripheral alimentary tract structures and the nerves innervating them, and observed the distribution of their terminals in NST. As pointed out by the authors, these studies demonstrated that the representation of the soft palate, pharynx, esoph-

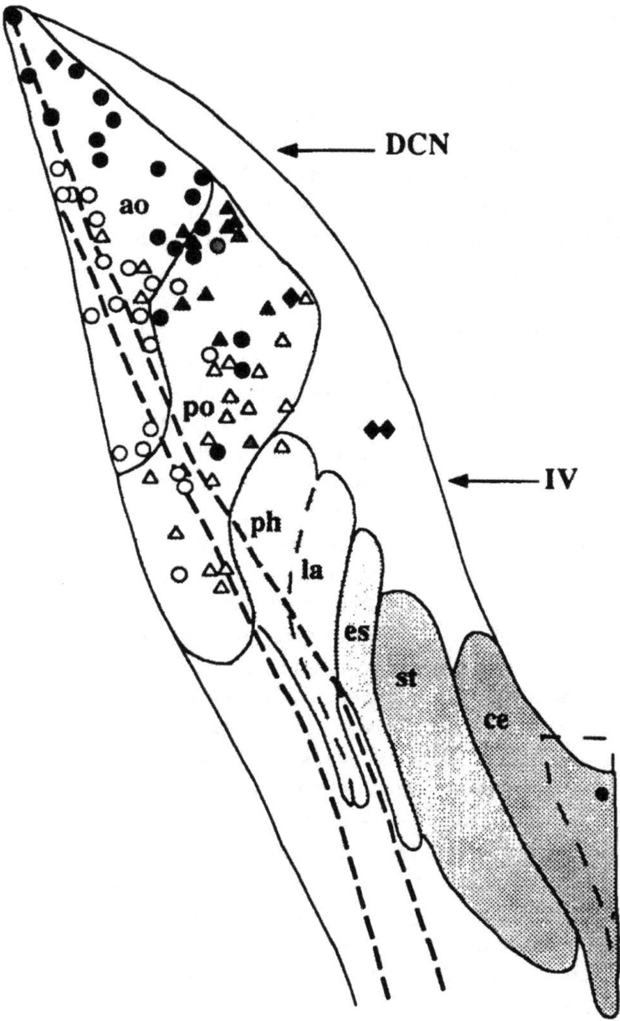

FIGURE 2. Schematic diagram of the NST viewed in the horizontal plane (outline modified from Reference 12), depicting alimentary tract representation in the nucleus. Rostral = top, medial = right. "DCN" indicates the level of NST that is coincident with the caudal limit of the dorsal cochlear nucleus; "IV" indicates where the medial border of the NST merges with the IVth ventricle. Successively more caudal NST regions represent successively more caudal alimentary tract structures: the anterior oral cavity (AO), posterior oral cavity (PO), pharynx (PH), larynx (LA), esophagus (ES), stomach (ST), and cecum (CE). The locations of the anterior and posterior oral cavity representations are based upon single-unit data collected in this laboratory.[55] The filled and open symbols depict neurons that responded to gustatory or only oral mechanical stimulation, respectively. Gustatory stimulation was accomplished by applying a mixture of the four standard taste stimuli (1.0 *M* sucrose, 0.3 *M* NaCl, 0.03 *M* HCl, and 0.01 *M* quinine HCl) to individual taste receptor subpopulations using a camel's hair brush and by

agus, stomach, and cecum occupy nested, elongated areas in progressively more medial and posterior regions in NST. Figure 2 summarizes our physiological data on oral gustatory and mechanical responses and the anatomical data of Altschuler and his colleagues[68,69] (see Figure 15,[68] Figure 6[69]), and also takes into consideration physiological data on distribution of epiglottal chemoresponsive neurons from Sweazey and Bradley[62,63] (see Figure 1,[62] Figure 1[63]) and anatomical data on the terminations of various branches of the subdiaphragmatic vagus from Norgren and Smith[70] (Figure 8[70]). Note that there appears to be a gap between the orally responsive gustatory region [closed symbols] and the region representing the larynx, which also contains chemoresponsive neurons.[62,63] It is possible that this gap is artifactual. In our studies of oral gustatory responses, we used techniques that may not have effectively stimulated the taste buds of the circumvallate papillae (the most posterior of the oral taste bud subpopulations). Similarly, the pharynx is known to contain taste buds,[71] but the anatomical techniques which revealed the pharyngeal representation cannot distinguish between the somatosensory and gustatory fibers. Thus, it is possible that gustatory responses span a continuous region that is medial to the NST somatosensory representation.

2. Convergence

a. Spatial

Several neurophysiological studies suggest that spatial integration of gustatory information is present at the level of the first central relay (Table 1). Indeed, such integration actually begins in the periphery. Individual chorda tympani fibers branch and innervate multiple fungiform papillae, and, correspondingly, a single fiber can respond to gustatory stimulation of multiple papillae.[72-75] Spatial integration continues in the NST and expands to encompass convergence of input from spatially separate taste bud subpopulations. When the anterior tongue was isolated within a tongue chamber, 64% of the gustatory neurons recorded in the hamster NST responded to independent stimulation of the anterior tongue as well as to stimulation of taste receptors located outside the tongue chamber.[45] This technique for stimulation provides

FIGURE 2 (continued).
flowing this mixture over the entire oral cavity. Mechanical stimulation consisted of stroking specified regions of the oral cavity using a blunt glass probe. Circles and triangles represent cells that responded maximally to AO or PO stimulation. For some gustatory neurons, however, the only effective gustatory stimulus was whole mouth stimulation. The latter cells are depicted by diamonds. The representations of the remainder of the alimentary tract are based upon the anterograde tracing studies of Altschuler and his colleagues,[68,69] and Norgren and Smith,[70] as well as the neurophysiological data of Sweazey and Bradley[62,63] on the location of epiglottal chemosensitive neurons (see text). The representations of the pharynx and larynx are separated only by a dotted line because they overlap greatly based on peripheral injections into these organs.[68] On the other hand, an anterograde tracing study by Sweazey and Bradley[15] suggests the IXth nerve is represented somewhat lateral to the Xth nerve at levels where the representations coexist in the NST. This distribution could reflect only the relative positions of oral vs. laryngeal responses,[62] but is also consistent with a similar relationship between pharyngeal and laryngeal responses.

TABLE 1
Convergence

Level	Reference	Spatial convergence, gustatory responses (%)	Gustatory/Mechanical Convergence (%)
NST	80	NR	62
NST	66	4.5	NR
NST	60	NR	64
NST	59	43	NR[a]
NST	52	53	NR
NST	45	64	NR
NST	76	54	NR
NST	55	49	73
Ave. NST		**41**	**64**
PBN	142	61	NR
PBN	176	50	69
PBN	143	24	NR
PBN	144	NR	84.5
Ave. PBN		**45**	**77**
VPMpc	168	29 (43 bilateral)	88
VPMpc	167	55	NR
Ave. VPMpc		**42**	**88**
Cortex[b]	187	NR (41 bilateral)	0
Cortex	192	NR	16
Cortex	190	NR	67
Ave. Cortex		**NR**	**28**

[a] "NR" indicates that the value was not reported.

[b] Cortical values are for primary gustatory cortex; i.e., insular cortex in rodents[187,190] and insular-opercular cortex in the monkey.[192] Except for the single study of the cortex in monkeys, the data in the rest of the table are from rodents.

limited resolution with regard to the contribution of individual receptor subpopulations to the observed convergence. Several subpopulations of taste buds are stimulated outside the tongue chamber, including taste buds located on posterior fungiform papillae, making it possible that the observed convergence was due simply to convergence between different fungiform papillae. More specific stimulating techniques, however, confirm convergence between spatially segregated receptor subpopulations.[52,55,59,66,67,76] Our investigations that separately stimulate individual oral taste bud subpopulations suggest that about 50% of the gustatory neurons in NST receive input from spatially discrete subpopulations, but that there appears to be an orderly pattern to this convergence.[52,55] As might be expected from the orotopic organization of gustatory responses in NST, more convergence appears to occur between receptor subpopulations within the AO (i.e., the anterior tongue, nasoincisor ducts, and sublingual organ) or PO (i.e., the foliate papillae, soft palate, and retromolar mucosa) than between AO and PO receptor subpopulations. In one study, in which the anterior tongue, nasoincisor ducts, foliate papillae, and soft palate were separately stimulated, convergence between the anterior tongue and nasoincisor ducts, which directly appose each other in the AO, accounted for 70% of the observed convergence.[52] Indeed, convergence between the anterior tongue and soft palate occurred less frequently than expected from chance alone, suggesting a possible bias against convergence of AO and PO input. In that study the portion of NST most extensively sampled was the rostral pole, and, accordingly, many more neurons were AO-responsive than PO-responsive. Thus, these data did not allow an adequate assessment of whether similar patterns of convergence occurred between PO taste bud groups. Our recent observations, drawn from a more evenly sampled NST population, suggest that this may be the case but emphasize the importance of quantitatively assessing the degree of convergence. In this study, 17/35 neurons with identified receptive fields received input from more than a single receptor subpopulation. If all responses reaching the threshold criterion were considered, about an equal number of neurons (9 vs. 8) received convergent input from subpopulations within the AO or PO, as between these 2 oral regions. On the other hand, if response magnitude was taken into account, a more orderly pattern was apparent. When only responses that were at least 20% of the response to the most effective receptor subpopulation were considered, fewer neurons received convergent input ($n = 11$), but only 2 of those cells received input from both the AO and PO. The remaining convergent cells were nearly equally divided between those neurons receiving convergent input from receptor subpopulations within the AO or PO. Similarly, the 2 most effective receptor subpopulations for a given convergent neuron nearly always (in 15 of 17 cases) were within the AO or PO. Figure 3 depicts responses from two convergent neurons in NST. Neuron 76 1-1-1 (left column) received input from the AO *and* PO (i.e., responses from both parts of the oral cavity reached criterion), but the AO inputs were much larger than those from the PO. Stimulation of the anterior tongue produced the largest response, which was ~1.8× as large as that produced by stimulating the nasoincisor ducts, but ~7.7× and ~10.8× as large as the responses produced by foliate and soft palate stimulation, respectively. Neuron 81 3-4-1 received all its inputs from PO structures, responding best to soft palate stimulation, but only a little less vigorously to foliate and about half as well to stimulation of the retromolar mucosa. Data from other laboratories also suggests that spatial convergence in NST is orderly. Sweazey and Bradley[63] recorded from several mechanically sensitive neurons in NST that received input from both the posterior tongue and palate. The lingual and palatal receptive fields for each of these convergent neurons exhibited a striking match, being in direct apposition to one another. In that same study, a population of NST neurons responsive to epiglottal stimulation were encountered. Interestingly, there was only a single instance of convergence between oral and epiglottal input, despite the fact that oral and epiglottal responses were near to one another in the

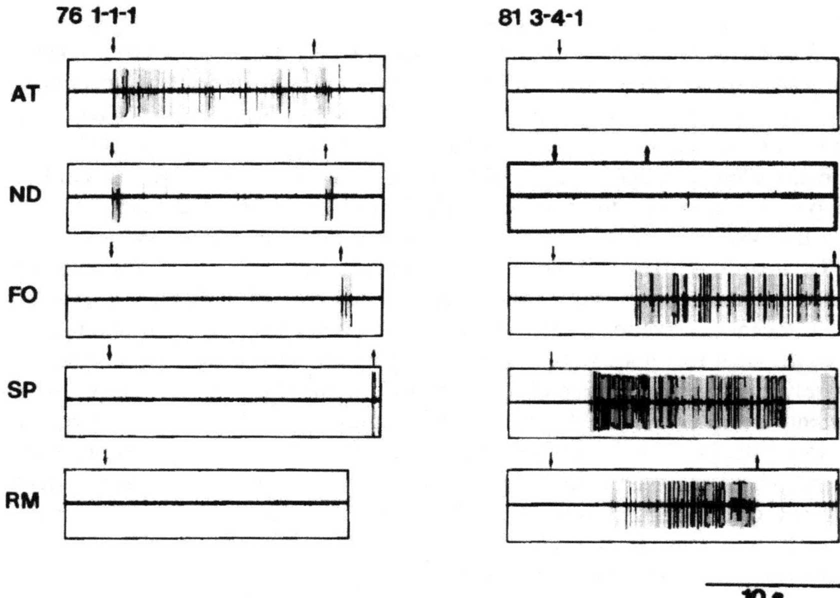

FIGURE 3. Oscillographic records for two NST neurons receiving convergent input. The neuron on the left (76 1-1-1) received its strongest input from the anterior oral cavity, the anterior tongue (AT), and nasoincisor ducts (ND); the neuron on the right (81 3-4-1) received input only from the posterior oral cavity, but from three different receptor subpopulations: the foliate papillae (FO), soft palate (SP), and retromolar mucosa (RM). Responses were evoked by independent stimulation of the AT, ND, FO, SP, and RM with a mixture of the four standard taste stimuli (1.0 *M* sucrose, 0.3 *M* NaCl, 0.03 *M* HCl, and 0.01 *M* quinine HCl). The arrows indicate the approximate times of stimulus onset (up arrow) and rinse (down arrow). Interestingly, the long latencies for the responses to FO and RM stimulation for neuron 81 3-4-1 were probably due to different mechanisms. The latency of the FO response probably was due to the time necessary for the stimulus to diffuse to the receptors, which are located within the FO trenches. The latency of the RM response appeared to be due to an initial mixture suppression, since the response to sucrose presented individually was robust and immediate.

nucleus. These data from the sheep support the idea that there are orderly organizational principles governing spatial convergence in NST. This spatial pattern of integration in NST appears to be appositional. The functional significance of this organization is probably related to the spatiotemporal pattern of stimulus contact with different taste bud groups during ingestion. Although the details of this pattern are not known, an anterior to posterior progression, paralleling the direction of intraoral transport, is the most likely possibility. During licking, sucking, or mastication, ingesta is not confined to the tongue or palate, but is manipulated between these structures in an anterior to posterior progression.[77,78] Thus, gustatory convergence would be likely to facilitate the detection of ingested substances either simply by increasing the probability of the interaction between tastant molecules and gustatory receptors or by actually summing afferent input from multiple sets of receptors. The spatial pattern of convergence appears to be designed to maximize detection for *limited*, functionally related segments of the oral cavity, i.e., those segments that would be likely to be nearly simultaneously stimulated at different times during the ingestive sequence. It has been pointed out that the different levels of the NST, which represent different regions of the oral cavity, appear to control distinct phases of the oropharyngeal consummatory sequence (reviewed in Reference 79).

b. Intermodal

Single NST gustatory neurons also sometimes respond to additional modalities. The most frequently observed instance is for gustatory and intraoral somatosensory stimulation to be effective in activating the same cell. Both thermal and mechanical stimulation are effective somatosensory submodalities and a majority of the taste-responsive neurons tested have been reported to respond to at least one of these two modalities.[55,60,63,67,80] It is unclear how much of this dual responsiveness arises from convergence of gustatory and somatosensory primary afferent input rather than simply reflecting the dual responsiveness of single primary afferent fibers. The thermal responsiveness of chemosensitive chorda tympani fibers (e.g., References 81,82), and the mechanical sensitivity of chemosensitive chorda tympani[75] and particularly of superior laryngeal nerve fibers,[83-85] however, have been well documented. Thus, a quantitative comparison of the responsiveness of elements at these two levels of the nervous system and selective stimulation/denervation experiments would help to establish the basis of the dual responsiveness of NST cells. Although the reported incidence of dual responsiveness to gustatory and somatosensory stimulation is high, in most studies the relative responsiveness of single neurons to the two modalities is not clear. This information is essential in establishing the functional significance of somatosensory responses in gustatory-responsive neurons and vice versa. Recent data suggest that the degree of relative responsiveness could depend, in part, upon the location of a neuron's receptive field. In the sheep NST, neurons that received input from the epiglottis were reported to be nearly equally responsive to gustatory and mechanical stimulation.[63] In contrast, our recent data for rat NST neurons show that those neurons that were dually responsive to oral taste and mechanical stimulation responded about twice as well to the gustatory stimulus.[55] Interestingly, in our sample, the proportion of gustatory-responsive neurons that were responsive to oral mechanical stimulation also depended on the gustatory receptive field: only about half of the AO gustatory-responsive neurons responded, versus *all* of the PO gustatory-responsive cells. Taken together, these data suggest an increasing tendency for NST taste neurons that represent successively more caudal peripheral regions to be increasingly responsive to somatosensory stimuli. In this light, it may be significant that taste fibers in the superior laryngeal nerve, unlike their counterparts in the chorda tympani and glossopharyngeal, cannot be classified into discrete types based upon their chemoresponsive profiles.[86] Thus, in contrast to oral chemosensitive neurons, laryngeal chemosensitive neurons do not appear to be concerned with differentiating among different chemical compounds, but merely with detecting their presence in the larynx. Their nearly equivalent dual responsiveness to chemical and mechanical stimulation would facilitate such a function.[63] The poorer responsiveness of oral gustatory neurons to mechanical stimuli would be consistent with a more specific role in gustatory discrimination for these cells. The function of somatosensory input in oral gustatory neurons is yet to be established, but it could serve mainly to enhance or otherwise modulate taste responses.

One report suggests that some gustatory cells can be coactivated by volatile chemical stimuli applied to the nasal mucosa.[87] The afferent pathways that contributed to the observed responses were not entirely clear, although the ethmoid nerve probably played some role. Because the stimulation procedure did not rigorously limit the area of contact to the nasal mucosa, it is also possible that some of these responses arose from stimulation of peripheral structures other than the nasal mucosa, such as the pharynx, larynx, or even the posterior tongue. Detailed information was not given in the report, but it would appear that the responses to the volatile chemicals were considerably smaller than the responses to taste stimulation. The substrate and significance of this intriguing phenomenon awaits further investigation.

Responses to gustatory stimulation in the rodent NST can be also be modulated by gastric distension,[88] intravenous glucose,[89] or insulin,[90] and sodium deprivation,[91] which suggests that gustatory-visceral interactions can also occur in the solitary nucleus. Electrical stimulation of the afferent fibers in the vagus nerve, however, appears to be ineffective for stimulating single gustatory neurons,[34] (but see Reference 92), so that the visceral interactions may be mediated via mechanisms other than the visceral afferent pathway. It should be noted that the modulation of gustatory responses in NST by homeostatic state appears to be species-specific. Thus, all the interactions noted above were demonstrated in experiments using the rat. When responses of gustatory neurons in the monkey NST were monitored while the monkey was fed to a state of satiety, no changes were noted.[93] Instead, such changes were limited to forebrain neurons in the limbic system or secondary gustatory cortex, suggesting an evolutionary trend for gustatory-visceral interactions to become encephalized.[94]

3. Descending Influences

As suggested by the anatomical data discussed earlier, gustatory afferent responses in NST can be modulated by forebrain influences. Thus far, interest in the effects of stimulation has centered largely upon the lateral hypothalamus,[95-97] a region long implicated in facilitating feeding behavior, and which reliably produces feeding behavior upon electrical stimulation.[98] Data from these studies demonstrate that neurons in the NST that can be activated by electrical stimulation of the chorda tympani nerve,[95] or oral gustatory stimulation,[96,97] also may respond to electrical stimulation of the lateral hypothalamus, but the incidence of coactivation in these studies varies from 25[97] to 62%.[96] Electrical stimulation of the hypothalamus appears to be capable of both increasing and decreasing the spontaneous activity of gustatory neurons, and, similarly, applying a conditioning stimulus to the hypothalamus is capable of both facilitating and depressing the response of a gustatory cell to subsequent electrical stimulation of the chorda tympani or tongue.[95,96] Effects of stimulating other forebrain structures having descending input to the rostral NST have not been directly assessed, but the overall influence of forebrain inputs on solitary nucleus neurons has been studied by comparing the properties of these neurons in intact vs. decerebrate animals.[59,99] The most compelling difference in the decerebrate state is a depression of the spontaneous and evoked firing rates of NST gustatory neurons, which suggests an overall facilitory input from the forebrain. Greater insight into the functional significance of forebrain-NST influences would be gained from a more detailed knowledge of the relationships between the origin and direction of various forebrain effects and the properties of the gustatory response profiles of the affected neurons. One logical prediction, for example, might be that the effects of lateral hypothalamic stimulation would vary according to the hedonic properties of the optimal gustatory stimuli for a given cell. Such specificity in descending modulatory influences would be consistent with the effects seen in other sensory systems, e.g., stimulation of the descending inputs originating in the central gray excites dorsal horn neurons responsive to low-threshold stimulation, but inhibits those responsive to high-threshold stimulation.[100]

D. EFFERENT PATHWAYS

The fundamental properties of the ascending gustatory pathway in rodents were elucidated in a classic series of experiments performed in rat by Norgren and his colleagues during the 1970s.[1-4] Neurons in the rostral, gustatory-responsive NST send axons into two main systems, an ascending pathway terminating in the medial half of the PBN of the pons, and a descending pathway, terminating in the more caudal solitary nucleus, parvicellular reticular formation, and lightly in the hypoglossal nucleus. Other investigators have confirmed and elaborated this pattern and have generalized the observations to hamsters.[33,53,101,102] Lagomorphs probably have a similar ascending organization since gustatory responses have been recorded in the PBN of the rabbit.[103-105] Primates, however, appear to be an exception. In the monkey, projections from the rostral pole of NST bypass the pontine relay and project directly to the thalamus.[106]

The parabrachial subnuclear location,[107] *(vide infra)* of NST terminations has recently been clarified.[101] Earlier studies demonstrated that these rostral

NST axons terminated in the medial half of both the classically described medial and lateral PBN (i.e., both dorsal and ventral to the brachium conjunctivum) in the caudal half of the nucleus.[1,2,4,53] The more recent data[101] delimits the medial portion of the PBN termination field to the medial and ventral lateral subnuclei[107] of the rat. Further, this latter study used WGA-HRP as an anterograde tracer, and thus was able to differentiate presumed terminal varicosities in the external medial subnucleus, which is located lateral to the brachium conjunctivum. Since the earlier anterograde tracing studies used autoradiographic or degeneration techniques, this labeling may have been presumed to represent fibers of passage, since many fibers that travel more medially do pass through this region.

Although the general organization of efferents arising from the rostral pole of the NST have been well defined, the intrinsic organization of these connections is only beginning to be determined. The overall efferent organization of the nucleus and the intrinsic organization of efferent projections from the caudal NST, however, suggests that the efferent organization of NST is heterogenous. Despite some overlap,[34] the rostral and caudal divisions of the NST make distinctive contributions to the ascending PBN projection. Input to the medial PBN arises nearly exclusively from the rostral NST, whereas input to the lateral PBN arises from both the rostral and caudal NST.[4,101,102] The input to the lateral PBN is further specialized, with the rostral NST projecting mainly to the ventral lateral subnucleus and the caudal NST projecting to other lateral subnuclei.[101] Recent studies have also demonstrated that the caudal NST itself is quite specifically organized with regard to both its ascending and descending efferent outflow.[101,108-111] Several elegant experiments make it apparent that the efferent connections of specific caudal NST regions are directly related to their afferent connections. For example, the central subnucleus[110] in the caudal NST receives afferent input mainly from the esophagus[68] and projects specifically to the compact formation of the nucleus ambiguus,[101,108] where the motoneurons innervating the esophageal musculature are located.[112]

Because the rostral NST evinces a heterogeneous afferent organization, our laboratory asked whether this organization is accompanied by a parallel heterogeneous efferent organization.[113,114] Injections of the sensitive anterograde tracer, biocytin,[115] made under electrophysiological guidance suggest that despite broad similarities, distinctions can be made between the efferent projections that arise from regions with distinctive sensory input. Injections centered in sites that responded predominantly to (1) gustatory stimulation of the AO, (2) gustatory and/or mechanical stimulation of the PO (mainly posterior tongue), or (3) exclusively to mechanical stimulation of the oral cavity all gave rise to ascending fibers that terminated in PBN, but distinctions between these projections were observed. The differences between the first two types of injections are depicted in Figure 4.

First, despite much overlap, the AO and PO gustatory projections appeared to maintain some orotopy in PBN, with the AO projection centered lateral and somewhat caudal to the PO projection. Second, the size of the fibers associated with the terminal fields arising from the AO and PO gustatory projections appeared different, with AO terminal endings noticeably larger than PO endings. Finally, judging by the density of the terminal field, the AO gustatory region projected more strongly to PBN than the PO gustatory region. Although much further work is necessary, we hypothesize that a greater proportion of AO than PO gustatory neurons contribute to the ascending processing "stream". Interestingly, on the basis of peripheral nerve recordings, Frank[64] has recently suggested that gustatory responses arising from the anterior tongue are more important in coding taste quality (presumably a function ultimately carried out by cortical forebrain neurons), whereas posterior tongue gustatory cells are more important in brainstem reflex circuits. Our observations regarding the descending projections of the AO vs. PO gustatory regions also lend some support to such a notion: the PO gustatory regions contributed more extensively to the descending intranuclear pathway. It is important to emphasize, however, that the AO *and* PO taste regions contributed to *both* the ascending and descending pathways. Interestingly, the region of NST that responded only to mechanical stimulation of the oral cavity projected *extremely* sparsely to PBN. A recent retrograde tracing study supports these findings. Very few NST-PBN projection neurons arose from the lateral area of NST that received overlapping input from the trigeminal and glossopharyngeal nerve,[116] which appears to be the same region in which we observe only oral mechanical responses (see Figure 2). Electrophysiological studies also document that few purely mechanically responsive neurons project to the PBN.[60,67] These studies suggest that the oral mechanical region of NST may play an almost exclusive role in local circuit function.

The differences in the efferent organization of AO and PO gustatory pathways in rodents will not be surprising to those familiar with comparative studies of gustatory anatomy and physiology. In fish, the somatotopic organization of the gustatory system is highly specialized, with morphologically distinct structures, the facial and vagal lobes, receiving input from extraoral and intraoral taste receptors, respectively. Elegant behavioral studies have demonstrated that these lobes are differentially important for distinct phases of the feeding cycle.[117] Correspondingly, the facial and vagal lobes give rise to distinctive efferent projections.[118,119] Both lobes contribute to the ascending pathway, terminating in the secondary gustatory nucleus in the pons, but their terminations are somatotopically organized. Further, only the facial lobe maintains direct projections to the forebrain, whereas projections from the vagal lobe have an obligatory relay in the secondary gustatory nucleus. Finally, the local connections made by axons arising from these two lobes are almost nonoverlapping. For example, the facial lobe makes direct connections with the sensory trigeminal complex, whereas the vagal lobe connects with the

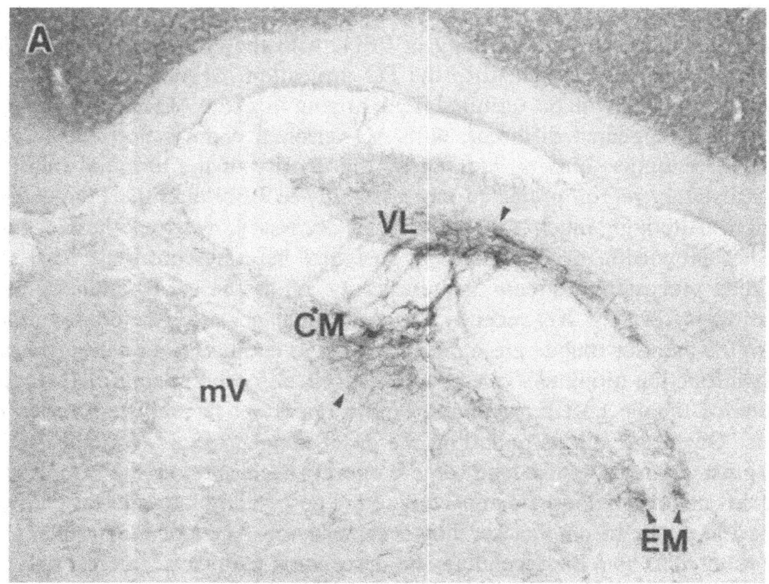

FIGURE 4. Light-field low-power photomicrographs of terminal labeling (indicated by the arrowheads) in the parabrachial nucleus (PBN) of the rat following biocytin injections into NST. Medial = left, "bc" = brachium conjunctivum, "mV" = tract of the mesencephalic nucleus of V. The location and density of the anterograde labeling in PBN varied, depending upon the orosensory properties recorded at the injection site in NST. The label in panel A was the result of an injection made into the rostral pole of NST at a site responsive to gustatory stimulation of the anterior tongue. The labeling in panel B resulted from an injection made in the NST at a site responsive to mechanical (and presumably gustatory) stimulation of the posterior tongue, which was located caudal and medial to the site injected for A. Both regions of NST project to the same three PBN subnuclei, the *ventral lateral* (VL), *central medial* (CM), and the *external medial* (EM). For both types of injection, the densest projection is to VL and CM, but the center of the projection from the anterior tongue is located more laterally and slightly more rostrally in these subnuclei. In addition, the overall density of the projection from the anterior tongue is noticeably greater. The asterisk over the "C" in the abbreviation for "C*M" indicates that this subnucleus has been delineated anatomically only in the hamster.[122] Interestingly, however, the boundaries of CM are roughly coincidental with the region ventral to the brachium conjunctivum that was occupied by the area covered by the label arising from both these injections. CM occupies a subset of the area occupied by the *medial* subnucleus in the rat,[107] which extends further laterally, approximately to the lateral border of the brachium.

nucleus ambiguus. These connections are consistent with the respective roles of the facial and vagal lobes in food selection and swallowing.[117]

Other studies suggest additional dimensions of the heterogeneous efferent organization of rostral NST. Interestingly, it appears possible that, at a given location in the horizontal plane, different populations of NST neurons give rise to the long ascending and descending pathways. J. Travers[53] compared the location of retrogradely labeled cells in the NST following injections made in the gustatory-responsive PBN or hypoglossal nucleus and adjacent reticular

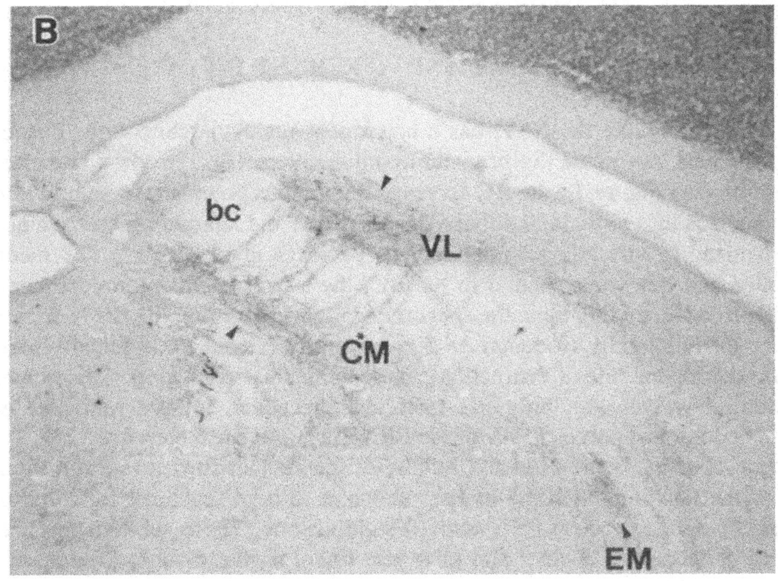

FIGURE 4B.

formation. Solitary nucleus neurons that projected rostrally were located dorsal to those projecting caudally. A recent analysis of the subnuclear location of NST-PBN projection neurons confirms that such cells are located dorsally in NST; specifically, the majority are in the rostral central subdivision.[102]

Physiological studies also support the notion that only a subset of NST neurons contribute to the ascending gustatory pathway. Indeed, a series of investigations by Ogawa and his colleagues suggest that only 34 to 38% of the gustatory-responsive NST neurons are pontine projection neurons.[60,67] Pontine projection neurons appear to have distinctive electrophysiological properties, being more responsive but less broadly tuned than neurons not projecting to PBN.[60] A more recent investigation[120] confirms that only a subset of NST gustatory neurons are pontine projection neurons, and that these tend to be more responsive, although the reported proportion of projection neurons was somewhat higher (49%) than that reported by the earlier studies. Additional information that correlates the sensory properties of gustatory neurons with their projection patterns should prove to be useful for understanding the functional roles of different taste neurons. Until recently, most hypotheses regarding the information processing capacities and coding strategies of gustatory neurons have revolved around assessing only their gustatory chemoresponsive profile. A knowledge of what anatomical connections a particular neuron makes, e.g., ascending, intranuclear, reticular, or direct motor, will facilitate the elucidation of the functional significance of different chemosensitive patterns, receptive field organizations, level of responsiveness, and neurotransmitter responsiveness.

III. PARABRACHIAL NUCLEUS

A. CYTOARCHITECTURE AND AFFERENT INPUT

1. Cytoarchitecture

The PBN, like the NST, has a heterogeneous cytoarchitecture. Because this nucleus surrounds the brachium conjunctivum (BC) in the pontine tegmentum (see Figure 4), the BC serves as an obvious boundary for subdividing the nucleus into two major subdivisions, medial and lateral, which have been recognized in virtually all mammalian species studied.[107,121-124] The medial subdivision lies ventromedial to the BC, the lateral subdivision dorsolateral to the BC. Depending upon the species under consideration, the BC is oriented more mediolaterally (rodents) or dorsoventrally (primates, especially man); thus, the medial/lateral distinction can be somewhat confusing. This nomenclature, however, not only has historical precedent but has remained entrenched in most parcellations of the PBN[107,122] (but see Reference 123). Two recent schemes, both in rodents, subdivide the nucleus further (consult Figure 4 for the locations of some of the subnuclei to be discussed). Fulwiler and Saper[107] divided the rat PBN into 10 subdivisions. These subdivisions were cytoarchitecturally distinct and gave rise to somewhat distinct efferent connections *(vide infra)*. Halsell and Frank[122] subdivided the hamster PBN similarly, but, in addition, directly related the nuclear subdivisions to the location of gustatory responses. Interestingly, the latter authors found that gustatory responses were restricted to only two of their nine subdivisions: the ventral lateral and central medial. These subdivisions are restricted to the caudal 2/3 of the PBN, abut the BC, and are found in the medial half (central medial) or middle third (ventral lateral) of the nucleus. Similar, albeit not identical, subdivisions have been identified in rat: the medial subnucleus of Fulwiler and Saper[107] includes the region analogous to the central medial subnucleus, but also areas lateral and slightly medial to it, whereas the ventral lateral subnucleus of the rat is located somewhat more medially than the analogous subdivision in the hamster. Although the cytoarchitectural subdivisions in the rat have not been directly related to the location of gustatory responses, published neurophysiological reports are consistent with the presence of taste responses in the general vicinity of the same subnuclei where they are found in the hamster. For the most part, the subnuclear location of gustatory responses in PBN agrees with the location of terminal endings arising from the rostral, gustatory pole of NST since the medial and ventral lateral subnuclei are major recipients of this input *(vide supra)*. Some uncertainty still exists with regard to the function of the external medial subnucleus (located at the lateral pole of the BC), since neuroanatomical studies suggest that this subnucleus receives afferents from the gustatory-responsive NST (see Figure 4),[101,113,114] but physiological studies have not thus far documented a gustatory function.

As is the case for gustatory NST subnuclei, the cell types found in the gustatory subnuclei of PBN are heterogeneous. Based on Nissl stains, the cells are of various sizes and are either round or elongated. The medial gustatory region appears somewhat more heterogeneous, since in both species, three cell types occur medially and only two laterally.[107,122] A more complete description of cellular architecture, based on Golgi staining, has recently become available for both these species.[125,126] Overall, the Golgi analyses are consistent with the Nissl analyses, although only 2 classes of neurons were delineated. In the rat, the analysis was done specifically for neurons in the "gustatory region", based upon previous neurophysiological studies, but the hamster analysis appeared to include the entire PBN. Nevertheless, the two analyses are in general agreement. In both species, one type of neuron, the "fusiform" neuron, is characterized by an elongated cell body and two to three primary dendrites, giving the cell a "bipolar" appearance. The other type is multipolar, characterized by a more spherical cell body and a larger number of neurons radiating in more than two directions. These types appear somewhat similar to the "elongate" vs. "stellate" or "tufted"[7] distinctions in NST.[125] In the hamster, direct comparisons made between NST and PBN indicate that PBN neurons are more branched, but less likely to show a preferred orientation than their NST counterparts.[125] The increased branching could indicate greater that the PBN is capable of more complex response processing than the NST.

2. Central Afferent Input

There are several sources of afferent input to PBN, other than that arising from NST, including projections from spinal cord, brainstem, and forebrain. Strong parabrachial projections, arising from the spinal cord and trigeminal nucleus, provide a substrate for convergence between special or general visceral and somatosensory input.[127-130] The afferent input from these sources is topographically arranged: spinal input is heaviest more anteriorly in PBN and projects nearly exclusively to the lateral subnuclei, whereas input from the trigeminal nuclei is heaviest more caudally and includes the medial and ventral lateral subnuclei, i.e., those associated with a gustatory function.[128] The origin of these inputs also displays considerable specificity. The spinal inputs arise from layer I of the spinal cord. The trigeminal input originates from layer I of subnucleus caudalis, and from a cell group intercalated among the spinal trigeminal fibers, called the "promontorium" or "paratrigeminal nucleus". It has been demonstrated that this intercalated cell group receives primary afferent input from the upper alimentary tract, including the oral mucosa. Transganglionic transport of anterograde tracers from peripheral nerve branches demonstrates input from oral trigeminal,[131] glossopharyngeal, superior laryngeal,[12,24] and ethmoid nerves.[24] Peripheral injections of WGA-HRP into the palate, pharynx, and esophagus document these projections and further specify the sources of afferent input into the paratrigeminal nucleus.[68] The outer layers of the medullary and spinal dorsal horn have long been associated with a nociceptive function, and a similar function has been suggested for the promontorium, based in part upon the histochemical staining properties of this region.[68] Thus, it appears possible that the somatosensory input converging at the level of the PBN is nociceptive. Since at least certain classes of nociceptors are chemosensitive,[132] the combined solitary and trigeminal input to PBN could provide a basis for integration of two types of oral chemosensory information, gustatory and common chemical.

Descending input from the cortex, hypothalamus, bed nucleus of the stria terminalis, and amygdala has been described in several species including the cat, rabbit, mouse, rat, and hamster.[133-141] Two of the more recent studies, using the rat[137] and hamster,[134] employed sensitive tracers and paid special attention to the parabrachial subnuclear location of these afferent inputs. Indeed, in the hamster study, some of the injections were electrophysiologically guided and placed into gustatory-responsive regions.[134] Thus, these latter studies provide a particularly good basis for inferring the relationship of forebrain afferent input to gustatory-responsive regions of PBN. In both the rat[137] and hamster,[134] there is a substantial projection from layers 5 and 6 of the insular cortex to PBN regions implicated in gustatory function. The projections may be organized topographically,[137] but the functional organization of this topography is not yet clear. There is also a diffuse projection from infralimbic cortex to several PBN regions, although the projection to gustatory PBN appears more moderate than that from insular cortex.[137] Strong projections to gustatory PBN also arise from cell populations in the bed nucleus of the stria terminalis and the central nucleus of the amygdala.[134,137] Several hypothalamic nuclei project strongly to the anterior and lateral PBN, i.e., the visceral zone, but projections from the hypothalamus are much lighter to gustatory PBN, and, when they exist, arise preferentially from the posterior lateral hypothalamus.[137,139]

B. PHYSIOLOGY

1. Topographic Organization

Neurophysiological studies indicate an orotopic arrangement of gustatory responses, as well as some segregation of gustatory and mechanical responses in PBN. The data, however, are not as complete as those describing these relationships in NST. Results of the first neurophysiological investigation of PBN (done in rat) demonstrated that the largest multiunit responses to sapid stimulation of the anterior tongue were located further ventrally than the largest responses to stimulating the remainder of the oral cavity.[142] Similarly, a more recent study reported that single neurons responding exclusively to the tongue (18/21 responded only to the anterior tongue) were located further ventrally than those responding only to the palate (either the nasoincisor ducts, soft palate, or both).[143] This latter study could mean that PBN orotopy is reorganized according to a tongue/palate dichotomy, in contrast to the AO/PO dichotomy observed in NST. It should be noted, however, that only a minority of the neurons in the PBN single-unit study received input from the PO, and thus the relative locations of AO and PO responses may have been difficult to determine. Our anatomical studies (*vide supra,* see Figure 4) suggest that a gustatory orotopy, based upon an AO/PO dichotomy, *is* likely to be preserved in PBN.

As observed in NST, there is a substantial population of PBN neurons that respond only to oral somatosensory stimulation, particularly to innocuous mechanical stimulation of the oral mucous membranes. Approximately 1/3 of a sample of almost 100 orally responsive cells responded exclusively to oral mechanical stimulation, whereas the remainder had gustatory (and often mechanical) inputs.[144] The location of the two neural populations in PBN overlapped considerably, but the purely mechanical neurons appeared to extend further laterally, particularly in the medial subnucleus (i.e., ventral to the BC). Although not noted by the authors, these cells appear to preferentially occupy an area corresponding to the ventral medial subnucleus in the hamster, i.e., a subnucleus in which gustatory responses were not found.[122] In the rat, this area spans the most lateral portion of the medial and the most medial portion of the external medial subnucleus.[107]

2. Convergence

a. Spatial

In agreement with what is observed in the solitary nucleus, single gustatory neurons in the PBN often respond to sapid stimulation of more than one receptor subpopulation. What is unclear at present, however, is whether or not more convergence is observed as a result of further processing at the pontine level. Table 1 tabulates the available data from investigations that address the question of the receptive field organization of central gustatory neurons. The reported incidence of convergent neurons, i.e., neurons with gustatory receptive fields that include two or more taste receptor subpopulations, spans a wide range, even *within* NST or PBN. In the NST, the reported proportion of covergent neurons ranges from 4.5 to 64%! This is undoubtedly due in part to the different techniques used to address this question. Sweazey and Smith[45] defined the gustatory receptive field for a neuron by enclosing the anterior tongue in a chamber and stimulating the isolated region and the remaining oral cavity separately, using all four of the standard taste stimuli. Some of the convergent neurons in this study thus may have been cells with receptive fields that were comprised only of fungiform papillae, albeit papillae both anterior *and* posterior to the tongue chamber. Such a procedure probably facilitates identification of "convergent" cells. On the other hand, in their several studies, Ogawa and his colleagues stimulated individual receptor subpopulations, but typically used only that neuron's "best stimulus" to define a receptive field.[59,66,143,144] This procedure would work against identification of convergent cells, especially when different stimuli are effective for the various receptor subpopulations comprising the receptive field.[52] Indeed, estimates using these two techniques result in the highest and lowest estimates of convergence, respectively. Keeping in mind the caveat that estimates of convergence at a single level vary (even within a single laboratory), it is possible to compare the incidence of convergence in different nuclei by calculating the average convergence across different studies. Comparing NST and PBN, only slightly more convergence is evident in the pontine (45%) than in the medullary (41%) relay. If the comparison is restricted to data collected in a single laboratory, and these are available from Ogawa and his colleagues, the increase in convergence is much more dramatic, 22% vs. 37%. A conclusion that convergence increases in PBN is thus a reasonable hypothesis, but one that requires further confirmation. Similarly, it is not yet clear whether there is any tendency for convergence in PBN neurons to be restricted to receptor subpopulations within the AO or PO, as has been observed in NST.[52,55]

b. Intermodal

A high proportion of PBN gustatory neurons, like their counterparts in NST, also respond to mechanical stimulation. Table 1 tabulates the incidence of gustatory/mechanical convergence at these two levels and suggests that the degree of convergence increases at the second synaptic relay, from ~65 to 77%. A similar phenomenon is observed for gustatory-visceral convergence: whereas no neurons in NST can be coactivated by gustatory stimulation and electrical stimulation of the cervical vagus nerve,[34] 31/47 gustatory-responsive neurons in the PBN also responded to vagal stimulation.[145] One might thus predict that body state changes, e.g., fluctuations in blood glucose or insulin levels or salt deprivation, would effect larger changes in gustatory responses in the PBN than they do in NST,[89-91] but such experiments have been performed only at the medullary level. Integration of responses to volatile chemical stimuli and sapid stimulation of the tongue also has been documented in a small population of single PBN gustatory-responsive neurons.[146] As was the case in NST,[87] however, the origin of the "olfactory" responses is uncertain. They may have arisen from olfactory receptor neurons, free nerve endings in the ethmoidal nerve, or perhaps even from sensory receptors in the pharynx.

3. Descending Influences

As noted above, there is much anatomical evidence indicating that there are descending influences from the cortex and limbic system on PBN gustatory neurons. The physiological nature of these influences has only begun to be investigated. Similarly to what is observed for neurons in the solitary nucleus, the most salient effect of decerebration is a diminution in the mean responses evoked by the most efficacious gustatory stimuli,[147] which suggests an overall facilitory effect of the forebrain on PBN gustatory neurons. In addition, "OFF" responses, i.e., increases in firing rate that occur after stimulus delivery, but before rinse onset, are observed in decerebrate, but not in intact rats. When the influence of only the gustatory cortex is removed using localized procaine infusions, similar "OFF" responses are observed, suggesting that the cortex is at least partially responsible for this effect.[148] Other effects of cortical removal, however, are more complex than those following decerebration. There is no change in any of the mean responses evoked by the four standard taste stimuli, but general increases or decreases in neural activity are observed for certain PBN neurons. For other PBN cells, only responses to certain stimuli are affected. It has been suggested that the specific nature of the effects of cortical input may allow "filtering" of gustatory signals, according to their salience, which changes as a function of learning or nutritional state.[148]

C. EFFERENT PATHWAYS

The efferent pathways arising from PBN are extensive and complex. They were first described in rodents[1-3,149] and data accumulated since that time has provided a fairly thorough account of the details of PBN efferent organization, particularly for these animals,[107,134,150-153] although some information on the rabbit,[154] cat,[155-157] and monkey (reviewed in References 158,159) is also available.

Descending pathways from PBN reach the spinal cord, as well as several regions of the brainstem, including the parvicellular medullary reticular formation and the raphe and oromotor nuclei. The descending pathways, however, appear to arise mostly from the region of the Kollicker-Fuse subnucleus,[107,151,155-157,160] a subnucleus which has not been implicated in oral sensation, but instead receives its strongest input from areas of the NST implicated in respiratory function.[101] Nevertheless, detailed investigations of orosensory responsiveness in Kollicker-Fuse have not been conducted, so such a function cannot be ruled out. In addition, the orally responsive PBN could influence Kollicker-Fuse via intranuclear connections.

Extensive ascending pathways from PBN have been also been described and the relationship of these pathways to orosensory function have become increasingly clarified, though not entirely resolved.[1-3,107,134,149-152,154] In rodents, ascending axons from PBN neurons take two major routes: one to the most medial, "parvi-", or "pauci-"cellular tip of ventroposteromedial nucleus of the thalamus (VPMpc, *vide infra*) and the other to several areas of the ventral forebrain, most notably the central nucleus of the amygdala, bed nucleus of the stria terminalis, hypothalamus, and substantia innominata.[1-3,107,134,149,151] In addition, there appears to be a direct PBN-insular cortex connection, at least in the rat[152] and mouse.[138] The efferent organization of PBN, however, is not homogeneous. Instead, different subnuclei, functional groups, and morphological cell types contribute differentially to the various pathways. In the opinion of some[153] but not other[107,134] authors, the thalamic projection from gustatory PBN is stronger than that from visceral PBN. It should be noted that there are also thalamic projections to the parafascicular and central lateral nuclei that do not arise from gustatory PBN, but specifically from the internal lateral subnucleus,[107,126] which apparently receives *no* afferent input from the NST.[101] It can also be concluded that the visceral PBN, especially the external and central lateral subnuclei, contributes more heavily to the ventral forebrain projections.[134,153,154] The medially located hypothalamic nuclei, such as the paraventricular or dorsomedial, in particular, receive most of their PBN connections from the lateral subnuclei.[107] The evidence for a heterogeneous efferent organization of PBN should not obscure the fact that these differences, for most of the projections, are relative, not absolute. This is certainly the case for PBN projections to the VB thalamus and central nucleus of the amygdala, where there is both anatomical[107,134,153] and neurophysiological[149,154] evidence for parallel gustatory and visceral pathways in both rodents and lagamorphs.

The ascending efferent organization of gustatory and visceral pathways in the primate is more divergent than in nonprimate species. It has already been mentioned that afferents arising from the rostral pole of the monkey NST bypass the PBN and project directly to the thalamus. Specifically, when autoradiographic injections were made at rostral NST sites, responsive to stimulation of the anterior tongue, no silver grains were observed at the pontine level. Projections to the PBN were observed, however, following injections into the caudal region of NST where responses to electrical stimulation of the vagus nerve were recorded.[106] Interestingly, there is convincing evidence for a strong PBN-ventral forebrain projection in the monkey, but the evidence for a PBN-thalamic projection is equivocal, suggesting that it is weak or lacking (see Reference 158). Thus, it appears that the differential contribution of gustatory and visceral afferent systems to forebrain projections, observed in lower mammals, becomes magnified in primates, actually arising from separate nuclei and having nearly nonoverlapping projections to the VB thalamus or ventral forebrain. One major caveat to this scenario, however, is that intermediate NST regions, occupied by primary afferent terminations of the glossopharyngeal nerve, also project to PBN.[106] This suggests the possibility that the monkey PBN processes posterior, but not anterior, tongue gustatory information, although the possibility has not yet been tested adequately using the necessary neurophysiological techniques (see Reference 158 for a discussion of this issue). If the PBN does indeed receive posterior tongue gustatory information, this suggests the intriguing possibility that these taste receptor subpopulations have privileged access to ventral forebrain regions, areas of the brain thought to be directly involved in emotional reactions to stimuli.

Neurophysiological data from antidromic stimulation experiments document the gustatory identity of parabrachio-thalamic[3,143,144] and parabrachio-ventral forebrain[3,149,154] pathways in rodents and lagamorphs. Although early neurophysiological data suggested that the *same* PBN cells often projected to both forebrain regions,[3,149] later neuroanatomical studies, utilizing retrograde tracing, double-labeling techniques, provided contrary evidence.[153] In addition, a recent study reported that the size and shape of the cell bodies of the two cellular populations were statistically different, although there was overlap between them.[134] It seems possible, therefore, that some dually driven neurons in the neurophysiological studies resulted from stimulating fibers of passage with the thalamic electrodes, as originally noted by the author.[149] Such an interpretation is consistent with the sizable populations of thalamic-only and dually driven neurons, compared to a small number of ventral-forebrain only cells, but is inconsistent with the fact that the most distant electrodes, i.e., those in the ventral forebrain, often drove PBN units with a shorter latency than did the thalamic electrode.[149] Regardless of whether there are some *dually* driven neurons or not, it appears likely that there are separate populations of PBN gustatory neurons with different projection patterns. There is also evidence that, as for NST neurons, cells in PBN with different projection patterns have different response properties.[143,144] Neurons that project bilaterally to the thalamus are more responsive to NaCl than those neurons with only ipsilateral projections. Indeed, parabrachial neurons that project to the thalamus are more responsive than those that do not. This was a consistent trend for each of the four standard stimuli, but only the differences for NaCl and HCl reached statistical significance.[144] The greater responsiveness of PBN-thalamic neurons suggests there may be an overall trend for elevated responsiveness in gustatory neurons that contribute to the ascending thalamo-cortical pathway since NST-PBN neurons are also more responsive than NST cells projecting locally.[60,120]

IV. VENTROPOSTEROMEDIAL THALAMIC NUCLEUS

A. CYTOARCHITECTURE AND AFFERENT INPUT

1. Cytoarchitecture

The thalamic relay for gustatory information lies just medial to the thalamic relay for oral somatosensory information in the ventroposteromedial nucleus (VPM). In several species, including rat,[161,162] cat, and primate (discussed in Reference 163), the medial VPM contains neurons that are smaller than those further lateral. As a consequence, the gustatory thalamic relay has been referred to as the parvicellular portion of the ventroposteromedial nucleus (VPMpc [e.g. Reference 32]) although other terminologies do not include this region as part of VPM (see Reference 163). A recent description of VPM cytoarchitecture in hamster, however, reported that the medial neurons were no smaller than those lying more laterally, although they were less densely packed. This suggests that VPMpc cytoarchitecture may not be as homogeneous as previously thought.[134] The name "paucicellular" was proposed to describe this region in hamster, so that the resulting abbreviation, VPMpc, is the same as for other species. The subnuclear organization of VPMpc has not been nearly so well described as for the NST and PBN. In the hamster, dorsal and ventral regions of the VPMpc could be differentiated based upon cellular morphology, as well as acetylcholinesterase and NADH dehydrogenase staining, suggesting some functional segregation. These two regions also received overlapping but differential projections from the lateral and medial PBN, so that these histochemically distinct regions *may* correspond to visceral and gustatory regions, respectively.[134] Such an arrangement of gustatory and visceral information, however, would be quite different from what has been proposed for the rat.[32] In the rat, visceral information has been proposed to be represented in a distinct subnucleus, the parvicellular portion of the ventral posterior lateral nucleus (VPLpc). This is a narrow strip of small neurons, morphologically similar to those in VPMpc, that are located lateral to VPMpc along the most basal portion of VPL. In the primate, there may be no, or only a weak, representation of visceral information in the VP, at least arising from the NST or PBN.[106,158] No systematic analyses of Golgi-stained material have been performed in VPMpc so that the dendritic morphology in this region has not been characterized.

2. Central Afferent Input

Unlike its brainstem counterparts, the VPMpc does not appear to receive a great variety of input from sources in addition to the main ascending pathway from PBN. Like most sensory thalamic relay nuclei, however, the VPMpc does receive descending input from the cortical gustatory relay in the insular cortex.[134,164] Perhaps surprisingly, there are still varied interpretations of the precise subnuclear origin of PBN neurons projecting to VPMpc. The disagreements are, to some extent, implied rather than overt debates, and are disagreements regarding the *relative* contribution of different sources, instead of debates about an exclusive site of origin. As discussed above, anatomical tracing and neurophysiological recording studies implicate the ventral lateral and (central) medial PBN subnuclei as the major site of gustatory processing in PBN. When electrophysiological guidance was used to target these PBN subnuclei for lesion placement or injection of autoradiographic tracers in these gustatory-responsive sites in the rat PBN, robust projections to VPMpc were demonstrated.[1,2] More recent studies, using both anterograde and retrograde tracing techniques, but not utilizing electrophysiologically guided injections, have argued that the external medial subnucleus is also a source of afferent input to VPMpc.[32,165,166] While the external medial subnucleus is not argued to be the exclusive source of thalamic input, it is argued to be the major one and, interestingly, is mostly contralateral. The most unsettling aspect of the notion that the external medial subnucleus, rather than the ventral lateral and central medial subnuclei, is the major source for the ascending gustatory pathway is its apparent conflict with the neurophysiological evidence that places the gustatory-responsive region of the PBN in the latter. On the other hand, although gustatory responses have not typically been recorded as far laterally as the external medial subnucleus, this region has not been thoroughly explored using neurophysiological techniques. Recent neuroanatomical studies suggest that it might well receive some gustatory input.[101,113,114] On the other hand, many neurons responsive to intraoral mechanical stimulation are located laterally in PBN, apparently near the external medial subnucleus.[144] Further neurophysiological studies of the external medial subnucleus are critical for a functional definition and a resolution of the origin of the gustatory PBN-thalamic pathway.

B. PHYSIOLOGY

1. Topographic Organization

Of all the "lemnsical" gustatory relays, at a neurophysiological level, the thalamus is probably the most enigmatic. Plagued by a paucity of studies and low response rates in anesthetized preparations (e.g., References 167 to 169), our knowledge of the gustatory diencephalic relay is meager. Most studies suggest that thalamic gustatory-responsive neurons are medial to those cells exclusively responsive to intraoral somatosensory information.[170-173] Studies from one laboratory, however, inconsistently report the two populations either to be intermingled[167] or separated along the anterior-posterior axis, with the gustatory neurons located further rostrally.[168] A similar discrepancy exists with regard to an orotopic organization of gustatory responses. The location of thalamic responses evoked by stimulation of the chorda tympani and glossopharyngeal nerves suggested that anterior and posterior oral cavity chorda tympani responses would be orotopically arranged in the thalamus,[174,175] but studies employing natural stimulation of the oral cavity suggest that there is no orotopic map, either for gustatory or intraoral somatosensory responses.[167,168] Unfortunately, the latter studies recorded very few gustatory neurons responsive to posterior oral cavity stimulation, so that any hypothesis regarding orotopy remains insufficiently tested.

2. Convergence

a. Spatial

Like gustatory neurons in the brainstem, those in the thalamus often receive input from multiple gustatory receptor subpopulations. As in the brainstem, however, it is difficult to accurately estimate the amount of convergence these neurons receive. There have been only two studies which address this question, both from the same laboratory, and the estimates of convergence ranged from 29[168] to 55%.[167] Averaged across both studies about 42% of the neurons receive input from multiple taste bud groups, a proportion no greater than observed in the brainstem. If *laterality* of receptive fields is taken into account, however, it is clear that receptive field size does increase in the thalamus. Parabrachial gustatory neurons have unilateral receptive fields,[176] but 43% of the gustatory thalamic neurons receive bilateral input.[168]

b. Intermodal

Consistent with what has been observed in the brainstem, there is evidence that gustatory-responsive thalamic neurons also respond to other modalities. The majority of thalamic gustatory neurons respond to mechanical stimulation of the oral cavity as well[168] (Table 1). The proportion of neurons that are dually responsive to taste and oral mechanical stimulation appears to be slightly higher in thalamus than in PBN, suggesting a trend for increasing convergence, but again PBN estimates vary considerably, making this conclusion tentative. One study also reports that thalamic neurons in the VPMpc can be dually driven by electrical stimulation of the lateral olfactory tract and tongue.[177] Of the 37 thalamic neurons recorded, 17 responded only to lingual stimulation, 18 responded to lingual and lateral olfactory tract stimulation, and 2 responded only to lateral olfactory tract stimulation. The lingual modalities involved were in most cases uncertain, but in one experiment (two cells) all of the tongue nerves except the chorda tympani were cut and dually responsive neurons were still observed. Convergence between gustatory and hepatic afferent inputs have been observed in a small population of thalamic neurons.[178]

3. Descending Influences

The physiological influence of cortico-thalamic pathways upon gustatory-responsive thalamic neurons has not received much attention. Two early studies which investigated the effects of cortical influence upon these cells did not use rigorous criteria to distinguish antidromic from orthodromic stimulation.[179,180] Nevertheless, these studies suggest that some thalamic gustatory cells are orthodromically affected by cortical stimulation. Both facilitory and inhibitory effects were observed, with the inhibitory effects more long-lasting and frequent.[180] A more recent study that did distinguish antidromic from orthodromic responses also reported orthodromic effects on these cells; again, inhibitory effects were more prominent.[167]

C. EFFERENT PATHWAYS

Similar to the organization of its afferent inputs, the organization of gustatory thalamic efferents is relatively simple compared to its brainstem counterparts. In several species, the major projection of the VPMpc is to the insular (rodents) or insular/opercular (cats and primates) cortex (hamster,[134] rat,[181-183] monkey,[184] cat[185,186]). Amygdalar projections have also been reported (hamster,[134] cat[186]), but this projection appears weaker. In the cat, additional projections are observed to perirhinal and infralimbic cortex.[186] In rodents and in cats, the thalamic projection to insular cortex appears to be to a single, contiguous region,[181-183,185,186] whereas in primates, VPMpc projects to two nearby but noncontiguous regions, one in "buried" insular-opercular cortex and the other on the lateral convexity, adjacent to somatosensory cortex.[184] Despite the rather simple nature of its efferent organization, there is some preliminary evidence for heterogeneity. In the one existing study, only 14/25 neurons tested could be antidromically invaded from the cortex.[167]

V. INSULAR/OPERACULAR CORTEX

A. CYTOARCHITECTURE AND AFFERENT INPUT

1. Cytoarchitecture

The major cytoarchitectural debate with regard to gustatory cortex has focused on the rodent and revolved around whether taste cortex is located in the granular, dysgranular, or agranular insular cortex. Initial descriptions, based upon evoked potential and single-unit recording studies, as well as anatomical and lesion studies (reviewed in Reference 187), placed taste cortex in the insular cortex, ventral to the somatosensory representation, but did not specifically note the cytoarchitectural features of the region. Recently, however, Kosar and her colleagues used electrophysiological techniques and cytoarchitectural analysis to specify precisely which oral modalities were associated with the cytoarchitecturally different regions of cortex.[188] Projections from gustatory-responsive regions of VPMpc were also studied using electrophysiologically guided injections of autoradiographic tracers.[181] These studies provided convincing evidence that the gustatory-responsive region was confined to the agranular insular cortex, albeit in the dorsal portion of this area (i.e., adjacent to granular cortex). A later study by this same author again found that gustatory neurons were primarily located in the dorsal agranular region.[189] Cechetto and Saper[32] redefined the boundaries of the insular region. They recorded from a small sample of gustatory cells in insular cortex and decided that taste neurons were in *dysgranular*, rather than agranular insular cortex. They regard any disagreement with Kosar et al.,[188] however, to be "primarily semantic". Regardless of whether this region is regarded as dygranular or agranular, the major point is that gustatory cortex is located in a region of insular cortex which is cytoarchitecturally distinct from the overlying granular region, which contains neurons responsive to oral somatosensory stimulation[188] *(vide infra)*. Most recently, however, the situation has become more confusing. Ogawa and his colleagues[190] have also studied gustatory responses in the insular cortex and find them to be located in *both* the granular and dysgranular cortex. Although differences in anesthetics or gustatory stimulating methods provide convenient explanations for the differences, a careful consideration of the data makes such arguments unconvincing. At present, the conflict remains unresolved. In the primate and cat, projections from VPMpc appear to reach granular and dysgranular regions, both in the buried insular/opercular cortex and on the lateral convexity.[184,186] Single-unit recording studies, however, indicate that gustatory-sensitive cells, at least in the primate, are mainly restricted to buried insular/opercular cortex[191,192] though they do seem to span both granular and dysgranular regions.[192]

2. Central Afferent Input

In addition to the input from VPMpc, the cortical gustatory area receives input from several other sources. In some rodents (mouse[138] and rat[152,193,194]) but not hamster,[134] the insular cortex receives direct input from PBN. In addition, insular cortex has been reported to receive direct input from the olfactory bulb,[195] lateral hypothalamus, and amygdala, as well as the prelimbic and infralimbic cortex.[196-198] The hypothalamic, amygdalar, and PBN projections appear to overlap extensively with projections from VPMpc, whereas the projections from infralimbic cortex and the mediodorsal thalamic nucleus terminate adjacent to the VPMpc projection in the agranular insular cortex.[196] Thus, it is likely that cortical gustatory neurons are influenced directly by only some of these projections, but they could be influenced indirectly, via local projections with adjacent regions, by the others. Despite this abundant anatomical evidence, only one study sheds some light on the function of these projections. In that study, stimulation of the amygdala resulted in both facilitory and inhibitory effects upon cortical gustatory neurons, with the inhibitory effects more prominent.[199] The multiple convergence of inputs at the cortical level is similar to that observed for hindbrain gustatory regions. Altogether, gustatory neurons in three gustatory relays, NST, PBN, and cortex, all can be affected by several sources of extragustatory input. Indeed, all three areas receive these inputs from many of the same, often limbic, sources. The gustatory thalamus appears somewhat unique, in being relatively free of these extrinsic influences.

B. PHYSIOLOGY

1. Topographic Organization

A topographic organization of oral modalities in the insular cortex was described by Kosar and her colleagues[188] in the study mentioned above. In addition to delimiting gustatory cortex to what these authors refer to as "agranular" insular cortex, thermal responses were found immediately dorsal to this region in their "dysgranular", and oro-mechanical responses were further dorsal in "granular" insular cortex. This systematic transition between oral modalities was observed without exception in the nearly 100 recording tracks reported. Such a progression from somatosensory to gustatory responses in insular cortex is consistent with the location of evoked potentials following stimulation of the lingual vs. the chorda tympani nerves (marmoset,[179] rat,[161,200,201] hamster,[202] rabbit[203]). Yamamoto and his colleagues also report a dorsal to ventral progression from mechanical to gustatory responses in the rat, both in acute[187] and chronic[204] preparations. A similar relationship between oral somatosensory and gustatory responses has recently been reported in primate, but with one major difference. Although gustatory responses were found much more commonly in buried insular cortex, neurons responding exclusively to oral mechanical stimulation were intermingled with gustatory-responsive neurons in this region, though they extended onto the exposed lateral convexity. In contrast to these many studies which suggest at least some separation of oral modalities, a recent report, in rat, claims that oral mechanical and gustatory neurons are entirely intermingled in granular and dysgranular insular cortex.[190]

Because most electrophysiological reports of the cortical gustatory area are either from studies employing only anterior tongue stimulation or using chronic, freely moving animals, data on the orotopic organization of cortical gustatory neurons are almost exclusively from evoked potential studies. Despite their limited resolution, these studies describe a consistent picture. The focus of evoked potentials elicited by stimulating of the chorda tympani nerve is anterior and somewhat dorsal to those elicited by stimulating the glossopharyngeal nerve in a variety of species (rabbit,[203] rat,[161,200,201] hamster[202]). These results suggest that the anterior-posterior axis of the oral cavity is represented along the same axis and, in a like manner, in cortex (although the monkey may be an exception[205]). Such an organization is consistent with the overall organization of visceral sensibilities within the insular cortical region, since neurons responsive to general visceral stimulation (mostly mediated by the Xth cranial nerve) are found posterior to gustatory neurons in the insular cortex.[32]

2. Convergence

Since convergence between modalities and between regions of the oral cavity is observed at lower levels of the gustatory neuraxis, it would be surprising if convergence did not occur in gustatory cortex. Indeed, both types of convergence are a feature of cortical gustatory neurons. Gustatory-responsive neurons also often respond to oral somatosensory stimulation, both thermal[206,207] and mechanical.[190,192] Again due to the paucity of data, the incidence of this convergence, relative to that observed at lower levels of the neuraxis, is rather difficult to compare. It would appear, however, that no major increase in gustatory-mechanical convergence occurs at the cortical level. Ogawa and his colleagues have tested the mechanical sensitivity of gustatory neurons at several levels of the neuraxis in the rat. The degree of convergence observed at the cortical level, 66.7%,[190,192] appears similar to what these investigators found in NST (see Table 1). It is informative that Yamamoto,[197] however, found *no* convergence between gustatory and mechanical responses in cortex. He attributes this to the fact that he used less intense stimulation than Ogawa's group (i.e., camel's hair brush vs. glass probe and forceps). It could be the case, then, that little or no convergence occurs at any level of the gustatory neuraxis between very low threshold trigeminal and gustatory afferents.

Convergence between gustatory and extraoral modalities is less frequently reported, albeit also less frequently investigated. One report found that 2/20 gustatory cortical neurons tested also responded to intraperitoneal injection of lithium chloride, suggesting convergence between gastrointestinal and gustatory signals.[207] When stomach afferents in this species were stimulated by inflation, however, no convergence was observed between these two types of responses in cortex.[32] It has been suggested that no interaction between gastrointestinal and gustatory signals occurs in the primate insular/opercular cortex, since taste neurons in this region did not alter their firing rates when the monkey was fed to satiety.[208,209] Similar to what has been reported for the brainstem, there is a small amount of data suggesting that a few (~6%) of gustatory cortical neurons also respond to odors.[207]

The receptive fields of cortical gustatory neurons have only been described in two studies. One study[187] stimulated only the anterior tongue, but did report a substantial proportion (41%) of neurons with bilateral receptive fields, a proportion similar to that observed in the thalamus. A second study stimulated all of the oral receptor subpopulations, but did not describe receptive field organization in detail.[190] Two examples of receptive fields are illustrated in the latter study, however, and it is clear that spatial convergence occurs in cortex. It is not clear, however, what the degree of this convergence is relative to that observed at lower levels.,

C. EFFERENT PATHWAYS

The efferent projections from the insular cortical region implicated in gustatory function fall into three major functional categories: (1) feedback projections to lower levels of the gustatory neuraxis, (2) subcortical motor and limbic areas, and (3) integrative regions of cortex. Projections to lower levels of the neuraxis have been discussed in previous sections and include projections to VPMpc,[30,134,152,183,186,194,195] PBN,[134,193,194,195,210] and NST.[30,193,195] Limbic and motor connections include projections to the amygdala,[134,186,194] mediodorsal[194] and centromedial,[30] thalamic nuclei, neostriatum,[186,194] and regions of the medullary reticular formation containing preganglionic parasympathetic neurons.[30,195] Intracortical projections include the infralimbic and perirhinal cortex in both the rat[194] and cat,[186] as well as the prefrontal cortex in the rat[193] and the caudolateral orbitofrontal cortex in the primate.[94] It should be noted that the functional nature of several of these cortical projections is not certain, i.e., no antidromic recording studies have documented their gustatory nature. It is highly likely, based upon the descriptions of injection and projection site sizes and locations, that at least some of the insular cortex projections to VPMpc, PBN, NST, and amygdala are gustatory. The gustatory nature of other projections, including those to the mediodorsal thalamic nucleus, infralimbic cortex, and striatum, however, is less definite. In the rat, for example, the reciprocal projections between the infralimbic and insular cortices appear to involve the most ventral portion of the agranular insular cortex, a region not implicated in gustatory function, albeit directly adjacent to the gustatory insular cortex.[197] Finally, responses to gustatory stimulation have been recorded in the caudolateral orbitofrontal cortical projection in the primate, indicating that this projection certainly conveys gustatory information.[94,211] Indeed, this region of cortex is considered to comprise a secondary cortical gustatory area in this species. This is a particularly interesting area, because, in contrast to the situation in rodents, it appears to be the first "lemniscal" site in the primate, where gustatory responses are modified according to need state.[94,211]

VI. CENTRAL NEURAL PROCESSING

This chapter has summarized evidence which details the complex nature of gustatory information processing. At every level of the neuraxis (with the possible exception of the thalamus), a salient feature of organization is the presence of multiple afferent and efferent connections. Afferent connections from different sensory modalities often converge upon gustatory-responsive neurons or reach areas adjacent to these cells. Although little is known about the details of the interaction between gustatory and other modalities, such inputs are likely to function in integrative processes essential for the synthesis of flavor. In addition, inputs from nonsensory, often limbic or cortical sources impinge upon these gustatory regions. Such inputs reflect the intimate relationship of the gustatory system with homeostatic mechanisms and the plasticity necessary for effective behavioral responses to gustatory stimuli in the face of a changing internal environment. Beginning with the NST, the presence of multiple efferent connections is also a ubiquitous feature of the neural organization of the gustatory system. These multiple outputs reflect the diverse functions of the gustatory system: discriminative, reflexive, and motivational.

The final section of this chapter will focus on a more specific feature of gustatory cells, the organization of chemosensitivities within individual neurons. The existence of gustatory neuron types, the breadth of tuning of individual gustatory neurons, and the relationship of chemosensitivity to anatomical organization will each be considered. In order to simplify matters, the discussion will largely be limited to the NST and insular or primary gustatory cortex, the two "extremes" of the central gustatory system.

A. GUSTATORY NEURON TYPES

1. Chemosensitive Response Profiles

Because of broad tuning and obvious heterogeneity between the chemosensitive profiles of individual gustatory neurons, the question of whether these cells may be grouped into a few distinct categories on the basis of their pattern of chemosensitivity has been a subject of debate.[212-215] Initial attempts to categorize neurons on this basis rested largely upon determining which of the four standard taste qualities was an optimal stimulus for a cell.[216] This practice was subject to the criticism that types were a given, since one stimulus is, by definition, the most effective. The neuron types obtained using this simple scheme, however, were noted to have orderly response profiles across all the other three standard taste stimuli. If the four standard stimuli were arranged in the order of behavioral preference, sucrose, NaCl, HCl, and quinine HCl (QHCl), the average responses of neurons in each of the categories showed only a *single* peak at the "best" stimulus, i.e., responses to the "sideband" stimulus diminished in an orderly fashion on either side of the peak. The orderliness of the sideband responses was not a foregone conclusion.[44,216] Over the last decade, neurons have also been categorized using the quantitative techniques of cluster analysis and multidimensional scaling to analyze the gustatory responses evoked by larger stimulus arrays.[61,91,99,191,207,214,217-222] These methods for classification have provided more objective evidence for a limited number of neuron types, and, interestingly, the groups are similar to those obtained merely using the "best stimulus" classification procedure. In addition, over the last several years, instead of stimulating only the anterior tongue, many studies have stimulated multiple groups of oral receptors,[45,52] the whole mouth,[60,67,99,217,218,219] or used chronic recording techniques.[191,207,208,211,220-225] These stimulating techniques provide an opportunity to observe the effects of spatial convergence *(vide supra)* upon the organization of chemosensitivities within individual cells. Since the taste qualities feeding into single neurons do not always match,[52] central convergence might be expected to blur distinctions between types. This does not appear to be the case, however. For example, in the rat NST, many neurons receive converging input from sweet receptors located on the nasoincisor ducts and salt receptors located on the anterior tongue. Even when both inputs are taken into account, however, orderly neuron types with "typical" response profiles are observed.[52] In addition, when the concept of neuron types is challenged with whole mouth stimulation[99,217-219] or chronic recording techniques,[207,211,220-224] the existence of a limited number of neuron types remains a viable proposition.

Although the above discussion suggests some agreement about existence of a limited number of gustatory neuron types, this agreement is not universal[212] and types are not always found.[215,225,226] Although the papers cited in the preceding paragraph all identify types, caveats concerning their discriminability are often included (e.g., References 220, 221). Indeed, cluster analysis and multidimensional scaling do not provide statistical probability statements for defining neuron types but merely provide visual and quantitative guidelines that are useful for making this determination.[227] When the data are examined closely and different data sets compared to each other, it is clear that some studies provide more equivocal evidence for clearly delineated gustatory neuron types than other studies, and that the characteristics of the observed types exhibit some variability. For example, in one acute study of NST neurons using whole mouth stimulation,[91] there was evidence for three well-delineated neuron clusters: (1) neurons optimally responsive to sucrose, (2) quinine HCl, or (3) salts and acids, with two subgroups either more or less specific to sodium salts, comprising this third cluster. In this study, the sucrose-sensitive cluster of neurons was very specifically tuned to sugars and distinct from the rest of the neurons. In the cluster analysis, sucrose-sensitive cells were intercorrelated relatively highly ($r \sim 0.6$), but correlated poorly with the remaining neurons ($r \sim 0.04$). In a study using comparable techniques,[218] four similar but not identical groups of neurons were delineated. In the latter study, three of the four groups actually responded optimally to NaCl, but each had distinct "sideband" sensitivities, which were similar to the optimal sensitivities for the neuron groups delineated in the first study. For example, in the latter study, the "sugar-sensitive" cells actually responded better to NaCl. Not surprisingly, they also formed a less distinct group than they did in the prior study. Although they were intercorrelated at about the same level as the analogous group in the first study ($r > 0.70$), this group of sugar-sensitive cells was correlated only a little less highly with the sodium-specific neurons ($r \sim 0.65$). Thus, from experiment to experiment, variations in sampling appear to produce similar but not identical types of neurons that exhibit varying specificity. Despite these caveats concerning neuron types, the facts that similar profiles recur and that certain combinations of sensitivities (e.g., sucrose and quinine) do not occur appear to reflect an underlying orderliness that determines the organization of chemosensitivities in individual neurons.[44,228]

2. Independent Evidence

The data described above provide only one criterion for establishing gustatory neuron types, differences in chemosensitive response profiles. Recent studies provide at least two kinds of independent evidence for establishing functional classes of neurons (see Reference 213 for a more complete discussion). First, when the salience of a gustatory stimulus is altered through learning or deprivation state, the neural responses evoked by that stimulus can change,[91,217] but the changes are limited to certain neuronal subsets. For example, when a conditioned taste aversion is given to a sweet stimulus, the responses of certain NST neurons are preferentially affected,[217] and the affected subset overlaps greatly with the sweet-sensitive neuron class, identified by chemosensitive response profiles. Second, when the passive sodium channel blocker, amiloride, is applied to the tongue, responses of single fibers to NaCl, both at the level of the peripheral nerve[229,230] and the NST,[218,231] are reduced. All neurons responding to NaCl, however, are not equally affected. Instead, neurons that respond well to sodium vs. nonsodium salts and acids are preferentially affected. Again, these neural groups can be identified solely on the basis of their chemosensitive response profiles.[216,232]

3. Breadth of Tuning

A further question with regard to gustatory neuron types is whether they persist and how they are altered through the several levels of processing that occur between the peripheral nerves and cortex. Because the notion of neuron types requires a certain amount of response specificity, one feature of interest has been the degree of response breadth that is characteristic of individual gustatory neurons and how this tuning changes at different levels of the neuraxis. Comparing the breadth of tuning at different levels can be measured in various ways, but one commonly used metric that has the advantage of requiring no response criterion is the breadth of tuning metric. This metric, originally used in information theory, was adapted for the analysis of gustatory data by Travers and Smith.[44] The measure ("entropy" or "H") varies from 0 to 1 and when calculated using responses to 4 stimuli (representatives of the "basic" taste qualities), "0" represents a cell that responds exclusively to one stimulus and "1" as a neuron responding equally well to all 4. In an earlier review,[228] we summarized evidence which suggested that response breadth increased from the periphery to the NST, and at least in certain classes of neurons, from the NST to PBN. If such a trend continued in the forebrain, this could result in extremely broad tuning in gustatory cortical neurons and it might therefore be difficult to classify them. Such an increase in breadth, however, does not appear to continue into the forebrain. In fact, it may be reversed! Table 2 extends our earlier summary of breadth of tuning to include values of "H" for more recent NST studies and several cortical studies as well. While the trend for increased breadth of tuning from periphery to brainstem is still obvious with the addition of the recent studies, an opposite trend actually occurs from the NST to the cortex. Across all the studies, H increases from 0.455 in the periphery to 0.735 in the NST, but decreases from 0.735 to 0.581 in the cortex. When the values from the different studies are considered, it is apparent that a great deal of variability occurs from study to study. Nevertheless, the periphery-NST difference is statistically significant ($p < 0.002$), as is the NST-cortex difference ($p < 0.03$). Thus, it may be true that the central processing actually sharpens response profiles, as has been suggested for central processing in the primate.[94] Because of the variability and limited amount of data available, however, such a conclusion is premature. In addition, the average "H" values for both the NST and cortex are indicative of neurons that are moderately broadly tuned. At any rate, continued increases in breadth seem very unlikely. Thus, it is perhaps not surprising that similar neuron types, equally differentiable from one another, are observed in both NST and cortex.

B. SPATIAL ORGANIZATION: CHEMOTOPY?

The notion of chemotopic organization has continued to intrigue gustatory researchers. In other sensory systems, examples of the topographic organization of submodalities or other nonspatial stimulus features are plentiful, and the evidence clear. For example, in the visual system there are ocular dominance columns, orientation columns, and color "blobs". In the auditory system, tonotopic organization is a persistent feature through several levels of the neuraxis.[233] Indeed, given the striking differential sensitivities of the peripheral gustatory afferent nerves (reviewed in Reference 228; also see Chapter 12) and the orotopic arrangement of gustatory responses at several

TABLE 2
Breadth of Tuning

Level	Reference	Species	Breadth of tuning (H)
Nerve (CT)	238	Mouse	.284
Nerve (IX)	239	Rat	.298(d)[a]
Nerve (CT)	238	Mouse	.327
Nerve (CT)	238	Mouse	.405
Nerve (CT)	232	Rat	.521
Nerve (CT)	81	Hamster	.523(d)
Nerve (CT)	81	Rat	.561(d)
Nerve (CT)	240	Monkey	.565(d)
Nerve (CT)	216	Hamster	.608(d)
Level Ave.			**.455**
NST	45	Hamster	.618
NST	67	Rat	.640
NST	52	Rat	.671(d)
NST	44	Hamster	.698
NST	91	Rat	.700
NST	220	Rat	.730
NST	218	Rat	.790
NST	225	Monkey	.821
NST	219	Rat	.830
NST	99	Rat	.860
Level Ave.			**.736**
Cortex[b]	241	Dog	.343
Cortex	207	Rat	.540
Cortex	187	Rat	.543
Cortex	222	Monkey	.560
Cortex	242	Monkey	.650
Cortex	225	Monkey	.670
Cortex	221	Monkey	.760
Level Ave.			**.581**

[a] "d" indicates that the value was derived from published figures.

[b] Cortical values are for primary gustatory cortex, i.e., insular cortex in rodents, and insular-opercular cortex in the monkey. There is a single study that reports breadth of tuning for the secondary (orbitofrontal) gustatory cortex in the monkey. The value was .390, indicating the possibility of sharpening at this level.[94,221]

levels of the neuraxis *(vide supra)*, a chemotopic organization would appear to be a logical consequence.[56] Unfortunately, the bulk of our information about the topographic organization of gustatory quality comes from studies that stimulate only the anterior tongue, providing less than optimal conditions for observing chemotopy. Nevertheless, some studies have stimulated multiple taste bud groups, and, even in those that do not, reports of chemotopy have recurred repeatedly and been observed at both extremes of the central gustatory neuraxis, NST and cortex.[57,61,191,207,225,234-237] In several studies of the NST, specific responses to NaCl appear to predominate anteriorly,[61,225,236] whereas responses to quinine HCl or NH_4Cl predominate more posteriorly.[57,61,225,236] Such an organization is consistent with peripheral nerve sensitivities[64,216,232] and the orotopic arrangement of gustatory responses in NST, despite the fact that some of these studies stimulated only the anterior tongue. In the gustatory cortex, responses to sucrose or glucose predominate anteriorly, responses to HCl or quinine, posteriorly.[191,207,238] Again this is consistent with the greater sensitivity of the glossopharyngeal nerve to HCl and QHCl,[64] and the representation of this nerve posteriorly in the cortex. Despite these reports of chemotopy, however, the organization of the gustatory system is not thought of as a clear exemplar of segregation by submodality. Even though some segregation of taste quality in central neural relays appears to exist, there is also much overlap between their representation (e.g. Reference 56). Thus, many studies do not comment on a chemotopic organization, and others comment on it and find it lacking.[221] The lack of entirely compelling evidence for or against chemotopy is probably due to several factors including imprecise histological analysis, inadequate stimulation of all oral taste receptors, and unequal sampling of neurons responsive to the full spectrum of taste qualities. These last two difficulties are probably related. In particular, only small numbers of central gustatory neurons responsive to posterior oral cavity stimulation have been reported, and neurons optimally responsive to bitter stimuli are also a rarity in central recordings. The question of whether strong chemotopy occurs and the details of its organization will become clarified only when investigations routinely are able to combine large and representative samples of gustatory neurons with precise histological analyses. Given recent advances in the neuroanatomical and neurochemical delineation of the gustatory nuclei, more comprehensive gustatory stimulating techniques, and increasing numbers of chronic recording studies, a resolution to this issue may be imminent.

ACKNOWLEDGMENTS

During the preparation of this chapter, the author was supported by a research grant from the National Institutes of Health (DC-00415). Ms. Lisa Akey provided able technical support for the unpublished electrophysiological experiments cited here. The neuroanatomical data on NST-PBN connections depicted in Figure 4 was collected by Mr. David Becker in partial fulfillment for the degree of Master of Science from the Department of Psychology at The Ohio State University (1992). The bibliographic assistance of Mrs. Faith Kelley is much appreciated. Dr. Joseph Travers made valuable comments on the manuscript.

REFERENCES

1. **Norgren, R. and Leonard, C. M.,** Taste pathways in rat brainstem, *Science*, 173, 1136, 1971.
2. **Norgren, R. and Leonard, C. M.,** Ascending central gustatory pathways, *J. Comp. Neurol.*, 150, 217, 1973.
3. **Norgren, R.,** Gustatory afferents to ventral forebrain, *Brain Res.*, 81, 285, 1974.
4. **Norgren, R.,** Projections from the nucleus of the solitary tract in the rat, *Neuroscience*, 3, 207, 1978.
5. **Davis, B. J.,** Computer-generated rotation analyses reveal a key three-dimensional feature of the nucleus of the solitary tract, *Brain Res. Bull.*, 20, 545, 1988.
6. **Davis, B. J. and Jang, T.,** A Golgi analysis of the gustatory zone of the nucleus of the solitary tract in the adult hamster, *J. Comp. Neurol.*, 278, 388, 1988.
7. **Whitehead, M. C.,** Neuronal architecture of the nucleus of the solitary tract in the hamster, *J. Comp. Neurol.*, 276, 547, 1988.
8. **Torvik, A.,** Afferent connections to the sensory trigeminal nuclei, the nucleus of the solitary tract and adjacent structures, *J. Comp. Neurol.*, 106, 51, 1956.
9. **Beckstead, R. M. and Norgren, R.,** An autoradiographic examination of the central distribution of the trigeminal, facial, glossopharyngeal, and vagal nerves in the monkey, *J. Comp. Neurol.*, 184, 455, 1979.
10. **Norgren, R.,** Central neural mechanisms of taste, in *Handbook of Physiology, The Nervous System III*, Brookhart, J. M. and Mountcastle, V. B., Eds., Williams & Wilkins, Baltimore, 1984, chap. 24.
11. **Miller, I. J., Jr. and Spangler, K. M.,** Taste bud distribution and innervation on the palate of the rat, *Chem. Senses*, 7, 99, 1982.
12. **Hamilton, R. B. and Norgren, R.,** Central projections of gustatory nerves in the rat, *J. Comp. Neurol.*, 222, 560, 1984.
13. **Whitcomb, M., Nicklas, K., and Travers, S.,** Innervation of extra-oral taste buds by the superior laryngeal nerve, *Chem. Senses*, 14, 760, 1989.
14. **Hanamori, T. and Smith, D. V.,** Gustatory innervation in the rabbit: central distribution of sensory and motor components of the chorda tympani, glossopharyngeal, and superior laryngeal nerves, *J. Comp. Neurol.*, 282, 1, 1989.
15. **Sweazey, R. D. and Bradley, R. M.,** Central connections of the lingual-tonsillar branch of the glossopharyngeal nerve and the superior laryngeal nerve in lamb, *J. Comp. Neurol.*, 245, 471, 1986.
16. **Astrom, K. E.,** On the central course of afferent fibers in the trigeminal, facial, glossopharyngeal, and vagal nerves and their nuclei in the mouse, *Acta Physiol. Scand.*, 106, 209, 1953.
17. **Contreras, R. J., Beckstead, R. M., and Norgren, R.,** The central projections of the trigeminal, facial, glossopharyngeal and vagus nerves: an autoradiographic study i the rat, *J. Auton. Nerv. Syst.*, 6, 303, 1982.
18. **Pfaller, K. and Arvidsson, J.,** Central distribution of trigeminal and upper cervical primary afferents in the rat studied by anterograde transport of horseradish peroxidase coupled to wheat germ agglutinin, *J. Comp. Neurol.*, 268, 91, 1988.
19. **Whitehead, M. C. and Frank, M. E.,** Anatomy of the gustatory system in the hamster: central projections of the chorda tympani and the lingual nerve, *J. Comp. Neurol.*, 220, 378, 1983.
20. **Shigenaga, Y., Doe, K., Suemune, S., Mitsuhiro, Y., Tsuru, K., Otani, K., Shirana, Y., Hosoi, M., Yoshida, A., and Kagawa, K.** Physiological and morphological characteristics of periodontal mesencephalic trigeminal neurons in the cat — intra-axonal staining with HRP, *Brain Res.*, 505, 91, 1989.

21. **Marfurt, C. F.,** The central projections of trigeminal primary afferent neurons in the cat as determined by the transgaglionic transport of horseradish peroxidase, *J. Comp. Neurol.*, 203, 785, 1981.
22. **Jacquin, M. F., Semba, K., Rhoades, R. W., and Egger, M. D.,** Trigeminal primary afferents project bilaterally to dorsal horn and ipsilaterally to cerebellum, reticular formation, and cuneate, solitary, supratrigeminal and vagal nuclei, *Brain Res.*, 246, 285, 1982.
23. **Anton, F. and Peppel, P.,** Central projections of trigeminal primary afferents innervating the nasal mucosa: a horseradish peroxidase study in the rat, *Neuroscience*, 41, 617, 1991.
24. **Panneton, W. M.,** Primary afferent projections from the upper respiratory tract in the muskrat, *J. Comp. Neurol.*, 308, 51, 1991.
25. **Whitehead, M. C.,** Anatomy of the gustatory system in the hamster: synaptology of facial afferent terminals in the solitary nucleus, *J. Comp. Neurol.*, 244, 72, 1986.
26. **Menetrey, D. and Basbaum, A. I.,** Spinal and trigeminal projections to the nucleus of the solitary tract: a possible substrate for somatovisceral and viscerovisceral reflex activation, *J. Comp. Neurol.*, 255, 439, 1987.
27. **Schaffar, N., Kessler, J. P., Bosler, O., and Jean, A.,** Central serotonergic projections to the nucleus tractus solitarii: evidence from a double labeling study in the rat, *Neuroscience*, 26, 951, 1988.
28. **Bandler, R. and Tork, I.,** Midbrain periaqueductal grey region in the cat has afferent and efferent connections with solitary tract nuclei, *Neurosci. Lett.*, 74, 1, 1987.
29. **Van Der Kooy, D., Koda, L. Y., McGinty, J. F., Gerfen, C. R., and Bloom, F. E.,** The organization of projections from the cortex, amygdala, and hypothalamus to the nucleus of the solitary tract in rat, *J. Comp. Neurol.*, 224, 1, 1984.
30. **Shipley, M. T.,** Insular cortex projection to the nucleus of the solitary tract and brainstem visceromotor regions in the mouse, *Brain Res. Bull.*, 8, 139, 1982.
31. **Yasui, Y., Ito, K., Kaneko, T., Shigemoto, R., and Mizuno, N.,** Topographical projections from the cerebral cortex to the nucleus of the solitary tract in the cat, *Exp. Brain Res.*, 85, 75, 1991.
32. **Cechetto, D. F. and Saper, C. B.,** Evidence for a viscerotopic sensory representation in the cortex and thalamus in the rat, *J. Comp. Neurol.*, 262, 27, 1987.
33. **Beckman, M. E. and Whitehead, M. C.,** Intramedullary connections of the rostral nucleus of the solitary tract in the hamster, *Brain Res.*, 557, 265, 1991.
34. **Hermann, G. E., Kohlerman, N. J., and Rogers, R. C.,** Hepatic-vagal and gustatory afferent interactions in the brainstem of the rat, *J. Auton. Nerv. Syst.*, 9, 477, 1983.
35. **Kruger, L. and Mantyh, P. W.,** Gustatory and related chemosensory systems, in *Handbook of Chemical Neuroanatomy*, Vol. 7., Bjorklund, A., Hokfelt T., and Swanson, L. W., Eds., Elsevier, Amsterdam, 1989, 323.
36. **Zardetto-Smith, A. M. and Gray, T. S.,** Organization of peptidergic and catecholaminergic efferents from the nucleus of the solitary tract to the rat amygdala, *Brain Res. Bull.*, 25, 875, 1990.
37. **Herbert, H. and Saper, C. B.,** Cholecystokinin-, galanin-, and corticotropin-releasing factor-like immunoreactive projections from the nucleus of the solitary tract to the parabrachial nucleus in the rat, *J. Comp. Neurol.*, 293, 581, 1990.
38. **Mantyh, P. W. and Hunt, S. P.,** Neuropeptides are present in projection neurones at all levels in visceral and taste pathways: from periphery to sensory cortex, *Brain Res.*, 299, 297, 1984.
39. **Wang, L., King, M., and Bradley, R. M.,** In vitro patch clamp analysis of substance P. Effect on neurons in the rostral nucleus tractus solitarius, Abstracts: The Fourteenth Annual Meeting of the Association for Chemoreception Sciences, #233, 1992.
40. **Cooper, J. R., Bloom, F. E., and Roth, R. H.,** *The Biochemical Basis of Neuropharmacology*, 6th ed., Oxford University Press, New York, 1991.

41. **Lasiter, P. S. and Kachele, D. L.,** Organization of GABA and GABA-transaminase containing neurons in the gustatory zone of the nucleus of the solitary tract, *Brain Res. Bull.*, 21, 623, 1988.
42. **Liu, H., Behbehani, M., and Smith, D. V.,** The effects of GABA in the gustatory portion of the hamster NST: a patch-clamp analysis of cells in a brainstem slice, Abstracts: The Fourteenth Annual Meeting of the Association for Chemoreception Sciences, #226, 1992.
43. **Wang, L. and Bradley, R. M.,** In vitro patch clamp analysis of GABA effects on neurons in the gustatory zone of the nucleus tractus solitarius, Abstracts: The Fourteenth Annual Meeting of the Association for Chemoreception Sciences, #265, 1992.
44. **Travers, J. B. and Smith, D. V.,** Gustatory sensitivities in neurons of the hamster nucleus tractus solitarius, *Sensory Processes*, 3, 1, 1979.
45. **Sweazey, R. D. and Smith, D. V.,** Convergence onto hamster medullary taste neurons, *Brain Res.*, 408, 173, 1987.
46. **Dietrich, W. D., Lowry, O. H., and Loewy, A. D.,** The distribution of glutamate, GABA and aspartate in the nucleus tractus solitarius of the cat, *Brain Res.*, 237, 254, 1982.
47. **Dockstader, K. C., Dunwiddie, T. V., and Finger, T. E.,** Glutamate is not the neurotransmitter of primary gustatory afferent fibers, *Chem. Senses*, 16, 514, 1991.
48. **Davis, B. J. and Jang, T.,** Tyrosine hydroxylase-like and dopamine β-hydroxylase-like immunoreactivity in the gustatory zone of the nucleus of the solitary tract in the hamster: light and electron-microscopic studies, *Neuroscience*, 27, 949, 1988.
49. **Davis, B. J. and Jang, T.,** The gustatory zone of the nucleus of the solitary tract in the hamster: light microscopic morphometric studies, *Chem. Senses*, 11, 213, 1986.
50. **Silverman, J. D. and Kruger, L.,** Analysis of taste bud innervation based on glycoconjugate and peptide neuronal markers, *J. Comp. Neurol.*, 292, 575, 1990.
51. **Lasiter, P. S. and Kachele, D. L.,** Elevated NADH-dehydrogenase activity characterizes the rostral gustatory zone of the solitary nucleus in rat, *Brain Res. Bull.*, 22, 777, 1989.
52. **Travers, S. P., Pfaffmann, C., and Norgren, R.,** Convergence of lingual and palatal gustatory neural activity in the nucleus of the solitary tract, *Brain Res.*, 365, 305, 1986.
53. **Travers, J. B.,** Efferent projections from the anterior nucleus of the solitary tract of the hamster, *Brain Res.*, 457, 1, 1988.
54. **Travers, S. P. and Norgren, R.,** Oral sensory responses in the nucleus of the solitary tract, *Soc. Neuro Abstr.*, 14, 1185, 1988.
55. **Travers, S. P., Norgren, R., and Akey, L.,** Functional organization of orosensory responses in the nucleus of the solitary tract, in preparation.
56. **Dickman, J. D. and Smith, D. V.,** Topographic distribution of taste responsiveness in the hamster medulla, *Chem. Senses*, 14, 231, 1989.
57. **Halpern, B. P. and Nelson, L. M.,** Bulbar gustatory responses to anterior and to posterior tongue stimulation in the rat, *Am. J. Physiol.*, 209, 105, 1965.
58. **Miller, I. J., Jr., Gomez, M. M., and Lubarsky, E. H.,** Distribution of the facial nerve to taste receptors in the rat, *Chem. Senses Flav.*, 3, 397, 1978.
59. **Hayama, T., Ito, S., and Ogawa, H.,** Responses of solitary tract nucleus neurons to taste and mechanical stimulations of the oral cavity in decerebrate rats, *Exp. Brain Res.*, 60, 235, 1985.
60. **Ogawa, H., Imoto, T., and Hayama, T.,** Responsiveness of solitario-parabrachial relay neurons to taste and mechanical stimulation applied to the oral cavity in rats, *Exp. Brain Res.*, 54, 349, 1984.
61. **McPheeters, M., Hettinger, T. P., Nuding, S. C., Savoy, L. D., Whitehead, M. C., and Frank, M. E.,** Taste-responsive neurons and their locations in the solitary nucleus of the hamster, *Neuroscience*, 34, 745, 1990.
62. **Sweazey, R. D. and Bradley, R. M.,** Responses of lamb nucleus of the solitary tract neurons to chemical stimulation of the epiglottis, *Brain Res.*, 439, 195, 1988.

63. **Sweazey, R. D. and Bradley, R. M.,** Responses of neurons in the lamb nucleus tractus solitarius to stimulation of the caudal oral cavity and epiglottis with different stimulus modalities, *Brain Res.*, 480, 133, 1989.
64. **Frank, M. E.,** Taste-responsive neurons of the glossopharyngeal nerve of the rat, *J. Neurophysiol.*, 65, 1452, 1991.
65. **Furusawa, K., Yamaoka, M., and Kumai, T.,** Properties of the lingual and LVP branches of the glossopharyngeal nerve, *Brain Res. Bull.*, 28, 1, 1991.
66. **Ogawa, H. and Hayama, T.,** Receptive fields of solitario-parabrachial relay neurons responsive to natural stimulation of the oral cavity in rats, *Exp. Brain Res.*, 54, 359, 1984.
67. **Ogawa, H., Hayama, T., and Yamashita, Y.,** Thermal sensitivity of neurons in a rostral part of the rat solitary tract nucleus, *Brain Res.*, 454, 321, 1988.
68. **Altschuler, S. M., Bao, X., Bieger, D., Hopkins, D. A., and Miselis, R. R.,** Viscerotopic representation of the upper alimentary tract in the rat: sensory ganglia and nuclei of the solitary and spinal trigeminal tracts, *J. Comp. Neurol.*, 283, 248, 1989.
69. **Altschuler, S. M., Ferenci, D. A., Lynn, R. B., and Miselis, R. R.,** Representation of the cecum in the lateral dorsal motor nucleus of the vagus nerve and commissural subnucleus of the nucleus tractus solitarii in rat, *J. Comp. Neurol.*, 304, 261, 1991.
70. **Norgren, R. and Smith, G. P.,** Central distribution of subdiaphragmatic vagal branches in the rat, *J. Comp. Neurol.*, 273, 207, 1988.
71. **Travers, S. P. and Nicklas, K.,** Taste bud distribution in the rat pharynx and larynx, *Anat. Rec.*, 227, 373, 1990.
72. **Miller, I. J.,** Peripheral interactions among single papilla inputs to gustatory nerve fibers, *J. Gen. Physiol.*, 57, 1, 1971.
73. **Nagai, T., Mistretta, C. M., and Bradley, R. M.,** Developmental decrease in size of peripheral receptive fields of single chorda tympani nerve fibers and relation to increasing NaCl taste sensitivity, *J. Neurosci.*, 8, 64, 1988.
74. **Oakley, B.,** Receptive fields of cat taste fibers, *Chem. Senses Flav.*, 1, 431, 1975.
75. **Robinson, P. P.,** The characteristics and regional distribution of afferent fibres in the chorda tympani of the cat, *J. Physiol.*, 406, 345, 1988.
76. **Travers, S. P. and Norgren, R.,** Coding the sweet taste in the nucleus of the solitary tract: differential roles for anterior tongue and nasoincisor duct gustatory receptors in the rat, *J. Neurophysiol.*, 65, 1372, 1991.
77. **Hiiemae, K. M. and Crompton, A. W.,** Mastication, food transport and swallowing, in *Functional Vertebrate Morphology*, Hildebrand, M., Bramble, D. M., and Liem, K. F., Wake, D. B., Eds., Harvard University Press, Cambridge, 1985, 262.
78. **Zeigler, H. P.,** Drinking in mammals: functional morphology, orosensory modulation and motor control, in *Thirst: Physiological and Psychological Aspects*, Ramsay, D. J. and Booth, D., Eds., Springer-Verlag, New York, 1991, chap. 15.
79. **Travers, J. B.,** Drinking: hindbrain sensorimotor neural organization, in *Thirst: Physiological and Psychological Aspects*, Ramsay, D. J. and Booth, D., Eds., Springer-Verlag, New York, 1991, chap. 16.
80. **Makous, W., Nord, S., Oakley, B., and Pfaffmann, C.,** The gustatory relay in the medulla, in *Olfaction and Taste*, Zotterman, Y., Ed., Pergamon Press, New York, 1963, 381.
81. **Ogawa, H., Sato, M., and Yamashita, S.,** Multiple sensitivity of chorda tympani fibres of the rat and hamster to gustatory and thermal stimuli, *J. Physiol.*, 199, 223, 1968.
82. **Yamashita, S., Ogawa, H., Kiyohara, T., and Sato, M.,** Modification by temperature change of gustatory impulse discharges in chorda tympani fibres of rats, *Jpn. J. Physiol.*, 20, 348, 1970.
83. **Bradley, R. M., Stedman, H. M., and Mistretta, C. M.,** Superior laryngeal nerve response patterns to chemical stimulation of sheep epiglottis, *Brain. Res.*, 276, 81, 1983.

84. **Harding, R., Johnson, P., and McClelland, M. E.,** Liquid-sensitive laryngeal receptors in the developing sheep, cat and monkey, *J. Physiol*, 277, 409, 1978.
85. **Storey, A. T. and Johnson, P.,** Laryngeal water receptors initiating apnea in the lamb, *Exp. Neurol.*, 47, 42, 1975.
86. **Smith, D. V. and Hanamori, T.,** Organization of gustatory sensitivities in hamster superior laryngeal nerve fibers, *J. Neurophysiol*, 65, 5, 1098, 1991.
87. **Van Buskirk, R. L. and Erickson, R. P.,** Odorant responses in taste neurons of the rat NTS, *Brain Res.*, 135, 287, 1977.
88. **Glenn, J. F. and Erickson, R. P.,** Gastric modulation of gustatory afferent activity, *Physiol. Behav.*, 16, 561, 1976.
89. **Giza, B. K. and Scott, T. R.,** Blood glucose selectively affects taste-evoked activity in rat nucleus tractus solitarius, *Physiol. Behav.*, 31, 643, 1983.
90. **Giza, B. K. and Scott, T. R.,** Intravenous insulin infusions in rats decrease gustatory-evoked responses to sugars, *Am. J. Physiol.*, 222, r994, 1987.
91. **Jacobs, K. M., Mark, G. P., and Scott, T. R.,** Taste responses in the nucleus tractus solitarius of sodium-deprived rats, *J. Physiol.*, 406, 393, 1988.
92. **Bereiter, D. A., Berthoud, H.-R., and Jeanrenaud, B.,** Chorda tympani and vagus nerve convergence onto caudal brain stem neurons in the rat, *Brain Res. Bull.*, 7, 261, 1981.
93. **Yaxley, S., Rolls, E. T., Sienkiewicz, Z. J., and Scott, T. R.,** Satiety does not affect gustatory activity in the nucleus of the solitary tract of the alert monkey, *Brain Res.*, 347, 85, 1985.
94. **Rolls, E. T.,** Information processing in the taste system of primates, *J. Exp. Biol.*, 146, 141, 1989.
95. **Bereiter, D., Berthoud, H. R., and Jeanrenaud, B.,** Hypothalamic input to brain stem neurons responsive to oropharyngeal stimulation, *Exp. Brain Res.*, 39, 33, 1980.
96. **Matsuo, R., Shimizu, N., and Kusano, K.,** Lateral hypothalamic modulation of oral sensory afferent activity in nucleus tractus solitarius neurons of rats, *J. Neurosci.*, 4, 1201, 1984.
97. **Murzi, E., Hernandez, L., and Baptista, T.,** Lateral hypothalamic sites eliciting eating affect medullary taste neurons in rats, *Physiol. Behav.*, 36, 829, 1986.
98. **Kupferman, I.,** Hypothalamus and limbic system, in *Principles of Neural Science*, 3rd ed., Kandel, E. R., Schwartz, J. H., and Jessel, T. M., Eds., Elsevier, New York, 1991, chap. 48.
99. **Mark, G. P., Scott, T. R., Chang, F-C. T., and Grill, H. J.,** Taste responses in the nucleus tractus solitarius of the chronic decerebrate rat, *Brain Res.*, 443, 137, 1988.
100. **Millar, J. and Williams, G. V.,** Effects of iontophoresis of noradrenaline and stimulation of the periaqueductal gray on single-unit activity in the rat superficial dorsal horn, *J. Comp. Neurol.*, 287, 119, 1989.
101. **Herbert, H., Moga, M. M., and Saper, C. B.,** Connections of the parabrachial nucleus with the nucleus of the solitary tract and the medullary reticular formation in the rat, *J. Comp. Neurol.*, 293, 540, 1990.
102. **Whitehead, M. C.,** Subdivisions and neuron types of the nucleus of the solitary tract that project to the parabrachial nucleus in the hamster, *J. Comp. Neurol.*, 301, 554, 1990.
103. **Di Lorenzo, P. M. and Schwartzbaum, J. S.,** Coding of gustatory information in the pontine parabrachial nuclei of the rabbit: magnitude of neural response, *Brain Res.*, 251, 229, 1982.
104. **Di Lorenzo, P. M. and Schwartzbaum, J. S.,** Coding of gustatory information in the pontine parabrachial nuclei of the rabbit: temporal patterns of neural response, *Brain Res.*, 251, 245, 1982.
105. **Schwartzbaum, J. S.,** Electrophysiology of taste-mediated functions in parabrachial nuclei of behaving rabbit, *Brain Res. Bull.*, 11, 61, 1983.

106. **Beckstead, R. M., Morse, J. R., and Norgren, R.,** The nucleus of the solitary tract in the monkey: projections to the thalamus and brain stem nuclei, *J. Comp. Neurol.*, 190, 259, 1980.
107. **Fulwiler, C. E. and Saper, C. B.,** Subnuclear organization of the efferent connections of the parabrachial nucleus in the rat, *Brain Res. Rev.*, 7, 229, 1984.
108. **Cunningham, E. T., Jr. and Sawchenko, P. E.,** A circumscribed projection from the nucleus of the solitary tract to the nucleus ambiguus in the rat: anatomical evidence for somatostatin-28-immunoreactive interneurons subserving reflex control of esophageal motility, *J. Neurosci.*, 9, 1668, 1989.
109. **Rinaman, L., Card, J. P., Schwaber, J. S., and Miselis, R. R.,** Ultrastructural demonstration of a gastric monosynaptic vagal circuit in the nucleus of the solitary tract in rat, *J. Neurosci.*, 9, 1985, 1989.
110. **Ross, C. A., Ruggiero, D. A., and Reis, D. J.,** Projections from the nucleus tractus solitarii to the rostral ventrolateral medulla, *J. Comp. Neurol.*, 242, 511, 1985.
111. **Shapiro, R. E. and Miselis, R. R.,** The central organization of the vagus nerve innervating the stomach of the rat, *J. Comp. Neurol.*, 238, 473, 1985.
112. **Bieger, D. and Hopkins, D. A.,** Viscerotopic representation of the upper alimentary tract in the medulla oblongata in the rat: the nucleus ambiguus, *J. Comp. Neurol.*, 262, 546, 1987.
113. **Becker, D. C.,** Efferent Projections of Electrophysiologically Identified Regions of the Rostral Nucleus of the Solitary Tract in the Rat, Thesis, Ohio State University, Columbus, 1992.
114. **Becker, D. and Travers, S. P.,** Functional heterogeneity in the efferent organization of the orally-responsive nucleus of the solitary tract, in preparation.
115. **King, M. A., Louis, P. M., Hunter, B. E., and Walker, D. W.,** Biocytin: a versatile anterograde tract-tracing alternative, *Brain Res.*, 497, 361, 1989.
116. **Lasiter, P. S.,** Postnatal development of gustatory recipient zones within the nucleus of the solitary tract, *Brain Res. Bull.*, 28, 667, 1992.
117. **Atema, J.,** Structures and functions of the sense of taste in the catfish (Ictalurus Natalis), *Brain Behav. Evol.*, 4, 273, 1971.
118. **Finger, T. E. and Morita, Y.,** Two gustatory systems: facial and vagal gustatory nuclei have different brainstem connections, *Science*, 227, 776, 1985.
119. **Morita, Y. and Finger, T. E.,** Reflex connections of the facial and vagal gustatory systems in the brainstem of the bullhead catfish, *Ictalurus nebulosus*, *J. Comp. Neurol.*, 231, 547, 1985.
120. **Monroe, S. and Di Lorenzo, P. M.,** Taste responses in units in the nucleus of the solitary tract that do and do not relay information to the parabrachial pons, Abstracts: The Thirteenth Annual Meeting of the Association for Chemoreception Sciences, #248, 1991.
121. **Block, C. H. and Estes, M. L.,** The cytoarchitectural organization of the human parabrachial nuclear complex, *Brain Res. Bull.*, 24, 617, 1990.
122. **Halsell, C. B. and Frank, M. E.,** Mapping study of the parabrachial taste-responsive area for the anterior tongue in the golden hamster, *J. Comp. Neurol.*, 306, 708, 1991.
123. **Kolesarova, D. and Petrovicky, P.,** Parabrachial nuclear complex in the rat (nuclei parabrachiales and nucleus Koelliker-Fuse). Detailed cytoarchitectonic division and connections compared, *J. Hirnforsch*, 28, 517, 1987.
124. **Petrovicky, P.,** The nucleus Koelliker-Fuse (K-F) and parabrachial nuclear complex (PBNC) in man. Location, cytoarchitectonics and terminology in embryonic and adult periods, and comparison with other mammals, *J. Hirnforsch*, 30, 551, 1989.
125. **Davis, B. J.,** The ascending gustatory pathway: a Golgi analysis of the medial and lateral parabrachial complex in the adult hamster, *Brain Res. Bull.*, 27, 63, 1991.
126. **Lasiter, P. S. and Kachele, D. L.,** Postnatal development of the parabrachial gustatory zone in rat: dendritic morphology and mitochondrial enzyme activity, *Brain Res. Bull.*, 21, 79, 1988.

127. **Berkley, K. J. and Scofield, S. L.,** Relays from the spinal cord and solitary nucleus through the parabrachial nucleus to the forebrain in the cat, *Brain Res.*, 529, 333, 1990.
128. **Cechetto, D. F., Standaert, D. G., and Saper, C. B.,** Spinal and trigeminal dorsal horn projections to the parabrachial nucleus in the rat, *J. Comp. Neurol.*, 240, 153, 1985.
129. **Ma, W. and Peschanski, M.,** Spinal and trigeminal projections to the parabrachial nucleus in the rat: electron-microscopic evidence of a spino-ponto-amygdalian somatosensory pathway, *Somatosensory Res.*, 5, 247, 1988.
130. **Nasution, I. D. and Shigenaga, Y.,** Ascending and descending internuclear projections within the trigeminal sensory nuclear complex, *Brain Res.*, 425, 234, 1987.
131. **Shigenaga, Y., Chen, I. C., Suemune, S., Nishimori, T., Nasution, I. D., Yoshida, A., Sato, H., Okamoto, T., Sera, M., and Hosoi, M.,** Oral and facial representation within the medullary and upper cervical dorsal horns in the cat, *J. Comp. Neurol.*, 243, 388, 1986.
132. **Jessell, T. M. and Kelley, D. D.,** Pain and analgesia, in *Principles of Neural Science*, 3rd ed., Kandel, E. R., Schwartz, J. H., and Jessel, T. M., Eds, Elsevier, New York, 1991, chap. 27.
133. **Berk, M. L. and Finkelstein, J. A.,** Efferent connections of the lateral hypothalamic area of the rat: an autoradiographic investigation, *Brain Res. Bull.*, 8, 511, 1982.
134. **Halsell, C. B.,** Organization of parabrachial nucleus efferents to the thalamus and amygdala in the golden hamster, *J. Comp. Neurol.*, 317, 57, 1992.
135. **Holstege, G., Meiners, L., and Tan, K.,** Projections of the bed nucleus of the stria terminalis to the mesencephalon, pons, and medulla oblongata in the cat, *Exp. Brain Res.*, 58, 379, 1985.
136. **Holstege, G.,** Some anatomical observations on the projections from the hypothalamus to brainstem and spinal cord: an HRP and autoradiographic tracing study in the cat, *J. Comp. Neurol.*, 260, 98, 1987.
137. **Moga, M. M., Herbert, H., Hurley, K. M., Yasui, Y., Gray, T. S., and Saper, C. B.,** Organization of cortical, basal forebrain, and hypothalamic afferents to the parabrachial nucleus in the rat, *J. Comp. Neurol.*, 295, 624, 1990.
138. **Shipley, M. T. and Sanders, M. S.,** Special senses are really special: evidence for a reciprocal, bilateral pathway between insular cortex and nucleus parabrachialis, *Brain Res. Bull.*, 8, 493, 1982.
139. **Touzani, K., Ferssiwi, A., and Velley, L.,** Localization of lateral hypothalamic neurons projecting to the medial part of the parabrachial area of the rat, *Neurosci. Lett.*, 114, 17, 1990.
140. **Takeuchi, Y. and Hopkins, D. A.,** Light and electron microscopic demonstration of hypothalamic projections to the parabrachial nuclei in the cat, *Neurosci. Lett.*, 46, 53, 1984.
141. **Takeuchi, Y., Mclean, J. H., and Hopkins, D. A.,** Reciprocal connections between the amygdala and parabrachial nuclei: ultrastructural demonstration by degeneration and axonal transport of horseradish peroxidase in the cat, *Brain Res.*, 239, 583, 1982.
142. **Norgren, R. and Pfaffmann, C.,** The pontine taste area in the rat, *Brain Res.*, 91, 99, 1975.
143. **Hayama, T., Ito, S., and Ogawa, H.,** Receptive field properties of the parabrachio-thalamic taste and mechanoreceptive neurons in rats, *Exp. Brain Res.*, 68, 458, 1987.
144. **Ogawa, H., Hayama, T., and Ito, S.,** Response properties of the parabrachio-thalamic taste and mechanoreceptive neurons in rats, *Exp. Brain Res.*, 68, 449, 1987.
145. **Hermann, G. E. and Rogers, R. C.,** Convergence of vagal and gustatory afferent input within the parabrachial nucleus of the rat, *J. Auton. Nerv. Syst.*, 13, 1, 1985.
146. **Di Lorenzo, P. M. and Garcia, J.,** Olfactory responses in the gustatory area of the parabrachial pons, *Brain Res. Bull.*, 15, 673, 1985.
147. **Di Lorenzo, P. M.,** Taste responses in the parabrachial pons of decerebrate rats, *J. Neurophysiol.*, 59, 1871, 1988.

148. **Di Lorenzo, P. M.,** Corticofugal influence on taste responses in the parabrachial pons of the rat, *Brain Res.*, 530, 73, 1990.
149. **Norgren, R.,** Taste pathways to hypothalamus and amygdala, *J. Comp. Neurol.*, 166, 17, 1976.
150. **Bernard, J. F., Carroue, J., and Besson, J. M.,** Efferent projections from the external parabrachial area to the forebrain: a *Phaseolus Vulgaris* leucoagglutinin study in the rat, *Neurosci., Lett.*, 122, 257, 1991.
151. **Saper, C. B. and Loewy, A. D.,** Efferent connections of the parabrachial nucleus in the rat, *Brain Res.*, 197, 291, 1980.
152. **Lasiter, P. S., Glanzman, D. L., and Mensah, P.,** Direct connectivity between pontine taste areas and gustatory neocortex in rat, *Brain Res.*, 234, 111, 1982.
153. **Voshart, K. and Van Der Kooy, D.,** The organization of the efferent projections of the parabrachial nucleus to the forebrain in the rat: a retrograde fluorescent double-labeling study, *Brain Res.*, 212, 271, 1981.
154. **Block, C. H. and Schwartzbaum, J. S.,** Ascending efferent projections of the gustatory parabrachial nuclei in the rabbit, *Brain Res.*, 259, 1, 1983.
155. **Gang, S., Mizuguchi, A., Kobayashi, N., and Aoki, M.,** Descending axonal projections from the medial parabrachial and Kolliker-Fuse nuclear complex to the nucleus raphe magnus in cats, *Neurosci. Lett.*, 118, 273, 1990.
156. **Holstege, G.,** Anatomical evidence for a strong ventral parabrachial projection to nucleus raphe magnus and adjacent tegmental field, *Brain Res.*, 447, 154, 1988.
157. **Stevens, R. T., Hodge, C. J., Jr., and Apkarian, A. V.,** Kolliker-Fuse nucleus: the principal source of pontine catecholaminergic cells projecting to the lumbar spinal cord of cat, *Brain Res.*, 239, 589, 1982.
158. **Norgren, R.,** Gustatory system, in *The Human Nervous System*, Paxinos, G., Ed., Academic Press, New York, 1990, chap. 25.
159. **Pritchard, T. C.,** The primate gustatory system, in *Smell and Taste in Health and Disease*, Getchell, T. V., Bartoshuk, L. M., Doty, R. L., and Snow, J. B., Eds., Raven Press, New York, 1991, chap. 7.
160. **Travers, J. B. and Norgren, R.,** Afferent projections to the oral motor nuclei in the rat, *J. Comp. Neurol.*, 220, 280, 1983.
161. **Benjamin, R. M. and Akert, K.,** Cortical and thalamic areas involved in taste discrimination in the albino rat, *J. Comp. Neurol.*, 3, 231, 1959.
162. **Faull, R. L. M. and Mehler, W. R.,** Thalamus, in *The Rat Nervous System*, Volume 1, Forebrain and Midbrain, Paxinos, G., Ed., Academic Press, New York, 1985, chap. 5.
163. **Jones, E. G.,** *The Thalamus*, Plenum Press, New York, 1985, chap. 7.
164. **Norgren, R. and Grill, H. J.,** Efferent distribution from the cortical gustatory area in rats, *Soc. Neurosci. Abstr.*, 176, 124, 1976.
165. **Cechetto, D. F. and Saper, C. B.,** Role of the cerebral cortex in autonomic function, in *Central Regulation of Autonomic Function*, Loewy, A. D. and Spyer, K. M., Eds., Oxford University Press, New York, 1990, chap. 12.
166. **Yasui, Y., Saper, C. B., and Cechetto, D. F.,** Calcitonin gene-related peptide immunoreactivity in the visceral sensory cortex, thalamus, and related pathways in the rat, *J. Comp. Neurol.*, 290, 487, 1989.
167. **Ogawa, H. and Nomura, T.,** Receptive field properties of thalamo-cortical taste relay neurons in the parvicellular part of the posteromedial ventral nucleus in rats, *Exp. Brain Res.*, 73, 364, 1988.
168. **Nomura, T. and Ogawa, H.,** The taste and mechanical response properties of neurons in the parvicellular part of the thalamic posteromedial ventral nucleus of the rat, *Neurosci. Res.*, 3, 91, 1985.
169. **Scott, T. R. and Erickson, R. P.,** Synaptic processing of taste-quality information in thalamus of the rat, *J. Neurophysiol.*, 34, 868, 1971.

170. **Benjamin, R. M.,** Some thalamic and cortical mechanisms of taste, in *Proceedings of the First International Symposium on Taste and Olfaction,* Zotterman, Y., Ed., Pergamon Press, Oxford, England, 1963, 309.
171. **Emmers, R.,** Separate relays of tactile, pressure, thermal, and gustatory modalities in the cat thalamus, *Proc. Soc. Exp. Biol. Med.,* 121, 527, 1966.
172. **Frommer, G. P.,** Gustatory afferent responses in the thalamus, in *The Physiological and Behavioral Aspects of Taste,* Kare, M. and Halpern, B. P., Eds., University of Chicago Press, Chicago, 1961, 50.
173. **Pritchard, T. C., Hamilton, R. B., and Norgren, R.,** Neural coding of gustatory information in the thalamus of *Macaca mulatta, J. Neurophysiol.,* 61, 1, 1989.
174. **Blomquist, A. J., Benjamin, R. M., and Emmers, R.,** Thalamic localization of afferents from the tongue in squirrel monkey *(Saimiri sciureus), J. Comp. Neurol.,* 118, 77, 1962.
175. **Emmers, R., Benjamin, R. M., and Blomquist, A.,** Thalamic localization of afferents from the tongue in albino rat, *J. Comp. Neurol.,* 118, 43, 1962.
176. **Ogawa, H., Hayama, T., and Ito, S.,** Convergence of input from tongue and palate to the parabrachial nucleus neurons of rats, *Neurosci. Lett.,* 28, 9, 1982.
177. **Giachetti, I. and MacLeod, P.,** Olfactory input to the thalamus: evidence for a ventroposteromedial projection, *Brain Res.,* 125, 166, 1977.
178. **Rogers, R. C., Novin, D., and Butcher, L. L.,** Electrophysiological and neuroanatomical studies of hepatic portal osmo- and sodium-receptive afferent projections within the brain, *J. Auton. Nerv. Syst.,* 1, 183, 1979.
179. **Ganchrow, D. and Erickson, R. P.,** Thalamocortical relations in gustation, *Brain Res.,* 36, 289, 1972.
180. **Yamamoto, T., Matsuo, R., and Kawamura, Y.,** Corticofugal effects on the activity of thalamic taste cells, *Brain Res.,* 193, 258, 1980.
181. **Kosar, E., Grill, H. J., and Norgren, R.,** Gustatory cortex in the rat. II. Thalamocortical projections, *Brain Res.,* 379, 342, 1986.
182. **Norgren, R. and Wolf, G.,** Projections of thalamic gustatory and lingual areas in the rat, *Brain Res.,* 92, 123, 1975.
183. **Wolf, G.,** Projections of thalamic and cortical gustatory areas in the rat, *J. Comp. Neurol.,* 132, 519, 1968.
184. **Pritchard, T. C., Hamilton, R. B., Morse, J. R., and Norgren, R.,** Projections of thalamic gustatory and lingual areas in the monkey, *Macaca fascicularis, J. Comp. Neurol.,* 244, 213, 1986.
185. **Ruderman, M. I., Morrison, A. R., and Hand, P. J.,** A solution to the problem of cerebral cortical localization of taste in the cat, *Exp. Neurol.,* 37, 522, 1972.
186. **Yasui, Y., Itoh, K., Sugimoto, T., Kaneko, T., and Mizuno, N.,** Thalamocortical and thalamo-amygdaloid projections from the parvicellular division of the posteromedial ventral nucleus in the cat, *J. Comp. Neurol.,* 257, 253, 1987.
187. **Yamamoto, T.,** Taste responses of cortical neurons, *Prog. Neurobiol.,* 23, 273, 1984.
188. **Kosar, E., Grill, H. J., and Norgren, R.,** Gustatory cortex in the rat. I. Physiological properties and cytoarchitecture, *Brain Res.,* 379, 329, 1986.
189. **Kosar, E. and Schwartz, G. J.,** Cortical unit responses to chemical stimulation of the oral cavity in the rat, *Brain Res.,* 513, 212, 1990.
190. **Ogawa, H., Ito, S., Murayama, N., and Hasegawa, K.,** Taste area in granular and dysgranular insular cortices in the rat identified by stimulation of the entire oral cavity, *Neurosci. Res.,* 9, 196, 1990.
191. **Scott, T. R., Yaxley, S., Sienkiewicz, Z. J., and Rolls, E. T.,** Gustatory responses in the frontal opercular cortex of the alert cynomolgus monkey, *J. Neurophysiol.,* 56, 876, 1986.
192. **Ogawa, H., Ito, S.-I., and Nomura, T.,** Oral cavity representation at the frontal operculum of macaque monkeys, *Neurosci. Res.,* 6. 283, 1989.

193. **Saper, C. B.,** Reciprocal parabrachial-cortical connections in the rat, *Brain Res.*, 242, 33, 1982.
194. **Saper, C. B.,** Convergence of autonomic and limbic connections in the insular cortex of the rat, *J. Comp. Neurol.*, 210, 163, 1982.
195. **Shipley, M. T. and Geinisman, Y.,** Anatomical evidence for convergence of olfactory, gustatory, and visceral afferent pathways in mouse cerebral cortex, *Brain Res. Bull.*, 12, 221, 1984.
196. **Allen, G. V., Saper, C. B., Hurley, K. M., and Cechetto, D. F.,** Organization of visceral and limbic connections in the insular cortex of the rat, *J. Comp. Neurol.*, 311, 1, 1991.
197. **Hurley, K. M., Herbert, H., Moga, M. M., and Saper, C. B.,** Efferent projections of the infralimbic cortex of the rat, *J. Comp. Neurol.*, 308, 249, 1991.
198. **Terreberry, R. R. and Neafsey, E. J.,** The rat medial frontal cortex projects directly to autonomic regions of the brainstem, *Brain Res. Bull.*, 19, 639, 1987.
199. **Yamamoto, T., Azuma, S., and Kawamura, Y.,** Functional relations between the cortical gustatory area and the amygdala: electrophysiological and behavioral studies in rats, *Exp. Brain Res.*, 56, 23, 1984.
200. **Benjamin, R. M. and Pfaffmann, C.,** Cortical localization of taste in the albino rat, *J. Neurophysiol.*, 18, 56, 1955.
201. **Yamamoto, T., Matsuo, R., and Kawamura, Y.,** Localization of cortical gustatory area in rats and its role in taste discrimination, *J. Neurophysiol.*, 44, 440, 1980.
202. **Yamamoto, T. and Kitamura, R.,** A search for the cortical gustatory area in the hamster, *Brain Res.*, 510, 309, 1990.
203. **Yamamoto, T. and Kawamura, Y.,** Cortical responses to electrical and gustatory stimuli in the rabbit, *Brain Res.*, 94, 447, 1975.
204. **Yamamoto, T., Matsuo, R., Kiyomitsu, Y., and Kitamura, R.,** Sensory inputs from the oral region to the cerebral cortex in behaving rats: an analysis of unit responses in cortical somatosensory and taste areas during ingestive behavior, *J. Neurophysiol.*, 60, 1303, 1988.
205. **Benjamin, R. M. and Burton, H.,** Projection of taste nerve afferents to anterior opercular-insular cortex in squirrel monkey *(Saimiri sciureus)*, *Brain Res.*, 7, 221, 1968.
206. **Yamamoto, T., Yuyama, N., and Kawamura, Y.,** Cortical neurons responding to tactile, thermal and taste stimulations of the rat's tongue, *Brain Res.*, 221, 202, 1981.
207. **Yamamoto, T., Matsuo, R., Kiyomitsu, Y., and Kitamura, R.,** Taste responses of cortical neurons in freely ingesting rats, *J. Neurophysiol.*, 61, 1244, 1989.
208. **Rolls, E. T., Scott, T. R., Sienkiewicz, Z. J., and Yaxley, S.,** The responsiveness of neurones in the frontal opercular gustatory cortex of the macaque monkey is independent of hunger, *J. Physiol.*, 397, 1, 1988.
209. **Yaxley, S., Rolls, E. T., and Sienkiewicz, Z. J.,** The responsiveness of neurons in the insular gustatory cortex of the macaque monkey is independent of hunger, *Physiol. Behav.*, 42, 223, 1988.
210. **Reep, R. L. and Winans, S. S.,** Efferent connections of dorsal and ventral agranular insular cortex in the hamster, *Mesocricetus auratus*, *Neuroscience*, 7, 2609, 1982.
211. **Rolls, E. T., Yaxley, S., and Sienkiewicz, Z. J.,** Gustatory responses of single neurons in the caudolateral orbitofrontal cortex of the macaque monkey, *J. Neurophysiol.*, 64, 1055, 1990.
212. **Erickson, R. P.,** Grouping in the chemical senses, *Chem. Senses*, 10, 333, 1985.
213. **Scott, T. R. and Plata-Salaman, C. R.,** Coding of taste quality, in *Smell and Taste in Health and Disease*, Getchell, T. V., Bartoshuk, L. M., Doty, R. L., and Snow, J. B., Eds., Raven Press, New York, 1991, chap. 18.
214. **Smith, D. V., Van Buskirk, R. L., Travers, J. B., and Bieber, S. L.,** Gustatory neuron types in hamster brain stem, *J. Neurophysiol.*, 50, 522, 1983.

215. **Woolston, D. C. and Erickson, R. P.,** Concept of neuron types in gustation in the rat, *J. Neurophysiol.*, 42, 1390, 1979.
216. **Frank, M.,** An analysis of hamster afferent taste nerve response functions, *J. Gen. Physiol.*, 61, 588, 1973.
217. **Chang, F.-C. and Scott, T. R.,** Conditioned taste aversions modify neural responses in the rat nucleus tractus solitarius, *J. Neurosci.*, 4, 1850, 1984.
218. **Giza, B. K. and Scott, T. R.,** The effect of amiloride on taste-evoked activity in the nucleus tractus solitarius of the rat, *Brain Res.*, 550, 247, 1991.
219. **Giza, B. K., Scott, T. R., Sclafani, A., and Antonucci, R. F.,** Polysaccharides as taste stimuli: their effect in the nucleus tractus solitarius of the rat, *Brain Res.*, 555, 1, 1991.
220. **Nakamura, K. and Norgren, R.,** Gustatory responses of neurons in the nucleus of the solitary tract of behaving rats, *J. Neurophysiol.*, 66, 1232, 1991.
221. **Smith-Swintosky, V. L., Plata-Salaman, C. R., and Scott, T. R.,** Gustatory neural coding in the monkey cortex: stimulus quality, *J. Neurophysiol.*, 66, 1156, 1991.
222. **Yaxley, S., Rolls, E. T., and Sienkiewicz, Z. J.,** Gustatory responses of single neurons in the insula of the macaque monkey, *J. Neurophysiol.*, 63, 689, 1990.
223. **Baylis, L. L. and Rolls, E. T.,** Responses of neurons in the primate taste cortex to glutamate, *Physiol. Behav.*, 49, 973, 1991.
224. **Nishijo, H. and Norgren, R.,** Responses from parabrachial gustatory neurons in behaving rats, *J. Neurophysiol.*, 63, 707, 1990.
225. **Scott, T. R., Yaxley, S., Sienkiewicz, Z. J., and Rolls, E. T.,** Gustatory responses in the nucleus tractus solitarius of the alert cynomolgus monkey, *J. Neurophysiol.*, 55, 182, 1986.
226. **Gill, J. M. and Erickson, R. P.,** Neural mass differences in gustation, *Chem. Senses*, 10, 531, 1985.
227. **Bieber, S. L. and Smith, D. V.,** Multivariate analysis of sensory data: a comparison of methods, *Chem. Senses*, 11, 19, 1986.
228. **Travers, J. B., Travers, S. P., and Norgren, R.,** Gustatory neural processing in the hindbrain, *Annu. Rev. Neurosci.*, 10, 595, 1987.
229. **Hettinger, T. P. and Frank, M. E.,** Specificity of amiloride inhibition of hamster taste responses, *Brain Res.*, 513, 24, 1990.
230. **Ninomiya, Y. and Funakoshi, M.,** Amiloride inhibition of responses of rat single chorda tympani fibers to chemical and electrical tongue stimulation, *Brain Res.*, 451, 319, 1988.
231. **Scott, T. R. and Giza, B. K.,** Coding channels in the taste system of the rat, *Science*, 249, 1585, 1990.
232. **Frank, M. E., Contreras, R. J., and Hettinger, T. P.,** Nerve fibers sensitive to ionic taste stimuli in chorda tympani of the rat, *J. Neurophysiol.*, 50, 941, 1983.
233. **Kelley, J. P.,** Hearing, in *Principles of Neural Science*, 3rd ed., Kandel, E. R., Schwartz, J. H., and Jessel, T. M., Eds., Elsevier, New York, 1991, chap. 32.
234. **Halpern, B. P.,** Chemotopic organization in the bulbar gustatory relay of the rat, *Nature*, 208, 393, 1965.
235. **Halpern, B. P.,** Chemotopic coding for sucrose and quinine hydrochloride in the nucleus of the fasciculus solitarius, in *Olfaction and Taste II*, Hayashi, T. H., Ed., Pergamon Press, New York, 1967, 549.
236. **Mistretta, C. M.,** How the anterior tongue and taste responses are mapped onto the NST in fetal and perinatal sheep, *Chem. Senses*, 13, 719, 1988.
237. **Yamamoto, T., Yuyama, N., Kato, T., and Kawamura, Y.,** Gustatory responses of cortical neurons in rats. II. Information processing of taste quality, *J. Neurophysiol.*, 53, 1356, 1985.
238. **Ninomiya, Y., Mizukosho, T., Higashi, T., Katsukawa, H., and Funakoshi, M.,** Gustatory neural responses in three different strains of mice, *Brain Res.*, 302, 304, 1984.

239. **Nowlis, G. H. and Frank, M. E.,** Quality coding in gustatory systems of rats and hamsters, in *Perception of Behavioral Chemicals*, D. M. Morris, Ed., Elsevier/North Holland, Amsterdam, 1981, 59.
240. **Sato, M., Ogawa, H., and Yamashita, S.,** Response properties of macaque monkey chorda tympani fibers, *J. Gen. Physiol.*, 66, 781, 1975.
241. **Funakoshi, M. and Ninomiya, Y.,** Relations between the spontaneous firing rate and taste responsiveness of the dog cortical neurons, *Brain Res.*, 262, 155, 1983.
242. **Scott, T. R., Plata-Salaman, C. R., Smith, V. L., and Giza, B. K.,** Gustatory coding in the monkey cortex: stimulus intensity, *J. Neurophysiol.*, 65, 76, 1991.

Chapter 14

PSYCHOPHYSICS: INSIGHTS INTO TRANSDUCTION MECHANISMS AND NEURAL CODING

Susan S. Schiffman and Robert P. Erickson

TABLE OF CONTENTS

0-8493-5341-6/93/$0.00 + $.50

The gustatory model commonly used today to represent the qualitative range of taste was proposed by Henning in 1916.[1] This model, which is based on the four primary tastes (sweet, sour, salty, and bitter), was reified in geometric form as a hollow tetrahedron with the four principal taste qualities located at the four corners. Intermediate sensations were located on the edges of the tetrahedron. Sensations produced by three primaries were located on the surfaces.

This closed-ended model which has guided most psychophysical and neurophysiological research has recently been found to be qualitatively incomplete. A variety of psychophysical methods including multidimensional scaling and sorting procedures have shown that the range of taste is far greater than the simple four proposed by Henning.[2-4] In this chapter representative psychophysical experiments will be examined from the viewpoint of modern biochemistry and neurophysiology. While there are many steps that occur between interaction of ions and molecules with taste membranes and the sensations measured in psychophysical studies (see Chapters 5, 6, 8, and 11), some inferences can be drawn about transduction and neural coding from psychophysical studies. The studies discussed below suggest that taste stimuli interact with a multiplicity of channels and receptors, and that the sense of taste is far more complex than the model proposed by Henning.

I. PSYCHOPHYSICAL IMPLICATIONS FOR RECEPTOR MECHANISMS

This section will describe implications for receptor mechanisms that derive from psychophysical studies. Mechanisms for salts, amino acids and peptides, acids, sweeteners, bitter, and astringent compounds will be reviewed.

A. SALTS

1. Amiloride-Sensitive Sodium Channels Transduce Salty Taste of Na^+ Salts

Psychophysical studies suggest that entry of ions into taste receptor cells via apical ion channels is an important early step in taste transduction of cations. Taste transduction of sodium salts has received the greatest experimental attention. The role of sodium transport in taste perception was discovered simultaneously by DeSimone et al.[5] in epithelial transport studies and Schiffman et al.[6] in psychophysical and electrophysiologic experiments. In both of these studies, application of the sodium transport inhibitor amiloride to the dorsal tongue surface inhibited responses to NaCl.

This was an important discovery because prior to 1981 the prevailing theory of taste transduction was that NaCl and other stimuli physically adsorbed to receptors in the membrane but did not penetrate the membrane itself. The studies of DeSimone et al. clearly indicated that ions penetrate lingual epithelium, and the psychophysical and neurophysiologic studies demonstrated directly that this process was related to the sense of taste.

The psychophysical studies to be described here (see Chapter 15) show that amiloride (*n*-amidino-3, 5-diamino-6-chloropyrazine carboxamide), which reduces sodium entry across the mucosal or outer surface of the epithelium in other tissues, also reduces the perceived taste intensity of sodium and lithium salts. This study indicates that sodium transport is directly involved in the perception of taste in humans.

The study was performed as follows. Amiloride HCl was delivered to the tongue at concentrations of 500 μM and 50 μM. The mode of application was via pieces of chromatography paper cut into the shape of half tongues. These "half tongues" were soaked in an amiloride solution or in a dionized water control for 10 min. Next, they were placed on the tongue; one impregnated with amiloride was applied to one side and one with the water control on the other. These were left in place for 2.5 min, removed, and a fresh set was then applied for a total application time of 5 min.

Then a standard concentration of a test stimulus impregnated in a 1.27-cm disc of chromatography paper was placed on the side of the tongue to

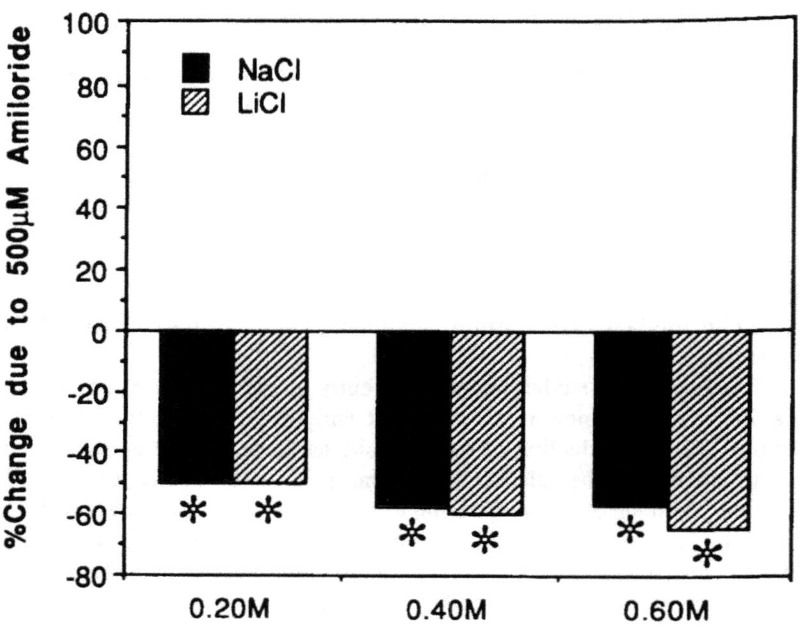

FIGURE 1. Percent suppression of the taste of NaCl and LiCl by 500 μM amiloride HCl in human subjects. * indicates that all decrements were statistically significant.

which the amiloride had been applied. Next, test stimuli were delivered to the untreated side, and the concentrations were adjusted to match the perceived intensity of the standard. The average concentration at which the subject judged the stimulus to be higher 50% of the time and lower 50% of the time was considered the concentration that matched the intensity of the standard. The results are shown for 500 μM amiloride in Figure 1. At 500 μM and 50 μM, amiloride reduced the perceived intensity of 0.6 M NaCl and LiCl by 50% or more. Amiloride had no effect on the perception of potassium or calcium salts nor did it reduce the perceived intensity of acids or bitter compounds. The psychophysical finding that amiloride blocks the taste of sodium and lithium salts with no effect on potassium salts is consistent with its action on a variety of cellular and epithelial transport systems.[6]

The finding that amiloride blocks responses to NaCl in humans[6] has been confirmed in several recent studies that utilize microdrop stimulation in conjunction with magnitude estimation. Tennissen[7] delivered taste stimuli to a set of papillae from 1-cc glass syringes mounted on a Harvard infusion pump. The taste stimuli were NaCl (0.4 M and 0.7 M) presented alone or mixed with 1 mM amiloride. She placed a drop of the test solution, small enough to cover one papilla, on the tongue. Then the drop was extended to cover neighboring papillae until the subject pushed a hand-held switch that terminated the stimulus delivery when he/she identified the taste quality. The intensity of the stimulus was then judged using the method of magnitude estimation. It was found that amiloride reduced the intensity of saltiness in most subjects. McCutcheon[8] used the same microdrop method and found that amiloride reduced the perceived intensity of NaCl but not citric acid. He also reported that half of his subjects required a large and significant increase in the number of papillae to detect saltiness in the presence of amiloride.

Amiloride-sensitive sodium pathways can be opened by quaternary ammonium compounds including the antifibrillary drug bretylium tosylate (BT) and the antiseptic cetylpyridinium chloride. BT has been shown to increase sodium transport through amiloride-sensitive channels in frog skin and has also been found to enhance taste responses to NaCl and LiCl.[9] In human psychophysical studies, 1 m*M* BT potentiated the taste of 0.2 *M* NaCl and 0.2 *M* LiCl by 33% and 12.5%, respectively, using the same testing procedure as Schiffman et al.[6] described above. These psychophysical findings were confirmed by electrophysiologic data from rat in which responses to NaCl and LiCl were potentiated by 30 to 40%. The potentiation induced by BT was reduced by amiloride, and, conversely, amiloride became ineffective in inhibiting taste responses to NaCl in the presence of BT. This suggests that these two drugs may be operating on the same channel in the taste cells.

Cetylpyridinium chloride (CPC) at low concentrations also enhanced the taste of 0.2 *M* NaCl[10] using the same experimental paradigm employed for amiloride and bretylium tosylate.[6,9] When the tongue was adapted to 0.1 m*M* CPC for 4 min, the taste of NaCl was significantly enhanced by 19.5% (Figure 2). Decrements in the intensity of NaCl occurred, however, at higher concentrations of CPC which were above its critical micelle concentration. When CPC was mixed in the salt solution, a similar pattern occurred with significant enhancement of NaCl by 21% at 0.1 m*M* CPC and decrements at higher concentrations of CPC. Some subjects reported increased persistence of the taste of NaCl on the side of the tongue exposed to 0.1 m*M* CPC. The psychophysical finding that CPC enhances the taste of 0.2 *M* NaCl is consistent with neurophysiological and epithelial transport studies. Application of 0.1 m*M* CPC for 5 min to the tongues of 4 rats resulted in a small increase of the multiunit taste activity recorded from nucleus of the solitary tract.[10] This increase was 10% for 0.3 *M* NaCl. CPC also increased Na^+ transport in isolated lingual epithelium when applied to the apical tongue surface.[11] The latter observation suggests that CPC opens amiloride-sensitive channels.

The psychophysical finding that an amiloride-blockable ion channel is involved in taste transduction of sodium salts has been confirmed by neurophysiologic investigations,[6] lingual epithelial ion transport studies,[5,12] and patch-clamp recordings from taste bud cells.[13] There are also some data which suggest that amiloride-insensitive Na^+ pathways may also exist in taste cell

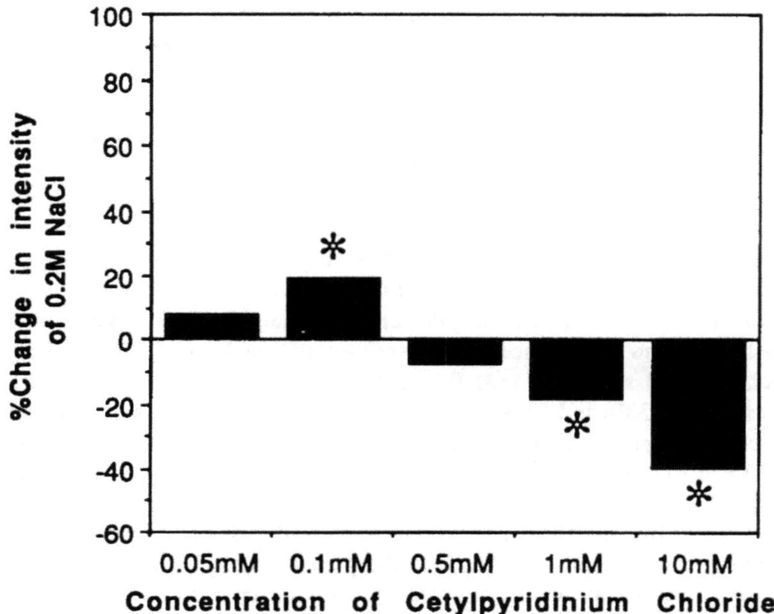

FIGURE 2. Percent change in saltiness of 0.2 *M* NaCl induced by cetylpyridinium chloride in human subjects. There is an enhancement at 0.1 m*M* which is below critical micelle concentration.

membranes.[14] However, Schiffman et al.[15] found that phenamil, a potent amiloride analog, blocked up to 98% of tonic taste responses to NaCl in the gerbil. This suggests that the amiloride-sensitive sodium channel alone may transduce salty taste in the gerbil although this may be species specific. Another possibility is that the tight junctions between taste cells are more cation-selective in the gerbil than in other species, and consequently the effectiveness of amiloride will be increased.[19]

2. Anions Modulate Taste Quality of Salts

Psychophysical studies indicate that anions modulate Na^+ taste. For many years it has been recognized that sodium salts have more taste qualities than saltiness alone which suggests that the anion must play a role (see Reference 16 for a review). Schiffman et al.[17] attempted to clarify the range of taste sensations for a set of sodium salts and to gain a greater understanding of the contribution of the cation and anion to taste quality. They found that the range could not be defined by the closed system of the four so-called primary or basic tastes.

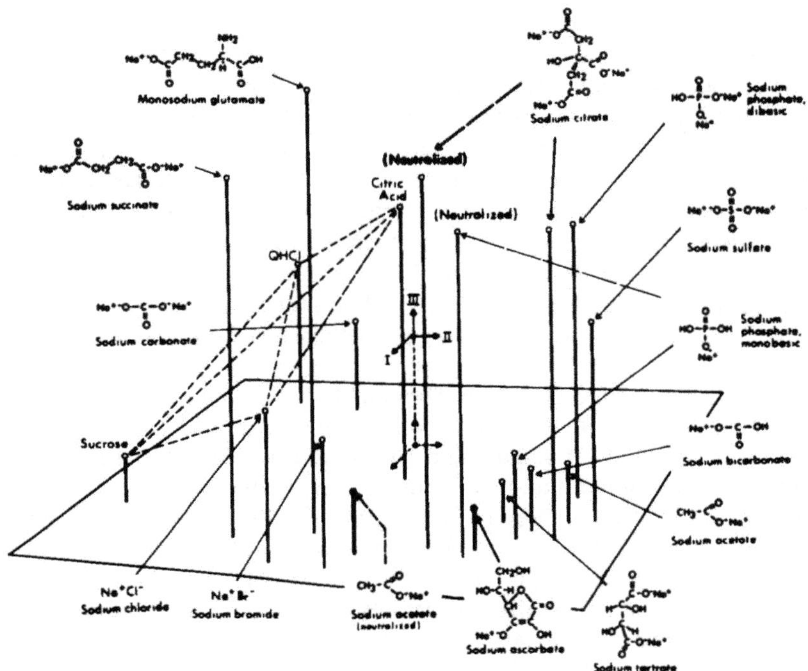

FIGURE 3. Arrangement of sodium salts and traditional standards derived by multidimensional scaling (from Reference 15). Most sodium salts are not contained within the closed system defined by salty (NaCl), sour (citric acid), bitter (quinine HCl), and sucrose (sweet).

Schiffman et al.[17] arranged sodium salts in a map on the basis of quantitative psychophysical similarity utilizing the mathematical technique of multidimensional scaling (MDS). In this method, stimuli rated similar in quality are located close to one another in the taste map while stimuli rated dissimilar are located distant from one another. Thirteen sodium salts as well as sucrose (sweet), citric acid (sour), and quinine HCl (bitter) stimuli were arranged in a three-dimensional map by the MDS procedure (see Figure 3). In the map the "primary tastes", that is, sucrose, citric acid, NaCl, and quinine HCl, fell toward one side of the space. Monosodium glutamate and sodium succinate fell apart from these stimuli toward the left of the "primaries", and the rest of the salts were arranged toward the right. Thus the spatial arrangement of sodium salts based on their similarities suggested that the taste qualities were far broader than the range of taste of traditional taste theory. Sodium salts of glutamate, acetate, ascorbate, bicarbonate, citrate, phosphate, succinate, sulfate, and tartrate each were found to have unique tastes of their own which are not confined to the "basic 4" tastes. Thus, the anion was shown to play a major role in determining the quality of each of the sodium salts.

It is not surprising that anions contribute to the taste of sodium salts because it has been well established that anions are transported by the lingual epithelium.[18] Recently, Ye et al.[19] found that anions affect taste reception through diffusion potentials across tight junctions between receptor cells. Both sodium ions and anions can permeate paracellular tight junctions; however, anions vary in their shunting capacity. For example, acetate and gluconate ions are less permeable across tight junctions than chloride ions by nature of their larger size; thus, they poorly compensate field potentials and result in hyperpolarization of the receptor cell (see Chapter 8 for additional details). This provides evidence that the functional unit for taste perception is not the single taste cell but also includes its paracellular microenvironment.

3. Anions Determine Detection Thresholds

The anion not only plays a major role in determining taste quality but also taste thresholds of sodium and potassium salts. Schiffman et al.[20] determined detection thresholds for sodium salts in young and elderly human subjects. The detection threshold was defined as the concentration at which the solution was distinguishable from deionized water. They found that the taste detection thresholds in humans for eight of the ten sodium salts were linearly correlated with molar conductivity values at infinite dilution of their ions. Molar conductivity (lambda) is directly proportional to both ionic mobility (u) and ionic charge (Z), i.e., molar conductivity is a measure of the electrical charge carried by the ion per unit time. Because molar conductivity is proportional to both ionic mobility and ionic charge, the correlation with threshold indicates that anionic charge mobility plays a role in determining detection threshold concentrations. That is, detection thresholds for sodium salts appear to be determined by the charge mobility of the anion. In addition, age-related losses in detection threshold are greatest for salts with anions having the greatest molar conductivity. A similar finding was made for potassium salts.[21]

4. Multiple Transduction Mechanisms Exist for Different Cations

The method of cross-adaptation has been used to investigate gustatory coding in humans.[22-24] Cross-adaptation experiments suggest a multiplicity of transduction mechanisms for cations.[22] When the human tongue is adapted to a variety of salts, the sensations to other salts are reduced. The degree of suppression differs with the cations that are cross-adapted, and suppression is strongest for salts with the same cation as the adapting salt and when the stimuli are judged equal in magnitude.

Other evidence for multiple transduction mechanisms for different ions comes from the fact that amiloride does not reduce the intensity of KCl in humans but does blunt the taste of NaCl. Thus Na^+ and K^+ involve different transduction mechanisms. In additional human studies, Schiffman[25] found that 1 m*M* trifluoperazine, which blocks calmodulin in a variety of cells,

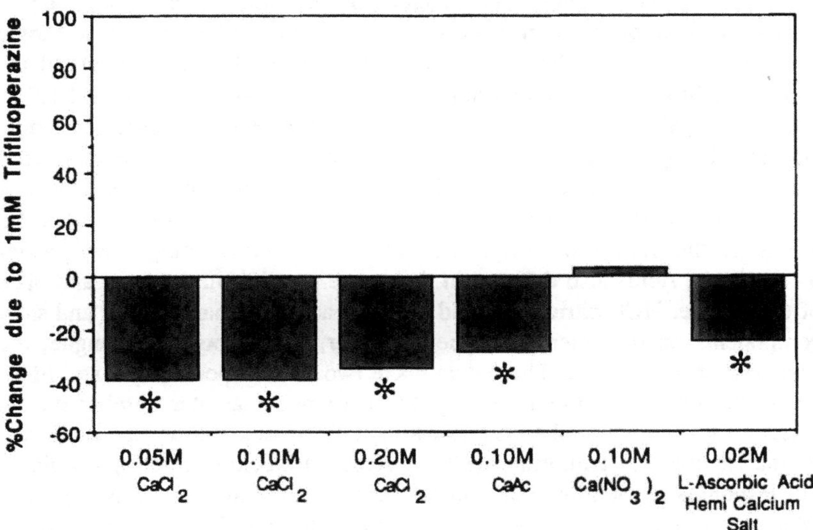

FIGURE 4. Percent suppression of calcium salts by 1 m*M* trifluoperazine in human subjects.

specifically blocks the taste of $CaCl_2$, CaAcetate, and L-ascorbic acid hemi calcium salt with no effect on sodium or potassium salts using the same procedure described for the amiloride experiment above (see Figure 4). This suggests that a mechanism for calcium, separate from those of sodium and potassium, exists in taste cells.

5. Concentration-Dependent Quality Changes May Be Due to Perturbation of Water Structure

The taste quality of suprathreshold inorganic salt solutions has been shown to change as a function of the concentration.[26-29] However, little is known about the mechanism by which these changes occur. Dzendolet[30] suggested that the sweet taste of weak concentrations of NaCl was due to localized hydrolysis which produces a structure in which the cations of the salt are surrounded by a shield of hydroxyl ions. These hydroxyl groups are capable of intermolecular hydrogen binding which is thought to be one of the mechanisms of binding for sweet tastants (to be discussed later). Other changes in taste quality with concentration may also be due to perturbation in water structure. Kemp et al.[31] have recently suggested that perturbation of water structure rather than stimulus molecular structure may play a role in taste thresholds.

6. Distribution of Channels Varies over the Tongue Surface

NaCl and other salts taste different on different parts of the tongue. Sandick and Cardello[32] compared the taste of three chloride salts and three acids on the front and the back of the tongue using 0.25 μl droplets of the stimulus solutions. On the anterior tongue, NaCl at both 250 m*M* and 1.0 *M* was rated predominantly salty with a sour component. However, on single circumvallate papillae on the posterior tongue, 250 m*M* was rated less salty than the front of the tongue and 1.0 *M* NaCl was rated more bitter and sour than salty. For KCl at the same concentrations, the salty ratings were far greater on the front of the tongue and the sour and bitter ratings were greater on the back. Acids also differed in their taste profiles on the front and back of the tongue. HCl, citric acid, and ascorbic acid had marked sour and salty components on the anterior tongue; however, on the posterior tongue, the sour taste predominated. These data make two points about salty taste. First, amiloride-sensitive components should be found in greater number on the anterior tongue than on the posterior tongue. This assertion is also supported by the finding that amiloride has little effect in blocking electrophysiologic salt responses in the glossopharyngeal nerve in rat.[33] Second, salts appear to interact with mechanisms for sour and bitter taste transduction as well as the amiloride-sensitive sodium channel.

B. PEPTIDES AND AMINO ACIDS

Amino acids and peptides have a wide range of taste qualities. Multidimensional scaling experiments reveal that amino acids have tastes that fall within and outside of the classical four tastes.[34] It is not surprising that amino acids and peptides induce tastes because high density neuropeptidergic fiber networks have been found around and within taste buds.[35,36] There must be receptors for peptides secreted from the taste neurons to the buds because these neuropeptides play a major role in the development and maintenance of taste buds (see Chapter 11).

1. Saltiness

A variety of peptides have been described as having salty tastes. For example, Shinoda et al.[37] and Tada et al.[38] reported synthesis of salty dipeptides without off-flavors. However, subsequent analyses at a number of food companies have found that most of the salty taste of one of these peptides, L-ornithyltaurine monohydrochloride, was simply due to the presence of NaCl as a byproduct of the synthesis. While amino acids and peptides may have salty components, no pure salty peptide similar to NaCl has been found.

2. Umami

The taste of glutamate salts is different from that of the classic four tastes[39] and this taste is termed "umami" in Japanese. Glutamate salts are dramatically enhanced in intensity by 5′-ribonucleotides;[39,40] however, these nucleotides have little effect on sweet, sour, salty, or bitter tastes which further suggests that glutamate taste operates by a different mechanism than the so-called primaries. Peptides as well as amino acids have umami tastes. For example, the peptide (Lys-Gly-Asp-Glu-Glu-Ser-Leu-Ala) found in beef soup has been reported to have an umami taste with a salty component.[41,42]

3. Bitterness and Hydrophobicity

Many peptides have bitter tastes, especially peptides formed by enzymatic hydrolysis of casein, lactalbumin, soy protein, ovalbumin, and gluten. Ney[43] and Bigelow[44] suggested that bitterness of these peptides could be predicted by the average hydrophobicities of the amino acid residues, and they utilized the transfer-free energies of the side chains of amino acids to calculate the average hydrophobicity. The transfer-free energies of the side chains of amino acids from water to nonpolar solvents were originally calculated by Tanford[45] and Nozaki and Tanford.[46] The specific sequence of the amino acids in a peptide had no influence on bitter taste.[43] Ney[47] calculated the average hydrophobicities for numerous proteins and found that all proteins with bitter tasting enzymatic hydrolysates have high average hydrophobicities. Ney's rule holds for the majority of bitter peptides known; exceptions including peptides containing glycine, however, do not obey the rule.[48,49]

Stereochemistry rather than hydrophobicity predicts the taste character and intensity of cyclic dipeptides, however.[50] The taste intensities of cyclic dipeptides that contain L-amino acid residues are slightly greater than those isomers containing D-amino acid residues. In addition, the configuration at the chiral centers of cylic dipeptides plays an important role in differentiation of taste properties.

C. SWEETNESS

Sweetness appears to have complicated taste properties that are transduced by multiple mechanisms. Psychophysical studies show that it is far more complex than a simple "primary" taste.

1. Multiple Sweet Receptor Mechanisms Transduce Sweet Taste

Seven lines of evidence from psychophysical experiments suggest that there are multiple receptor mechanisms for sweetness. The first line of evidence derives from experiments that reveal nonhomogeneous variability among sweeteners in individual subjects for thresholds, intensity ratings, and the effect of pronase E. Faurion et al.[51] found that the thresholds for sweeteners varied independently from subject to subject. It was not possible to predict a subject's threshold sensitivity to one sweetener given his or her response to another sweetener. This suggests that the relative numbers of receptor sites for each sweetener is unique for a given individual because sweet thresholds are independent of each other. Suprathreshold intensity ratings exhibited the same variability. Faurion et al.[51] trained subjects to recognize 7 concentrations of sucrose and then asked subjects to match 7 concentrations of 12 other sweeteners against the standards. Noncovariance between individual intensity functions were found for the 12 sweeteners which suggests that a unique peripheral chemoreception mechanism cannot account for sweetness. Schiffman et al.[52] reached the same conclusions as Faurion et al. using a different methodology. The relative reduction of sweetness by pronase E in humans also shows different profiles for individual subjects[53] which further supports the position that there are multiple sweet receptor types.

The second line of evidence derives from cross-adaptation experiments. If all sweet compounds equally cross-adapt, one would expect a single receptor site type. This does not occur, however, and there are strong differences in the degree of cross-adaptation among sweeteners. Schiffman et al.[54] found that the degree of cross-adaptation among sweet-tasting compounds was related to the possible AH-B configurations in the molecules. A prominent theory of sweet taste (Chapter 16) suggests that a sweet molecule possess an AH-B configuration where A and B are electronegative atoms separated by 2.5 to 4 angstroms, and H is a hydrogen atom that is part of a polarized system A-H. A complementary AH-B group in the receptor is presumed to interact with the AH-B site in the stimulus molecule to form two simultaneous hydrogen bonds. Schiffman et al.[54] found that stimuli which show the greatest cross-adaptation such as sodium saccharin and acesulfame-K have identical AH-B configurations and thus presumably share the same receptor sites. Neohesperidin dihydrochalcone and acesulfame-K which do not cross adapt have totally different AH-B types.

A third line of evidence for multiple sweet receptors comes from multidimensional scaling of sweeteners.[55] Sugars were found to be distant from large proteins such as thaumatin in the three-dimensional space based on sweet quality. Thus, sweeteners vary tremendously in the type or nature of the sweet sensation they impart which suggests "sweetness" itself is not a single or unitary quality. In addition, sweet tasting proteins such as thaumatin produce more intense sweetness at the edges of the tongue while sucrose is more intense at the tip of the tongue.[56] Also, the sweetness for thaumatin and monellin develops more slowly and has a longer duration than the sweet sensation of sucrose. These differences in sensory characteristics between sweet proteins and sucrose suggest that they engage different taste receptors on the tongue.

A fourth line of evidence is that the dose-response curves are dramatically different for different sweeteners. Dubois et al.[57] constructed dose-response curves for a wide range of sweeteners from suprathreshold intensity judgments made by a trained taste panel. Six concentrations of sucrose (2, 5, 7.5, 19, 12, and 16%) were used to standardize sweetness intensity ratings of the taste panel on a 15-cm line scale. The panel was trained to assign intensity values of 2, 5, 7.5, 10, 12, and 15 to these concentrations, respectively. All standards were dissolved in deionized water. The dose-response curves for sucrose, fructose, and sugar alcohols continued to increase in intensity beyond the equivalent sweetness of 15% sucrose. Aspartame and alitame never got much sweeter than 15 to 16% sucrose even at maximum solubility. Thaumatin never got sweeter than a 9% sucrose equivalent.

A fifth line of evidence for multiple receptors comes from suprathreshold intensity data on sweeteners derived by the method of magnitude estimation.[52] Loss in perceived intensity with age is not uniform across sweeteners. Rather, the greatest loss was for large molecules such as thaumatin, rebaudioside, and neohesperidin dihydrochalcone which have the greatest number of possible AH-B systems. This nonuniformity of age-related loss suggests multiple receptors.

A sixth line of evidence comes from cooling studies of the tongue. Green and Frankmann[58] used the method of magnitude estimation to determine the effect of cooling the tongue on the perceived intensity of taste. They found that when the tongue was cooled to 20°C, the sweetness of sucrose and the bitterness of caffeine were reduced in intensity; the sourness of citric acid and the saltiness of sodium chloride were unaffected by cooling. A further study showed that cooling the tongue did not reduce the sweetness of all sweet compounds.[59] While fructose and glucose had temperature sensitivities similar to sucrose, saccharin did not. This suggests that multiple rather than a single receptor mechanism underlies sweet taste perception.

The last line of evidence comes from studies using methyl xanthines to modulate taste. Schiffman et al.[60] found that caffeine enhances the taste of some sweeteners including thaumatin, stevioside, sodium saccharin, acesulfame-K, neohesperidin dihydrochalcone, and D-tryptophan with no effect on other sweeteners such as aspartame, sucrose, fructose, and calcium cyclamate. This finding again emphasizes that multiple mechanisms for sweetness occur in the taste cell.

2. Degree of Hydrogen Bonding Predicts Detection Thresholds of Sweeteners

Support for the AH-B hypothesis of sweet taste comes from threshold studies.[52] There is a strong relationship between the number of possible types of systems for hydrogen bonding (i.e., AH-B systems) and mean thresholds. The lowest thresholds are for the sweeteners monellin and thaumatin which have many possible types of AH-B systems. The highest thresholds are for sweeteners with only one type.

3. Amiloride-Sensitive Sodium Channels Are Involved in Sweet Taste in Humans

Psychophysical studies suggest that entry of sodium ions into taste receptor cells via apical sodium is an important early step in taste transduction of sweeteners in humans as well as NaCl. Schiffman et al.[6] not only found that amiloride blocked responses to NaCl but also blocked responses to every sweetener tested (stevioside, acesulfame-K, sucrose, thaumatin, glucose, calcium cyclamate, rebaudioside-A, apartame, neohesperidin dihydrochalcone, and fructose). The reduction in sweetness by amiloride was confirmed in a recent study that utilized microdrop stimulation in conjunction with magnitude estimation. Tennissen[7] delivered sucrose (0.3 *M*, 0.5 *M*, 0.7 *M*, 1.0 *M*) stimuli to a set of papillae from 1-cc glass syringes mounted on a Harvard infusion pump. The taste stimuli were presented alone or mixed with 1 m*M* amiloride. She found that amiloride reduced the intensity of sweetness in some but not all subjects.

4. Adaptation Suggests a Desensitization Mechanism for the Taste Receptors Similar To That of the Beta Adrenergic Receptor

Decreased sensitivity to a taste stimulus after repeated tasting is found for all sweeteners. Schiffman[25] has tested sweetness intensity of a variety of sweeteners with repeated tasting using the scaling method of DuBois et al.[57] described above. Panelists were presented with 10-ml samples of the same concentration of a given sweetener every 30 s with 2-min breaks after 4 samples at the same concentration. For the disaccharide sucrose at a weak concentration (3%), there was approximately a 30% decrement from the first to the fourth sip; for strong concentrations (12%) the decrement was only 6%. For the monosaccharide fructose, the decrement was 42% after 4 sips of 3% fructose but only 9% after 12% fructose. Similar patterns of adaptation were found for glucose, mannitol, and sorbitol; that is, there was far greater adaptation at low concentrations than high concentrations. For sodium saccharin, however, the degree of adaptation after four sips was similar at both high and low concentrations. At 250 ppm, there was 46% adaptation from the first to the fourth sip; at 1000 ppm, the adaptation was 50%. This pattern in which the degree of adaptation was similar over a range of concentrations was also found for other high potency sweeteners including rebaudioside-A, stevioside, and aspartame. This adaptation with repeated sips may be due to desensitization of the receptor similar to that for beta adrenergic receptors or rhodopsin.[61]

D. ACIDS

The taste of acids is not confined to the quality of sourness, but extends beyond the closed-ended model of Henning. Schiffman[25] performed a multidimensional scaling of 11 acids along with sucrose (sweet), citric acid (sour), and quinine HCl (bitter). Many acids fell outside the range defined by the primary tastes when scaled in a three-dimensional map. Thus, the anion plays a major role in the taste quality of acids. In addition, sour taste is not quantitatively proportional to pH. Both the anion and undissociated acid molecules play a role. Moskowitz[62] found that the pH of an acid does not directly predict its sourness. Makhlouf and Blum[63] and Price and DeSimone[64] proposed theories about this phenomenon. Basically, it is possible that the anion indirectly facilitates the sour taste by decreasing the positive charge produced by the binding of hydrogen ions to the taste cell membrane.

Ganzevles and Kroeze[65] studied the sour taste of acids and investigated the role of the hydrogen ion and the undissociated acid as sour agents. Carboxylic acids (weak acids that are slightly dissociated in water) taste more sour than HCl (a strong acid that is highly dissociated in water) when the concentrations are matched for pH.[66] Thus the sour taste of weak acids is not entirely due to the H_3O^+ ion concentration.

In order to determine if the transduction processes by which H^+ and undissociated acid molecules (HA) elicit sour taste are the same or different, Ganzevles and Kroeze[65] used a cross-adaptation paradigm to investigate adaptation and cross-adaptation of HCl and the following carboxylic acids: tartaric acid, acetic acid, and lactic acid. If the transduction mechanisms are the same, then HCl and weak carboxylic acids should cross adapt. However, their findings suggest otherwise. When HCl was the adapting stimulus, it did not cross adapt with the carboxylic acids. However, mutual cross-adaptation did occur among tartaric, acetic, and lactic acids. These results suggest that the tastes of HCl and carboxylic acids are elicited by different receptor processes even though both have a sour component.

E. BITTER COMPOUNDS

Transduction of bitter taste is just as complex as that for salts, peptides, acids, and sweet compounds.

1. Multiple Mechanisms Transduce Bitter Taste

There is clearly a multiplicity of mechanisms for bitter tastants (see Chapter 10). Cross-adaptation studies showed that multiple receptor types for bitter compounds exist[23] because bitter compounds do not equally cross-adapt. Lush[67] suggested multiple bitter site types because there is genetic dimorphism with respect to the ability to detect the bitter tastes of strychnine and sucrose octaacetate in mice.

2. Hydrophobicity Determines Bitterness Thresholds

Hydrophobicity as determined by octanol/water partition coefficients has been shown to be highly correlated with the taste detection threshold of bitter compounds.[68] This may be due to the fact that bitter compounds bind to hydrophobic amino acids on the receptors or that penetration of the bilipid layer is necessary for detection of bitter taste.

3. Modulation of Bitter Taste

Bitter taste can be enhanced by other components in a mixture. For example, Keil et al.[69] found that theobromine enhanced the bitter taste of cyclic dipeptides in roasted cocoa. Schiffman et al.[60] also found that treating the tongue with methyl xanthines including theobromine enhanced certain bitter tastes including quinine HCl and sweeteners with a bitter component. Schiffman suggested that this may be due to modulation of bitter taste via interaction of methyl xanthines with an adenosine receptor.

The antagonism between bitter and sweet taste has been shown with mixtures of neohesperidin dihydrochalcone and naringin.[70] The sweetness of neohesperidin dihydrochalcone (NHDC) has a long persistence. However, when it is mixed with naringin, the bitter flavone analog of NHDC, the sweetness persistence is reduced.

Schiffman et al.[71] have shown that mixing sweet compounds with bitter compounds elevates thresholds for bitter tastes to various degrees. One reason for the antagonism between bitter and sweet tastes may be due to the fact that they have opposite reactions to the elevation of cAMP within a cell.[72] When membrane permeable forms of cAMP are applied to the tongue of a gerbil, electrophysiologic responses to sweeteners are enhanced and responses to bitter compounds are suppressed. Thus, the adenylate cyclase system which is known to exist in taste papillae[73-77] may play a role in the suppression of bitter taste by sweeteners.

The sodium salts of guanosine-5′-monophosphate and inosine-5′-monophosphate have been reported to block the persistent aftertaste of monoammonium glycyrrhizzinate,[78] and a soluble natural carbohydrate from lach wood, arabinogalactan, has been found to remarkably reduce the persistent aftertaste of the protein sweetener thaumatin.[79] Hence, modulation of second messenger systems or channels by one stimulus can alter the taste of another.

F. ASTRINGENT COMPOUNDS

Many foods such as fruits and vegetables produce astringent sensations when introduced into the oral cavity. These sensations are described as having long-lasting puckering and drying components that involve the tongue and membranes of the mouth.[80-82] A wide variety of compounds induce these sensations including vegetable tannins (polyphenolic compounds such as tannic acid), some salts of multivalent cations (aluminum, chromium, zinc, and boron), mineral acids, and dehydrating agents such as ethyl alcohol.

Historically there has been controversy whether astringency is a fundamental taste property analogous to sweet, sour, salty, and bitter (stimulating the taste nerves) or is a tactile sensation that involves the trigeminal nerve. Until recently, the consensus was that astringency stimulated the sense of touch (and hence the trigeminal nerve) by coagulating proteins of the saliva and mucous epithelium.[81,83,84] However, a pioneering neurophysiological study by Kawamura et al.[85] found that tannic acid stimulated the taste nerves (chorda tympani and glossopharyngeal nerves) but did not stimulate the trigeminal nerve in rat.

Recent psychophysical and electrophysiologic studies suggest that astringency is indeed a taste and not a tactile sensation.[82,86,87] Schiffman et al.[86] found that astringent compounds evoked significant responses in human subjects when the stimuli came in contact with the tongue, but not when they came in contact with the inside of the cheek which is devoid of taste buds. Fifty milliliters of tannic acid (0.26 *M* and 0.94 *M*), tartaric acid (0.03 *M* and 0.4 *M*), or deionized water were delivered in randomized order to the inside of the cheek on the buccal membrane. The subjects were tested in a reclining position so that the liquid would not reach the tongue. Between delivery of each tastant, there was a thorough rinse and a long interstimulus interval. Deionized water was just as likely to elicit "sour", "dry", "tingly", or "astringent" responses as were tannic and tartaric acid solutions after 5 and 25 s. However, when subjects touched their anterior tongues to the stimulated area, tannic and tartaric acids (especially the higher concentrations) elicited significant "dry", "tingly", "rough", and "astringent" responses. The fact that the tip of the tongue must be stimulated to perceive astringent sensations from tannic and tartaric acids delivered to the cheek suggests that astringency may involve the sense of taste and not touch.

Electrophysiological studies in rodents confirm these results.[82,87] Schiffman et al.[82] recorded from the chorda tympani nerve which transmits taste information from the anterior 2/3 of the tongue and from the lingual nerve which transmits tactile, thermal, and pain sensations. All five astringent compounds tested (tannic acid, tartaric acid, gallic acid, aluminum ammonium sulfate, and aluminum potassium sulfate) rapidly and reversibly (at lower concentrations) stimulated the chorda tympani nerve over the pH range from 1.98 to 6. None of these compounds stimulated the lingual nerve at comparable pH values. These data suggest that astringency is a taste quality in rodents.

The rapid and reversible responses to astringent compounds found by Schiffman et al.[82] suggest that these stimuli interact with membrane-bound proteins associated with ion transport in taste cells rather than precipitating water-soluble proteins in the saliva. Thus, the transduction mechanism for astringent taste is probably similar to that for other taste stimuli known to involve ion transport (see Chapters 8 and 11). Support for this hypothesis comes from Schiffman et al.[87] Electrophysiologic recordings were made from the chorda tympani nerve of gerbils to determine the interactive effect of astringent-tasting molecules with a broad spectrum of tastants. Astringent compounds (both tannic acid and aluminum ammonium sulfate) markedly inhibited taste responses elicited by mono- and divalent salts, bitter compounds, acids, and sweeteners. This suggests that astringent compounds elicit a variety of transport pathways and receptors in taste cells and that these pathways may contribute to the astringent sensation.

G. CONCLUSIONS ABOUT TRANSDUCTION MECHANISMS FOR TASTE

The sense of taste is far more complex than four primary sensations. Many stimuli have sensations that fall outside of these traditional four qualities in multidimensional scaling experiments. Sodium, potassium, and calcium channels probably exist in taste cells; anions interact with the taste cell complex, and glutamate and amino acid receptors are implicated by psychophysical experiments. Astringent compounds also appear to interact with channels and receptors in taste cell membranes. Multiple mechanisms for sweetness, bitterness, and sourness are also suggested by psychophysical experiments. Paracellular currents also play a role in taste transduction.

II. PSYCHOPHYSICAL IMPLICATIONS FOR NEURAL ORGANIZATION

In addition to implications for receptor processes, psychophysical data have been used to understand neural organization. This section will examine a basic issue in neural organization; that is, whether taste consists of a continuum, or, on the other hand, of several discrete and independent mechanisms. The discrete view will be referred to as the "quadripartite" position whether proponents of this view advocate 3, 4, 5, or some other small number of tastes.

A. EXPERIMENTAL EVIDENCE FOR CONTINUOUS VS. QUADRIPARTITE ORGANIZATIONS OF NEURAL CODING

The key concepts of these two points of view (continua vs. quadripartite) are illustrated with Henning's taste tetrahedron.[1] The two views of this model are as follows: both would accept the four corners of the tetrahedron as representing four "basic tastes", sweet, sour, salty, and bitter. The profound and important differences between these views concern the meaning of the continua connecting the corners. For Henning (see Reference 88), a point halfway between the salty and sour corners would be a *singular* taste equally similar to both salty and sour; but for Öhrwall (see Reference 88) this point would not be singular but would have to represent *two* tastes, one salty and the other sour.

Henning's view suggests that the dimensions of taste, like color vision, are continua of many different singular sensations. Henning's model is also consistent with the psychophysics of synthesis, and the Across-Fiber Patterning neural coding theory (to be discussed below). While Henning's prism needs to be expanded to include tastes that do not fall into the domain of the four primaries, the model is historically important because it first introduced the concept of taste as a continuum. Öhrwall's view has led to the position that the tastes of all substances, including mixtures, are composites of a small number of tastes. His view is consistent with the psychophysics of analysis and its attendant concepts of suppression and the Labeled-Line neural coding theory. Taste psychophysics and neural organization must be intimately related and should provide insights into each other.

1. Taste Mixture Data

Several seminal papers by Bartoshuk[89,90] addressed fundamental issues in the psychophysics of taste intensity of mixtures. She investigated the mathematics of the "compressed" concentration/response functions for mixtures of similar and different stimuli. Using equally "intense" concentrations of NaCl, HCl, sucrose, and QHCl, each was reduced in intensity in mixtures with the others, with the amount of "suppression" of each being an inverse function of the amount of compression shown by each stimulus in its own concentration/intensity function.

Bartoshuk's studies of taste mixtures, like most that preceded and followed, have made the assumption basic to the quadripartite position that the tastes of the components change in intensity in the mixture, while their qualities remain unchanged. These studies did not evaluate whether the data were best interpreted from the quadrapartite or the continuous point of view. In fact, studies of individual as well as mixed stimuli back to Öhrwall were not designed to test the premise that there are only four tastes. As examples, Kroeze,[91] following Öhrwall's assumption that any variations in the tastes of simple substances such as salts were due to "side-tastes" of other primary tastes, showed that the "suppressive" effects of NaCl on sucrose were in part a function of the "sweet" side-taste of the NaCl which contributes to the total sweetness of the mixture. That NaCl might be a singular taste very close to "salty" along the "salty-sweet" Henning continuum was not considered although it would lead to Kroeze's conclusions.

Smith and Theodore[92] showed that rats will avoid a binary mixture in accordance with the intensities of the negatively trained component. Thus, rats trained to avoid sucrose will also avoid a mixture of sucrose and NaCl, HCl, or QHCl to the degree that the mixture contains sucrose. The conclusion is offered[93] that the rats were identifying the sucrose in the mixture. However, the same results would be obtained if these mixtures had singular tastes occupying points along Henning's "sweet-salty", "sweet-sour", or "sweet-bitter" continua, those with more sucrose being situated closer to the negative stimulus sucrose, and thus being more avoided.

Recently, studies have appeared which were designed to address the issue (quadrapartite vs. continuous) more directly. In these experiments it is clear that instead of only the intensities of the component sensations changing (prothetic effects) as required in "suppression", the results usually indicate some qualitative (metathetic) interactions between stimuli including suggestions that tastes other than the four "basics" are formed. For example, in a study in which sodium deficient, adrenalectomized rats drank a mixture of NaCl and HCl not preferred by normal rats, McCutcheon[94] concluded that since the motivational properties of the NaCl were reduced by the nonaversive concentration of HCl, the two components did not remain separate but interacted in some manner which he termed a "sensory composite" as distinct from synthesis, analysis, or fusion; however, qualitative interaction was suggested.

Kuznicki and Ashbaugh[95] concluded that mixture components formed a perceptual blend of the elements to which human subjects were incapable of selectively attending. They then showed[96] that this blend was present only when the components were presented together, were attenuated when separated spatially, and absent when presented at different times. Stimuli in mixtures could also sort themselves out over time;[97] that is, although NaCl and QHCl formed a blend, the different latencies of the components permitted their separate discrimination. Kuznicki et al.[98] also showed that tastes are more difficult to sort in mixtures, the idea being that instead of two independent tastes, the accompanying taste may have qualitatively interfered with the target taste to make identification of that taste cognitively difficult.

Moskowitz[99] found that sweet (glucose or fructose) mixed with sour (citric acid) or bitter (quinine sulfate) blended such that the sweet "flavor" in the mixture differed from that of simple sugar sweetness. He concluded that the presence of the sour or bitter taste modified the qualitative aspect of sweetness.

Further evidence that qualitative changes occur in mixtures was reported by DiLorenzo et al.[100] They showed that alcohol is apparently experienced by rats as similar to a mixture of sucrose and quinine, but is not like either of these stimuli alone. Thus sucrose and quinine must form an alcohol-like taste which is not like either of the components, suggesting that a new separate taste was formed. This was elaborated by Kiefer and Lawrence[101] and Kiefer et al.[102] in similar studies in which rats responded in a similar way to alcohol and to a mixture of sucrose and quinine, but this was not generalizable to sucrose or quinine alone or to NaCl or HCl. They concluded that "the mixture of the sweet and bitter taste components must synthesize a new taste not characterized by a simple sum of the individual components". Together these studies suggest that in mixtures component tastes may not always remain qualitatively (as distinct from intensively) unchanged or separate.

The basic point concerning these data on which the two positions differ is very simple. If the quadripartite position is correct, then a mixture of two stimuli must be perceived as two stimuli, one or both of which are weakened while remaining qualitatively unchanged. On the other hand, if the continuous position is correct the components in a mixture could change qualitatively, being perceived as one stimulus (synthesis), as in color mixtures. If human subjects are asked whether lights, tones, colors, and tastes are "singular" or "more than one",[103,104] a point of light and a single tone are reported as singular, and all colors were experienced as singular. On the other hand, in simple visual figures of more than one component, or with several simultaneous tones, the stimuli were reported as "more than one".

With these data as reference points, Erickson[103] and Erickson and Covey[104] designed experiments to determine if the taste system is analytic or continuous. Subjects were presented with simple solutions of "basic tastes" as well as mixtures to determine whether the mixtures are perceived as singular and intermediate to the components, or whether they are perceived as more than one as required by the analytic, quadripartite point of view. It is important to note an important procedural difference between this method and those typically used. Subjects were not asked to use taste names because in such a situation, clearly, they could only resort to the common four whether there were just four or a continuum.

Interestingly, no taste was entirely singular. The most singular "basic tastes" were rated as unitary on about 80% of the trials. However, mixtures of several "basic" tastes were judged as singular, or more singular, than the components themselves, suggesting that the individual components were fused into one sensation in the mixtures. Together these results indicate that in mixtures a new taste was formed which was as singular as, but different from, the simplest "basic" taste solution. These different tastes were presumably intermediate but similar to both the components, as a blue-green is singular and similar to, but distinct from, its blue and green components. These tastes would be different from the end point tastes, but similar to them as indicated by their placement on the continua. Such fusion of multiple inputs into a singular sensation has been termed "synthesis", with "analysis" defining instances of components being kept separated.[105,106] Mixtures of electrolytes were found in these studies to fuse or synthesize. In distinction to mixtures of electrolytes, the mixtures of electrolytes with sugars appeared to be analytic as these were typically judged as "more-than-one".

In a similar study, comparing the relative "mixedness" ("more-than-oneness") of taste stimuli, O'Mahoney et al.[107] confirmed that no stimuli were entirely singular, and possibly that none were absolutely "mixed" (analyzed), although the methodology of comparing mixedness with that of quinine HCl (QHCl) leaves unclear the issue of just how "unmixed" they were. A mixture of fructose and citric acid was second only to QHCl in singularity, which certainly requires synthesis.

Bartoshuk[108] stated that the sensory singularity of these mixtures, rather than instances of qualitative change or synthesis, is due to "suppression" of one of the sensations by the other. If Bartoshuk is correct that the mixture is singular because one component has been erased by suppression, the resultant sensation must be identical with the component which "suppressed" the other; i.e., if two components were still perceived it would not be singular. This was tested experimentally by Erickson et al.;[106] they found that the singular mixture had not become identical to the "suppressor" stimulus in that it remained *clearly* distinguishable from *both* of its components.

Thus, the facts of "suppression" in all important aspects are consonant with descriptions of a generalization gradient of singular sensations, including those resulting from stimulus mixtures, along a continuum. If the psychophysical task is to rate mixtures in terms of their components, of necessity they will be rated as "weaker" on each, if their mixture has resulted in synthesis of a new singular taste as is true with colors.

2. Implications of Continuous and Quadripartite Organizations for Neural Coding

The mutually exclusive continuous and quadripartite theories of taste organization have different implications for neural organization. The concept of a continuum of tastes, with "basic" nodal points and other tastes between these arranged throughout in terms of similarities, e.g., Henning's taste tetrahedron, requires many singular tastes, with the tastes of many mixtures also being singular. The "Across-Fiber Pattern" theory is required both for the continuum and the multiplicity of tastes. In this theory, individual neurons are not exclusively labeled for a particular sensation but cooperate with the others in the ensemble, implying interactions among its elements as in color vision[3] which allows such interactive effects as have been presented above (Section II.A.1).

The AFP theory was originally formulated in the model of cooperation among neuron groupings (see Reference 105 for overview) with color vision as a primary example. Later, this was extended to nongrouped systems and topographic modalities such as somesthetic localization.[105] AFP theory is equally applicable whether taste is considered to be grouped or not.

On the other hand, the psychophysical concept of a few independent and exclusive tastes implies the psychophysical concept of analysis in which all taste sensations of whatever complexity are analyzed into differing amounts of the few ("basic") tastes. This in turn requires neural groups to support the coding theory of "Labeled-Lines" which indicates a psychoneural identity between a neural group and a sensory function, i.e., one group of neurons signals "salty", another separate group signals "sweet", etc. This model of separate and noninteractive groups of neurons is not compatible with the idea of a taste continuum, nor with interactions among the elements such as mixture suppression.[3]

The only clear test to differentiate between these theories appears to exist in the kinds of psychophysical studies of mixtures mentioned earlier, which in some way directly examine whether there are only a few tastes, or many.

B. CRITICISMS OF THE CONTINUOUS POSITION

The main psychophysical and neural data that have been used to support the quadripartite position have been summarized by McBurney[109] and Scott and Plata-Salaman.[110] The fact that these data are not clear tests of this theory, being as acceptable for the continuous as for the quadripartite position, is usually not noted (see Reference 3 for a review). These arguments will be very briefly reviewed below.

1. Structure-Activity Relations

It is argued that each of the four tastes has a separate and distinct receptor mechanism. As discussed in the first part of this chapter, there appear to be many transduction mechanisms, certainly more than four, which would suggest more than four tastes. Further, it is evident that each receptor cell, and the afferents they serve, is typically responsive to several or all of the "basic" tastes; these facts are compatible with the continuous position, while ruling heavily against the exclusive labeling of any receptor cell or afferent neuron for one "taste", and the related idea of a correspondingly few number of tastes.

2. Which Neurons Carry the Code?

In the context of the labeled-line position, it has been argued[110] that it is contrary to AFP theory for the neural code for a particular taste to be restricted to less than the total population of neurons. However, it has been pointed out that within AFP theory, some neurons play little or no role in the encoding of certain stimuli. The import of a particular neuron's activity[111,112] may be negligible for some stimuli, i.e., not all neurons are involved with all stimuli. The coding role of a particular neuron varies,[113] taking a greater or lesser, or indeed different, role in encoding different stimuli. Even in very clearly continuous models such as the auditory system, it is clear that not all neurons respond to and encode all stimuli equivalently; a low-frequency neuron may not be involved in the encoding of high tones, but this does not require a grouping model for audition.

3. Ethological Considerations

As suggested by McBurney,[109] the quadripartite position suggests four independent taste systems, each with the specific task of encoding one distinct physiologic function. The taste system certainly was formed by evolutionary pressures, but whether these are understood well enough to provide conclusions about the structure of taste is problematic. Such simplicity evokes the idea of four pheromone-like systems with very specifiable stimulus-behavior connections. With only one message each, such systems would not require large amounts of neural machinery and would not allow much complexity for the perceptual world of taste and taste behavior. The increasing understanding of the anatomical complexity and mass of the gustatory system as well as the complexity of behaviors and perceptions mediated by taste seem to argue against this model.

4. The Nature of Continua

In connection with discussion of the remaining criticisms, a primary characteristic of continua must be recalled.[3] All continua, by definition, vary in their characteristics from one end to the other. Continuously organized sensations or neurons would be expected to be sensitive to experimental manipulation in a way that would show such differences along their continua; discussion of the most commonly used examples follows.

5. Chemotopic Organization

The oldest and most primary argument for four tastes is based on the fact that taste sensitivity is roughly laid out on the tongue (and in the nervous system) so that one place is more responsive to a given taste than another. The fact that taste sensitivities are laid out in a literally continuously manner, in a gently graded and overlapping way across space, rather than each taste being restricted to one area, might be taken as a primary argument that taste is a continuum rather than four separate entities. That is, the continuous spread in sensitivities across space does not strictly limit one "basic" sensation to the tip of the tongue, and another to the sides or back, the argument being equivalent in central taste areas. This is analogous to the fact that the auditory system is tonotopically organized so that a high tone is more effective at a different place of the basilar membrane than a low tone; this is a basic characteristic of a continuous system. Although at best equivocal, and perhaps more supportive of a continuous than quadripartite position, chemotopic organization has always been accepted as one of the stronger supports for the four separate taste model.

As with chemotopic organization, the following kinds of effects are all to be expected in continuous systems in which stimulus characteristics at one part of the continuum differ from those at other parts of the continuum. The response/concentration functions differ in slope between different stimuli. The time courses of the response evoked by each taste differ neurally and perceptually. The thresholds of various stimuli vary with temperature. (That temperature variations have effects on tactile afferents is of a different, lesser order than the rather profound effects of one tastant on another or one color on another.) Taste is similar to other continuous systems in that cross adaptation is a function of the proximity of the stimulus pairs in question, i.e., remote stimuli, such as the "basic" tastes, may have little effect on each other. Anesthetics and taste modifiers should have effects selective to relatively discrete portions of a continuum; that is, it is not to be expected in a continuous system that all taste modifying substances would have equal effect on all parts of the continuum. These issues are all discussed in detail in Schiffman and Erickson.[3]

Two other major arguments have been used in opposition to the idea of a taste continuum.

6. Taste Names

It has been suggested[108] that probably no more than a few tastes exist since no other good taste names have been proposed beyond the basic four. The existence of taste continua with many tastes does not require the discovery of new taste names. The continua involved in Henning's tetrahedron, describing many intermediate tastes, do not require any taste names in addition to the four at the corners. By analogy, all colors are describable in terms of a few color names.

7. Filling of a Continuum

Scott and Plata-Salaman[110] have averred that a continuum, to exist, must be evenly filled, and certainly no multi-dimensional scaling solution has shown an even filling of the space. The concept of a continuum on the one hand is distinct from the idea of a continuous filling of a continuum on the other, i.e., a map of the U.S. is continuous, but is not evenly filled with cities or people.

C. CONCLUSIONS ABOUT NEURAL ORGANIZATION

Following the above discussion, it is concluded here that the issue of the basic organization of taste is still not well understood, and the possibility that taste is continuously organized is strong. More studies need to be conducted which are relevant to this issue. Nearly all studies to date have utilized a limited number of stimuli and have been interpreted in terms of four tastes. Given the complexity of taste sensations, it is unlikely that the quadripartite position is a correct one.

III. GENERAL SUMMARY

Taste sensations are far broader than the traditional four qualities of sweet, sour, salty, and bitter. These sensations are produced by a large variety of transduction mechanisms including ion channels, receptors, and transepithelial currents. The relative activation of these mechanisms by a taste stimulus determines its quality and thus defines its location on taste continua.

REFERENCES

1. **Henning, H.,** Die Qualitatenreihe des Geschmacks, *Z. Psychol.*, 74, 203, 1916.
2. **Schiffman, S. S. and Erickson, R. P.,** A psychophysical model for gustatory quality, *Physiol. Behav.*, 7, 617, 1971.
3. **Schiffman, S. S. and Erickson, R. P.,** The issue of primary tastes versus a taste continuum, *Neurosci. Biobehav. Rev.*, 4, 109, 1980.
4. **Ishii, R. and O'Mahony, M.,** Taste sorting and naming: can taste concepts be misrepresented by traditional psychophysical labelling systems?, *Chem. Senses*, 12, 37, 1987.
5. **DeSimone, J. A., Heck, G. L., and DeSimone, S. K.,** Active ion transport in dog tongue: a possible role in taste, *Science*, 214, 1039, 1981.
6. **Schiffman, S. S., Lockhead, E., and Maes, F. W.,** Amiloride reduces the taste intensity of Na^+ and Li^+ salts and sweeteners, *Proc. Natl. Acad. Sci. U.S.A.*, 80, 6136, 1983.
7. **Tennissen, A. M.,** Amiloride reduces intensity responses of human fungiform papillae, *Physiol. Behav.*, 51, 1061, 1992.
8. **McCutcheon, N. B.,** Human psychophysical studies of saltiness suppression by amiloride, *Physiol. Behav.*, 51, 1069, 1992.
9. **Schiffman, S. S., Simon, S. A., Gill, J. M., and Beeker, T. G.,** Bretylium tosylate enhances salt taste, *Physiol. Behav.*, 36, 1129, 1986.

10. **Schiffman, S. S. and Graham, B.,** unpublished data.
11. **DeSimone, J. A.,** personal communication.
12. **Simon, S. A. and Garvin, J. L.,** Salt and acid studies on canine lingual eptihelium, *Am. J. Physiol.*, 249, C398, 1985.
13. **Avenet, P. and Lindemann, B.,** Amiloride-blockable sodium currents in isolated taste receptor cells, *J. Membr. Biol.*, 105, 245, 1988.
14. **Ninomiya, Y., Sako, N., and Funakoshi, M.,** Strain differences in amiloride inhibition of NaCl responses in mice, Mus musculus, *J. Comp. Physiol. A*, 166, 1, 1989.
15. **Schiffman, S. S., Suggs, M. S., Cragoe, E. J. Jr., and Erickson, R. P.,** Inhibition of taste responses to Na^+ salts by epithelial Na^+ channel blockers in gerbil, *Physiol. Behav.*, 47, 455, 1990.
16. **Schiffman, S. S.,** Contribution of the anion to the taste quality of sodium salts, in Kare, M. R., Fregley, M. J., and Bernard, R. A., Eds., *Biological and Behavioral Aspects of NaCl Intake*, Nutrition Foundation Monograph Series, Academic Press, New York, 1980, 99.
17. **Schiffman, S. S., McElroy, A. E., and Erickson, R. P.,** The range of taste quality of sodium salts, *Physiol. Behav.*, 24, 217, 1980.
18. **Mierson, S., Heck, G. L., DeSimone, S. K., Biber, T. U. L., and DeSimone, J. A.,** The identity of the current carriers in canine lingual epithelium *in vitro, Biochim. Biophys. Acta*, 816, 283, 293, 1985.
19. **Ye, Q., Heck, G. L., and DeSimone, J. A.,** The anion paradox in sodium taste reception: resolution by voltage-clamp studies, *Science*, 254, 724, 1992.
20. **Schiffman, S. S., Crumbliss, A. L., Warwick, Z. S., and Graham, B. G.,** Thresholds for sodium salts in young and elderly subjects: correlation with molar conductivity of anion, *Chem. Senses*, 15, 671, 1990.
21. **Schiffman, S. S., Crumbliss, A. L., and Warwick, Z. S.,** Detection thresholds of potassium salts are related to the molar conductivity of the anion, *Chem. Senses*, 16, 574, 1991.
22. **McBurney, D. H. and Lucas, J. A.,** Gustatory cross adaptation between salts, *Psychonom. Sci.*, 4, 301, 1966.
23. **McBurney, D. H., Smith, D. V., and Shick, T. R.,** Gustatory cross-adaptation: sourness and bitterness, *Percept. Psychophys.*, 22, 228, 1972.
24. **Smith, D. V. and McBurney, D. H.,** Gustatory cross-adaptation: does a single mechanism code the salty taste?, *J. Exp. Psychol.*, 80, 101, 1969.
25. **Schiffman, S. S.,** unpublished data.
26. **Hober, R. and Kiesow, F.,** Ueber den Geschmack von Salzen and Laugen, *Z. Phys. Chem.*, 27, 601, 1898.
27. **Renqvist, Y.,** Ueber den Geschmack, *Skand. Arch. Physiol.*, 38, 97, 1919.
28. **Dzendolet, E. and Meiselman, H.,** Gustatory quality changes as a function of solution concentration, *Percept. Psychophys.*, 2, 29, 1967.
29. **Cardello, A. V. and Murphy, C.,** Magnitude estimates of gustatory quality changes as a function of solution concentration of simple salts, *Chem. Senses*, 2, 327, 1977.
30. **Dzendolet, E.,** A structure common to sweet-evoking compounds, *Percept. Psychophys.*, 3, 65, 1968.
31. **Kemp, S. E., Grigor, J. M., and Birch, G. G.,** Do taste receptors respond to perturbation of water structure?, *Experientia*, in press.
32. **Sandick, B. and Cardello, A. V.,** Tastes of salts and acids on circumvallate papillae and anterior tongue, *Chem. Senses*, 8, 59, 1983.
33. **Formaker, B. K. and Hill, D. L.,** Lack of amiloride sensitivity in SHR and WKY glossopharyngeal taste responses to NaCl, *Physiol. Behav.*, 50, 765, 1991.
34. **Schiffman, S. S. and Dackis, C.,** Taste of nutrients: amino acids, vitamins, and fatty acids, *Percept. Psychophys.*, 17, 140, 1975.

35. **Luts, A., Montavon, P., Lindstrand, K., and Sundler, F.,** Peptide-containing nerve fibers in the circumvallate papillae, *Reg. Peptides*, 27, 209, 1990.
36. **Montavon, P., Lindstrand, K., Luts, A., and Sundler, F.,** Peptide-containing nerve fibers in the fungiform papillae of pigs and rats, *Reg. Peptides*, 32, 141, 1991.
37. **Shinoda, I., Tada, M., and Okai, H.,** A new salty peptide, ornithyl-B-alanine hydrochloride, in *Peptide Chemistry*, Munekata, E., Ed., Protein Research Foundation, Osaka, 1984.
38. **Tada, M., Shinoda, I., and Okai, H.,** L-ornithyltaurine, a new salty peptide, *J. Agr. Food Chem.*, 32, 992, 1984.
39. **Schiffman, S. S. and Gill, J. M.,** Psychophysical and neurophysiological taste responses to glutamate and purinergic compounds, in *Umami: A Basic Taste*, Kawamura, Y. and Kare, M. R., Eds., Marcel Dekker, New York, 1987, 271.
40. **Schiffman, S. S., Frey, A. E., Luboski, J. A., Foster, M. A., and Erickson, R. P.,** Taste of glutamate salts in young and elderly subjects: role of inosine 5'-monophosphate and ions, *Physiol. Behav.*, 49, 843, 1991.
41. **Yamasaki, Y. and Maekawa, K.,** A peptide with delicious taste, *Agr. Biol. Chem.*, 42, 1761, 1978.
42. **Tamura, M., Nakatsuka, T., Tada, M., Kawasaki, Y., Kikuchi, E., and Okai, H.,** The relationship between taste and primary structure of "delicious peptide" (Lys-Gly-Asp-Glu-Glu-Ser-Leu-Ala) from beef soup, *Agric. Biol. Chem.*, 53, 319, 1989.
43. **Ney, K. H.,** Voraussage der Bitterkeit von Peptiden aus deren Aminosaurezusammensetzung, *Z. Lebensm. Unters. Forsch.*, 147, 64, 1971.
44. **Bigelow, C. C.,** On the average hydrophobicity of proteins and the relation between it and protein structure, *J. Theor. Biol.*, 16, 187, 1967.
45. **Tanford, C.,** Contribution of hydrophobic interactions to the stability of globular conformation of proteins, *J. Am. Chem. Soc.*, 84, 4240, 1962.
46. **Nozaki, Y. and Tanford, C.,** The solubility of amino acids and related compounds in aqueous urea solutions, *J. Biol. Chem.*, 238, 4074, 1963.
47. **Ney, K. H.,** Aminosaure-Zusammensetzung von Proteinen und die Bitterkeit ihrer Peptide, *Z. Lebensm. Unters. Forsch.*, 149, 321, 1972.
48. **Guigoz, Y. and Solms, J.,** Bitter peptides, occurrence and structure, *Chem. Senses*, 2, 71, 1976.
49. **Schiffman, S. S. and Engelhard, H. H.,** Taste of dipeptides, *Physiol. Behav.*, 17, 523, 1976.
50. **Siemion, I. Z., Kolasa, T., and Paradowski, A.,** On the taste of stereoisomeric cyclic dipeptides containing a proline residue, *Chem. Senses*, 4, 127, 1979.
51. **Faurion, A., Saito, S., and MacLeod, P.,** Sweet taste involves several distinct receptor mechanisms, *Chem. Senses*, 5, 107, 1980.
52. **Schiffman, S. S., Lindley, M. G., Clark, T. B., and Makino, C.,** Molecular mechanism of sweet taste: relationship of hydrogen bonding to taste sensitivity for both young and elderly, *Neurobiol. Aging*, 2, 173, 1981.
53. **Faurion, A., Bonaventure, L., Bertrand, B., and MacLeod, P.,** Multiple approach for the sweet taste sensory continuum: psychophysical and electrophysiological data, in *Olfaction and Taste VII*, Van der Starre, H., Ed., IRL Press, London, 1980.
54. **Schiffman, S. S., Cahn, H., and Lindley, M. G.,** Multiple receptor sites mediate sweetness: evidence from cross adaptation, *Pharmacol. Biochem. Behav.*, 15, 377, 1981.
55. **Schiffman, S. S., Reilly, D. A., and Clark, T. B.,** Qualitative differences among sweeteners, *Physiol. Behav.*, 23, 1, 1979.
56. **Van der Wel, H. and Arvidson, K.,** Qualitative psychophysical studies on the gustatory effects of the sweet tasting proteins thaumatin and monellin, *Chem. Senses*, 3, 291, 1978.

57. **DuBois, G. E., Walters, D. E., Schiffman, S. S., Warwick, Z. S., Booth, B. J., Pecore, S. D., Gibes, K., Carr, B. T., and Brands, L. M.,** Concentration-response relationships of sweeteners: a systematic study, in *Sweeteners Discovery, Molecular Design, and Chemoreception*, Walters, D. E., Orthoefer, F. T., and DuBois, G. E., Eds., ACS Symposium Series 450, American Chemical Society, Washington, D.C., 1991, 261.
58. **Green, B. G. and Frankmann, S. P.,** The effect of cooling the tongue on the perceived intensity of taste, *Chem. Senses*, 12, 609, 1987.
59. **Frankmann, S. P. and Green, B. G.,** Effect of cooling on the sweetness of natural and artificial sweeteners, *Chem. Senses*, 12, 656, 1987.
60. **Schiffman, S. S., Diaz, C., and Beeker, T. G.,** Caffeine intensifies taste of certain sweeteners: role of adenosine receptor, *Pharmacol. Biochem. Behav.*, 24, 429, 1986.
61. **Lorenz, W., Inglese, J., Palczewski, K., Onorato, J. J., Caron, M. G., and Lefkowitz, R. J.,** The receptor kinase family: primary structure of rhodopsin kinase reveals similarities to the beta-adrenergic receptor kinase, *Proc. Natl. Acad. Sci. U.S.A.*, 88, 8715, 1991.
62. **Moskowitz, H. R.,** Ratio scales of acid sourness, *Percept. Psychophys.*, 9, 371, 1971.
63. **Makhlouf, G. M. and Blum A. L.,** Kinetics of taste response to chemical stimulation: a theory of acid taste in man, *Gastroenterology*, 63, 67, 1972.
64. **Price, S. and DeSimone, J. A.,** Models of taste receptor cell stimulation, *Chem. Senses Flav.*, 2, 427, 1977.
65. **Ganzevles, P. G. J. and Kroeze, J. H. A.,** The sour taste of acids. The hydrogen ion and the undissociated acid as sour agents, *Chem. Senses*, 12, 563, 1987.
66. **Ganzevles, P. G. J. and Kroeze, J. H. A.,** Acid sourness, the effect on taste intensity of adaptation and cross-adaptation to common ions, *Physiol. Behav.*, 40, 641, 1987.
67. **Lush, I. E.,** The genetics of tasting in mice. II. Strychnine, *Chem. Senses*, 7, 93, 1982.
68. **Schiffman, S. S., Gatlin, L. A., Frey, A. E., Heiman, S. A., Stagner, W. C., and Cooper, D. C.,** Taste perception of bitter compounds in young and elderly persons: relation to lipophilicity of bitter compounds, *Neurobiol. Aging*, in press.
69. **Pickenhagen, W., Dietrich, P., Keil, B., Polonsky, J., Nouaille, F., and Lederer, E.,** Identification of the bitter principle of cocoa, *Helv. Chim. Acta*, 58, 1078, 1975.
70. **Naim, M., Dukan, E., Yaron, L., Levinson, M., and Zehavi, U.,** Effects of the bitter additives naringin and sucrose octaacetate on sweet persistence and sweet quality of neohesperidin dihydrochalcone, *Chem. Senses*, 11, 571, 1986.
71. **Schiffman, S. S., Gatlin, L. A., Sattely-Miller, E. A., Graham, B. G., Heiman, S. A., Stagner, W. C., and Erickson, R. P.,** The effect of sweeteners on bitter taste thresholds in young and elderly subjects, *Physiol. Behav.*, in press.
72. **Schiffman, S. S., Gatlin, L. A., Suggs, M. S., Heiman, S. A., Stagner, W. C., and Erickson, R. P.,** Effect of modulators of the adenylate cyclase system on electrophysiological taste responses in the gerbil, *Pharmacol. Biochem. Behav.*, in press.
73. **Asanuma, N. and Nomura, H.,** Histochemical localization of adenylate cyclase and phosphodiesterase activities in the foliate papillae of the rabbit. II. Electron microscopic observations, *Chem. Senses*, 7, 1, 1982.
74. **Kurihara, I.,** Inhibition of cyclic 3′-5′-nucleotide phosphodiesterase in bovine taste papillae by bitter taste stimuli, *FEBS Lett.*, 27, 270, 1972.
75. **Kurihara, K. and Koyama, N.,** High activity of adenyl cyclase in olfactory and gustatory organs, *Biochem. Biophys. Res. Commun.*, 48, 30, 1972.
76. **Price, S.,** Phosphodiesterase in tongue epithelium: activation by bitter taste stimuli, *Nature*, 241, 54, 1973.
77. **Cagan, R. H.,** Biochemical studies of taste sensation. II. Labeling of cyclic AMP of bovine taste papillae in response to sweet and bitter stimuli, *J. Neurosci. Res.*, 2, 363, 1976.
78. **MacAndrews and Forbes Company,** A sweetening agent comprising ammoniated glycyrrhizin and 5′-nucleotide, Israeli Patent 42, 862, 1975.

79. **Burge, M. L. E. and Nechutny, Z. L. A. Z.,** Sweetening compositions containing arabinogalactan, U. S. Patent 4, 228, 1980.
80. **Haslam, D. and Lilley, T. H.,** Natural astringency in foodstuffs: a molecular interpretation, *CRC Crit. Rev. Food Sci. Nutr.*, 27, 1, 1988.
81. **Joslyn, M. S. and Goldstein, J. L.,** Astringency of fruits and fruit products in relation to phenolic content, *Adv. Food Res.*, 13, 179, 1964.
82. **Schiffman, S. S., Suggs, M. S., Sostman, A. L., and Simon, S. A.,** Chorda tympani and lingual nerve responses to astringent compounds in rodents, *Physiol. Behav.*, 51, 55, 1991.
83. **Bates-Smith, E. C.,** Haemanalysis of tannins: the concept of relative astringency, *Phytochemistry*, 12, 907, 1973.
84. **Lyman, B. J. and Green, B. G.,** Oral astringency: effects of repeated exposure and interactions with sweeteners, *Chem. Senses*, 15, 151, 1990.
85. **Kawamura, Y., Funakoshi, M., Kasahara, Y., and Yamamoto, T.,** A neurophysiological study on astringent taste, *Jpn. J. Physiol.*, 19, 851, 1969.
86. **Schiffman, S. S., Simon, S. A., and Graham, B. G.,** Regional differences in sensitivity to astringent compounds in the oral cavity of humans, *Chem. Senses*, 16, 575, 1991.
87. **Schiffman, S. S., Suggs, M. S., and Simon, S. A.,** Astringent compounds suppress taste responses in gerbil, *Brain Res.*, in press.
88. **Erickson, R. P.** Öhrwall, Henning, and von Skramlik: The foundations of the four primary position in taste, *Neurosci. Biobehav. Rev.*, 8, 105, 1984.
89. **Bartoshuk, L. M.,** Taste mixtures. Is mixture suppression related to compression?, *Physiol. Behav.*, 14, 643, 1975.
90. **Bartoshuk, L. M.,** Psychophysical studies of taste mixtures, in *Olfaction and Taste VI*, LeMagnen, J. and MacLeod, P., Eds., 1977, 377.
91. **Kroeze, J. H. A.,** The relationship between the side tastes of masking stimuli and masking in binary mixtures, *Chem. Senses*, 7, 23, 1982.
92. **Smith, D. V. and Theodore, R. M.,** Conditioned taste aversions: generalization to taste mixtures, *Physiol. Behav.*, 32, 983, 1984.
93. **Smith, D. V.,** Neural and behavioral mechanisms of taste mixture perception in mammals, in *Perception of Complex Smells and Tastes*, Academic Press, Australia, 1989, 149.
94. **McCutcheon, B.,** Response to NaCl taste in mixture with HCl by sodium deficient rats, *Physiol. Behav.*, 34, 97, 1985.
95. **Kuznicki, J. T. and Ashbaugh, N.,** Taste quality differences within the sweet and salty taste categories, *Sensory Processes*, 3, 157, 1979.
96. **Kuznicki, J. T. and Ashbaugh, N.,** Space and time separation of taste mixture components, *Chem. Senses*, 7, 39, 1982.
97. **Kuznicki, J. T. and Turner, L. S.,** Temporal dissociation of taste mixture components, *Chem. Senses*, 13, 45, 1988.
98. **Kuznicki, J. T., Hayward, M., and Schultz, J.,** Perceptual processing of taste quality, *Chem. Senses*, 7, 273, 1983.
99. **Moskowitz, H. R.,** Perceptual changes in taste mixtures, *Percept. Psychophys.*, 11, 257, 1972.
100. **DiLorenzo, P. M., Kiefer, S. W., Rice, A. G., and Garcia, J.,** Neural and behavioral responsivity to ethyl alcohol as a tastant, *Alcohol*, 3, 55, 1986.
101. **Kiefer, S. W. and Lawrence, G. J.,** The sweet-bitter taste of alcohol: aversion generalization to various sweet-quinine mixtures in the rat, *Chem. Senses*, 13, 633, 1988.
102. **Kiefer, S. W., Bice, P. J., Orr, M. R., and Dopp, J. M.,** Similarity of taste reactivity responses to alcohol and sucrose mixtures in rats, *Alcohol*, 7, 115, 1990.
103. **Erickson, R. P.,** Studies on the perception of taste: do primaries exist?, *Physiol. Behav.*, 28, 57, 1982.
104. **Erickson, R. P. and Covey, E.,** On the singularity of taste sensations: what is a taste primary?, *Physiol. Behav.*, 25, 527, 1980.

105. **Erickson, R. P.**, Stimulus coding in topographic and nontopographic sensory modalities: on the significance of the activity of individual sensory neurons., *Psych. Rev.*, 75, 447, 1968.
106. **Erickson, R. P., Priolo, C. V., Warwick, Z. S., and Schiffman, S. S.**, Synthesis of tastes other than the 'primaries': Implications for neural coding theories and the concept of 'suppression', *Chem. Senses*, 15, 495, 1990.
107. **O'Mahoney, M., Atassi-Sheldon, S., Rothman, L., and Murphy-Ellison, T.**, Relative singularity/mixedness judgements for selected taste stimuli, *Physiol. Behav.*, 31, 749, 1983.
108. **Bartoshuk, L. M.**, Taste, in *Stevens' Handbook of Experimental Psychology I*, Atkinson, R. C., Herrnstein, R. J., Lindzey, G. and Luce, R. D., Eds., 2nd ed., John Wiley & Sons, New York, 461, 1988.
109. **McBurney, D. H.**, Are there primary tastes for man?, *Chem. Senses Flav.*, 1, 17, 1974.
110. **Scott, T. R. and Plata-Salaman, C. R.**, Coding of taste quality, in *Smell and Taste in Health and Disease*, Getchell, T. V., et al., Eds., Raven Press, New York, 1991, 345.
111. **Erickson, R. P.**, A neural metric, *Neurosci. Biobehav. Rev.*, 10, 377, 1986.
112. **Gill, J. M., II and Erickson, R. P.**, Neural mass differences in gustation, *Chem. Senses*, 10, 531, 1985.
113. **Smith, D. V., VanBuskirk, R. L., Travers, J. B., and Bieber, S. L.**, Coding of taste stimuli by hamster brain stem neurons, *J. Neurophysiol.*, 50, 541, 1983.

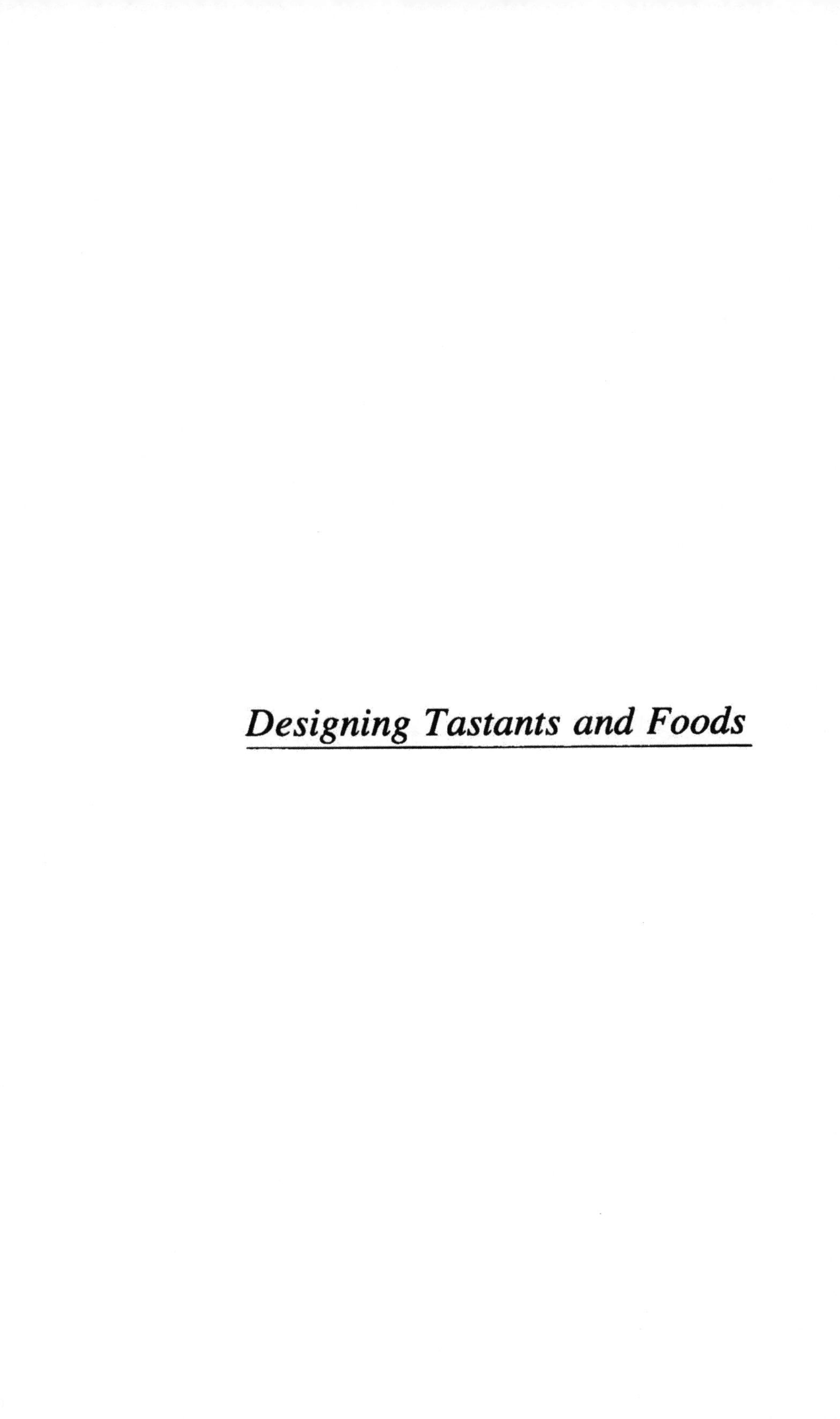

Designing Tastants and Foods

Chapter 15

MOLECULAR MODELING STUDIES OF AMILORIDE ANALOGS

Carol A. Venanzi and Thomas J. Venanzi

TABLE OF CONTENTS

0-8493-5341-6/93/$0.00 + $.50

I. INTRODUCTION

A. FROM PHARMACOPHORE TO GUSTAPHORE: THE USE OF MOLECULAR MODELING TECHNIQUES TO INTERPRET HOST-GUEST MOLECULAR RECOGNITION

The goal of our work in both biomimetic chemistry[1-10] and chemoreception[11-20] is to study the relationship of molecular structure and molecular properties to chemical and biological activity in order to "tune" the molecular design of particular host-guest complexes for optimal activity. To that end, we use molecular modeling techniques, a combination of computational chemistry and computer graphics, to analyze the nature of the molecular interactions in the host-guest complexes. We study two types of hosts: (1) relatively small macrocyclic receptors which act as cation-selective complexation sites[3,8] and/or enzyme mimics[1,2,6,7,9,10] and (2) macromolecules, such as enzymes or receptor proteins, which bind substrates, drugs, or tastants,[11-20] causing a cascade of biochemical events.[21] In both biomimetic and tastant systems, molecular recognition of the guest by the host is the first step towards binding and activity. Understanding of the molecular recognition process derives from knowledge of the intra- and intermolecular forces involved in the host-guest interaction. Compared to guest-receptor complexes, the study of enzyme-substrate or enzyme mimic-substrate systems is made easier by the fact that the molecular structure of both host and guest are known. For example, the X-ray structure of enzyme-inhibitor complexes can be used to provide information on steric and electrostatic aspects of molecular recognition of the guest by the enzyme.[22,23] This can provide a general starting point for the understanding of the nature of protein (receptor) recognition of small guests.[24] In the drug-receptor or tastant-receptor interaction, however, only the structure of the guest is known because of the difficulties involved in the crystallization of membrane-bound proteins. Therefore, the guest must be used as a probe of the steric and electrostatic features of its binding site on the host. In this chapter, we use our work on amiloride, a potent sodium channel blocker, to illustrate the use of molecular modeling to study the tastant-receptor interaction.

Many of the computational techniques we use are those which have been historically used to interpret drug-receptor interactions.[25,26] Dean[27] has given an extensive presentation of the use of pharmacological and molecular modeling techniques in drug design, while Lenz et al.[28] have included a survey of molecular modeling applications in their monograph on opiates. Boyd[29] has summarized some particularly successful uses of molecular modeling in drug and agrochemical design, and Weinstein et al.[30] have demonstrated the use of the *ab initio* molecular electrostatic potential to predict the biological activity of congeners of 5-hydroxytryptamine.

Computational analysis of such host-guest systems, in which the molecular structure of the host is unknown, uses as a starting point a set of structure-activity data that relates changes in the molecular structure and molecular properties of the guest to changes in the resultant biological activity of the host-guest system. In structure-activity studies, a series of analogs of a guest is synthesized and tested to determine their relative binding affinity and/or biological activity. The purpose of the studies is to define a "pharmacophore" for biological activity. This pharmacophore consists of a particular spatial arrangement of steric and electrostatic components of the guest that results in optimal biological activity of the host-guest system. Since guests of totally unrelated molecular structure may act at the same receptor site, the pharmacophore is determined through an analysis of molecular properties rather than molecular substructure. Guests whose binding to the receptor results ultimately in biological activity are known as agonists; guests whose binding does not lead to biological activity are antagonists. Comparison of the steric and electrostatic features of agonists and antagonists which act at the same receptor can lead to the identification of the common components of the pharmacophore which control the recognition and binding event of the complexation process. Analysis of the differences in the steric and electrostatic features of the agonist and antagonist can lead to a clear and definite identification of those features of the agonist which may be implicated in biological activity.

For example, acetylcholine and phencyclidine, two structurally unrelated molecules, act as an agonist and an antagonist, respectively, at the muscarinic cholinergic receptor. Weinstein and co-workers[31,32] identified a common spatial orientation of elements in the quantum mechanical molecular electrostatic (MEP) patterns of acetylcholine and phencyclidine. They defined these commonalities as the components of the interaction pharmacophore for the recognition and binding of a guest by the receptor. In this way, they explained the recognition of the synthetic drug, phencyclidine, and the endogenous neurotransmitter, acetylcholine, by the muscarinic cholinergic receptor in terms of the similarity in the pharmacophores of the two molecules. Conformational analysis of a series of phencyclidine derivatives was able to further pinpoint the stereochemical requirements of the interaction pharmacophore for phencyclidine.[33] In addition, Weinstein and co-workers[32] have also suggested that the anticholinergic activity of phencyclidine is due to the fact that its rigid molecular framework prevents the conformational rearrangement of the drug-receptor complex required for activation. In this way, comparison of the relative conformational flexibility of the agonist and antagonist has led to identification of the particular steric features of the agonist responsible for its cholinergic activity.

Many of the computational and experimental techniques which have been developed for drug design are now being applied to elucidation of the tastant-receptor interaction and to the design of more potent artificial tastants. In Chapter 16, Walters, Dubois, and Kellogg describe much of the experimental structure-activity studies and computational studies that have led to an understanding of the molecular interactions involved in sweet and bitter taste. Much of our work in this area,[11-14] which follows the methodology set out below, is summarized in a recent monograph on sweet taste.[15]

B. THE UNDERLYING ASSUMPTIONS OF THE MOLECULAR MODELING APPROACH TO TASTANT-RECEPTOR BINDING

The molecular modeling approach combines the use of computational chemistry and computer graphics to investigate the host-guest binding event. In the ideal situation, the molecular modeler will work along with the synthetic chemist and the taste experimentalist in a collaborative effort to design and test a series of tastant analogs. First, a small set of experimental structure-activity data is needed from which can be formulated a hypothesis concerning the steric and electrostatic features of the guest that lead to recognition of the guest by the host and, ultimately, to the taste response. Molecular modeling techniques applied to analysis of this set of data can provide a qualitative and quantitative evaluation of the steric and electrostatic components of the "gustaphore", a concept analogous to the pharmacophore described above. Then analogs of differing steric and electrostatic properties can be designed and synthesized to test the limits of the proposed gustaphore. In the case of amiloride (described below), which acts as an antagonist of sodium transport across taste cells, we use the term gustaphore to describe those features which lead to the formation of a stable blocking complex with the ion channel.

In using molecular modeling techniques to investigate the tastant-receptor interaction, we make several assumptions:

1. *The receptor provides a binding site which is complementary in both molecular shape and molecular electrostatic potential to that of the*

guest. Therefore, the proposed gustaphore defines the steric and electrostatic components of the tastant which are optimal in complementarity to those of the receptor binding site.

2. *The most potent analog provides the steric and electrostatic template for the gustaphore*. Comparison of the steric and electrostatic features of the other analogs to that of the most potent analog leads to an identification of the features which are important to recognition and binding and, ultimately, to optimal activity.
3. *Electrostatic forces are dominant in directing the tastant into the receptor site*. In the classical description of the molecular interaction of two charged particles i and j, where i and j can be considered atoms of the tastant and receptor, respectively, with fractional point charges q_i and q_j, the coulombic term which describes the electrostatic interaction energy between the two particles is given by:

$$q_i q_j/(r_{ij}\ \epsilon) \qquad (1)$$

where r_{ij} is the distance between i and j and ϵ is the effective dielectric constant of the medium. In contrast, the Lennard-Jones term, which describes the van der Waals (or dispersion) interaction between i and j, is given by

$$A_{ij}/r_{ij}^{12} - B_{ij}/r_{ij}^{6} \qquad (2)$$

where A_{ij} and B_{ij} are constants. Comparison of Equations 1 and 2 shows that when the tastant is far from the receptor (i.e., i is far from j and r_{ij} is large), then $1/r_{ij}$ is much larger than $1/r_{ij}^6$ or $1/r_{ij}^{12}$.[12] Thus, the electrostatic interaction between the tastant and receptor dominates at long range, whereas the dispersion interaction becomes more important at short range. Therefore, tastant-receptor electrostatic interactions can be assumed to direct and orient the tastant into the binding site where short range steric and van der Waals interactions come into play. In physiological solutions, the Debye length is about 10 Å; so, this is the range of electrostatic interactions.
4. *Low energy conformations of the tastant, not just the global energy minimum conformation calculated in the gas phase, should be considered as templates for the gustaphore*. As described below, conformational analysis of tastants by computational methods is generally carried out in the gas phase. Boltzmann statistics[34] show that in a distribution of molecules of different energy, those states of lower energy will be more highly populated. Since interactions with the receptor or solvent[35] could stabilize these conformations by several kcal/mol, altering their relative energy, several gas phase, low energy conformations of the tastant should be considered as potential binding conformations. Alter-

natively, the specific effect of solvent on conformation and energy can be assessed by molecular dynamics simulation of the tastant in solution, as described below.

II. COMPUTATIONAL METHODOLOGY

A. REVIEWS AND SOURCES

Every computational technique involves some degree of approximation. It is important to use a computational technique of sufficient accuracy to be able to answer the scientific question being posed. To be guided in the selection of the appropriate technique, it is crucial to know the level of approximation inherent in each method. It is impossible in this short chapter to give an overview of quantum mechanical and classical mechanical computational techniques. However, several excellent books are available. Boyd[36] gives an excellent short introduction to *ab initio* and semiempirical molecular orbital techniques for the nonspecialist. He also gives a listing of sources for standard computational and graphics packages.[37] Clark[38] compares the classical molecular mechanics method to quantum mechanical molecular orbital techniques in a volume which also discusses input and output to standard computational programs. Sadlej[39] surveys semiempirical methods in quantum chemistry. Hehre et al.[40] describe different types of *ab initio* basis sets and evaluate their choice by comparison of the results of theoretical calculations of molecular structure and properties to experiment. Burkert and Allinger[41] give an introduction to molecular mechanics methods.

A survey of molecular dynamics simulation methods, with applications to proteins and nucleic acids, is given by McCammon and Harvey[42] and by Brooks et al.[43] Algorithms used in molecular dynamics simulation are discussed by Berendsen and van Gunsteren[44] and Clementi et al.[45] Many applications of molecular dynamics simulation to problems of chemical and biological significance are collected by Beveridge and Jorgensen[46] and Goodfellow.[47] Applications of molecular mechanics and dynamics to conformational analysis of carbohydrates are presented by Madsen et al.,[48] Tran and Brady,[49] and Grigera.[50]

In addition, the use of molecular shape similarity in molecular recognition is discussed by Hopfinger and Burke[51] and Dean.[52] The concept of the quantum mechanical molecular electrostatic potential and its use as a tool to interpreting molecular reactivity is given by Scrocco and Tomasi.[53] Several applications are given in the book by Politzer and Truhlar.[54]

B. INTRODUCTION TO TECHNIQUES USED FOR AMILORIDE

We have chosen to study tastants of relatively small size and some conformational rigidity[11-16] such as amiloride[17-20], **1a**,

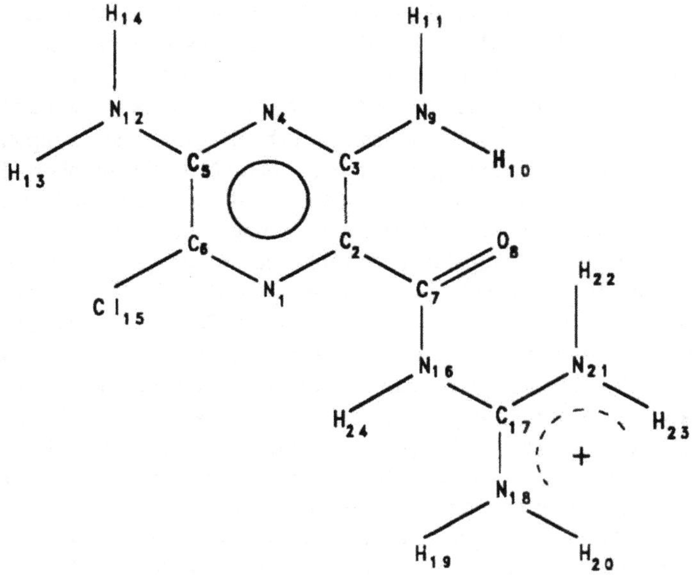

STRUCTURE 1a.

a bitter-tasting potassium-sparing acylguanidinium diuretic. We have selected semi-rigid tastants for analysis because in these compounds the steric and electrostatic components of the gustaphore are "frozen out" and can more easily be discerned than in conformationally flexible molecules for which a wide range of binding conformations need be considered. We have chosen to study small molecules because this allows us to treat the tastants by more exact computational techniques. In this way, the results are as accurate as possible and the calculations are tractable from a computer resource standpoint.

Below, we give a short introduction to the computational techniques we used in our studies of amiloride. In Section III.C, we describe our calculations on amiloride using these techniques and evaluate whether, for the amiloride analogs, use of the more exact techniques is required for accurate calculation of molecular properties.

1. Conformational Analysis by Quantum Mechanics

Ab initio molecular orbital theory describes the electronic structure of a molecule from "first principles" by constructing a molecular wavefunction from a weighted set (or linear combination) of one-electron orbitals called basis functions. The electronic energy of the molecule is determined by finding the set of weighting factors for the basis functions which minimizes the energy. The basis functions are mathematical functions of a particular form. The use of basis functions speeds up the computation by allowing the energy of the molecule to be calculated analytically rather than by numerical integration. For example, Gaussian-type functions consist of powers of x, y, and z multiplied by an exponential factor, exp $(-\alpha r^2)$. They can be combined to give functions similar to *s, p, d,* etc. atomic orbitals. Since the computational expense of the calculations goes as N^4, where N is the number of basis functions in the set, the choice of basis set size is a compromise between the efficiency of the calculation and the accuracy with which the molecular property of interest can be calculated. A split valence basis set is one in which two basis functions, instead of one, are centered on each atomic nucleus and are used to describe each valence atomic orbital. Only one basis function is used for the inner shell atomic orbitals. A polarization basis set, in contrast, allows for displacement of a charge away from the atomic center and is necessary for describing the charge polarization that occurs in highly polar molecules. For example, although the ground electronic state of chlorine has electrons in only the *2s* and *2p* orbitals, a polarization basis set for chlorine includes functions of the form of *d* orbitals, as well.

Computational techniques are available that are both cheaper and faster than the *ab initio* molecular orbital method. However, these techniques must be evaluated in terms of their accuracy for the system of interest. Both semiempirical molecular orbital and molecular mechanics calculations can be carried out in a fraction of the time needed for an *ab initio* calculation. But, these methods depend on empirically derived parameters and are, therefore, only as accurate as the parameter sets they employ. If a parameter set has not yet been devised for atoms of the type found in the molecule of interest, then use of these methods may lead to inaccurate results for molecular geometries and/or properties.

The semiempirical quantum mechanical approach proceeds in the same fashion as the *ab initio* method except that several integrals which are components of the energy are assumed to be small and are, therefore, neglected. Other integrals are "tuned" to give agreement with the molecular properties of a group of molecules of a particular type, resulting in a set of empirical parameters for atoms in this set. Only the valence electrons of each atom are explicitly treated in the calculation and a restricted basis set of only one *s* and three *p* orbitals is used per atom. As a result, this technique is considerably cheaper and faster computationally than the *ab initio* method. However, its accuracy may be limited because the empirical parameters have been tuned to reproduce the properties of certain types of molecules. If a new molecule does not fit that category, then the predetermined parameter set may be of little relevance to the description of its molecular properties. For example, in Sections III.C.1 and III.C.4, we describe how we tested the semiempirical technique against the *ab initio* approach for calculation of the molecular geometries and barriers to internal rotation of the amiloride analogs.

2. Molecular Electrostatic Potential Patterns

The molecular electrostatic potential at point $\vec{r}$, $V(\vec{r})$, is given by[55]

$$V(\vec{r}) = \sum_{A} Z_A/|\vec{R}_A - \vec{r}| - \int \rho(\vec{r}')\, d\vec{r}'/|\vec{r}' - \vec{r}| \tag{3}$$

where Z_A is the atomic number of atom A, $\vec{R}_A$ is the location of that atom from the origin of the coordinate system, and $\rho(\vec{r}')$ is the electron density function of the molecule calculated from the quantum mechanical wavefunction. The first term in Equation 4 gives the nuclear contribution to the molecular electrostatic potential; the second term gives the electronic contribution. $V(\vec{r})$ is a real, physical property. The product of $V(\vec{r})$ times the absolute value of the charge on the electron is equal to the interaction energy between the static charge distribution of the molecule and a positive unit point charge located at position $\vec{r}$. This energy is the quantum mechanical equivalent of the classical expression given in Equation 1 and is derived from the electrostatic forces operating on the system (see assumption 3 in Section I.B). Since the MEP is calculated from the molecular wavefunction, which in turn is composed of a sum of atomic basis functions, the accuracy of the MEP can depend upon the basis set chosen for the calculation.

The molecular electrostatic potential is sometimes approximated by

$$V(\vec{r}) = \sum_{A} Q_A/|\vec{R}_A - \vec{r}| \tag{4}$$

where Q_A is a Mulliken- [56] or potential-derived[57] atomic point charge for nucleus A. Since the atomic point charges do not represent a physical observable, the MEP calculated in this fashion is not an equivalent representation of Equation 3. In Section III.C.2, we describe how we tested the point charge representation of the MEP against the *ab initio* MEP for several tastants.

3. Solvent Simulation by Molecular Dynamics

Molecular dynamics simulation is a technique which can be used to determine the effect of solute-solvent interactions on molecular conformation and energetics. In a molecular dynamics simulation, the atoms of the solute and solvent are allowed to change their position and velocity with time. At each new time step, the interaction energy between each atom of the solute and all the solvent atoms must be recalculated. Because of the large number of atoms in the solute-solvent systems and the number of times the energy need be recalculated, it is not computationally feasible to use quantum mechanical molecular orbital techniques to calculate the solute-solvent interaction energy in a molecular dynamics simulation. Instead a classical mechanical empirical force field approach, called molecular mechanics, is used. The molecular mechanics approach, the basis of the molecular dynamics simulation technique, is a classical description of the internal potential energy of a molecule in which the electrons are not treated explicitly as in quantum mechanics. Instead, the energy of the system is given as the sum of bond stretching, angle bending, torsional, coulombic, and Lennard-Jones contributions from each atom:

$$
\begin{aligned}
E = & \sum_{\text{bonds}} K_r (r - r_{eq})^2 + \sum_{\text{angles}} K_\theta (\theta - \theta_{eq})^2 \\
& + \sum_{\text{dihedrals}} V_n/2 \, [1 + \cos(n\theta - \gamma)] \\
& + \sum_{i<j} [A_{ij}/R_{ij}^{12} - B_{ij}/R_{ij}^{6} + q_i q_j/(\epsilon R_{ij})] \\
& + \sum_{\text{hydrogen bonds}} [C_{ij}/R_{ij}^{12} - D_{ij}/R_{ij}^{10}]
\end{aligned}
\tag{5}
$$

where the variable r is the bond length, θ is the bond angle, r_{ij} is the distance between i and j, and ϕ is the torsional angle. The parameters K_r, K_θ, V_n are force constants for bond stretching, and bending, and torsional motions, respectively; r_{eq} and θ_{eq} represent equilibrium bond lengths and equilibrium angles, respectively; A_{ij}, B_{ij}, C_{ij}, and D_{ij} are parameters defining the Lennard-Jones and hydrogen bonding interactions between atoms i and j; q_i and q_j are the atomic point charges of atoms i and j, n and γ are phase factors, and ϵ is the effective dielectric constant of the medium. Alternatively, the effective dielectric constant can be set equal to one and explicit waters of hydration can be included in the calculation. The parameters in Equation 5 are determined empirically from experimental data for a set of molecules of a particular type or they are derived from *ab initio* calculations of molecular properties such as torsional barriers. The energy is determined by finding the set of atomic coordinates which minimize Equation 5.

Since the energy is a sum of analytical potential functions, calculation of the energy is computationally quite efficient. Again, as with the semiempirical molecular orbital method, the results of a molecular mechanics calculation will only be accurate to the extent that the molecule of interest is chemically similar to the set of molecules for which the force field parameters have been derived. Some molecular mechanics packages have been parameterized for proteins and nucleic acids[58-60] or for hydrocarbons and other small molecules.[61,62]

In the molecular dynamics simulation technique, Newton's classical equations of motion for the solute-solvent system are numerically integrated over a certain time period. Knowledge of the atomic position and velocity at a point in time, combined with knowledge of the forces acting upon that atom, can be used to predict the position and velocity of the atom at a future point in time. Since the force is defined as the negative gradient of the potential energy, it can be calculated analytically from Equation 5. The position x after a short time interval Δ t is, therefore, given by

$$x(t + \Delta t) = x(t) + v(t)\,\Delta t + a(t)(\Delta t)^2/2 + \ldots \tag{6}$$

where $v(t)$ is the velocity at time t, $a(t)$ is the acceleration, calculated by Newton's second Law of Motion (force = mass × acceleration), and higher order terms in t are ignored. The resultant trajectory gives the position and velocity of the interacting atoms at each point in time. Since the simulations are carried out over several picoseconds (ps) of time, requiring many hours of computer time, the results provide abundant data which can be analyzed statistically to provide average thermodynamic and geometric quantities for the system of interest. In Section III.C.3, we describe the results of simulation of the free base and protonated forms of amiloride in water. From these calculations, we determine the average conformation of amiloride in solution and the relative energy (i.e., enthalpy) of the *A1* and *A4* conformers.

III. AMILORIDE: POTENT BLOCKER OF SODIUM TRANSPORT

A. EXPERIMENTAL BACKGROUND AND STRUCTURE-ACTIVITY STUDIES

Amiloride, **1a**, has been shown[63-66] to be a potent inhibitor of sodium transport in a variety of cellular and epithelial transport systems, such as the epithelial sodium channel, the Na^+/H^+ antiporter, the Na^+/Ca^{+2} exchanger, voltage-gated Na^+ and Ca^{+2} channels, a K^+ channel, the nicotinic acetylcholine receptor, and Na^+/K^+-ATPase. In Chapter 8, DeSimone and Heck describes how amiloride has been used to probe the mechanism of taste transduction.[67-79]

Although radio-ligand binding studies[80,81] have located the amiloride binding site on sodium channels purified from bovine renal papilla and amphibian cultured A6 cells, no data is yet available about the molecular structure of the epitheliel sodium channel. Therefore, structure-activity studies involving amiloride analogs provide a means of probing the steric and electrostatic requirements of the amiloride binding site. Much of the structure-activity data that correlates modifications of the structure of amiloride to changes in its efficacy as an ion transport inhibitor has been reviewed.[63-66] In summary, the studies show that amiloride analogs exhibit relatively similar behavior for binding to the Na^+ channel and the Na^+/Ca^{+2} antiporter. Modification of the terminal guanidine nitrogen by substitution with hydrophobic groups enhances the inhibition of the epitheliel Na^+ channel and the Na^+/Ca^{+2} exchanger, while decreasing inhibition of the Na^+/H^+ antiporter. Substitution of the chlorine at position 6 of the pyrazine ring with hydrogen or fluorine leads to decreased inhibition of all three ion transport systems. In contrast, substitution with bromine or iodine results in decreased inhibition of the Na^+ channel and the Na^+/Ca^+ antiporter, but increased inhibition of the Na^+/H^+ antiporter. Alkyl substitution in place of the amino group at position 5 of the pyrazine ring decreases inhibition of the Na^+ channel, but increases inhibition of the two antiporters.

Using a set of 30 amiloride analogs, Li et al.[82,83] have carried out a series of electrophysiological studies on the apical sodium channels of the abdominal

TABLE 1
Structure-Activity Relationships for Selected Amiloride Analogs[a]

Analog	Substituent at Position 5	Substituent at Position 6	k_{on} (s^{-1} μM^{-1})	k_{off} (s^{-1})	Block time (msec)
1	$-NH_2$	-Cl	13.17 ± 0.25	3.93 ± 0.19	255
2	$-NH_2$	-Br	14.19 ± 1.09	5.58 ± 0.92	179
3	$-NH_2$	-I	11.43 ± 0.90	17.41 ± 0.40	57
4	$-NH_2$	-F	13.54 ± 0.65	32.20 ± 1.57	31
5	$-NH_2$	-H	14.47 ± 0.68	176.25 ± 17.73	6
6	-H	-Cl	3.32 ± 0.44 (3.42)	10.89 ± 1.35	92
7	-Cl	-Cl	5.16 ± 0.46 (5.57)	151.10 ± 16.48	7

Analog	Elongation with	k_{on} (s^{-1} μM^{-1})	k_{off} (s^{-1})	Block time (msec)
18	-O-	1.22 ± 0.07 (13.4)	20.67 ± 3.72	48
19	-NH-	2.16 ± 0.11	3.41 ± 0.55	293

[a] Data from References 82 and 83.

skin of *Rana ridibunda*. These experiments relate differences in the microscopic rate constants for the binding of the analog to alterations in the molecular structure of the pyrazine ring[82] or to alterations in the acylguanidinium side chain[83] bonded to the pyrazine ring at position 2. Table 1 summarizes the results for the analogs which we have selected for molecular modeling analysis. The analogs are identified by the numbering scheme used by Li et al.[82,83] The numbers in parentheses are the pK_a-corrected values which are based on the concentration of only the protonated species. The structure-activity data in Table 1 correlate changes in the substituent pattern at positions 5 and 6 of the pyrazine ring or changes in the side chain to differences in the microscopic association constant, k_{on}, and the dissociation constant, k_{off}. Table 1 shows that substitution of bromine, iodine, fluorine, or hydrogen for chlorine at position 6 of the pyrazine ring has little effect on k_{on}. However, k_{off} for the series steadily increases to the point that analog **5** forms a significantly less stable blocking complex with the channel than does amiloride. Conversely, substitution of hydrogen (analog **6**) or chlorine (analog **7**) for the amino group at position 5 of the pyrazine ring affects both k_{on} and k_{off}. For **7**, the increase in k_{off}, which implies a less stable blocking complex, is particularly notable. The decrease in k_{on} for **6** and **7** (as well as for **18** and **19**) indicates that the k_{on} rates for **1** to **5** are not diffusion limited.

Analogs **18** and **19** are examples of the set of compounds with elongated side chains. Table 1 shows that **18** has a pK_a-adjusted k_{on} close to that of amiloride, while that of **19** is much smaller. This indicates that **18** and **19**

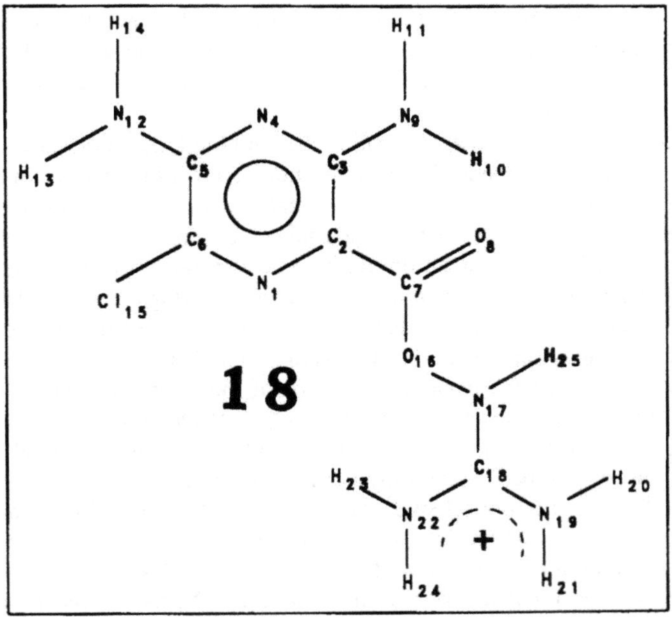

18

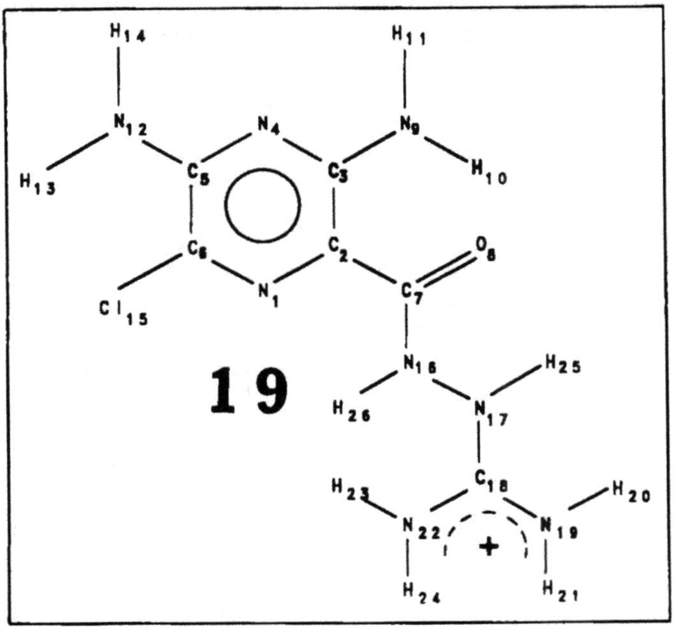

19

have different initial molecular interactions with the ion channel.[83] Comparison of the values of k_{off} for **18** and **19** show that they are almost the same order of magnitude: **19** is a slightly better blocker than amiloride, with a block time of 293 msec, while **18,** with a block time of 48 msec, is closer to analogs **3** and **4.**

From these data, Li et al.[82,83] following the plug-type model of Cuthbert[84] postulated a two-step model for the amiloride-channel interaction. In the first step, the guanidinium side chain of the analog invades the channel entrance and interacts with an anionic site on the channel to form an encounter complex. Then the chlorine atom of amiloride binds to an electropositive site on the channel, leading to the formation of a stable blocking complex. Since the molecular structure of the channel is not known, the specific molecular interactions involved in the amiloride-channel complex have not been further identified. We used the basic assumptions of the Li model (binding of the guanidinium group to an anionic site on the channel and binding of the chlorine to an unspecified channel site) as a starting point for our computational studies.

Prior to our work,[17-20] the only computational study of amiloride was a CNDO/2[85] molecular orbital study of several conformers and tautomers of **1** carried out in conjunction with a series of ^{1}H, ^{13}C, and ^{15}N NMR studies.[86] This work attempted to identify the preferred ground state structure of the free base and protonated forms of amiloride. The NMR results showed that the free base form of amiloride exists primarily in the *A* tautomer, **1b,** while the conjugate acid is found in the *F1* form, **1a.** The gas phase CNDO/2 calculations, however, predicted the isoimino *E* tautomer, **1c,**

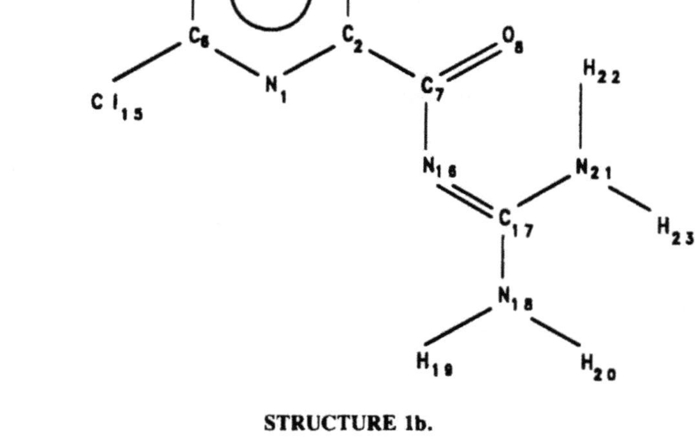

H14
N12
H13
N4
C5
C3
N9
H11
H10
C6
C2
N1
Cl15
C7
O8
N16
C17
N18
H19
H20
N21
H22
H23

STRUCTURE 1b.

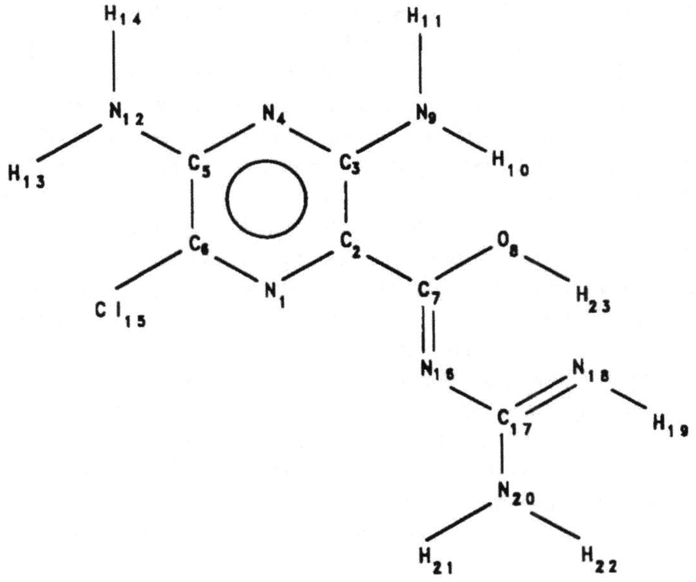

STRUCTURE 1c.

to be the more stable free base form found in solution, in disagreement with the NMR results. The NMR studies were not able to distinguish between the *A1* conformer ($O_8C_7C_2N_1$ = 180°) and the *A4* conformer ($O_8C_7C_2N_1$ = 0°) of the free base form. The CNDO/2 calculations, carried out in the vacuum phase, predicted the *A1* conformer to be 0.88 kcal/mol more stable than the *A4* conformer, with a barrier to rotation around the C_2–C_7 bond of 6.02 kcal/mol. The calculations, although nearly state-of-the-art at the time they were performed, suffer from several unavoidable deficiencies due to the relatively large size of the molecules and the resultant limitation of computer facilities at that time: (1) the CNDO/2 semiempirical molecular orbital technique was used, and (2) geometry optimization of the molecules was not carried out. Instead the molecular structure was built from standard (average) bond angles and bond lengths. These were held fixed as the torsional angles changed, i.e., from *A1* to *A4*. It has been noted[87] that semiempirical calculations, even with full geometry optimization, often give a poor estimate of relative energies for tautomeric pairs. For this reason, we decided to use the more exact molecular orbital techniques described in Sections III.C.1 and III.C.2 to clarify the relative energies of the free base and protonated species and ultimately to interpret the structure-activity data of Li et al.[82,83] in terms of molecular recognition of the analogs by the epithelial sodium channel.

B. COMPUTATIONAL GOALS AND ASSUMPTIONS

It is the goal of work in this laboratory to interpret the activity of the amiloride analogs at the molecular level by an analysis of the steric and electrostatic features which determine the optimal interactions of amiloride with the channel binding site and thereby lead to the formation of a stable blocking complex. As a first step in that direction, we chose to study the set of analogs with pyrazine ring modifications.[82] Since these analogs are, in general, sterically similar to amiloride, this enabled us to focus primarily on the electrostatic features of molecular recognition, which we determined by comparison of *ab initio* molecular electrostatic potential patterns of the amiloride analogs. In this way, we were able to relate differences in the MEP patterns of the analogs to differences in the kinetic data[82,83] for formation of the encounter and blocking complexes. From this data, we developed a hypothesis describing the important molecular recognition features which determine the stability of the blocking complex formed between the channel and the analogs with pyrazine ring modifications. In this way, we attempted to define a model gustaphore for amiloride that describes the steric and electrostatic features which modulate its function as an antagonist of sodium transport.

In the next step, which involves work still in progress, we applied a similar type of MEP analysis to analogs **18** and **19**, as representative analogs with elongated side chains. Comparison of the MEP patterns and steric volumes occupied by the pyrazine ring and side chain analogs is leading to a model of the blocking complex and to refinement of the gustaphore for amiloride.

C. RESULTS

Below, we give a description of the computational techniques we used in our studies of amiloride, along with an explanation of why certain techniques were chosen over others, and some caveats about approximate methods in light of the subsequent results. We focus on the results for the analogs with pyrazine ring modifications in Sections 1 to 3 below, and describe our work in progress on the analogs with side chain modifications in Section 4.

1. Selection of Conformers

In this section, we compare the results of conformational analysis of amiloride by *ab initio* and semiempirical molecular orbital techniques. We evaluate the transferability of amiloride torsional barriers to other analogs.

Since it is the protonated form of amiloride which acts as a sodium channel blocker, we wanted to clarify the results of Smith et al.[86] in terms of which species would be found in the protonated form in solution. Using the GAUSSIAN[88] computer package, we carried out an *ab initio* molecular orbital study, with geometry optimization, to determine the relative energy of the *A* and *E* tautomers.[17] Since amiloride has a chlorine substituent at position 6 of the pyrazine ring, we chose the 3-21G(*) basis set[89] because it is the smallest split-valence basis set with polarization that could provide reliable relative energies and geometries for such a large molecule. Others have shown the reliability of the 3-21G(*) basis set in calculations on similar types of molecules. For example, the 3-21G(*) basis set was able to correctly predict the relative stability of a series of pyridone tautomers[90] and to give good results for molecular geometries, relative energy differences, and protonation energies for heterocyclic molecules such as pyrazine.[91] For amiloride, we found[17] that the geometry calculated in the 3-21G(*) basis set was in excellent agreement with gas phase electron diffraction and X-ray crystal studies of similar systems.

Our results, in agreement with the NMR study, showed that the *A* tautomer, **1b**, is significantly lower in energy and, therefore, more likely than the *E* tautomer, **1c**, to be found in the protonated form in solution. We then calculated[17] the relative energy and torsional barrier to *A1*/*A4* rotation, finding *A1* to be 2.50 kcal/mol more stable than *A4* and the barrier to rotation to be 19 kcal/mol.

We then tested the AM1[92] semiempirical molecular orbital method against the 3-21G(*) basis set in terms of its ability to calculate accurate molecular geometries and relative energies for *A1*, *A4*, *E1*, and *F1*.[17] We also studied the torsional barriers for rotation around C_2-C_7 in the *A* tautomer of amiloride. We found that, although the geometries and relative energies of the planar conformations, **1a**, **1b**, and **1c**, were in good agreement with the *ab initio* and experimental results, the energy of *A* at $O_8C_7C_2N_1 = 90°$ was inaccurate. This seems to be due to the difficulty of the AM1 method to accurately reproduce torsional energies for rotation around partial double bonds. For example, in amiloride, because of conjugation between the pyrazine ring and acylguanidinium side chain, the C_2-C_7 bond has partial double bond character. Our results are in line with those of Fabian,[93] who found several examples in which the AM1 method gives poor results for rotational barriers around partial double bonds. Therefore, since the semiempirical technique could not produce accurate torsional barriers for the amiloride analogs, we chose to use the *ab initio* method for our study. Similar poor results were obtained with other semiempirical techniques for torsional rotations in the side chains of **18** and **19**. These results are described in Section 4.

Since the other analogs were modified at positions 5 and 6 of the pyrazine ring, far from the acylguanidinium side chain, we assumed that these modifications did not affect the height of the $O_8C_7C_2N_1$ torsional barrier. Therefore, we assumed that the analogs with pyrazine ring modifications adopt a planar, *F1*-like conformation in solution.

2. Molecular Electrostatic Potential Analysis

In this section, we outline our stepwise approach for using the molecular electrostatic potential to interpret structure-activity data. We compare the accuracy of point charge representations of the MEP to the more exact MEP calculated from Equation 4. In Section d and subsequent sections, analog **1** refers to the protonated form of amiloride, **1a.** All MEP maps were calculated *in vacuo*.

a. General Procedure

Although the formation of the blocking complex may be a concerted procedure, in order to approach the problem computationally, we used a stepwise protocol:

1. The MEP patterns of the unprotonated analogs were calculated and inspected in order to determine the site of protonation.
2. Then, the protonated forms of the analogs were constructed and their MEP patterns were determined. These were inspected and compared to differences in the value of k_{on} in order to identify features in the MEP pattern of the protonated analog that might be related to the formation of the encounter complex.
3. Finally, the MEP maps of simple model encounter complexes (protonated analog and formic acid anion) were calculated and compared to differences in k_{off} in order to identify features in the MEP pattern of the analog that might be related to the formation of a stable blocking complex. The simple encounter complex was not meant to model the complete analog-channel interaction. Rather, it was used to illustrate the change in the MEP pattern of the analog that occurs upon binding to a putative anionic site, such as that assumed in the models of Li et al.[82,83] and Cuthbert,[84] and to thereby highlight the molecular features of the bound analog that might be implicated in the formation of a stable blocking complex. From these results, we formulated a gustaphore for amiloride activity as an antagonist of sodium transport.

b. Choice of Basis Set

The molecular electrostatic potential patterns were used: (1) to determine the site of protonation of the free base forms of the analogs and (2) to identify sites which might interact with complementary sites on the ion channel. To that end, we needed to have an accurate representation of the location, extent, and depth of the minima in the MEP. Previously, in studies of the perillartine sweeteners[94] we found that the 3-21G basis set was able to accurately reproduce the position and depth of MEP minima calculated in the 4-31G and 6-31G(*) basis sets. Therefore, we selected the 3-21G(*) basis set, with polarization functions on the chlorine atom, as a reasonable choice for calculation of the MEPs of the amiloride analogs.[18] Since the model encounter complex (protonated analog plus formate anion) is an even larger molecular system, computer disk space limitations at that time forced us to use a smaller basis set to study the complexes. Therefore, we tested[18] the STO-3G basis set against the 3-21G(*) in its ability to represent the MEP of *A1*. We found that the STO-3G map was able to reproduce the general location but not the depth of the minima noted in the 3-21G(*) map. Specifically, the STO-3G basis set showed the minimum off the oxygen to be less negative and the minimum off the chlorine to be more negative than the respective 3-21G(*) minima. Since these are relatively small differences which would be constant for the series of analogs to be studied, we decided to use the STO-3G basis for the analysis of the model encounter complexes. Thus, in comparing the MEP patterns of the amiloride analogs calculated with a single basis set (see Section d below), what is important is not the sign of the minima (negative in the case of chlorine, positive in the case of bromine), but rather the relative trends within a series of substitutents.

It is interesting to note that in a similar comparison for the molecular electrostatic potential patterns of the sweet-tasting compounds acesulfame and saccharin, we found essentially no difference in the MEP calculated by the STO-3G and 3-21G basis sets.[12] This indicates that differences in the MEP due to the choice of basis set should be carefully evaluated for each system studied.

c. Atomic Point Charge MEP Maps

s135We have compared the molecular electrostatic potential patterns of amiloride and perillartine sweeteners calculated from the exact and approximate definitions of $V(\vec{r})$ given in Equations 3 and 4, respectively. We have shown[95] that MEP maps calculated from 3-21G(*) Mulliken charges, AM1 charges, and AM1 potential derived charges do not correctly reproduce the location or relative depth of the minima found in the MEP calculated with the 3-21G(*) basis set. Therefore, although the point charge representation is a cheap method for calculating the MEP, it is not necessarily sufficiently accurate to be used with confidence when investigating questions of molecular reactivity. In contrast, it is interesting to note that the conformational energy given by Equation 5 seems to be much less sensitive to the atomic point charges used in the calculation. For example, we have shown[1] that the main features of the conformational potential energy surface of β-cyclodextrin are independent of the choice of atomic point charge set. This indicates that molecular reactivity is much more sensitive than molecular conformation to the correct representation of the molecular electrostatic potential.

d. Ab Initio *MEP Maps*

We calculated the molecular electrostatic potential maps for the free base, protonated, and model encounter complex forms of analogs **1** through **7** using the 3-21G(*) basis set.[18] The maps were calculated in a plane 1.3 Å from the plane of the molecule. At each point on the two-dimensional grid, the interaction energy between a positive point charge located at that position and the static charge distribution of the molecule, represented by Equation 3, was calculated. Points of equal energy (in kcal/mol) were connected and displayed with the ARCHEM[96] contouring program. The resulting contour map is similar in principle to contour maps used by hikers to indicate changes in elevation.

In the MEP case, however, the maxima in the map correspond to the regions of high positive electrostatic potential in the molecule. The negative minima in the MEP map correspond to regions of negative electrostatic potential. These minima are particularly important indicators of reactive sites in the molecule. They identify sites which may interact with electrophilic sites on other molecules.

We illustrate the use of the molecular electrostatic potential maps in formulating the model of the gustaphore by focusing here on the results for the protonated forms and model encounter complexes of **1**, **5**, and **6**.

i. Molecular Electrostatic Potentials of the Protonated Analogs

Figure 1 gives the molecular electrostatic potential maps of the protonated analogs **1**, **5**, and **6**. We used the maps of analogs **1** through **7** to determine which molecular recognition features of the analogs are involved in the formation of the encounter complex. All the maps, including those not shown here, exhibit a broad maximum in the MEP over the guanidinium side chain and show that the positive potential is spread out over the molecule. This is in contrast to schematic diagrams of amiloride like **1** in which the charge is localized near the amino groups. Figure 1a shows that the MEP of protonated amiloride has local minima in the positive potential off the carbonyl oxygen, the amino groups at positions 3 and 5 of the pyrazine ring, and at N_4 of the pyrazine ring, the location of the global minimum. The map also exhibits a local minimum off the chlorine atom. Figure 1b shows the MEP map of **5**, in which hydrogen has replaced chlorine at position 6 of the pyrazine ring. The positions of the local minima are the same as for **1**, with the exception of the minimum off position 6 which is missing in **5**. The spatial extent of the MEP maximum over the guanidinium group is similar in **1** and **5**. Figure 1c shows the MEP map of **6**. The replacement of the electron-donating amino group of **1** by a hydrogen at position 5 has significantly altered the MEP pattern of the molecule. Neither the carbonyl oxygen nor positions 3, 4, and 5 of the pyrazine ring show minima of the depth of **1** and **5**. In addition, the broad MEP maximum over the guanidinium side chain extends into the ring region of **6**.

In comparing the MEP maps to the k_{on} values in Table 1, we noted that **1**, **4**, and **5** have similar values of k_{on} for the formation of the encounter complex and have very similar maps, except in the region of position 6 of the pyrazine ring. Since the value of k_{on} is an indication of the initial molecular interactions between the analog and the channel, this seemed to indicate that position 6 of the pyrazine ring is not involved in the formation of the encounter complex. Rather, the positively charged side chain most likely initiates the electrostatic interaction with the channel. Since **1**, **4** (map not shown), and **5** have very similar minima off the carbonyl oxygen and positions 3, 4, and 5 of the pyrazine ring, it may be that the functional groups at these positions also interact with the channel to form the encounter complex. Or, alternatively, they may function to control the electrostatic features over the ring which are

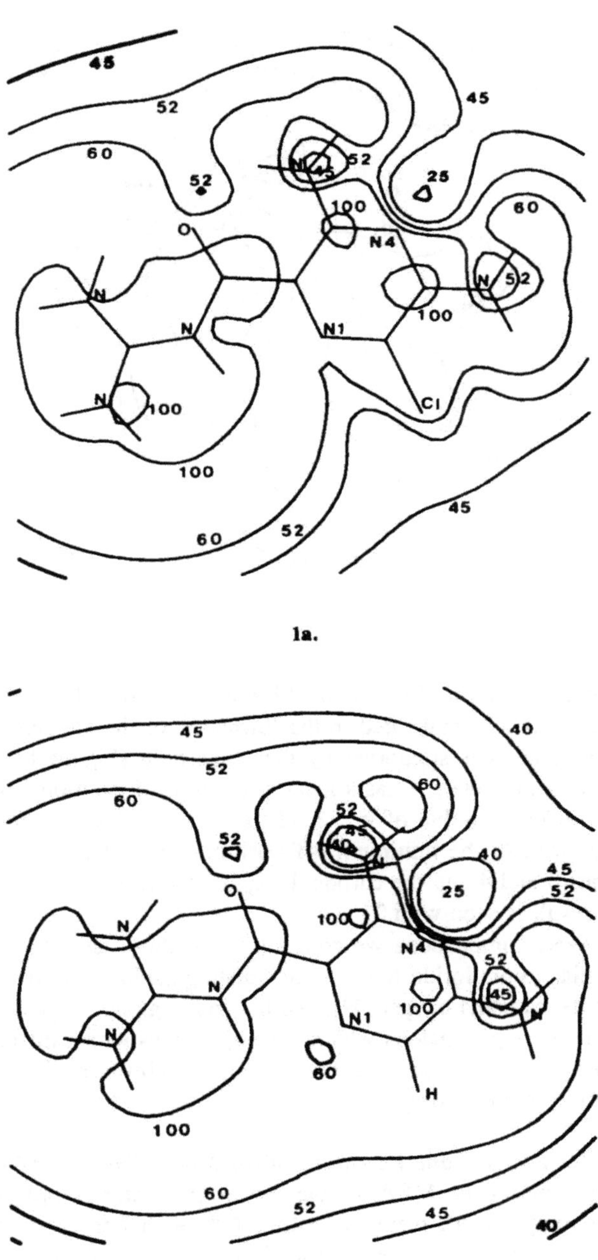

FIGURE 1. The molecular electrostatic potential patterns of the protonated amiloride analogs: (a) **1**, (b) **5**, (c) **6**. Contour levels are given in units of kcal/mol. (From Venanzi, C. A., Plant, C., and Venanzi, T. J., *J. Med. Chem.*, 35, 1646, 1992. With permission.)

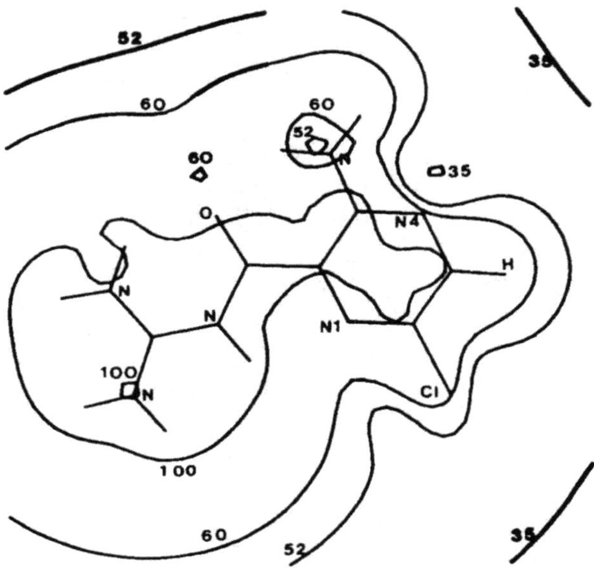

FIGURE 1c.

complementary to those of the channel binding site. In either case, it is clear that position 6 is not implicated in the formation of the encounter complex.

This hypothesis is supported by the maps of **6** (Figure 1c) and **7** (not shown). Although analog **6** has a local minimum off position 6 as does **1**, it has a very different value of k_{on}: 3.32 vs. 13.17 s^{-1} μM^{-1} for **1**. On the other hand, with **6**, the minimum off position 5 no longer exists and those off positions 3 and 4 and the carbonyl oxygen are less deep than those of **1**. The same result is seen with **7**.

From these comparisons we concluded that those analogs that have k_{on} values similar to amiloride have strong, distinguishing minima in the MEP pattern off the carbonyl oxygen, N_4, and the amino groups at positions 3 and 5 of the pyrazine ring. Analogs which have k_{on} values which differ from amiloride lack two or more of these features and exhibit a much more positive pattern on the pyrazine ring.

ii. Molecular Electrostatic Potentials of the Model Encounter Complexes

Figure 2 shows the MEPs of the model encounter complexes of **1**, **5**, and **6**. We compared the maps of analogs **1** through **7** in order to determine which molecular recognition features are important to the formation of a stable blocking complex. Figure 2a shows the MEP pattern of the model encounter complex formed by **1** and the formate anion. The region around the anion is negative, while most of the region around **1** is positive. However, two negative minima are associated with the pyrazine ring: the minimum off N_4 and the

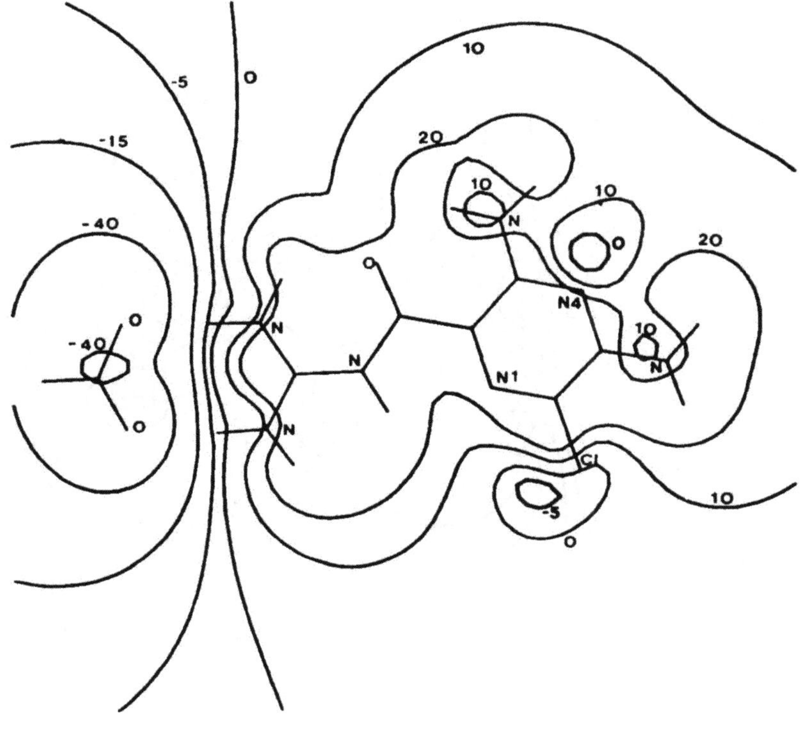

2a.

FIGURE 2. The molecular electrostatic potential patterns of the model encounter complexes formed by the amiloride analogs and the formate anion: (a) **1**, (b) **5**, (c) **6**. Contour levels are given in units of kcal/mol. (From Venanzi, C. A., Plant, C., and Venanzi, T. J., *J. Med. Chem.*, 35, 1647, 1992. With permission.)

minimum off the chlorine atom at position 6 of the pyrazine ring. Figure 2b shows the MEP of the model encounter complexes with **5.** The pattern is very similar to that of **1,** with the exception that there is no negative minimum off the atom at position 6. In fact, **5,** of all the analogs with halogen substituents at position 6, shows the most positive MEP pattern in this region of the map.

As an example of an analog with substitution at position 5, Figure 2c shows that in the MEP of the model encounter complex with **6,** the minimum off N_4 is missing, but the minimum off position 6 is similar to that of **1.** For **6** as well as **7** (not shown), the minima off the N_4 are much less distinctive than for **1** through **5.** All the model encounter complexes exhibit a broad positive region extending from the guanidinium side chain to cover the ring.

From this we concluded that if, as hypothesized,[82,83] the chlorine atom of amiloride interacts with a site on the channel, then the MEP pattern of the model encounter complex in Figure 2a defines the electrostatic requirements

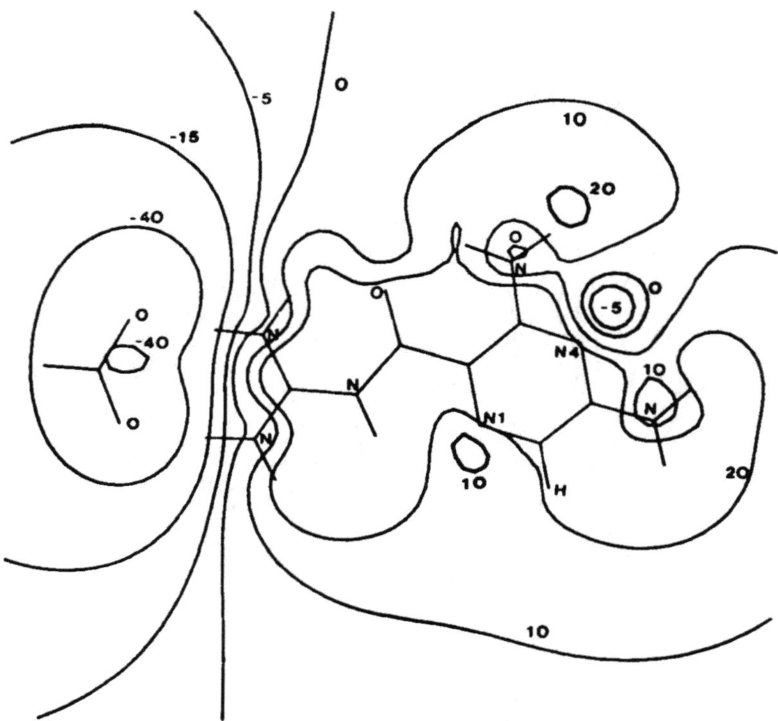

FIGURE 2b.

for formation of a stable analog-channel blocking complex. Table 1 shows that the values for k_{off} for **1** and **5** are 3.93 s^{-1} and 176.25 s^{-1}, respectively. Since the residence time for the analog on the channel is inversely proportional to k_{off}, this indicates that **5** forms a very poor blocking complex compared to amiloride. Comparison of Figures 2a and 2b shows that both molecules exhibit minima off N_4, but only amiloride has an additional minimum off the chlorine atom at position 6 of the pyrazine ring. Therefore, electrostatic interactions between the channel and position 6, rather than position 4, are more important to the stability of the blocking complex. In the case of **5**, which exhibits a broad, positive region off position 6 rather than the localized minimum of **1**, this electrostatic molecular recognition feature is missing, leading to a poor interaction with a complementary site on the channel.

The MEPs of **2**, **3**, and **4** (not shown) taken together with those of **1** and **5** show that the efficacy of an analog as a sodium transport blocker is related to the depth of the minimum off position 6. As the minimum becomes progressively less negative, the block time becomes shorter. This supports the hypothesis of Li et al.[82,83] that the ligand at the 6 position binds to an electropositive site on the channel. Li et al.[82] also noted that formation of a long-

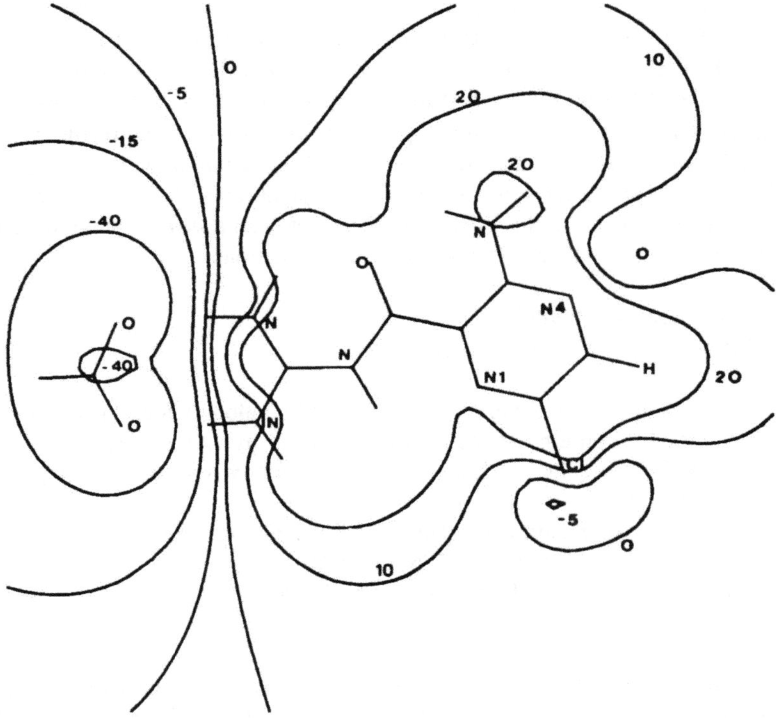

FIGURE 2c.

lived blocking complex requires a highly electronegative substituent at position 6. But they did not address the fact that, although fluorine is the most electronegative atom in the series, **4** is not the best blocker. The MEP maps, however, clarify this point. In contrast to the deep minimum off the chlorine in **1**, **4** has a very positive MEP pattern in this region. This is probably due to the strong hydrogen bonding interaction between fluorine and the amino group at position 5 noted in the geometry optimization of **4**. Therefore, the most electronegative atom does not exhibit the deepest MEP minimum off position 6.

In addition, comparison of the MEP maps of **2** and **3** (not shown) shows that they are very similar and yet the values of k_{off} are slightly different: 5.58 s^{-1} for **2** and 17.41 s^{-1} for **3**. Since iodine has a much larger van der Waals radius than bromine, this suggests that steric factors also influence the binding of the 6-position ligand to the ion channel. We decided to investigate the importance of steric interactions in more detail through a study of analogs with elongated side chains, as discussed in Section 4 below.

The importance of a deep, localized minimum off position 6 is further supported by the model encounter complex with **6** (see Figure 2c). Although the minimum off N_4 is negligible in this complex, the minimum off the chlorine at position 6 is similar to that of **1**. Since the k_{off} of **6** ($10.89\ s^{-1}$) is similar to that of amiloride, the presence of this minimum at position 6 seems to be the determining feature in the formation of a stable blocking complex. Further support for this hypothesis was shown by the results for **7**, which does not form a particularly stable blocking complex with the channel and does not have a localized minimum off position 6. Rather, **7** shows a broad negative region in the MEP spread out between positions 5 and 6, with the global minimum located between the two positions.

e. The Gustaphore Model

The results of our study indicated that:

1. A stable blocking complex is formed with analogs which have a deep, localized minimum off the 6 position of the pyrazine ring;
2. The stability of the blocking complex is directly related to the depth of that minimum;
3. Substitution at position 5 affects not only the depth, but also the location and size of the minimum off position 6;
4. Steric factors may influence the optimal binding of the 6-position ligand to the ion channel.

Although steric factors will be investigated more extensively in the study of the analogs with side chain modifications (described in Section 4 below), the pyrazine ring analogs provide some information about the spatial requirements of the channel binding site. The work of Li et al.[82,83] indicates that both the guanidinium group and the chlorine atom are implicated in the binding of amiloride to the channel. Our studies have identified the size, depth, and location of the MEP minimum off the chlorine as features which seem to be related to the stability of the blocking complex. It is possible that both the guanidinium group and the chlorine atom are involved in hydrogen bonding interactions with separate sites on the channel. In the formulation of the model encounter complex, we assumed that the guanidinium group binds to the formate anion in a chelate-type orientation. Therefore, the position of the minimum off the chlorine atom of amiloride, taken together with the positions of the proton donors, H_{20} and H_{23}, may identify the relative spatial location of complementary sites on the channel. Therefore, since **1** is the most potent analog in the series of analogs with pyrazine ring modifications, we defined the steric and electrostatic components of the gustaphore in terms of the template provided by **1**. In summary, the gustaphore for amiloride activity as an antagonist of sodium transport is defined by:

1. Strong, distinguishing minima in the MEP pattern off the carbonyl oxygen and positions 3, 4, and 5 of the pyrazine ring.

2. A broad positive MEP maximum localized over the side chain.
3. A deep, localized MEP minimum off the 6 position of the pyrazine ring.
4. A fixed distance between the proton donors of the chelating guanidinium group and the minimum off position 6 of the pyrazine ring.

Features 1 and 2 define the requirements for the initial molecular interaction with the channel, whereas features 3 and 4 define the requirements for formation of a stable blocking complex with the channel. Feature 4, in particular, indicates that the size of the analog, specifically the distance between the proton donors and the minimum off the chlorine, may be an important steric requirement for binding. In order to probe the steric limitations of the proposed gustaphore, we are studying analogs with elongated side chains. Our work in progress on **18** and **19** is described below in Section 4.

As illustrated here, the MEP method has been extremely useful in identifying electrostatic recognition features of the analogs that can be correlated to values of k_{on} and k_{off}. On the other hand, the method cannot distinguish whether the carbonyl oxygen and pyrazine ring substituents actually interact with the ion channel directly to form the encounter complex, or if they in some way control the interaction in a more global fashion by their effect on the overall molecular electrostatic potential pattern of the analog. This latter effect may be important, for example, to the formation of a stacking complex with the channel.

3. Investigation of the Solvent Effect on Conformation and Energetics

The gustaphore assumes that the analogs interact with the channel in planar, *F1*-like conformations which exhibit intramolecular hydrogen bonding interactions similar to **1**. However, it may be possible that interaction of the analog with discrete solvent molecules disrupts the intramolecular hydrogen bonding pattern to such an extent that the analog is no longer planar. Although, to a significant degree, the torsional barrier for rotation around the C_2–C_7 bond determines the planarity of the amiloride pyrazine ring with respect to the side chain, it may be possible that solute-solvent interactions stabilize slightly nonplanar conformations of the solute. Since, upon binding to the ion channel, the attractive tastant-solvent interactions may be replaced by attractive tastant-channel interactions, it would seem that any stable nonplanar conformations identified through the solvent simulation should be considered as likely models from which to extract additional information about the gustaphore.

In order to investigate this solvent effect, we used the molecular dynamics simulation method to analyze which conformers of the free base and protonated forms of amiloride might be stable in aqueous solution. Previously, the *A4* conformer was shown[17,86] to have a higher dipole moment than *A1*. This indicates that the *A4* form may be stabilized by solvent interactions to a greater degree than the *A1* form, such that *A4* becomes the more stable conformer in solution. In order to clarify which free base form is likely to predominate in solution and to determine potential modes of binding of amiloride to the ion channel, we carried out molecular dynamics simulations of the *A1*, *A4*, and *F1* species in water.[20] However, in order to accomplish this, we first had to derive molecular mechanics force field parameters to describe amiloride since none of the available molecular dynamics simulation packages have been parameterized for small, acylguanidinium heterocycles. The chief deficiency in the available force fields was the lack of appropriate torsional potential functions. We, therefore, used *ab initio* quantum mechanics to calculate the torsional barriers for *A1/A4*[17] and *F1/F4*[19] rotation. The results indicated that there is a large energy penalty for both the free base and protonated species to adopt nonplanar conformations. In addition, we found the *F4* conformation to be 33 kcal/mol less stable than *F1* due to hydrogen-hydrogen repulsion between H_{10} and H_{24}.[19] We then used the method of Hopfinger and Pearlstein[97] to fit the variable parameters in the analytical expression for the torsional potential function in Equation 5 to reproduce the *ab initio* results. We used this information to parameterize the GROMOS (GROningen MOlecular Simulation)[98] molecular dynamics force field for amiloride.

We carried out molecular dynamics simulations of 30 ps in length for the *A1*, *A4*, and *F1* conformers of amiloride.[20] Each conformer was surrounded by a bath of approximately 400 water molecules. We found that the large torsional barrier to interconversion between the *A1* and *A4* conformers constrains the average structure of each conformer to a near planar conformation. Comparison of the relative internal energy of the *A1* and *A4* conformers showed the *A1* conformer to be more stable than the *A4*, in agreement with the *ab initio* results. Comparison of the solvent-solute interaction energy showed the *A4*-solvent system to be more stable than the *A1*-solvent system, a trend expected from the higher dipole moment of *A4*. However, the more favorable solvent-solute interaction energy of *A4* was not sufficient to offset the large difference in internal energy of *A1* and *A4*. Therefore, the results predicted the *A1* conformer to be more stable in solution than *A4*.

We also found that the high torsional barrier for *F1/F4* interconversion constrains the protonated species to the planar, *F1*, conformation. Solvent-solute interactions were not found to significantly stabilize nonplanar conformations. These results support our assumption of a planar *F1*-like conformation of amiloride as the template from which the gustaphore is defined.

4. Work in Progress: Analogs with Side Chain Modifications

Since analogs **18** and **19** have several rotatable bonds which may affect their mode of binding to the ion channel, we are in the process of carrying out a study of the torsional barriers for rotation around the C_2–C_7, C_7–O_{16} or C_7–N_{16}, and O_{16}–N_{17} or N_{16}–N_{17} bonds of these analogs using both the 3-21G(*) basis set and the AM1, PM3,[99] and MNDO[100] semiempirical techniques. We used the MOPAC 6.0 molecular orbital package[101] for the semiempirical calculations. The side chain rotations involve either rotation around a partial double bond or change in the $O_{16} \ldots N_1$ lone pair-lone pair interaction in **18** or the $H_{26} \ldots N_1$ hydrogen-lone pair interaction in **19.** Our preliminary results[19] show that, again, the semiempirical method fails to reproduce accurate torsional barriers. Our results agree with those of Fabian,[93] who showed several examples in which the AM1 method gives poor results for conformational energies which are influenced by lone pair-lone pair or lone pair-hydrogen interactions.

We have calculated the 3-21G(*) MEP maps of analogs **18** and **19** in the orientations shown above. The MEPs of the analogs have many of the same electrostatic features as amiloride. The encounter complexes of **18** and **19** show a deep, localized MEP minimum off the chlorine atom and a positive maximum in the MEP over the guanidinium group. However, superposition of the structures and maps of **18** and **19** onto those of **1** shows that, because the side chain analogs are longer than amiloride, the locations of these features in **18** and **19** do not coincide with those in **1.** For example, superposition of the encounter complexes by atom-to-atom matching of the pyrazine ring fragments shows that the guanidinium groups of **18** and **19** occupy a very different region of space than that of **1.** Alternatively, superposition of the guanidinium fragments of **18** and **19** onto that of **1** shows that the minimum off the chlorine atom in **18** and **19** is located in a different region of space than that of **1.** If, as proposed in the gustaphore, the distance between the deep, localized MEP minima off the chlorine and the proton donors of the chelating guanidinium moiety defines the spatial relationship between the complementary binding sites on the channel, then it is difficult to see how **18** and **19** could fit this template. The elongated side chains would seem to prevent the guanidinium moiety and chlorine atom of **18** and **19** from simultaneously being in the proper regions of space to interact with the complementary sites on the channel.

However, it may be that analogs **18** and **19** do not bind in the conformations shown here. From our torsional studies of **18** and **19,**[19] we have already located other planar conformations that are only a few kcal/mol higher in energy than the conformations shown here. In these conformations, the analogs take on a more compact form such that the location and orientation of the guanidinium group with respect to the chlorine is more similar to that in amiloride. It is possible that interaction with the solvent or the ion channel could stabilize these conformations, making them energetically more favorable. In addition, the full torsional study of **18** and **19,** which includes rotation around the C_2–C_7, C_7–O_{16} or C_7–N_{16}, and O_{16}–N_{17} or N_{16}–N_{17} bonds, may reveal nonplanar conformations of relatively low energy that should also be considered as candidates that would fit the proposed gustaphore. These alternative conformations will also be investigated by molecular dynamics simulation of **18** and **19** in water.

In our formulation of the model encounter complex, we have assumed that the analogs bind to the formate anion in a chelate-type interaction, which is the most typical orientation of the guanidinium and carboxylate moieties found in the X-ray crystal of peptides and proteins.[102-106] If, in order to fit into the channel binding site, the analogs with elongated side chains bind in the compact form described above, then the chelate-type interaction may no longer be possible. For example, it may be that H_{22} and H_{23}, rather than H_{20} and H_{23}, fall in the neighborhood of the putative anionic site. We are, therefore, investigating alternative possibilities for the formulation of the model encounter complex for analogs **18** and **19**.

IV. FUTURE DIRECTIONS

As emphasized in the Introduction, the molecular modeling approach builds upon a set of well-defined, experimental structure-activity data. We continue to seek such data for other tastant-receptor systems. In addition to our work on amiloride, we are studying some of the smaller saccharin-like tastants which appear to act at the same receptor site in the gerbil.[107] We are also collaborating with Bryant and coworkers[108] on the definition of a gustaphore for L-alanine taste receptors in the channel catfish. In these studies, we apply the techniques described here to interpret the structure-activity data on tastant-receptor binding and activation.

ACKNOWLEDGMENTS

This work was supported by grants to C.A.V. from the Campbell Institute for Research and Technology, the New Jersey Commission on Science and Technology, the Hoffmann LaRoche Corporation, the Pittsburgh Supercomputing Center, and New Jersey Institute of Technology. The authors would like to thank Ronald A. Buono for helpful comments on the manuscript, William J. Skawinski for assistance with bibliographic searching, N. U. M. Shyamantha for help in preparation of the structures and figures, and John DeSimone for bringing the work of Li, Lindemann, and Cragoe to their attention.

REFERENCES

1. **Wertz, D. A., Shi, C.-X., and Venanzi, C. A.,** A comparison of distance geometry and molecular dynamics simulation techniques for conformational analysis of β-cyclodextrin, *J. Comp. Chem.*, 13, 41, 1992.
2. **Venanzi, C. A., Canzius, P. M., Zhang, Z., and Bunce, J. D.,** A molecular mechanics analysis of molecular recognition by cyclodextrin mimics of α-chymotrypsin, *J. Comp. Chem.*, 10, 1038, 1989.

3. **Maye, P. V. and Venanzi, C. A.**, Host-guest preorganization and complementarity: a molecular mechanics and molecular dynamics study of cationic complexes of a cyclic urea-anisole spherand, *J. Comp. Chem.*, 12, 994, 1991.
4. **Maye, P. V. and Venanzi, C. A.**, An ab initio study of the geometry and rotational barrier of 4-phenylimidazole, *Structural Chem.*, 1, 517, 1990.
5. **Venanzi, C. A. and Maye, P. V.**, An ab initio study of the geometries and rotational barriers of 1-, 2-, and 5-phenylimidazole, *Structural Chem.*, 2, 493, 1991.
6. **Venanzi, C. A. and Namboodiri, K.**, Electrostatic features of molecular recognition by cyclic urea mimics of chymotrypsin, *Anal. Chim. Acta*, 210, 151, 1988.
7. **Venanzi, C. A. and Bunce, J. D.**, Molecular design of cyclic urea mimics of α-chymotrypsin, *Enzyme*, 36, 79, 1986.
8. **Venanzi, C. A. and Bunce, J. D.**, Molecular recognition by artificial enzymes: cyclic urea mimetics of α-chymotrypsin, *Int. J. Quantum Chem.*, Quantum Biology Symposium, 12, 69, 1986.
9. **Venanzi, C. A. and Bunce, J. D.**, Molecular recognition by cyclic urea mimetics of alpha-chymotrypsin, *Ann. N.Y. Acad. Sci.*, 471, 318, 1986.
10. **Venanzi, C. A.**, Computational analysis of molecular recognition by artificial enzymes, in *Environmental Influences and Recognition in Enzyme Chemistry*, Vol. 10, *Molecular Structure and Energetics*, Liebman, J. F. and Greenberg, A., Eds., VCH Publishers, New York, 1988, 251.
11. **Venanzi, T. J. and Venanzi, C. A.**, A conformational study of a biologically active conjugated *syn*-oxime, *J. Comp. Chem.*, 9, 67, 1988.
12. **Venanzi, T. J. and Venanzi, C. A.**, Acesulfame: an ab initio molecular electrostatic potential study, *Anal. Chim. Acta*, 210, 213, 1988.
13. **Venanzi, T. J. and Venanzi, C. A.**, Ab initio molecular electrostatic potentials of perillartine analogs: implication for sweet taste receptor recognition, *J. Med. Chem.*, 31, 1879, 1988.
14. **Venanzi, T. J. and Venanzi, C. A.**, Sweet taste receptor recognition of the conformationally flexible aldoxime molecule, in *QSAR: Quantitative Structure-Activity Relationships in Drug Design*, Fauchère, J.-L., Ed., Alan R. Liss, New York, 1989, 321.
15. **Venanzi, T. J. and Venanzi, C. A.**, Electrostatic recognition patterns of sweet-tasting compounds, in *Sweeteners: Discovery, Molecular Design, and Chemoreception*, ACS Symposium Series, Vol. 450, Walters, D. E., Jr., Orthoefer, F. T., and DuBois, G. E., Eds., American Chemical Society, Washington, D.C., 1990, 193.
16. **Venanzi, T. J. and Venanzi, C. A.**, Barriers to C-C rotation in molecules with C=C and C=N conjugated bonds. A molecular orbital study, *Chem. Phys. Lett.*, 192, 469, 1992.
17. **Venanzi, C. A., Plant, C., and Venanzi, T. J.**, A molecular orbital study of amiloride, *J. Comp. Chem.*, 12, 850, 1991.
18. **Venanzi, C. A., Plant, C., and Venanzi, T. J.**, Molecular recognition of amiloride analogs: a molecular electrostatic potential analysis. I. Pyrazine ring modifications, *J. Med. Chem.*, 35, 1643, 1992.
19. **Venanzi, C. A., Skawinski, W. J., and Venanzi, T. J.**, A molecular orbital study of torsional barriers of protonated amiloride analogs, manuscript in preparation.
20. **Buono, R. A., Venanzi, C. A., and Venanzi, T. J.**, Molecular dynamics simulation of solvated amiloride conformers, manuscript in preparation.
21. **Bruch, R. C.**, Signal transduction in olfaction and taste, in *G Proteins*, Iyengar, R. and Birnbaumer, L., Eds., Academic Press, New York, 1990, 411.
22. **Liebman, M. N., Venanzi, C. A., and Weinstein, H.**, Structural analysis of carboxypeptidase A and its complexes with inhibitors as a basis for modeling enzyme recognition and specificity, *Biopolymers*, 24, 1721, 1985.

23. **Weinstein, H., Osman, R., Topiol, S., and Venanzi, C. A.,** Molecular determinants for biological mechanisms: model studies of interactions in carboxypeptidase, in *Quantitative Approaches to Drug Design*, Deardon, J. C., Ed., Elsevier, The Netherlands, 1983, 81.
24. **Liebman, M. N., Venanzi, C. A., and Weinstein, H.,** Theoretical principles of drug action: the use of enzymes to model receptor recognition and activity, in *New Methods in Drug Research*, Makriyannis, A., Ed., J. R. Prous, Barcelona, 1985, 233.
25. **Weinstein, H. and Green, J. P., Eds.,** *Quantum Chemistry in the Biomedical Sciences*, Ann. N.Y. Acad. Sci., Vol. 367, 1981.
26. **Silipo, C. and Vittoria, A., Eds.,** *QSAR: Rational Approaches to the Design of Bioactive Compounds*, Vol. 16, Pharmacochemistry Library, Elsevier, New York, 1991.
27. **Dean, P. M.,** *Molecular Foundations of Drug-Receptor Interactions*, Cambridge University Press, New York, 1987.
28. **Lenz, G. R., Evans, S. M., Walters, D. E., and Hopfinger, A. J.,** *Opiates*, Academic Press, New York, 1986.
29. **Boyd, D. B.,** Successes of computer-assisted molecular design, in *Reviews in Computational Chemistry*, Vol. 1, Lipkowitz, K. B. and Boyd, D. B., Eds., VCH Publishers, New York, 1990, 355.
30. **Weinstein, H., Osman, R., Green, J. P., and Topiol, S.,** Electrostatic potentials as descriptors of molecular reactivity: the basis for some successful predictions of biological activity, in *Chemical Applications of Atomic and Molecular Electrostatic Potentials*, Politzer, P. and Truhlar, D. G., Eds., Plenum Press, New York, 1981, 309.
31. **Weinstein, H., Maayani, S., Srebrenik, S., Cohen, S., and Sokolovsky, M.,** Psychotomimetic drugs as anticholinergic agents. II. Quantum mechanical study of molecular interaction potentials of 1-cyclohexylpiperidine derivatives with the cholinergic receptor, *Mol. Pharmacol.*, 9, 820, 1973.
32. **Weinstein, H.,** New quantum mechanical procedures for the analysis of drug-receptor interactions, *Int. J. Quantum Chem.*, Quantum Biology Symposium 2, 59, 1975.
33. **Weinstein, H., Maayani, S., Pazhenchevsky, B., Venanzi, C. A., and Osman, R.,** Molecular determinants for recognition of phencyclidine derivatives at muscarinic cholinergic receptors, *Int. J. Quantum Chem.*, Quantum Biology Symposium 10, 309, 1983.
34. **McQuarrie, D. M.,** *Statistical Mechanics*, Harper & Row, New York, 1976, 72.
35. **Némethy, G. and Scheraga, H. A.,** The structure of water and hydrophobic bonding in proteins. III. The thermodynamic properties of hydrophobic bonds in proteins, *J. Phys. Chem.*, 66, 1773, 1962.
36. **Boyd, D. B.,** Aspects of molecular modeling, in *Reviews in Computational Chemistry*, Vol. 1, Lipkowitz, K. B. and Boyd, D. B., Eds., VCH Publishers, New York, 1990, 321.
37. **Boyd, D. B.,** Compendium of software for molecular modeling, in *Reviews in Computational Chemistry*, Vol. 1, Lipkowitz, K. B. and Boyd, D. B., Eds., VCH Publishers, New York, 1990, 383.
38. **Clark, T.,** *A Handbook of Computational Chemistry*, John Wiley & Sons, New York, 1985.
39. **Sadlej, J.,** *Semi-Empirical Methods of Quantum Chemistry*, Halsted Press, New York, 1985.
40. **Hehre, W. J., Radom, L., v. R. Schleyer, P., and Pople, J. A.,** *Ab Initio Molecular Orbital Theory*, John Wiley & Sons, New York, 1986.
41. **Burkert, U. and Allinger, N. L.,** *Molecular Mechanics*, ACS Monograph, Vol. 177, American Chemical Society, Washington, D.C., 1982.
42. **McCammon, J. A. and Harvey, S. C.,** *Dynamics of Proteins and Nucleic Acids*, Cambridge University, New York, 1987.
43. **Brooks, C. L., III, Karplus, M., and Pettitt, B. M.,** *Proteins: A Theoretical Perspective of Dynamics, Structure, and Thermodynamics*, John Wiley & Sons, New York, 1988.

44. **Berendsen, H. J. C. and van Gunsteren, W. F.,** Practical algorithms for dynamics simulations, in *Molecular Dynamics Simulation of Statistical Mechanical Systems,* North-Holland, New York, 1986, 43.
45. **Clementi, E., Corongiu, G., Aida, M., Niesar, U., and Kneller, G.,** Monte Carlo and molecular dynamics simulations in *Modern Techniques, in Computational Chemistry: MOTECC-90,* Clementi, E., Ed. Escom, Leiden, 1990, 805.
46. *Computer Simulation of Chemical and Biochemical Systems,* Beveridge, D. L. and Jorgensen, W. L., Eds., Ann. N.Y. Acad. Sci., Vol. 482, New York Academy of Sciences, New York, 1986.
47. *Molecular Dynamics: Applications in Molecular Biology,* Goodfellow, J. M., Ed., CRC Press, Boca Raton, FL, 1990.
48. **Madsen, L. J., Ha, S. N., Tran, V. H., and Brady, J. W.,** Molecular dynamics simulations of carbohydrates and their solvation, in *Computer Modeling of Carbohydrate Molecules,* French, A. D. and Brady, J. W., Eds., ACS Symposium Series, Vol. 430, American Chemical Society, Washington, D.C., 1990, 69.
49. **Tran, V. H. and Brady, J. W.,** Conformational flexibility of sucrose: static and dynamical modeling, in *Computer Modeling of Carbohydrate Molecules,* French, A. D. and Brady, J. W., Eds., ACS Symposium Series, Vol. 430, American Chemical Society, Washington, D.C., 1990, 213.
50. **Grigera, J. R.,** Solvent effects on conformation of carbohydrates: molecular dynamics simulation of sorbitol, mannitol, and methoxytetrahydropyran, in *Computer Modeling of Carbohydrate Molecules,* French, A. D. and Brady, J. W., Eds., ACS Symposium Series, Vol. 430, American Chemical Society, Washington, D.C., 1990, 152.
51. **Hopfinger, A. J. and Burke, B. J.,** Molecular shape analysis: a formalism to quantitatively establish spatial molecular similarity, in *Concepts and Applications of Molecular Similarity,* Johnson, M. A. and Maggiora, G. M., Eds., John Wiley & Sons, New York, 1990, 173.
52. **Dean, P. M.,** Molecular recognition: the measurement and search for molecular similarity in ligand-receptor interaction, in *Concepts and Applications of Molecular Similarity,* Johnson, M. A. and Maggiora, G. M., Eds., John Wiley & Sons, New York, 1990, 211.
53. **Scrocco, E. and Tomasi, J.,** Electronic molecular structure, reactivity, and intermolecular forces: an euristic interpretation by means of electrostatic molecular potentials, in *Advances in Quantum Chemistry,* Vol. 2, Lowdin, P.-O., Ed., Academic Press, New York, 1978, 115.
54. *Chemical Applications of Atomic and Molecular Electrostatic Potentials,* Politzer, P. and Truhlar, D. G., Eds., Plenum Press, New York, 1981.
55. **Politzer, P. and Truhlar, D. G.,** Introduction: the role of the electrostatic potential in chemistry, in *Chemical Applications of Atomic and Molecular Electrostatic Potentials,* Politzer, P. and Truhlar, D. G., Eds., Plenum Press, New York, 1981, 1.
56. **Mulliken, R. S.,** Electronic population analysis on LCAO-MO molecular wavefunctions. II. Overlap populations, bond order, and covalent bond energies, *J. Chem. Phys.,* 23, 1841, 1955.

57a. **Singh, U. C. and Kollman, P. A.,** An approach to computing electrostatic charges for molecules, *J. Comp. Chem.,* 5, 129, 1984.

57b. **Chirlian, L. E. and Francl, M. M.,** Atomic charges derived from electrostatic potentials: a detailed study, *J. Comp. Chem.,* 8, 894, 1987.

57c. **Besler, B. H., Merz, K. M., Jr., and Kollman, P. A.,** Atomic charges derived from semiempirical methods, *J. Comp. Chem.,* 11, 431, 1990.

58. **Weiner, S. J., Kollman, P. A., Case, D. A., Singh, U. C., Ghio, C., Alagona, G., Profeta, S., Jr., and Weiner, P.,** A new force field for molecular mechanical simulation of nucleic acids and proteins, *J. Am. Chem. Soc.,* 106, 765, 1984.

59a. **Brooks, B. R., Bruccoleri, R. E., Olafson, B. D., States, D. J., Swaminathan, S., and Karplus, M.,** CHARMM: a program for macromolecular energy, minimization, and dynamics calculations, *J. Comp. Chem.*, 4, 187, 1983.

59b. **Nilsson, L. and Karplus, M.,** Empirical energy functions for energy minimization and dynamics of nucleic acids, *J. Comp. Chem.*, 7, 591, 1986.

60a. **Momamy, F. A., McGuire, R. F., Burgess, A. W., and Scheraga, H. A.,** Energy parameters in polypeptides. VII. Geometric parameters, partial atomic charges, nonbonded interactions, hydrogen bond interactions, and intrinsic torsional potentials for the naturally occurring amino acids, *J. Phys. Chem.*, 79, 2361, 1975.

60b. **Sippl, M. J., Némethy, G., and Scheraga, H. A.,** Intermolecular potentials from crystal data. VI. Determination of empirical potentials for O–H . . . O=C hydrogen bonds from packing configurations, *J. Phys. Chem.*, 88, 6231, 1984.

60c. **Némethy, G., Pottle, M. S., and Scheraga, H. A.,** Energy parameters in polypeptides. IX. Updating of geometrical parameters, nonbonded interactions, and hydrogen bond interactions for the naturally occurring amino acids, *J. Phys. Chem.*, 87, 1883, 1983.

61a. **Allinger, N. L., Yuh, Y.-H., and Lii, J.-H.,** Molecular mechanics. The MM3 force field for hydrocarbons, *J. Am. Chem. Soc.*, 111, 8551, 1989.

61b. **Allinger, N. L., Li, F., and Yan, L.,** Molecular mechanics. The MM3 force field for alkenes, *J. Comp. Chem.*, 11, 848, 1990.

61c. **Allinger, N. L., Rahman, M., and Lii, J.-H.,** A molecular mechanics force field (MM3) for alcohols and ethers, *J. Am. Chem. Soc.*, 112, 8293, 1990.

62a. **Bowen, J. P., Pathiaseril, A., Profeta, S., Jr., and Allinger, N. L.,** New molecular mechanics (MM2) parameters for ketones and aldehydes, *J. Org. Chem.*, 52, 5162, 1987.

62b. **Bowen, J. P., Reddy, V. V., Patterson, D. G., Jr., and Allinger, N. L.,** Molecular mechanics (MM2) parameters for divinyl ethers and aromatic halide derivatives, *J. Org. Chem.*, 53, 5471, 1988.

62c. **Bowen, J. P. and Allinger, N. L.,** Molecular mechanics treatment of β-heteroatom-substituted cyclohexanones, *J. Org. Chem.*, 52, 1830, 1987.

62d. **Bowen, J. P. and Allinger, N. L.,** Molecular mechanics parameters for organophosphines, *J. Org. Chem.*, 52, 2937, 1987.

63. **Garty, H. and Benos, D. J.,** Characteristics and regulatory mechanisms of the amiloride-blockable sodium channel, *Physiol. Rev.*, 68, 309, 1988.

64. **Benos, D. J.,** Amiloride: chemistry, kinetics, and structure-activity relationships, in *Na^+/H^+ Exchange*, Grinstein, S., Ed., CRC Press, Boca Raton, FL, 1988, 121.

65. **Kleyman, T. R. and Cragoe, E. J., Jr.,** Cation transport probes: the amiloride series, in *Methods in Enzymology*, Fleischer, S., Ed., Academic Press, New York, 191, 1990, 739.

66. **Kleyman, T. R. and Cragoe, E. J., Jr.,** Amiloride and its analogs as tools in the study of ion transport, *J. Membr. Biol.*, 105, 1, 1988.

67. **DeSimone, J. A., Heck, G. L., and DeSimone, S. K.,** Active transport in dog tongue: a possible role in taste, *Science*, 214, 1039, 1981.

68. **DeSimone, J. A., Heck, G. L., Mierson, S., and DeSimone, S. K.,** The active transport properties of canine lingual epithelia in vitro, *J. Gen. Physiol.*, 83, 633, 1984.

69. **Heck, G. L., Mierson, S., and DeSimone, J. A.,** Salt taste transduction occurs through an amiloride-sensitive sodium transport pathway, *Science*, 223, 403, 1984.

70. **DeSimone, J. A. and Ferrell, F.,** Analysis of amiloride inhibition of chorda tympani taste response of rat to NaCl, *Am. Physiol. Soc.*, R52, 1985.

71. **Heck, G. L., Persaud, K. C., and DeSimone, J.,** Direct measurement of translingual epithelial NaCl and KCl currents during the chorda tympani taste response, *Biophys. J.*, 55, 843, 1989.

72. **DeSimone, J. A., Heck, G. L., Persaud, K. C., and Mierson, S.,** Stimulus-evoked transepithelial lingual currents and gustatory neural response, *Chem. Senses*, 1, 13, 1989.

73. **Ye, Q., Heck, G. L., and DeSimone, J.,** The anion paradox in sodium taste reception: resolution by voltage clamp studies, *Science*, 254, 724, 1991.
74. **Avenet, P. and Lindemann, B.,** Amiloride-blockable sodium currents in isolated taste receptor cells, *J. Membr. Biol.*, 105, 245, 1988.
75. **Schiffman, S. S., Lockhead, E., and Maes, F. W.,** Amiloride reduces the taste intensity of Na^+ and Li^+ salts and sweeteners, *Proc. Natl. Acad. Sci. U.S.A.*, 80, 6136, 1983.
76. **Schiffman, S. S., Simon, S. A., Gill, J. M., and Beeker, T. G.,** Bretylium tosylate enhances salt taste, *Physiol. Behav.*, 36, 1129, 1986.
77. **Schiffman, S. S., Suggs, M. S., Cragoe, E. J., Jr., and Erickson, R. P.,** Inhibition of taste responses to Na^+ salts by epithelial Na^+ channel blockers in gerbil, *Physiol. Behav.*, 47, 455, 1990.
78. **Schiffman, S. S., Frey, A. E., Suggs, M. S., Cragoe, E. J., Jr., and Erickson, R. P.,** The effect of amiloride analogs on taste response in gerbil, *Physiol. Behav.*, 47, 435, 1990.
79. **Desor, J. A. and Finn, J.,** The effect of amiloride on salt taste in humans, *Chem. Senses*, 14, 793, 1989.
80. **Benos, D. J., Saccomani, G., and Sariban-Sohraby, S.,** The epithelial sodium channel: subunit number and location of the amiloride binding site, *J. Biol. Chem.*, 262, 10613, 1987.
81. **Tousson, A., Alley, C. D., Sorscher, E. J., Brinkley, B. R., and Benos, D. J.,** Immunochemical localization of amiloride-sensitive sodium channels in sodium-transporting epithelia, *J. Cell Sci.*, 93, 349, 1989.
82. **Li, J. H.-Y., Cragoe, E. J., Jr., and Lindemann, B.,** Structure-activity relationship of amiloride analogs as blockers of epithelial Na channels. I. Pyrazine-ring modifications, *J. Membr. Biol.*, 83, 45, 1985.
83. **Li, J. H.-Y., Cragoe, E. J., Jr., and Lindemann, B.,** Structure-activity relationship of amiloride analogs as blockers of epithelial Na channels. II. Side-chain modifications, *J. Membr. Biol.*, 95, 171, 1987.
84. **Cuthbert, A. W.,** Similarities between sodium channels in excitable membranes and in epithelia, *Experientia*, 32, 1321, 1976.
85. **Pople, J. A. and Segal, G. A.,** Approximate self-consistent molecular orbital theory. III. CNDO results for AB_2 and AB_3 systems, *J. Chem. Phys.*, 44, 3289, 1966.
86. **Smith, R. L., Cochran, D. W., Gund, P., and Cragoe, E. J., Jr.,** Proton, carbon-13, and nitrogen-15 nuclear magnetic resonance and CNDO/2 studies on the tautomerism and conformation of amiloride, a novel acylguanidine, *J. Am. Chem. Soc.*, 101, 191, 1979.
87. **Kwiatkowski, J. S., Zielinski, T. J., and Rein, R.,** Quantum mechanical prediction of tautomeric equilibria, *Adv. Quantum Chem.*, 18, 85, 1986.
88. **Frisch, M. J., Head-Gordon, M., Trucks, G. W., Foresman, J. B., Schlegel, H. B., Raghavachari, K., Robb, M. A., Binkley, J. S., Gonzalez, C., Defrees, D. J., Fox, D. J., Whiteside, R. A., Seeger, R., Melius, C. F., Baker, J., Martin, R. L., Kahn, L. R., Stewart, J. J. P., Topiol, S., and Pople, J. A.,** GAUSSIAN90 is available from Gaussian, Inc., Pittsburgh, PA, 1990.
89. **Pietro, W. J., Francl, M. M., Hehre, W. J., DeFrees, D. J., Pople, J. A., and Binkley, J. S.,** Self-consistent molecular orbital methods. XXIV. Supplemented small split-valence basis sets for second row elements, *J. Am. Chem. Soc.*, 104, 5039, 1982.

90a. **Schlegel, H. B., Gund, P., and Fluder, E.,** Tautomerization of formamide, 2-pyridone, and 4-pyridone: an ab initio study, *J. Am. Chem. Soc.*, 104, 5347, 1982.

90b. **Scanlan, M. J., Hillier, I. H., and MacDowell, A. A.,** Theoretical studies of the tautomeric equilibria and isomeric energetics of 2-, 3-, and 4-hydroxypyridine, *J. Am. Chem. Soc.*, 105, 3568, 1983.

91. **Mo, O., DePaz, J. L. G., and Yanez,** Protonation of azines: an ab initio molecular orbital study, *J. Mol. Struct. (Theochem.)*, 150, 135, 1987.

92. **Dewar, M. J. S., Zoebisch, E. G., Healy, E. F., and Stewart, J. J. P.,** AM1: a new general purpose quantum mechanical molecular model, *J. Am. Chem. Soc.*, 107, 3902, 1985.
93. **Fabian, W. M. F.,** AM1 calculations of rotation around essential single bonds and preferred conformations in conjugated molecules, *J. Comp. Chem.*, 9, 369, 1988.
94. **Venanzi, C. A. and Venanzi, T. J.,** unpublished results.
95. **Shyamantha, N. U. M., Plant, C., Venanzi, C. A., and Venanzi, T. J.,** unpublished results.
96a. **Hermsmeier, M. and Gund, T. A.,** A graphical representation of the electrostatic potential and electric field on a molecular surface, *J. Mol. Graphics*, 7, 150, 1989.
96b. **Shukla, K.,** Modification of ARCHEM for representation of two-dimensional molecular electrostatic potential patterns, unpublished program.
97. **Hopfinger, A. J. and Pearlstein, R. A.,** Molecular mechanics force field parameterization procedures, *J. Comp. Chem.*, 5, 486, 1984.
98a. **van Gunsteren, W. F., Berendsen, H. J. C., Hermans, J., Hol, W. G. J., and Postma, J. P. M.,** Computer simulation of the dynamics of hydrated protein crystals and its comparison with X-ray data, *Proc. Natl. Acad. Sci. U.S.A.*, 80, 4315, 1983, and references therein.
98b. GROMOS, available from Biomos, B. V., Groningen, The Netherlands.
99a. **Stewart, J. J. P.,** Optimization of parameters for semiempirical methods. I. Methods, *J. Comp. Chem.*, 10, 209, 1989.
99b. **Stewart, J. J. P.,** Optimization of parameters for semiempirical methods. II. Applications, *J. Comp. Chem.*, 10, 221, 1989.
100. **Dewar, M. J. S. and Thiel, W.,** Ground states of molecules. XXXVIII. The MNDO method. Approximations and parameters, *J. Am. Chem. Soc.*, 99, 4899, 1977.
101a. **Stewart, J. J. P.,** MOPAC: A semiempirical molecular orbital program, *J. Comput. Aided Mol. Des.*, 4, 1, 1990.
101b. MOPAC 6.0 is available from Quantum Chemistry Program Exchange, Indiana University, Bloomington, Indiana.
102. **Salunke, D. M. and Vijayan, M.,** Specific interactions involving guanidyl group observed in crystal structures, *Int. J. Peptide Protein Res.* 18, 348, 1981.
103. **Singh, J., Thornton, J. M., Snarey, M., and Campbell, S. F.,** The geometries of interacting arginine-carboxyls in proteins, *FEBS Lett.*, 224, 161, 1987.
104. **Tintelot, M. and Andrews, P.,** Geometries of functional group interactions in enzyme-ligand complexes: guides for receptor modelling, *J. Comput. Aided Mol. Des.*, 3, 67, 1989.
105. **Eggleston, D. S. and Hogson, D. J.,** Guanidyl-carboxylate interactions: crystal structures of arginine dipeptides, *Int. J. Peptide Protein Res.* 25, 242, 1985.
106. **Milburn, M. V., Privé, G. G., Milligan, D. L., Scott, W. G., Yeh, J., Jancarik, J., Koshland, D., Jr., and Kim, S.-H.,** Three dimensional structures of the ligand binding domain of the bacterial aspartate receptor with and without a ligand, *Science*, 254, 1342, 1991.
107a. **Jakinovitch, W., Jr.,** Stimulation of the gerbil's gustatory receptors by saccharin *J. Neurosci.*, 2, 49, 1982.
107b. **Jakinovitch, W., Jr.,** Stimulation of the gerbil's gustatory receptors by artificial sweeteners, *Brain Res.*, 210, 19, 1981.
108. **Bryant, B. P., Leftheris, K., Quinn, J. V., and Brand, J. G.,** Molecular structural requirements for binding and activation of L-Alanine taste receptors, *Amino Acids*, in press.

Chapter 16

DESIGN OF SWEET AND BITTER TASTANTS

D. Eric Walters, Grant E. DuBois, and Michael S. Kellogg

TABLE OF CONTENTS

0-8493-5341-6/93/$0.00 + $.50

I. INTRODUCTION

The relationship between chemical structure and sweet or bitter taste is far more complex and more difficult to unravel than the corresponding relationships for salty and sour tastes. Literally thousands of structurally diverse chemicals are known to have sweet or bitter tastes. Cohn,[1] Moncrieff,[2] and Beets[3] have published extensive compilations of chemical structures and their tastes. In order to understand and rationally design sweet or bitter tastants, either we must have a clear understanding of the relevant receptors and transduction systems, or we must have receptor models which are consistent with known structure-taste relationships. To date, no one has isolated and characterized either sweet or bitter taste receptors. Moreover, the structural nature of the molecular recognition units which we call receptors is not known. Perhaps they are proteins of the ubiquitous G-linked receptor family, having seven transmembrane domains; receptors of this type have now been identified and sequenced in the olfactory system.[4] In addition, mechanisms of transduction are only partially elucidated. It is not entirely clear whether we are dealing with a single receptor type or multiple receptor types. However, many receptor models have been proposed, and some of these have been used successfully in the design of new tastants. In this chapter we discuss the following aspects of sweet and bitter tastant design:

- The question of multiple receptor types for sweet and bitter taste
- The relationship between sweet and bitter taste
- Modeling and design of sweet tastants
- Modeling of bitter tastants

II. MULTIPLE RECEPTOR TYPES FOR SWEET AND BITTER TASTE?

A key question which must be considered in understanding structure-taste relationships is whether there are multiple receptor types mediating sweet taste and bitter taste. Multiple receptors are clearly the case for olfactory chemoreception; Buck and Axel[4] have reported sequences for 18 different olfactory receptors and propose that there may be as many as 100 to 200 different receptor types mediating olfaction.

A. SWEETENER RECEPTORS

DuBois et al.[5] have reviewed the evidence for multiple vs. single sweetener receptor(s). Most of the evidence points toward multiple receptor types for sweet taste. In cross-adaptation studies,[6-8] the quantitative taste response to a sweetener is measured alone and immediately after exposure to a second sweetener; in some cases the second sweetener causes a significant decrease in response to the first sweetener, so it is presumed that the two sweeteners trigger a common receptor. Sweetness synergy (higher than predicted potency when two sweeteners are combined) has been observed and has also been taken as evidence for multiple receptors.[9] Faurion and co-workers[10] have studied interpanelist response differences and have found that saccharin sweetness detection thresholds do not correlate with aspartame sweetness detection thresholds from one panelist to the next, again suggesting different receptors. DuBois and co-workers[11] found that concentration-response relationships for high-potency sweeteners are substantially different from those for polyol sweeteners; this may indicate different receptors or a fundamentally different mechanism for these two broad classes of sweeteners. Finally, single nerve fiber electrophysiological recordings in rhesus monkey by Hellekant and Ninomiya[12] unequivocally argue for multiple sweetener receptor types. In this work it was found that, among a group of sweeteners judged in behavioral and electrophysiological experiments to be sucrose-like, there were single nerve fibers which responded to some sweeteners but not others.

In contrast to the experiments described above, *in vivo* photo-deactivation and sweet taste antagonist studies show no selectivity among sweeteners, suggesting a single transduction system. Nagarajan et al.[13] were able to induce irreversible inhibition of all sweeteners tested by photolysis of a high-potency sweetener containing a photolabile substituent; presumably, the photolyzed compound becomes covalently bound to the receptor site and prevents other sweeteners from acting. In this work, the experimental protocol did not allow for an assessment of whether or not the photoinactivation caused parallel shifts in concentration-response functions for all sweeteners. Sweet taste antagonists have shown little or no selectivity for different classes of sweeteners. These include gymnemic acid,[14] hodulcin,[15] lactisole,[16] an arylurea-sulfonic acid,[17] ziziphin,[18] methyl 4,6-dichloro-4,6-dideoxy-α-D-galactopyranoside,[19] alkyl and aryl sulfonates,[20,21] and benzoic acid derivatives.[22] Sweetness inhibitors discovered to date have all had one or more drawbacks which limit their usefulness for the study of taste mechanisms. All are relatively low in potency and therefore are likely to exhibit broad activity among members of a receptor family. Many have an accompanying bitter taste which makes it difficult to quantitate sweetness. Some of these inhibitors have structural similarities to sweet compounds, suggesting that they may act as competitive inhibitors. However, among compounds which have been suggested to be competitive inhibitors, only lactisole has been studied in a rigorous protocol which looks for parallel shifts in concentration-response functions.[23]

Thus, while many experiments point toward multiple receptors for sweet taste, there has been no selectivity observed in reversible or irreversible inhibition studies. The apparent contradiction between experiments suggesting multiple receptors and those suggesting a single pathway can be rationalized as follows. Recent evidence indicates that the sweet taste transduction pathway involves receptor/G protein activation of adenylyl cyclase[24,25] leading to kinase-controlled gating of potassium channels.[26] If the transduction pathway

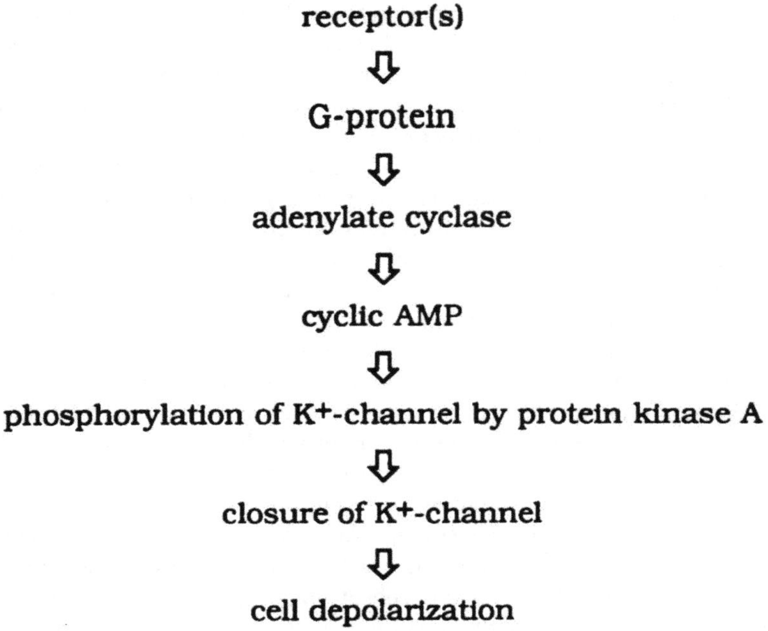

receptor(s)

⇩

G-protein

⇩

adenylate cyclase

⇩

cyclic AMP

⇩

phosphorylation of K^+-channel by protein kinase A

⇩

closure of K^+-channel

⇩

cell depolarization

FIGURE 1. Probable transduction pathway for sweet taste chemoreception.

involves one or more receptor proteins which activate a G protein/cyclic AMP cascade as shown in Figure 1, then different sweeteners could trigger this system by binding to different receptors, by binding to different sites on a common receptor, by interaction with intermediate stages (G protein, adenylate cyclase, cyclic AMP, protein kinase), or by opening K^+ channels. Conversely, photo-deactivation probes and inhibitors which act at a late step in this pathway would inhibit the responses of all sweeteners which initiate response at any earlier step. It is possible that carbohydrates and other polyol sweeteners, which are taste-active only at concentrations in the one tenth molar range and show no stereospecificity,[27] do not directly bind to receptors but have a nonspecific (perhaps osmotic) effect on taste cell membranes which indirectly alters the conformation of receptor protein(s) or ion channels.

B. BITTERNESS RECEPTORS

Bitter taste is not as thoroughly studied from the standpoint of sensory experiments (cross-adaptation, synergy, antagonism, concentration-response relationships, electrophysiology), but it is clear that there are multiple types of bitter tastant receptors or multiple transduction systems or both. In mice, ability to taste quinine appears to be controlled by a different gene than the ability to taste sucrose octaacetate, strychnine, and denatonium.[28] In humans, there are clear dichotomies in the ability to detect bitter taste in PTC[29] and saccharin.[30] There are also selectivities among bitter taste inhibitors; these have been reviewed by Roy.[31] For example, rhoifolin[32] and neodiosmin[33] inhibit the bitterness of limonin and naringin (bitter components of citrus). The bitterness of peptides and protein hydrolysates can be blocked by various glutamic dipeptides.[34]

At least some bitter taste may be induced by interaction of bitter tastants with membrane lipids. Kurihara and co-workers[35,36] have done extensive work on the effect of bitter tastants on membrane potentials in model systems. As in the case of sweeteners, bitter taste is probably transduced by a multistep pathway, and different bitter compounds may act at different steps along the pathway. Some bitter compounds may have specific interactions with receptor proteins, and others may act via relatively nonspecific interactions with membrane lipids.

It may well be the case that we simply have not paid enough attention to bitter taste to develop an adequate vocabulary for describing different bitter tastes; if we dislike a taste, we call it "bitter" and avoid tasting it further. An example is the off-taste which some people detect in saccharin.[29] It is commonly called "bitter," but experienced tasters can easily distinguish it from a caffeine- or quinine-like bitter taste; they refer to it as "metallic."

III. RELATIONSHIP BETWEEN SWEET AND BITTER TASTE

Are sweet taste and bitter taste related? Or, in view of the evidence for multiple types of receptors, are some kinds of sweet taste and some kinds of bitter taste related? Certainly the taste qualities (and subjective human reactions to them) are very different. The spatial distribution of sweet and bitter taste perception (more sweetness perception on the front of the tongue, more bitterness perception toward the back) also suggests that sweet and bitter are quite distinct. However, at least two sweetness inhibitors (an arylurea-sulfonic acid described by Muller et al.[17] and lactisole[30]) have been reported to selectively inhibit some bitter tastes. In addition, there are numerous instances in which small alterations in chemical structure convert a sweet compound to a bitter one. Mazur and co-workers[37] showed that the dipeptide ester aspartame (L-Asp-L-Phe-OMe) is sweet, while many analogs (e.g., L-Asp-D-Phe-OMe, L-Asp-L-LeuOMe, L-Asp-L-Trp-OMe, L-Asp-L-Ala-OMe) are bitter. Workers at Queen Elizabeth College and at Tate & Lyle[38] found that substitution of halogens for hydroxyl moieties of sucrose at positions 4, 1′, 4′, and 6′ resulted in compounds with sweetness potencies up to 7500 times sucrose, while chlorination at the 2-position resulted in a compound with bitterness potency comparable to quinine. Some aminoacyl-sugars are sweet,[39] while others (including acylated sugars such as sucrose octaacetate) are bitter.[40] Acton and Stone[41] reported structure-taste relationships for 51 oximes; some were sweet, some were bitter, and some were both sweet and bitter. Glycine, alanine,

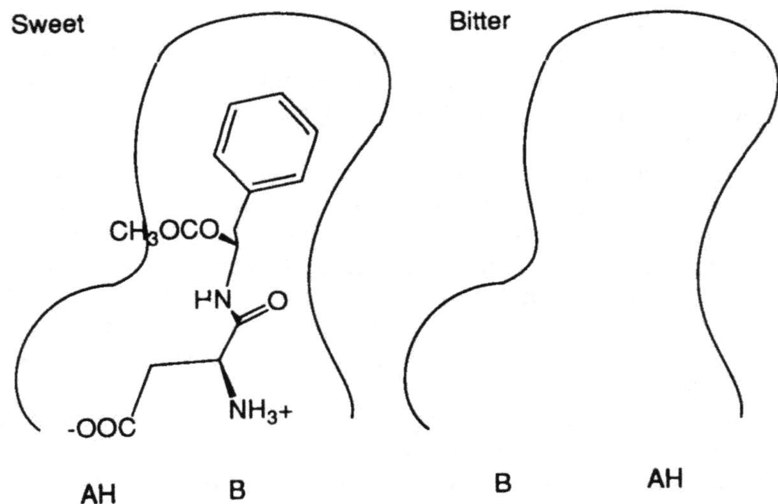

FIGURE 2. Schematic representation of sweet and bitter taste receptor models proposed by Temussi and co-workers.

and many hydrophobic D-amino acids taste sweet, while many L-amino acids have a bitter taste.[42] Numerous bitter and sweet flavonoid derivatives are known; for example, naringin is bitter, while naringin dihydrochalcone (cleavage of a single oxygen-carbon bond) is sweet.[43]

Several proposals have been made to relate sweet and bitter tastes. Belitz and co-workers[44] have proposed a receptor model which could accommodate both sweet and bitter compounds in partially overlapping regions of a single site. Temussi and co-workers[45] have proposed sweet and bitter receptor site models which are essentially mirror images with respect to the placement of hydrophobic and charged functional groups (Figure 2). We may speculate that sweet and bitter receptor proteins are evolutionarily related; if this is the case, there may be overall similarities in receptor structure, with differences in the specificities of the ligand binding sites due to differences of just a few amino acids. Such a relationship could account for the close structural relationships seen between some classes of sweet and bitter compounds. It could also rationalize the observation that some sweetness inhibitors also inhibit some bitter tastants. Alternatively, sweetness and bitterness could utilize a common receptor and be related as agonist and inverse agonist; such a situation has been observed, for instance, in the case of benzodiazepine receptors.[46]

IV. MODELING AND DESIGN OF SWEET TASTANTS

A. SINGLE RECEPTOR TYPE MODELS

Numerous models have been proposed to relate chemical structure to sweet taste. Many of these models begin with the implicit or explicit assumption that all sweeteners act at a common site. In his monumental work on the tastes of organic compounds, Cohn[1] proposed that sweet taste may be dependent on certain functional groups (e.g., hydroxyl and amino groups) which he called "sapophores." He also noted that these sapophores often occur in pairs. Oertly and Myers[47] further developed this idea by dividing sapophores into two subgroups: "glucophores", which are groups capable of imparting sweet taste to otherwise tasteless structures, and "auxoglucs", which are groups that, when combined with a glucophore, yield sweet compounds. Shallenberger and co-workers[48] made the first attempt to relate three-dimensional structure to sweet taste; they proposed that the minimal requirement for sweet taste is a hydrogen bond donor and a hydrogen bond acceptor (AH-B) separated by a distance of about 3 Å. Such a distance can be achieved by adjacent hydroxyl groups (e.g., in sugars) in gauche (but not extended) conformations. Shallenberger and co-workers[27] later added a "steric barrier" to their model to account for the observation that many D-amino acids are sweet, while the corresponding L-amino acids are not. Kier[49] added to Shallenberger's model a "dispersion binding" or hydrophobic site, 3.5 Å from the AH group and 5.5 Å from the B group. The most comprehensive of the single-receptor models is that developed by Tinti and Nofre.[50] This model incorporates eight interaction points (including AH and B sites) in a three-dimensional spatial arrangement.

B. MULTIPLE RECEPTOR TYPE MODELS

Some workers have taken the AH-B concept as a starting point for the development of multiple models, to account for the diversity of sweet-tasting chemical structures. Alternatively, some models have been developed for limited classes of sweeteners. Van der Heijden and co-workers[51,52] evaluated distances from the hydrophobic site to the AH-B pair for ten structural classes of sweeteners; they reported optimum distances ranging from 5.6 to 10.8 Å for the classes of compounds examined. Belitz and co-workers[53] replaced the AH-B groups with a more general electrophile/nucleophile (e/n) system. In a study of twelve structural classes of sweeteners, they found two broad groups, with e/n distances of 3 Å and 8 Å, respectively.

The discovery of aspartame by Mazur and Schlatter[37] touched off an extensive series of structure-taste studies, in laboratories all over the world, which has produced over 1000 aspartame analogs.[54] Along with this mass of structure-taste data came a number of models to rationalize the accumulated data. Mazur and co-workers[55] noted that, for aspartic acid-derived sweeteners, sweetness potency depends on the stereochemistry and size of the substituents on the carbon atom attached to the amide nitrogen. Maximal potencies occur when the largest substituent is hydrophobic and polarizable. Based on a study of space-filling models, Brussel et al.[56] deduced that the length of the hydrophobic group should be between 4.8 and 8.8 Å (optimally about 8 Å). Fujino and co-workers[57] further characterized the shape of the hydrophobic site, based on a series of 21 L-aspartyl-aminomalonic diesters. They proposed a site with approximate dimensions of 7 × 7 × 6 Å. Iwamura[58] used quantitative structure-activity relationships (QSAR) to study 217 aspartic amides; he derived a view of the hydrophobic site as an asymmetric space, with an optimum length and width of about 5.5 Å.

Conformational analysis of sweet aspartic derivatives has been used in the development of several three-dimensional models for sweet taste chemoreception. Temussi and co-workers,[59,60] van der Heijden and co-workers,[61] and Goodman and co-workers[62] have all used NMR and computational chemistry methods to postulate "active conformations" of aspartic derivatives, from which receptor models were then derived. Unfortunately, aspartic derivatives have several conformations which are relatively close in energy, so that assumptions must be made in order to select one "active conformation". Compared to recently discovered high-potency sweeteners,[63] aspartame is a relatively low-potency sweetener (apparent kDa = 1.9 m*M*,[11] compared to compounds which are taste-active at hundreds of n*M*); the "active conformation" may well be far from the global minimum energy conformation.

C. CULBERSON AND WALTERS MODEL

Culberson and Walters[64] developed a three-dimensional model based on the assumption that aspartic derivatives, arylurea-dipeptides,[65] and arylguanidines[63] act at a common receptor site. This assumption is based on the observation that these three classes of sweeteners have similar functional group requirements (carboxylate group, polar NH groups, large hydrophobic substituent). A four-step strategy was used in development of this model:

- Selection of appropriate compounds for modeling studies
- Conformational analysis of chosen compounds
- Calculation and mapping of electrostatic potential fields for the chosen compounds in appropriate low-energy conformations
- Identification of low-energy conformations which allow good superposition of steric and electronic features of the active compounds

The five compounds used to develop this model are shown in Figure 3. Aspartame, **1**, was chosen as the prototype compound in this series because of its clean sweet taste quality; its potency on a weight basis relative to 2% sucrose, $P_w(2)$, is 180 times sucrose. The aspartyl-aminomalonyl-diester **2** was selected because it is one of the most potently sweet aspartic dipeptides known, $P_w(2)$ = 50,000.[57] Addition of a cyanophenylurea moiety to aspartame was found by Tinti and Nofre[65] to produce a 40-fold increase in potency (compound **3**, $P_w(2)$ = 7800). Compounds **4** and **5** represent high-potency guanidine-acetic acid derivatives discovered by Nofre and co-workers;[63] both have potencies $P_w(2) > 100{,}000$ times sucrose.

Conformational analysis was carried out using the MMFF force field of Chemlab[66] and the MacroModel program.[67] Force field parameters for urea

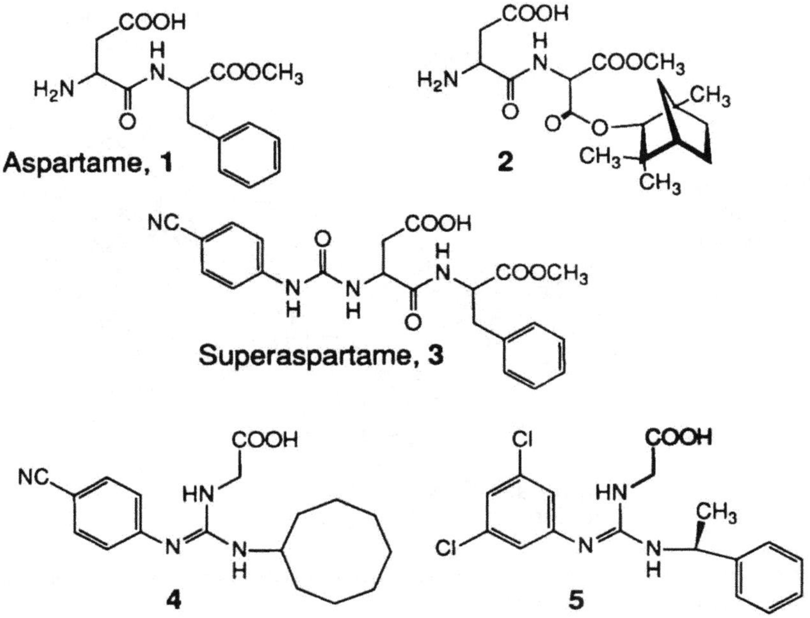

FIGURE 3. Structures of the five compounds used in development of the Culberson & Walters model for sweet taste receptor site.

and guanidinium groups were generated by the method of Hopfinger and Pearlstein,[68] using PRDDO calculations[69] to estimate torsional barriers and optimal geometries. While the aspartic derivatives were found to have as many as 150 low-energy conformations, the aryl-guanidines were found to have only 4 to 6 low energy conformations each, due to conjugation and steric bulk of the trisubstituted arylguanidinium system. Electrostatic potential surfaces were calculated for compounds in low-energy conformations, using point charges calculated by the INDO/S method.[70]

The surfaces were superimposed so as to maximize overlap of steric and electronic features. As shown in Figure 4, the superposition of the five compounds places all of the carboxylate groups in one region. All compounds have a polar N-H group in another region 4 to 6 Å away (carboxylate-to-proton distance). All compounds have a hydrophobic group in a third region, 7 to 9 Å from the carboxylate and 6 to 8 Å from the N-H group. Additionally, three of the most potently sweet compounds have an electron-deficient aromatic ring in a fourth region. The resulting receptor model is a van der Waals surface over the five superimposed compounds; onto this surface is mapped a composite electrostatic potential. The model incorporates distance-between-functional group information as well as detailed three-dimensional steric shape and electronic properties.

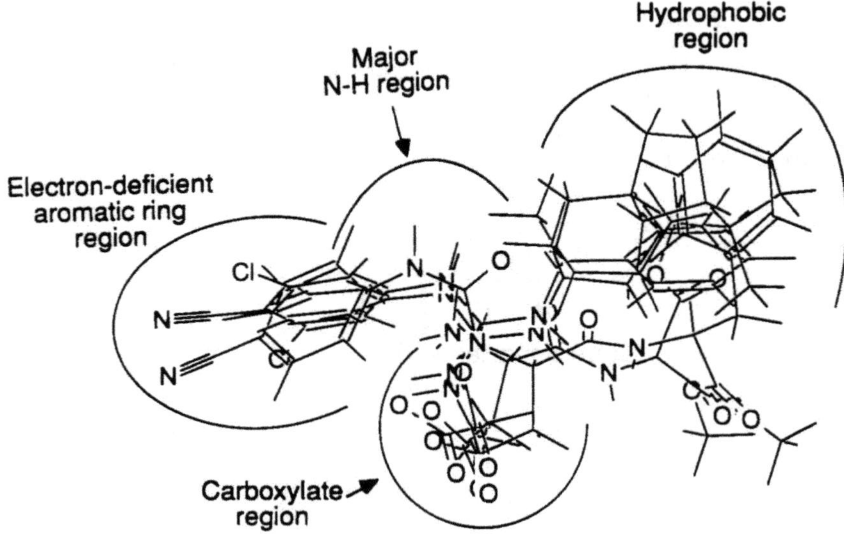

FIGURE 4. Superimposition of the five compounds used in development of the Culberson & Walters model for sweet taste receptor site.

This model was successfully used to rationalize existing structure-taste relationships for carboxylate-containing sweeteners and their nonsweet analogs. Essentially any compound which can fit within the boundaries of the surface and which has appropriate distribution of negative and positive charge (in the carboxylate and N-H regions, respectively) is very likely to taste sweet. Increasing occupation of the hydrophobic region results in increased potency; compounds which extend beyond the boundary of the hydrophobic pocket have low potencies or no sweet taste at all. Compounds which place an electron-deficient aromatic substituent in the fourth region generally have an additional order of magnitude increase in potency. Compounds which have partial negative charge in the main N-H region have highly potent bitter taste.

This model led to the design of arylurea-sulfonic acid inhibitors of sweet taste[17] such as compound **6.** It also led to the successful design of new classes of sweeteners such as arylurea derivatives of substituted β-amino acids.[71] When the sweetener suosan[72] is placed in the receptor model (Figure 5), it is apparent that a substituent in the β-position could place a hydrophobic substituent in a favorable region of the model; the model predicts that this should result in increased potency. This prediction was shown to be correct when aryl-substituents were placed at this position;[71] compounds with aryl-substituents in the β-position have $P_w(2)$ ranging up to 25,000 times sucrose, compared to a potency of about 400 times sucrose for suosan. The model also correctly predicts that only the stereoisomer with S absolute configuration should have high-potency sweet taste; the S-isomer has twice the potency of

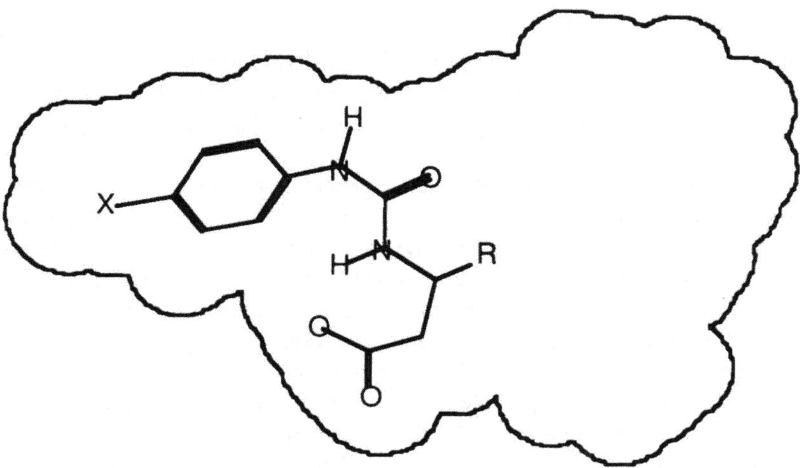

FIGURE 5. Orientation of suosan (R = H, X = NO_2) and higher potency analogs (R = aryl, X = CN) in the Culberson & Walters model for sweet taste receptor site.

the racemate, and the sweetness potency of the R-isomer can be accounted for by the amount of contaminating S-isomer.

V. MODELING OF BITTER TASTANTS

There is broad diversity in the chemical structures of bitter-tasting compounds: inorganic cations (Cs^{2+}, Ca^{2+}), small peptides (Arg-Pro-Gly), sugar derivatives (sucrose octaacetate), alkaloids (strychnine, caffeine), and many other classes have bitter taste. This structural diversity, as well as studies described previously in Section II.B, points to the likelihood that bitter taste is mediated by several (perhaps many) receptor types or mechanisms. In contrast to the extensive efforts which have gone into modeling and designing sweet tastants, bitter tastants have received relatively little attention. There is little economic incentive to discover new bitter tasting compounds; most of the work reported has attempted to understand undesired bitterness which may be present in some food systems. Models which have been proposed for bitter taste receptors are generally less detailed than those recently developed for sweet taste.

Some of the models proposed have been derived by modification of sweetener receptor models. Kubota and Kubo[73] proposed that bitterness is dependent on hydrogen bond donor and acceptor groups separated by a distance of 1.5 Å (compared to a distance of about 3 Å in Shallenberger's sweetener model). Belitz and co-workers have carried out extensive studies of the relationship between structure and bitter taste;[74] they proposed a bitterness receptor model[43] analogous to their sweetener receptor model (see above), with comparable distances between electrophilic and nucleophilic sites, and with a different orientation for the hydrophobic region. Temussi and co-workers[75] proposed a bitter receptor model essentially identical to their sweetener receptor model, except that hydrogen bond donor and acceptor sites are reversed.

Many investigators have observed a general correlation between hydrophobic character and bitter taste. Ney[76] proposed the Q-value as a measure of peptide hydrophobicity and described a relationship between Q-value and bitterness for peptides of molecular weight less than 6000. Belitz and co-workers[77] showed an inverse correlation between bitterness threshold values and octanol/water partition coefficients for series of alkylamines, cycloalkylamines, and azacycloalkanes. Iwamura and co-workers[78] carried out QSAR studies of 93 amino acids, peptides, and peptide derivatives, finding statistical correlations between bitterness threshold concentrations and partition coefficients and molecular size parameters. Okai's group,[39] in a study of *O*-aminoacyl sugars, observed an inverse correlation between bitterness thresholds and hydrophobicity of the amino acid side chains. They proposed a model for a bitterness receptor having a pocket 15 Å wide, with two determinant sites (a "binding unit" and a "stimulating unit") at the base of the pocket, and hydrophobic interactions lining the sides of the pocket. The determinant sites were postulated to be 4.1 Å apart.

VI. CONCLUSION

Models have been useful in understanding and utilizing structure-taste relationships. These models have led to the successful design of new sweet compounds and have provided insight into the ways in which sweet and bitter compounds may trigger taste responses. Ongoing studies of the biochemical mechanisms of sweet and bitter taste transduction may soon provide additional information about the number of receptor types involved and the pathways through which sweet and bitter compounds act. The discovery of taste inhibitors and the design of high affinity compounds which can covalently label receptors should contribute to the biochemical studies of these systems.

REFERENCES

1. **Cohn, G.,** *Die Organischen Geschmacksstoffe*, F. Siemenroth, Berlin, 1914.
2. **Moncrieff, R. W.,** *The Chemical Senses*, John Wiley &, Sons, New York, 1946.
3. **Beets, M. G. J.,** *Structure-Activity Relationships in Human Chemoreception*, Applied Science Publishers, London, 1978.
4. **Buck, L. and Axel, R.,** A novel multigene family may encode odorant receptors: a molecular basis for odor recognition, *Cell*, 65, 175, 1991.

5. **DuBois, G. E., Walters, D. E., and Kellogg, M. S.,** The rational design of ultra-high potency sweeteners, in *Sweet Taste Reception,* Birch, G. and Mathlouthi, M., Eds., Parthenon, 1992.
6. **McBurney, D. H.,** Gustatory cross adaptation between sweet-tasting compounds, *Percept. Psychophys.,* 11, 225, 1972.
7. **Schiffman, S. S., Cahn, H., and Lindley, M. G.,** Multiple receptor sites mediate sweetness: evidence from cross adaptation, *Pharmacol. Biochem. Behav.,* 15, 377, 1981.
8. **Lawless, H. T. and Stevens, D. A.,** Cross adaptation of sucrose and intensive sweeteners, *Chem. Senses,* 7, 309, 1983.
9. **DuBois, G. E., Bittman, R., Carr, B. T., and Pecore, S. D.,** presented at Am. Chem. Soc. Great Lakes Regional Meeting, Chicago, June 11, 1987.
10. **Faurion, A., Saito, S., and Mac Leod, P.,** Sweet taste involves several distinct receptor mechanisms, *Chem. Senses,* 5, 107, 1980.
11. **DuBois, G. E., Walters, D. E., Schiffman, S. S., Warwick, Z. S., Booth, B. J., Pecore, S. D., Gibes, K., Carr, B. T., and Brands, L. M.,** Concentration-response relationships of sweeteners: a systematic study, in *Sweeteners: Discovery, Molecular Design, and Chemoreception,* Walters, D. E., Orthoefer, F. T., and DuBois, G. E., Eds., American Chemical Society, Washington, D.C., 1991, 261.
12. **Hellekant, G. and Ninomiya, Y.,** On the taste of umami in chimpanzee, *Physiol. Behav.,* 49, 927, 1991.
13. **Nagarajan, S., DuBois, G. E., Kellogg, M. S., and Hellekant, G.,** Design and synthesis of super potent guanidines as probes for the sweet taste receptor, paper presented at the 199th Natl. Meeting, American Chemical Society, Boston, April 22 to 27, 1990, AGFD 61.

14a. **Kennedy, L. M.,** Gymnemic acids: specificity and competitive inhibition, *Chem. Senses,* 14, 853, 1989.

14b. **Jakinovich, W., Jr.,** Specificity and mechanisms of gymnemic acid antagonism (inhibition), *Chem. Senses,* 14, 859, 1989.

15. **Kennedy, L. M., Saul, L. R., Sefecka, R., and Stevens, D. A.,** *Chem. Senses,* 13, 529, 1988.
16. **Lindley, M. G.,** Phenoxyalkanoic acid sweetness inhibitors, in *Sweeteners: Discovery, Molecular Design, and Chemoreception,* Walters, D. E., Orthoefer, F. T., and DuBois, G. E., Eds., American Chemical Society, Washington, D.C., 1991, 251.
17. **Muller, G. W., Culberson, J. C., Roy, G. M., Ziegler, J., Walters, D. E., Kellogg, M. S., Schiffman, S. S., and Warwick, Z. S.,** Carboxylic acid replacement structure-activity relationships in suosan type sweeteners. A sweet taste antagonist, *J. Med. Chem.,* 35, 1747, 1992.
18. **Kurihara, Y., Ookubo, K., Tasaki, H., Kodama, H., Akiyama, Y., Yagi, A., and Halpern, B.,** Studies on the taste modifiers. I. Purification and structure determination of sweetness inhibiting substance in leaves of *Ziziphus jujuba, Tetrahedron,* 44, 61, 1988.
19. **Jakinovich, W., Jr.,** Methyl 4,6-dichloro-4,6-dideoxy-α-D-galactopyranoside: an inhibitor of sweet taste response in gerbils, *Science,* 219, 408, 1983.
20. **Palmer, M. D., Hickernell, G. L., and Zanno, P. R.,** An ingestible product containing a sweetness inhibitor and process for inhibiting sweetness in an ingestible product, Eur. Pat. Appl. 86109045.4, 1986.
21. **Barnett, R. E. and Yarger, R. G.,** Foodstuffs containing 3-aminobenzenesulfonic acid as a sweetener inhibitor, U.S. Pat. 4,642,240, 1987.
22. **Barnett, R. E. and Yarger, R. G.,** Sweetening composition and process for modifying the sweetness perception of a foodstuff, Eur. Pat. Appl. 83103881.2, 1983.
23. **Lindley, M. G.,** Phenoxyalkanoic acid sweetness inhibitors, in *Sweeteners: Discovery, Molecular Design, and Chemoreception,* Walters, D. E., Orthoefer, F. T., and DuBois, G. E., Eds., American Chemical Society, Washington, D.C., 1991, 251.
24. **Avenet, P., Hofmann, F., and Lindemann, B.,** Transduction in taste receptor cells requires cAMP-dependent protein kinase, *Nature,* 331, 351, 1988.

25. **Striem, B. J., Pace, U., Zehavi, U., Naim, M., and Lancet, D.,** Sweet tastants stimulate adenylate cyclase coupled to GTP-binding protein in rat tongue membranes, *Biochem. J.*, 260, 121, 1989.
26. **Tonosaki, K. and Funakoshi, M.,** Cyclic nucleotides may mediate taste transduction, *Nature*, 331, 354, 1988.
27. **Shallenberger, R. S., Acree, T. E., and Lee, C. Y.,** Sweet taste of D and L-sugars and amino-acids and the steric nature of their chemo-receptor site, *Nature*, 221, 555, 1969.
28. **Whitney, G., Maggio, J. C., and Harder, D. B.,** Manifestations of the major gene influencing sucrose octaacetate (SOA) tasting among mice: classic taste qualities, *Chem. Senses*, 15, 243, 1990.
29a. **Fox, A. L.,** Tasteblindness, *Science*, 73, 14, 1931.
29b. **Reddy, B. M. and Rao, D. C.,** Phenylthiocarbamide taste sensitivity revisited: complete sorting test supports residual family resemblance, *Genet. Epidemiol.*, 6, 413, 1989.
29c. **Olson, J. M., Boehnke, M., Neiswanger, K., Roche, A. F., and Siervogel, R. M.,** Alternative genetic models for the inheritance of the phenylthiocarbamide taste deficiency, *Genet. Epidemiol.*, 6, 423, 1989.
30. **Helgren, F. J., Lynch, M. J., and Kirchmeyer, F. J.,** A taste panel study of the saccharin "off-taste," *J. Am. Pharm. Assoc. Sci. Ed.*, 44, 353, 1955.
31. **Roy, G. M.,** The applications and future implications of bitterness reduction and inhibition in food products, *Crit. Rev. Food Sci. Nutr.*, 29, 59, 1990.
32. **Kuang-chih, T. and Hua-zhong, H.,** Structural theories applied to taste chemistry, *J. Chem. Ed.*, 64, 1003, 1987.
33. **Guadagni, D. G., Maier, V. P., and Turnbaugh, J. G.,** Effect of neodiosmin on threshold and bitterness of limonin in water and orange juice, *J. Food Sci.*, 41, 681, 1976.
34. **Arai, S.,** The bitter flavor due to peptides or protein hydrolysates and its control by bitterness-masking with acidic oligopeptides, in *The Analysis and Control of Less Desirable Flavors in Foods and Beverages*, Academic Press, New York, 1980, 133.
35. **Kumazawa, T., Nomura, T., and Kurihara, K.,** Liposomes as model for taste cells: receptor sites for bitter substances including N–C=S substances and mechanism of membrane potential changes, *Biochemistry*, 27, 1239, 1988.
36. **Kumazawa, T., Kashiwayanagi, M., and Kurihara, K.,** Contribution of electrostatic and hydrophobic interactions of bitter substances with taste receptor membranes to generation of receptor potentials, *Biochim. Biophys. Acta*, 888, 62, 1986.
37. **Mazur, R. H., Schlatter, J. M., and Goldkamp, A. H.,** Structure-taste relationships of some dipeptides, *J. Am. Chem. Soc.*, 91, 2684, 1969.
38. **Jenner, M. R.,** Sucralose: how to make sugar sweeter, in *Sweeteners: Discovery, Molecular Design, and Chemoreception*, Walters, D. E., Orthoefer, F. T., and DuBois, G. E., Eds., American Chemical Society, Washington, D.C., 1991, 68.
39. **Tamura, M., Nakatsuka, T., Kinomura, K., Okai, H., and Fukui, S.,** Methyl 2,3-di-*O*-(L-alanyl)-α-D-glucopyranoside, a new sweet substance, *Agric. Biol. Chem.*, 49, 891, 1985.
40. **Tamura, M., Miyoshi, T., Mori, N., Kinomura, K., Kawaguchi, M., Ishibashi, N., and Okai, H.,** Mechanism for the bitter tasting potency of peptides using *O*-aminoacyl sugars as model compounds, *Agric. Biol. Chem.*, 54, 1401, 1990.
41. **Acton, E. M. and Stone, H.,** Potential new artificial sweetener from study of structure-taste relationships, *Science*, 193, 584, 1976.
42. **Berg, C. P.,** Physiology of the D-amino acids, *Physiol. Rev.*, 33, 145, 1953.
43. **Horowitz, R. M.,** Taste effects of flavonoids, in *Plant Flavonoids in Biology and Medicine: Biochemical, Pharmacological, and Structure-Activity Relationships*, Alan R. Liss, 1986, 163.

44. **Belitz, H.-D., Chen, W., Jugel, H., Treleano, R., Wieser, H., Gasteiger, J., and Marsili, M.,** Sweet and bitter compounds: structure and taste relationship, in *Food Taste Chemistry,* Boudreau, J. C., Ed., American Chemical Society, Washington, D.C., 1979, 93.
45. **Lelj, F., Tancredi, T., Temussi, P. A., and Toniolo, C.,** Interaction of α-L-aspartyl-D-phenylalanine methyl ester with the receptor site of the bitter taste, *Farmaco Ed. Sci.,* 35, 988, 1980.
46. **Villar, H. O., Uyeno, E. T., Toll, L., Polgar, W., Davies, M. F., and Loew, G. H.,** Molecular determinants of benzodiazepine receptor affinities and anticonvulsant activities, *Mol. Pharmacol.,* 36, 589, 1989.
47. **Oertly, E. and Myers, R. G.,** A new theory relating constitution to taste. Simple relations between the constitution of aliphatic compounds and their sweet taste, *J. Am. Chem. Soc.,* 41, 855, 1919.
48. **Shallenberger, R. S. and Acree, T. E.,** Molecular theory of sweet taste, *Nature,* 216, 480, 1967.
49. **Kier, L. B.,** A molecular theory of sweet taste, *J. Pharm. Sci.,* 61, 1394, 1972.
50. **Tinti, J.-M. and Nofre, C.,** Why does a sweetener taste sweet? A new model, in *Sweeteners: Discovery, Molecular Design, and Chemoreception,* Walters, D. E., Orthoefer, F. T., and DuBois, G. E., Eds., American Chemical Society, Washington, D.C., 1991, 206.
51. **van der Heijden, A., van der Wel, H., and Peer, H. G.,** Structure-activity relationships in sweeteners. I. Nitroanilines, sulphamates, oximes, isocoumarins and dipeptides, *Chem. Senses,* 10, 57, 1985.
52. **van der Heijden, A., van der Wel, H., and Peer, H. G.,** Structure-activity relationships in sweeteners. II. Saccharins, acesulfames, chlorosugars, tryptophans and ureas, *Chem. Senses,* 10, 73, 1985.
53. **Rohse, H. and Belitz, H.-D.,** Shape of sweet receptors studied by computer modeling, in *Sweeteners: Discovery, Molecular Design, and Chemoreception,* Walters, D. E., Orthoefer, F. T., and DuBois, G. E., Eds., American Chemical Society, Washington, D. C., 1991, 176.
54. **Janusz, J. M.,** Peptide sweeteners beyond aspartame, in *Progress in Sweeteners,* Grenby, T. H., Ed., Elsevier Applied Science, London, 1989, 1.
55. **Mazur, R. H., Reuter, J. A., Swiatek, K. A., and Schlatter, J. M.,** Synthetic sweeteners. III. Aspartyl dipeptide esters from L- and D-alkylglycines, *J. Med. Chem.,* 16, 1284, 1973.
56. **Brussel, L. B. P., Peer, H. G., and van der Heijden, A.,** Structure-taste relationship of some sweet-tasting dipeptide esters, *Z. Lebensm. Unters. Forsch.,* 159, 337, 1975.
57. **Fujino, M., Wakimasu, M., Mano, M., Tanaka, K., Nakajima, N., and Aoki, H.,** Structure-taste relationships of L-aspartyl-aminomalonic acid diesters, *Chem. Pharm. Bull.,* 24, 2112, 1976.
58. **Iwamura, H.,** Structure-sweetness relationship of L-aspartyl dipeptide analogues. A receptor site topology, *J. Med. Chem.,* 24, 572, 1981.
59. **Lelj, F., Tancredi, T., Temussi, P. A., and Toniolo, C.,** Interaction of α-L-aspartyl-L-phenylalanine methyl ester with the receptor site of the sweet taste bud, *J. Am. Chem. Soc.,* 98, 6669, 1976.
60. **Castiglione-Morelli, M. A., Lelj, F., Naider, F., Tallon, M., Tancredi, T., and Temussi, P. A.,** Conformation-activity relationship of sweet molecules. Comparison of aspartame and naphthimidazolesulfonic acids, *J. Med. Chem.,* 33, 514, 1990.
61. **van der Heijden, A., Brussel, L. B. P., and Peer, H. G.,** Chemoreception of sweet-tasting dipeptide esters: a third binding site, *Food Chem.,* 3, 207, 1978.
62. **Goodman, M., Coddington, J., Mierke, D. F., and Fuller, W. D.,** A model for the sweet taste of stereoisomeric retro-inverso and dipeptide amides, *J. Am. Chem. Soc.,* 109, 4712, 1987.

63. **Nofre, C., Tinti, J.-M., and Chatzopoulos-Ouar, F.,** Preparation of (phenylguanidino)- and [[1-(phenylamino)ethyl]amino]acetic acids as sweeteners, Eur. Pat. Appl. EP 241,395, 1987; *Chem. Abstr.*, 109, 190047k, 1988.
64. **Culberson, J. C. and Walters, D. E.,** Three-dimensional model for the sweet taste receptor. Development and use, in *Sweeteners: Discovery, Molecular Design, and Chemoreception*, Walters, D. E., Orthoefer, F. T., and DuBois, G. E., Eds., American Chemical Society, Washington, D.C., 1991, 214.
65. **Tinti, J.-M. and Nofre, C.,** Synthetic sweeteners, Fr. Demande FR 2,533,210, 1984; *Chem. Abstr.*, 101, 152354k, 1984.
66. **Pearlstein, R. A., Malhotra, D., Orchard, B. J., Tripathy, S. K., Potenzone, R., Jr., Grigoras, S., Koehler, M., Mabilia, M., Walters, D. E., Doherty, D., Harr, R., and Hopfinger, A. J.,** in *New Methods in Drug Research*, Makriyannis, A., Ed., J. R. Prous, Barcelona, 1988, 147.
67. **Mohamadi, F., Richards, N. G. J., Guida, W. C., Liskamp, R., Lipton, M., Caufield, C., Chang, G., Hendrickson, T., and Still, W. C.,** MacroModel — an integrated software system for modeling organic and bioorganic molecules using molecular mechanics, *J. Comp. Chem.*, 11, 440, 1990.
68. **Hopfinger, A. J. and Pearlstein, R. A.,** Molecular mechanics force-field parameterization procedures, *J. Comp. Chem.*, 5, 486, 1984.
69. **Halgren, T. A., Kleier, D. A., Hall, J. H., Jr., Brown, L. D., and Lipscomb, W. N.,** Speed and accuracy in molecular orbital calculations. A comparison of CNDO/2, INDO, PRDDO, STO-3G, and other methods, including AAMOM, VRDDO, and ESE MO, *J. Am. Chem. Soc.*, 100, 6595, 1978.
70. **Ridley, J. E. and Zerner, M. C.,** An intermediate neglect of differential overlap technique for spectroscopy: pyrrole and azines, *Theor. Chim. Acta (Berlin)*, 32, 111, 1973.
71. **Muller, G. W., Madigan, D. L., Culberson, J. C., Walters, D. E., Carter, J. C., Klade, C. A., DuBois, G. E., and Kellogg, M. S.,** High-potency sweeteners derived from β-amino acids, in *Sweeteners: Discovery, Molecular Design, and Chemoreception*, Walters, D. E., Orthoefer, F. T., and DuBois, G. E., Eds., American Chemical Society, Washington, D.C., 1991, 113.
72. **Petersen, S. and Müller, E.,** Über eine neue Gruppe von Süßstoffen, *Chem. Ber.*, 81, 31, 1948.
73. **Kubota, T. and Kubo, I.,** Bitterness and chemical structure, *Nature*, 223, 97, 1969.
74. **Belitz, H.-D. and Wieser, H.,** Bitter compounds: occurrence and structure-activity relationships, *Food Rev. Intl.*, 1, 271, 1985.
75. **Tancredi, T., Lelj, F., and Temussi, P. A.,** Three-dimensional mapping of the bitter taste receptor site, *Chem. Senses*, 4, 259, 1979.
76. **Ney, K. H.,** in *Food Taste Chemistry*, Boudreau, J. C., Ed., American Chemical Society, Washington, D. C., 1979, 149.
77. **Belitz, H.-D., Chen, W., Jugel, H., Stempfl, H., Treleano, R., and Wieser, H.,** QSAR of bitter tasting compounds, *Chem. Ind. (London)*, 23, 1986.
78. **Asao, M., Iwamura, H., Akamatsu, M., and Fujita, T.,** Quantitative structure-activity relationships of the bitter thresholds of amino acids, peptides, and their derivatives, *J. Med. Chem.*, 30, 1873, 1987.

Chapter 17

DESIGN OF NEW FOODS

H. I. Frier

TABLE OF CONTENTS

0-8493-5341-6/93/$0.00 + $.50

I. INTRODUCTION

A. HEALTH DIET LINK

The present knowledge of nutrition has taken a new turn in recent years. Instead of the old saw of moderation, balance, and variety, which was an attempt to rid society of outright nutrient deficiencies, the health community is pinpointing specific disease states to components of the diet and educating the public on the consequences of misusing them. Epidemiological relationships for cardiovascular disease, still the number one killer in the U.S., cancer of various tissue origins, and the cascade of diseases brought about by obesity is being elucidated.[1-3] Hence, American consumers are receiving information to reduce their caloric intake and more importantly to do it by reducing their fat intake.

Because of the tremendous growth in the area of designing new foods that require both hedonic and health considerations, this chapter can only touch on some of the key aspects of this very interesting area. The need to educate the public on the use of these novel foods, implications on consumer behavior, alterations in individual food patterning and nutrient intake, and the changes that will occur in public health policy and regulation are just a few of the more interesting areas that will need intellectual debate. This is an area where the technology will far outstrip our understanding of the outcome.

In considering this chapter for *Mechanisms of Taste Transduction*, two distinct areas become apparent: ingredients which provide functionality and the finished product containing these ingredients. The following sections will try to provide an understanding of the transitions occurring in the food industry today. It will also provide the reader with an introduction to some of the newer and unique ingredients being used in low fat food products.

B. CONSUMER FOOD INDUSTRY

In 1991 the food industry introduced nearly 1600 new food products.[4] Introductions of new or partially new brand names grew from 1249 in 1987 (23% of total introductions) to 2331 in 1992 (36% of all food introductions).[5] It would be difficult, given the space provided, to do justice to the contributions of the food industry, but it is important to provide the reader with an understanding of the scope of the activities occurring in this industry.

With the increased significance of dietary fat intake and public health, the consumer is being bombarded with information on the importance of reducing fat intake. The food industry has capitalized on growing trends and increased product introductions which are low or no fat. As an example, sales of nonfat dairy products soared 33% ($1.3 billion) in 1990 comprising 7%

of dairy sales. Low fat product sales increased 13% ($5.7 billion) during this same period.[6] Today, no or low fat dairy products represents 63% of dairy department sales. It seems that consumers have demanded a change in the composition of products they buy and the food industry has adjusted in meeting those demands.

The food industry, despite the consolidation in the past 10 years, has remained active in the area of technology development. Between 1988 to 1990 459 patents were issued; Nabisco was granted 156, Kraft General Foods 104, and Proctor and Gamble 65.[7] In a majority of cases these patents reduce concepts to real products.

It has been argued that new product introductions lack innovation and point of difference from existing brands on the shelves. However, the major food manufacturers account for only 20% of line extension introductions with new products being introduced under existing well known trademark or branded lines.[5] This strategy allows the majors to reduce the cost of a new product roll out by linking the product with a consumer recognized brand.

C. FOOD INGREDIENTS INDUSTRY

Due to the unparalleled success with the introduction of aspartame in the early 1980s, the food ingredient business has taken on a new significance relative to the profits that can be made by opening up new markets. With the newer heat stable, high intensity sweeteners, there is an expanded science of application in baking and thermoprocessing development. In order to provide similar consistencies to reduced calorie products that have the attributes that sucrose provides beyond sweetness many more components, in increasing complex food systems, must be modified. The reader is referred to more detailed reviews on the overlapping technologies involving high intensity sweeteners and the requirements for bulking agents.[8,9]

Due to a shift in research and technology development emphasis in the large food companies, the food ingredient industry has taken on the role of providing the new and unique ingredients and their application to specific food systems.

With the increased importance of reducing dietary fat there has been a shift in the present technological drive. There is a high level of activity occurring in the food ingredient industry today to provide ingredients that allow for finished consumer products with reduced calories by targeting the fat component of the diet. In the past 2 years this area has seen an explosion of ingredient development and their methods of use. The balance of this chapter will be dedicated to reviewing the newer and technically unique materials that have regulatory approval and are finding their way into consumer food products.

II. STRATEGIES FOR CALORIC REDUCTION

"Ye shall eat no manner of fat of oxen, or of sheep or of goat."

Leviticus 7:23

A. GENERAL CONSIDERATIONS

Moses was not the only one who heard the Lord's protestations against eating fat. The food industry has also received this message and is beginning to penetrate the market with reduced fat products. Baked goods and frozen desserts have experienced significant growth over the last 2 years as depicted in Table 1 which lists the share and dollar size of the U.S. low fat food category.

A number of approaches to designing fat-reduced foods exists. The simplest is the removal of fat from the food item. The dairy industry has had considerable success by using skim or low fat milk in the production of various dairy items, e.g., yogurt, cottage cheese, and ice milk. Applications that require the fat for its physical attributes, i.e., whipping cream or butter, are severely restricted in quality and functionality of the product.

A second approach is the use of low fat food ingredients in conjunction with gums and small amounts of emulsifiers to give specific components a fatty mouthfeel. The development of the low calorie portioned frozen food entree is a successful example of this approach.

The above two approaches do not provide a parity with full fat products but are used as an alternative to that market segment more concerned with fat and caloric content than organoleptic qualities. There is a tradeoff for this market segment: a willingness to sacrifice taste for a perception of healthy eating. Exceptions do exist. The growth in skim milk and Diet Coke® consumption is based on the taste of superior alternatives rather than health.

Another means of providing products that match the traditional full fat versions is with the use of fat substitutes or mimetics. These fat replacers must provide the functionality of the fat being replaced, be nontoxic, produce no toxic metabolites, and have a minimal caloric reduction of at least 50%.[10]

Qualitative physical criteria required for an acceptable agent has been described as a (1) proper viscosity — a thickness having a specific body or fullness specific for that food system; (2) lubricity — an acceptable level of creaminess or smoothness; and (3) absorption/adsorption — a physical or chemical effect on the tastebud.[11] Additional characteristics that have been described but not quantified are specific degrees of cohesiveness, mouthcoat, mouthmelt (coolness), and waxiness.

The task of simulating the many characteristics of fats when their levels have been reduced or eliminated is most difficult. It is obvious from reviewing a number of the agents used for fat replacement that no universal fat replacer exists. It is presently believed that the most effective way of formulating reduced fat products is with a combination of ingredients that contribute to

TABLE 1
The Market for Reduced Fat and Reduced Calorie Food Products in the U.S., 1991

Category	Retail Sales ($ MM)	% Shares of reduced fat/low cal products		Retail sales reduced fat/low cal products ($ MM)
		1989	1991	
Bakery products		8		
Bread & related	11,100		10	1,110
Cookies & crackers	5,900		1	59
Sweet goods	3,800		5	190
Frozen baked goods	1,500		5	75
Processed packaged meats	19,000	<5	<5	950
Snack foods	10,300	<5	<3	309
Cheese	7,000	3	2[b]	140
Ice cream and frozen desserts[a]	6,600	26	33	2,178
Margarines and spreads	2,200	33	31	682
Mayonnaise and spoonable salad dressings	1,400	23	27	378
Pourable salad dressings	1,000	27	30	300
Total selected segments	52,700			6,370

[a] Includes products that are low fat by definition, i.e., frozen yogurt and ice milk.
[b] Next 5 years expected to be pivotal for low fat cheese, which may reach 11% of total category by 1995.

Reproduced with permission from A. D. Little.

the various textural and functional attributes of the missing fat. The food scientist must be aware of the myriad of ingredients available, their specific characteristics, and techniques for their use.

B. FAT SUBSTITUTES

1. Caprenin

Recently Proctor & Gamble filed a GRAS (generally recognized as safe) petition with the Food and Drug Administration for the use of a synthetic triglyceride, caprenin (caprocaprylobehenin acid), as a fat substitute in confectionery coatings for candies, wafers, granola type bars, snack cakes and cookies, fruits, nuts and nut products, and other snack foods.

Caprenin is a structured lipid formed by the random esterification of glycerol with three fatty acids normally found in foods. Two of these fatty acids are of medium chain length (MCFA [C8:0, caprylic acid; C10:0, capric acid]) and three fatty acids of very long chain lengths (VLCA [predominately

C22:0, behenic acid; with lesser amounts of C20:0 and C24:0, arachic and lignoceric acids]).[12] This unusual triglyceride is digested and absorbed similarly to other lipids.[13] It is a slower rate of absorption of the VLCAs and a unique metabolic pathway of the MCFAs that reduce caloric availability. In man, caprenin has been estimated to provide 5 kcal/g rather than 9.[14-16]

Caprenin has the functionality of cocoa butter and, due to its degree of saturation contributed by the MCFAs, has a lower melting point. By manipulating the composition with different combinations of MCFAs, it is taught that the physical properties of the fat can be altered for different food applications.[17] It is further taught that these fats are good tasting, have comparable mouthmelt, texture, and flavor display as the fat in chocolate.

The first introduction of a consumer product was launched by M&M Mars in the spring of 1992. Caprenin will be part of the calorie-reduced matrix in combination with polydextrose to provide 8 grams of fat and 190 calories in a 2.05-oz version of the Milky Way® candy bar. The present full-fat Milky Way® has 280 calories and contains 11 grams of fat in 2.15 ounces.

The reduction of 3 grams of fat is not sufficient to account for the 90 calorie reduction in the total product, assuming 9 calories/g of fat. The discrepancy can be accounted for by the replacement of sucrose with polydextrose, a synthetic bulking agent that provides one fourth the calories of sucrose.[9,19]

According to Pfizer, polydextrose provides the mouthfeel and texture of fats and has applications in specific food systems such as frostings and salad dressings.[18] Since polydextrose and a recently improved version, Litesse,™ mainly function in replacing the humectancy, bulk, and water activity of sugar, it will not be reviewed in this chapter. The reader is referred to two recent chapters on application and functionality.[19,20]

2. Prime O Lean™

Analyses of the 1977 to 1978 Nationwide Food Consumption Survey (NFCS) and the 1985 to 1986 Continuing Survey of Food Intakes by Individuals (CSFII) suggested a downward shift in total fat intake and a shift from concentrations of higher fat beef to a lower content.[21,22] The largest contributor of dietary fat in the U.S. diet is the ground beef category which, based on the National Health and Nutrition Survey II, accounts for 7% of total fat intake.[23] This is not surprising; over 7 billion lbs of ground beef products are consumed annually in the U.S. and account for 44% of total fresh beef consumption.[24]

The meat industry has responded by introducing newer versions of ground meat ranging from 5 to 10% fat. Typically, ground beef contains 20 to 30% fat. Consumer sensory paneling of ground beef demonstrated peak acceptability at 20% fat content and was strongly correlated with beef flavor intensity.[24] Pork sausage,[25] in similar paneling studies, was most acceptable at 50% fat, similar but lower in acceptability between 20 to 40%, and least acceptable at 10%.

Prime O Lean™ was developed to provide the flavor, juiciness, fluidity, and mouthfeel of 20% fat ground beef in the drier 90 to 95% lean versions that are presently being marketed. Prime O Lean™ is a matrix of water, partially hydrogenated canola oil, hydrolyzed beef plasma, tapioca flour, sodium alginate, colors, and flavorings.

The functionality of this fat substitute is based on the ability of the matrix to hold high levels of moisture and not impart that water to the meat during processing or cooking.[26] The low level of canola oil (8 to 10% of the matrix) provides the fatty mouthfeel expected in a 20% fat hamburger and, due to the emulsifying properties of the beef plasma, does not separate from the moisture. The sodium alginate binds the ingredients in the matrix and aids in binding lean ground meat which tends to crumble and lose form when cooked.

Low fat ground meat containing this lipid matrix has a 50% reduction in saturated fats and cholesterol and a 30% reduction in calories. The very stable water-holding properties of the matrix produces a finished product with less shrinkage and fluid loss than regular hamburger. The finished product has a high eating quality when compared to a 20% fat hamburger.

Newly introduced meat products, McLean Deluxe™ (McDonald Corp.) and Healthy Choice (ConAgra), made with carrageenan and oat, do not replace the fat but mimic the fatty mouthfeel by their interaction with the moisture added to the formulations. Further details of the carrageenan product will be discussed in the next section on mimetics.

The oat product, oatrim, is too newly introduced to comment at length. The fat-like properties can be attributed to the soluble gums that are inherent in oat starch and the hydrolyzed maltodextrins. The degree of polymerization of the maltodextrins produced is important in determining the functionality of the oat material. Maltodextrins of lower DE are most suitable for use as a fat replacer, due to their water-holding and swelling characteristics. A 25% dispersion of oatrim in water forms a fat-like gel which provides 1 cal/g in product application. Oatrim, potentially, has broad applications but has only been used in the ground beef category.[27]

C. FAT MIMETICS

1. Carrageenans

Carrageenan is classified under the broad category of edible hydrocolloids and is derived from several sources of red seaweed.[11] The many seaweed sources that are approved extracts for food use contain varying amounts of the three types of carrageenan, kappa, iota, and lambda. Carrageenan is a high molecular weight polysaccharide made up of repeating galactose units and anhydrogalactose units, some being sulfated, that are joined by alternating alpha 1-3 and beta 1-4 glycosidic linkages.[28]

The gelation requirements differ for the three carrageenans. All have differing properties when conditions of shear, freezing and thawing, and exposure to different salts are varied.[29] Creating the specific reduced fat meat product, e.g., hamburger, pork sausage, and meat batters for frankfurters, dictates which carrageenan can be used to provide the desired characteristics.

The most successful and market-visible use of the carrageenan in meat application is McDonald's McLean Deluxe™ which is a modification of the Auburn University (AU) formulation.[24] The AU formulation was developed by first identifying a suitable "gold standard" to make comparisons, establishing palatability and flavor parameters, and by testing agents to provide the juiciness and tenderness qualities similar to the 20% fat gold standard.[30]

The AU researchers found that iota carrageenan provided the necessary properties to make a reduced fat hamburger that could impart the functionality and sensory qualities of 20% fat ground meat. This specific carrageenan provided a high degree of moisture retention due to its elastic gelling characteristics. The ability to hold moisture in the cooked state was key in providing a hamburger that was rated higher in juiciness and tenderness than the 20% fat hamburger (panel ratings on an 8-point scale for juiciness and tenderness with 1 = extremely dry and tough and 8 = extremely juicy and tender; carrageenan = 6.7 and 6.6 when compared to 20% fat 5.8 and 5.6).[24]

Morphologically, the carrageenan gels maintained their polymorphic structure after cooking and appeared similar in size and shape to the lipid droplets in the 20% fat ground meat.[24] It was suggested that the fat mimetic quality of the carrageenan was due to the similarity of the hydrated carrageenan droplets to lipid droplets in the full fat control. Additionally, the gel strength of the carrageenan was thermostable which aided in retention of moisture within the system.

Palatability and flavor issues were overcome by size of grind and use of a salt and hydrolyzed vegetable protein blend. Combining these three key attributes, the researchers at AU had developed a reduced fat hamburger (8.2% fat content upon cooking) which rated similarly to hamburgers containing greater amounts of fat. The recommended caloric contribution of fat to the diet is 30% of calories;[1] the AU formula, although missing the target, provides 43% of calories as fat and is a major improvement over the standard at 64%.

2. Pectin

Like carrageenan, pectin is a hydrocolloid extracted from plants and is abundant in fruits and vegetables. Pectin is a complex, high molecular weight polysaccharide and consists mainly of partial methyl esters of polygalacturonic acid. The acid groups are partly neutralized by sodium, calcium, potassium, and ammonium ions.[31]

Pectin can form stable gels and increase the viscosity of acidic sugar-containing solutions, e.g., jams and jellies. Viscosity and gelling capacities are dependent on the concentration of polyvalent ions and the degree of esterification. Calcium is usually an integral part of the gelling system. The higher the degree of esterification the stronger the gel and lesser dependence on low acidity. Other factors that influence the gelling mechanisms are temperature and sugar concentration.[32]

In the past the food applications for pectin were functional. It has been used to stabilize food systems, add thickness to fruit beverages, prevent coagulation of casein, and improve the shelf life of pasteurized sour milk products.[31,32] Hence, its utility has been as a food modifier requiring small amounts to produce the desired functionality.

Recently, a specific pectin, Slendid™ (Hercules, Inc.), was introduced with proprietary methods for its preparation and use as a fat replacer. This ingredient can be hydrated separately and added as a gel to other ingredients or mixed with dry components, hydrated, and allowed to gel during processing. In applications requiring a creaminess similar to that delivered by fat, e.g., cheesecake, a homogenization step is required to shear the gel into smaller particles, mimicking the physical and organoleptic qualities of emulsified fat globules.[33]

The Slendid™ pectin, as with other hydrocolloids, derives its functionality by its ability to gel and retain large amounts of moisture. Based on product literature for a low fat baked cheesecake formulation, the manufacturer recommends 1.7 g of Slendid™ for a 100-g (3.53 oz) serving.[34] The specialized pectin is wet prepared with water (96 g water:3.72 g Slendid™), homogenized, and added to the formulation at 46% of total solids. The resultant fat and caloric content (wet basis) of the formula is 5.2 g and 149 cal/100 g serving compared to a standard recipe where fat is 19 g and calories 284. This ingredient imparts a moist and fat-like mouthfeel similar to the standard cheesecake and provides an acceptable healthy alternative with fat and calories reduced by 80 and 52%.

Slendid™ does not impart flavor, interact with other flavors in food systems, and has no discernible aftertaste. Therefore, it provides the food technologist/flavorist with an excellent base for adding flavor systems. The flavor neutrality and high temperature stability of the formed gels allows for a wide range of low fat applications, e.g., baked goods, dairy products, creamy soups, frozen desserts, and whipped toppings.

This ingredient meets the regulatory requirements for pectin and is affirmed GRAS. Because of its natural positioning, i.e., the ingredient is essentially fruit pectin, the food processor will, in all probability, label the ingredient as pectin or fruit pectins and emphasize the reduction of fat in the finished product. It is unclear whether the manufacturer intends to market Slendid™ as a branded ingredient; hence, the consumer may not be able to identify or differentiate this ingredient from other pectins.[35]

3. Corn Starch

Stellar™ is a newly introduced and unique modified food starch, having several significant features besides its fat mimicry. This "particulate gel" acts as an antistaling agent in baked goods and reduces dough toughening in microwave applications. Staling of commercial baked goods is caused by starch and protein interactions as water migrates out of the product over time.

Stellar's ability to bind water prevents this interaction and adds shelf life to baked goods.[36]

The primary chemical structure of this starch is a branched chain amylodextrin comprised of anhydro-D-glucopyranose units with alphas 1-4 (linear) and 1-6 (branched) linkages. This food ingredient is a dry, white powder that is 70% insoluble in water.[37] The functional form of the carbohydrate crystallite is produced by high pressure shearing of the powder in an aqueous system (mixture of 25% crystallite with 75% water).

Particle diameter of the creme is extremely small, 0.02 μ, but the surface area is large when compared to the parent cornstarch or Stellar™ powder. Small angle X-ray scattering and transmission EM suggests that the creme is a particulate gel comprised of aggregated submicron particles.[37] This unique physical structure imparts a high degree of water immobilization (moisture retention) to the creme. The combination of particle networks with large surface areas and high moisture content provides this ingredient with a short texture (similar to plasticized shortening), tenderness, lubricity, and a smooth fat-like mouthfeel.

The processing of the Stellar™ powder into the creme form provides a caloric value of 1. A Stellar™ creme reduced fat cheesecake, incorporating 42% creme as suggested in the product literature,[38] would provide a 60% fat-reduced product containing 30% fewer calories than a full fat version. Stellar™ is affirmed GRAS and appears in the Code of Food Regulations under the classification of food starch, modified. Similar to the pectin product, the consumer will not be able to identify with this material when reading from the food product's list of ingredients.

III. SUMMARY AND CONCLUSIONS

There seems to be no end in sight for the introduction of new fat replacers to the food industry. National Starch and Chemical Company recently announced the introduction of a new line of five fat replacers under the N-lite brand. Since no single fat replacer satisfies all food systems, this company's strategy was to provide the food manufacturer with a class of fat-like compounds with narrowed functionality to fit specific food applications. Protein-based fat replacers, Simplesse 100 and 300, are on the market in frozen deserts such as Simple Pleasures.® Konjac flour (Nutricol®, FMC Corp.) has been introduced as a fat replacer in the ground meat category.

Arguably, as long as public health policy emphasizes fat reduction as part of a healthy lifestyle, the demand for reduced fat foods will grow. Limitations to the success of these ingredients will be their cost to the food industry, the ease in which these ingredients can be applied to existing formulas and plant operations, and end product eating quality. It is the consumer, balanced by how well the fat-reduced version replaces the full fat product and the retail cost, who will make the final judgment.

Introductions of new ingredients will be limited by the ease and length of time required to achieve government approval for the safety of an ingredient. Food ingredient companies are attempting to avoid this by redeveloping existing recognized safe ingredients and self-affirming their GRAS status.

In the majority of cases, the technologies used today have been around for a number of years. It is the prospect of opening new markets that is pulling the technology rather than a new technology driving the market. Existing food ingredients are being modified by new processes and applied in new and unique ways to target the requirements of a reduced fat food. The majority of the fat replacers are based on technologies to enhance their water-holding capacities. The moisture retained may or may not interact with the material itself to impart lubricity and a fatty mouthfeel.

Beyond its physical attributes, fat imparts the majority of the flavor to foods. Flavor profiles and onset of flavor perception, as a function of time, are important attributes of fat. Consideration must be given to chemical changes of fat replacers and flavors. As an example, amylose portions of carbohydrate-based fat replacers can form helices and bind lipophilic flavors reducing the expected flavor impact. Understanding the nature of the fat replacer and the interaction with flavor systems will be key to the success of the low fat food market.

The consumers desire to reduce fat intake is not a fad. The market for new and novel low and no fat products will continue to expand and will only be limited by the food industry's inability to provide parity products at an acceptable cost.

REFERENCES

1. U.S. Department of Health and Human Servic, The Surgeon General's Report on Nutrition and Health, Summary and Recommendations, DHHS(PHS) Publ. No. 88-50211, U.S. Government Printing Office, Washington, D.C., 1988.
2. WHO Study Group, Diet, nutrition and the prevention of chronic diseases, World Health Organization Technical Report Series 797, World Health Organization, Geneva, 1990, chap. 3.
3. National Research Council, Obesity and eating disorders, in *Diet and Health*, National Academy Press, Washington, D.C., 1989, chap. 21.
4. Gorman's New Product News, Gorman's Publishing Co., Chicago, IL, December 1991, 8.
5. **Lawrence, R.,** Fresh perspectives on line extensions, *Food Bus.*, 12, 48, 1991.
6. **McNair, R.,** No-fat dairy items hit $1.3 billion, presented at Food MegaTrends Conference, Chicago, October 26–30, 1991.
7. Food Industry Report, Sunvale Publ. Co., Inc., 3(8), 1991.
8. **Lindley, M. G.,** Sweetener markets, marketing and product development, in *Handbook of Sweeteners*, Marie, S. and Piggott, J. R., Eds., AVI, New York, 1991, chap. 7.
9. **Various authors,** in *Alternative Sweeteners*, 2nd ed., Nabors, L. O. and Gerlardi, R. C., Eds., Marcel Dekker, New York, 1991, chaps. 2, 3, 4, 7, and 10.

10. **Singhal, R. S., Gupta, A. K., and Kulkarni, P. R.,** Low-calorie fat substitutes, *Food Sci. Technol.*, 2, 235, 1991.
11. **Glicksman, M.,** Hydrocolloids and the search for the "oily grail", *Food Technol.*, 45(10), 94, 1991.
12. **Stipp, G. K. and Yang, D. K.,** Selective esterification of long chain fatty acid monoglycerides with medium chain fatty acids, W.O. patent 91/09099, 1991.
13. **Webb, D. R. and Sanders, R. A.,** Caprenin 1. Digestion, absorption and rearrangement in thoracic duct-cannulated rats, *J. Am. Coll. Tox.*, 10, 323, 1991.
14. **Bach, A. C. and Babayan, V. K.,** Medium-chain triglycerides: an update, *Am. J. Clin. Nutr.*, 36, 950, 1982.
15. **Webb, D. R., Peters, J. C., Jandacek, R. J., and Fortier, N. E.,** Caprenin 2. Short-term safety and metabolism in rats and hamsters, *J. Am. Coll. Toxicol.*, 10, 325, 1991.
16. **Peters, J. C., Holcombe, B. N., Hiller, L. K., and Webb, D. R.,** Caprenin 3. Absorption and caloric value in adult humans, *J. Am. Coll. Toxicol.*, 10, 341, 1991.
17. **Mohlenkamp, M. J., Jr. and Pflaumer, P. F.,** Food compositions containing reduced calorie fats and reduced calorie sugars, W.O. patent 91/09537, 1991.
18. **Haumann, B. F.,** Getting the fat out, *J. Am. Oil Chem. Soc.*, 63, 278, 1986.
19. **Moppett, F. K.,** Polydextrose, in *Alternative Sweeteners*, 2nd ed., Nabors L. O. and Gerlardi, R. C., Eds., Marcel Dekker, New York, 1991, chap. 21.
20. **Smiles, R. E.,** The functional applications of polydextrose, in *Chemistry of Food and Beverages: Recent Developments*, Academic Press, Orlando, FL, 1982, 305.
21. **Stephen, A. M. and Wald, N. J.,** Trends in individual consumption of dietary fat in the United States, 1920–1984, *Am. J. Clin. Nutr.*, 52, 457, 1990.
22. **Popkin, B. M., Haines, P. S., and Reidy, K. C.,** Food consumption trends of U.S. women: patterns and determinants between 1977 and 1985, *Am. J. Clin. Nutr.*, 49, 1307, 1989.
23. **Block, G., Dresser, C. M., Hartman, A. M., and Carroll, M. D.,** Nutrient sources in the American diet: quantitative data from the NHANES II survey. II. Macronutrients and fats, *Am. J. Epidemiol.*, 122, 27, 1985.
24. **Egbert, W. R., Huffman, D. L., Chen, C. M., and Dylewski, D. P.,** Development of low-fat ground beef, *Food Technol.*, 45(6), 64, 1991.
25. **Egbert, W. R., Huffman, D. L., and Reeves, J. C.,** Fat content is major factor in acceptability of fresh pork sausage, *Ala. Agric. Exp. Stn.*, 37, 15, 1990.
26. **Duxbury, D. D.,** New fat replacers sourced from GRAS ingredients, *Food Proc.*, 52(9), 86, 1991.
27. **Inglett, G. E.,** USDA's oatrim replaces fat in many food products, *Food Technol.*, 44, 100, 1990.
28. **Howse, G. T. and Gingras, J.,** Carrageenan — a novel meat product ingredient, *Natl. Prov.*, 196, 48, 1987.
29. FMC Corporation, Marine Colloids, Introductory Bulletin A-1, 1991.
30. **Huffman, D. L. and Egbert, W. R.,** Advances in lean ground beef production, *Ala. Agric. Exp. Sta. Bull.*, No. 606, Auburn Univ., Ala., 1990.
31. **Christensen, S. H.,** Pectins, in *Food Hydrocolloids*, vol. 3, Glicksman, M., CRC Press, Inc., Boca Raton, FL, 1986, chap. 9.
32. **Dreher, M. L.,** in *Handbook of Dietary Fiber*, Marcel Dekker, New York, 1987, 416.
33. **Pszcola, D. E.,** Pectin's functionality finds use in fat-replacer market, *Food Technol.*, 45(12), 116, 1991.
34. Hercules Technical Report, Slendid™ — Nature's Fat Replacer, Product Brochure, Hercules, Inc., Wilmington, DE, 1991.
35. **Duxbury, D. D.,** Pectin on your label can take fat off your table, *Food Proc.*, 52(12), 54, 1991.
36. **Swientek, R. J.,** New fat replacer — creme of the crop?, *Food Proc.*, 52(8), 48, 1991.
37. Staley Technical Data, Stellar™ fat replacer, A. E. Staley, Decatur, IL, TIB 29 195060, 1991.

Index

INDEX

A

B

D

E

N

O

Q

U

V

W

X

Z

Milton Keynes UK
Ingram Content Group UK Ltd.
UKHW021919071024
449327UK00022B/1685